ANALYTICAL CHEMISTRY

OF

AEROSOLS

Edited by

Kvetoslav Rudolf Spurny

Schmallenberg, Germany

CRC Press
Taylor & Francis Group
Boca Raton London New York

CRC Press is an imprint of the
Taylor & Francis Group, an **informa** business

CRC Press
Taylor & Francis Group
6000 Broken Sound Parkway NW, Suite 300
Boca Raton, FL 33487-2742

First issued in paperback 2019

© 1999 by Taylor & Francis Group, LLC
CRC Press is an imprint of Taylor & Francis Group, an Informa business

No claim to original U.S. Government works

ISBN-13: 978-1-56670-040-5 (hbk)
ISBN-13: 978-0-367-39980-1 (pbk)
Library of Congress Card Number 98-54990

Library of Congress Cataloging-in-Publication Data

Analytical chemistry of aerosols / edited by K.R. Spurny.
　　　　p.　　cm.
　　　Includes bibliographical references and index.
　　　ISBN 1-56670-040-X (acid-free paper)
　　　1. Aerosols—Analysis. 2. Chemistry, Analytic. I. Spurny, Kvetoslav.
　　QC882.46.A5 1999
　　551.51'13—dc21
98-549900
CIP

Visit the Taylor & Francis Web site at
http://www.taylorandfrancis.com

and the CRC Press Web site at
http://www.crcpress.com

The Editor

Professor Dr. Kvetoslav Rudolf Spurny was Head of the Department of Aerosol Chemistry at the Fraunhofer Institute for Environmental Chemistry and Ecotoxicology in Germany from 1972 to 1988. He has retired but is still working as an aerosol chemist. Prior to 1972, he was an environmental chemist at the Institute for Occupational Hygiene in Prague (1952 to 1956) and Head of the Department of Aerosol Sciences at the Czechoslovak Academy of Sciences in Prague (1957 to 1972). He was a visiting scientist at the National Center for Atmospheric Research, Boulder, CO, U.S.A. (1966 to 1967) and visiting scientist at the Nuclear Research Center, Fontenay aux Roses, France, in 1969.

Dr. Spurny obtained his Diplomate in Physics and Chemistry from Charles University, Prague, in 1948; a Ph.D. in chemistry at the same university in 1952; and a C.Sc. as a Candidate of Chemical Sciences at the Czechoslovak Academy of Sciences in Prague in 1964.

Professor Spurny is a member of the American Chemical Society, American Association for the Advancement of Science, American Association of Aerosol Research, British Occupational Hygiene Society, the New York Academy of Sciences; and was president of the Association for Aerosol Research from 1983 to 1984. He has written six books on aerosols and over 150 original publications in aerosol physics and chemistry. He was recipient of the David Sinclair Award in Aerosol Sciences in 1989.

Contributors

Ian Colbeck
University of Essex
Department of Biological and Chemical
 Sciences
Colchester, England

Lieve A. De Bock
University of Antwerpen
Chemical Department
Wilrijk, Belgium

Vladimir V. Gridin
Department of Chemistry
TECHNION, Israel Institute of Technology
Haifa, Israel

Suneetha Indurthy
McNeese State University
Department of Chemistry
Lake Charles, Louisiana

Mikio Kasahara
Kyoto University
Institute of Atomic Energy
Kyoto, Japan

A-M. N. Kitto
University of Southern California
Department of Civil Engineering
Los Angeles, California

B. Kopcewicz
Institute of Geophysics
Polish Academy of Science
Warsaw, Poland

Michal Kopcewicz
Institute of Electronic Materials Technology
Warsaw, Poland

Yong-III Lee
McNeese State University
Department of Chemistry
Lake Charles, Louisiana

Christopher A. Noble
University of California
Department of Chemistry
Riverside, California

Ivo Orlić
The National University of Singapore
Department of Physics
Singapore

Josef Podzimek
524 Northern Oaks Dr.
Groveland, Illinois

Miroslava Podzimek
524 Northern Oaks Dr.
Groveland, Illinois

Kimberly A. Prather
University of California
Department of Chemistry
Riverside, California

Israel Schechter
Department of Chemistry
TECHNION, Israel Institute of Technology
Haifa, Israel

Václav Potoček
Charles University
Prague, Czech Republic

Gustav Schweiger
Ruhr University
Department of Lase Technique
Bochum, Germany

Mark V. Smith
McNeese State University
Department of Chemistry
Lake Charles, Louisiana

Joseph Sneddon
McNeese State University
Department of Chemistry
Lake Charles, Louisiana

Kvetoslav R. Spurny
Aerosol Chemistry
Schmallenberg, Germany

Rene E. Van Grieken
University of Antwerpen
Chemical Department
Wilrijk, Belgium

Nicola D. Yordanov
Bulgarian Academy of Science
Institute of Catalysis
Sofia, Bulgaria

Table of Contents

Section III: Single Particle Analysis

Section IV: Special Systems

Acknowledgments

The editor is extremely grateful to all the contributors to this book. Their reviews build a mosaic together that illustrates the state of art in the methods for chemical analysis of aerosols at the end of the 20th century.

He would also like to thank the publishers, who have, with great perseverance, enthusiasm, and patience, brought this book to fruition. In doing so, they have contributed to the development of atmospheric environmental science.

What Is an Aerosol?

An assembly of liquid or solid particles suspended in a gaseous medium long enough to be observed and measured; generally about 0.001 to 100 μm in size (K. Willeke and P. A. Baron in *Aerosol Measurement*, Van Nostrand, Reinhold, New York, 1993).

The designation *aerosol* was coined by F. G. Donnan in about 1918 and introduced in the meteorological literature in 1920 by A. Schmauss (The chemistry of fog, clouds, and rain. *Umschau* 24: 61–63, 1920).

Introduction

During the "classical" period of aerosol science, i.e., before the 1960s, only physical properties of aerodispersed systems were of interest and only physical parameters of particle clouds and of single particles were measured. Only later, mainly since the 1980s, has interest in physicochemical and chemical properties of aerosols increased.

There were several reasons for this development. In the fields of environmental hygiene, medicine, and toxicology, the importance of the chemical composition and chemical properties and interactions of inhaled "bad" aerosols had been recognized. In the various fields of high-technology, which use aerosol systems and their reactivities to produce important materials, the importance of aerosol chemistry and chemical aerosol analysis has been discovered.

Furthermore, the chemistry and chemical composition of aerosols are of great, and often basic, importance in several other fields. The chemistry of atmospheric aerosols is involved in cloud physics processes, such as condensation, evaporation, ice crystal formation, etc. Modern clean-room technologies require almost "aerosol-free" atmospheres. Even very low concentrations of very small particles are "dangerous" for production processes. The knowledge of the chemical composition of these particulate air contaminants is helpful in estimating and finding air contamination sources.

Other fields in which chemical composition of aerosols plays an important role are the applications of medical aerosols and basic research in animal inhalation toxicology. Basic as well as applied research and control strategies in indoor and outdoor atmospheric environments could not progress without well developed and effective methods for chemical aerosol analysis.

The great progress in chemical aerosol analysis was enabled by a fast, intensive, and successful development and improvement of analytical chemistry. Modern physical, physicochemical, biochemical, and biophysical analytical procedures have made it possible to analyze ppm, ppb, ppt, and even smaller amounts of substances in aerosol samples, and even single aerosol analysis has become realistic and useful.

Generally speaking, there exists only one analytical discipline — analytical chemistry. Its application is very broad, and for several applications, smaller or greater methodological sampling and other modifications are necessary. For this reason, it seems reasonable to use the term *chemical analysis of aerosols* and/or *analytical chemistry of aerosols*.

Methods of chemical analysis of aerosols are continuously developing and improving. This is probably why it is difficult to compile a handbook or a textbook dealing with the analytical chemistry of aerosols. Nevertheless, it is worth mentioning some of the important monographs in this field, such as the book by Malissa and Robinson, *Analysis of Aerosol Particles by Physical Methods*, published by CRC Press in 1978, the monograph by Spurny, *Physical and Chemical Characterization of Individual Airborne Particles*, published by Ellis Horwood, Chichester, in 1986, and the important book by J. P. Lodge, *Methods of Air Sampling and Analysis*, published in 1988 by CRC Press.

I have tried in this book to encourage several well-known colleagues from the field of aerosol analysis to contribute their work. Our task was to show the existing procedures and trends in the measurement and analysis of atmospheric aerosols. We hope our aim has been at least partially fulfilled and that we have also presented some new or experimental methods for chemical aerosol analysis. The fact that the authors are from different continents, demonstrates that modern analytical techniques are now available for aerosol analysis in all technically well-developed countries.

Kvetoslav R. Spurny
Schmallenberg Germany
November 1998

Section I

General Approach

1 Methods of Aerosol Measurement Before the 1960s

Kvetoslav R. Spurny

CONTENTS

1.1   INTRODUCTION

The period of classical aerosol physics (Spurny, 1993) was characterized by the use and exploitation of measurement and experimental techniques common during that time. In my opinion, the classical period of aerosol science research lasted approximately until the middle of the 20th century, ending with the publication of the *Mechanics of Aerosols* (Fuchs 1955, 1964). No lasers, no computers, and no spectroscopic analytical tools were available during this period.

1.2   THE EARLY DAYS

The existence of unpleasant and harmful particles in the outdoor and indoor atmosphere was referred to in very early literature. For example, the Romans complained of the foul air in ancient Rome. Serious particulate air pollution led to the prohibition of coal burning in London in 1273, followed

FUMIFUGIUM:

OR,

The Inconvenience of the A E R,

A N D

SMOAKE of LONDON

DISSIPATED

TOGETHER

With some REMEDIES humbly proposed

By John Evelyn Esq;

To His Sacred MAJESTIE,

A N D

To the PARLIAMENT now Assembled.

Published by His Majesty's Command.

TO   THE KINGS MOST SACRED

M A J E S T Y.

SIR,

I T was one day, as I was Walking in Your MAJESTIES Palace at WHITE–HALL, (where I have sometimes the honour to refresh myself with the Sight of Your Illustrious Presence, which is the Joy of Your Peoples hearts) that a presumptuous Smoake issuing from one or two Tunnels neer Northumberland-house, and not far from Scotland-yard, did so invade the Court; that all the Rooms, Galleries, and Places about it were filled and infested with it; and that to such a degree, as Men could hardly discern one another for the Clowd, and none could support, without manifest Inconveniency.

FIGURE 1.1   The title page of John Evelyn's *Fumifugium.*

by a Royal Proclamation by Edward I in 1306. In 1661, the first major tract regarding particulate air pollution was submitted to Charles II by John Evelyn (Lodge, 1969). His *Fumifugium...* (Figure 1.1) contained a graphic description of pollution in London. However, the very birth of aerosol science and aerosol measurement did not occur until the second half of the 19th century. It was closely combined with initial developments in colloid chemistry. During this period, the first observations and simple measurement of fine particles in the atmosphere occurred (Podzimek, 1985, 1989).

Following Podzimek's review (1985) H. Becquerel hypothesized about the existence of fine particles in the air, the *condensation nuclei*, in 1847. Their existence was confirmed about 30 years later by the experiments of Coulier (Coulier, 1875). Between 1880 and 1890, John Aitken made several observations that demonstrated the fundamental role of dust particles in the formation of clouds and fog. He evaluated Coulier's experiments on condensation phenomena and his condensation nuclei hypothesis and recognized Coulier as the first to show the important part played by nuclei in the cloudy condensation of water (Aitken, 1880, 1881, 1888, 1889). Aitken developed the first expansion-type dust chamber, and in 1887 also developed the first out-of-pocket condensation nuclei counter. J. Aitken (Figure 1.2), who was born in Falkirk, Scotland in 1839 and died in 1919, built this instrument himself. It is interesting that the next generation of Aitken instruments were not produced and commonly used until the 1930s in Germany (Scholz, 1931).

Of equal importance in the middle of the 19th century were the observations and simple measurements performed by John Tyndall, who was born in 1820 in Ireland and died in 1893 in England. Tyndall studied in Marburg, Germany with W. Bunsen, and later worked with Michael Faraday, eventually becoming his successor. His observation, that dust and smoke in a room are easily detectable from scattered light when a beam of sunlight enters the room, was used in 1856 by Faraday to indicate the presence of colloidal particles in liquids. A decade later, Tyndall extended the method to detect aerosols and first applied it to the detection of the particulate pollution in London air (Tyndall, 1871; Gentry, 1996). Tyndall was not only the father of tyndallometers and nephelometers, of ultramicroscopes and optical particle counters, but also the indirect inventor of thermal precipitators. In 1870 he reported the observation of a narrow region above a heated body

FIGURE 1.2 John Aitken (1839–1919). The founder of atmospheric aerosol science and aerosol measurement techniques.

in a dusty atmosphere. Several years later Lord Rayleigh (1882), Lodge (1883), Lodge and Clark (1894), as well as J. Aitken (1894) observed that a dark space completely surrounded a heated body. This was the discovery of the thermophoretic effect (see also Fuchs, 1971).

Some early measurements exist, prior to the 1900s, of microbiological aerosol particles in room air (Singerson, 1870–1874; Preining, 1996). However, the broader development of aerosol measurement methods and equipment is dated after 1900 and primarily after 1920. During this period the negative health effects of industrial aerosols and dusts were recognized (Sinclair, 1950; Davies, 1954; Drinker and Hatch, 1954). The measurement of aerosols in general, and specifically industrial dust, can be made while particles are airborne, or particles can be collected on a surface and measured physically or chemically. In the early 1920s, as well as during the entire period before the 1960s, these were the preferred collection methods in the industrial hygiene field.

1.3 THE MEASUREMENT PHILOSOPHY

I began my aerosol measurement work at the end of the 1940s and can remember very well the philosophy of dust measurement at that time. The most important reason for dust measurement in the workplace was the high incidence of silicosis in industry and in the mines. An important observation of the high mortality of hard-rock miners, accredited to Agricola (George Bauer, *De Re Metalica*), first appeared in 16th century literature (Drinker and Hatch, 1954). The recognition that silica (quartz dust) produces the characteristic pulmonary diseases of pneumoconiosis and silicosis dates to the latter 1920s (Collis, 1926).

A broad need for the measurement of industrial dust in the workplace was recognized before, but primarily after, the Second World War. What kinds of physical methods for dust sampling and for sample evaluation were available at that time? Generally speaking, knowledge of inertia particle separation, filtration, thermophoresis, and, somewhat later, particle separation in electrostatic fields already existed. However, very few sample evaluation methods existed, even though light micros-copy methods that could be used for particle counting and sizing were available. Therefore, the

FIGURE 1.3 Circular konimeter with microscope.

procedures of choice were particle sampling on plain, smooth transparent surfaces or in liquids, and particle counting and sizing by light microscopy methods.

1.4 AEROSOL SAMPLING METHODS

Gravity is not fast enough to separate respirable dust and aerosols from air samples. Inertia and thermal and electrostatic forces must be applied to speed particle deposition, or an efficient filtration method must be used. All of these methods are suitable for sampling aerosols to estimate the particle numbers or particle mass concentrations (Spurny et al., 1961).

1.4.1 KONIMETERS

The term *konimeter* was used before the 1960s to designate the one-stage impactor. The first konimeter was constructed by Sir Robert Kotze in 1919 in the Union of South Africa. Dust particles were collected by impaction on a glass plate covered with a thin film of petroleum or glycerin jelly, which trapped and retained the dust. In the U.K., the Owens Jet Dust Counter (Owens, 1922) was used for a long time. It was similar to the Kotze konimeter and contained an entrance chamber in front of the rectangular nozzle. No adhesive substance was used on the impaction glass surface. Instead, the entrance chamber was lined with moistened blotting paper to ensure humidification of the sampled air volume.

Commercial konimeters produced in the U.K. and in Germany were later available. An English konimeter, the Bausch and Lomb counter (Gurney et al., 1938), was an improved Owens konimeter. It could collect 12 samples on a circular glass plate. This instrument included a light microscope, similar to the later Zeiss konimeter. The Zeiss konimeter (Zeiss, 1950), shown in Figure 1.3, could collect 30 dust samples on a single glass disk, which was rotated to permit the immediate examination of the spots under the built-in microscope.

1.4.2 CASCADE IMPACTORS

An important improvement in the field of dust sampling was provided by an instrument consisting of four impaction stages. It was developed by May (1945) in England. Four jets were arranged in a series, and the dust particles were collected on adhesive-coated microscope slides. The May

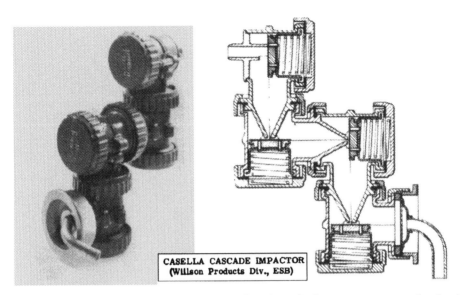

FIGURE 1.4   The Casella Cascade Impactor and its schematic diagram arrangement showing jets.

cascade impactor was produced commercially in England by the Casella company. This apparatus as well as the pictures of four fractions of a liquid aerosol are illustrated in Figure 1.4. At the end of the 1950s the Andersen cascade impactor, consisting of six stages, was introduced. The development of more sophisticated cascade impactors began after the 1960s (Mercer, 1973).

1.4.3   IMPINGERS

Impingers were similar to konimeters. The only difference between them was that, with the impinger, particle impaction was combined with subsequent collision of dust particles in water or alcohol. The impinger, shown schematically in Figure 1.5, operates like a konimeter (impactor) except that the jet is immersed in the liquid. In operation, particles larger than about 1 μm are captured by inertial mechanisms and end up suspended in the liquid. The collection efficiency drops off rapidly for particles less than 1 μm. After an air sample of known volume is taken, dust particle concentration is evaluated by light microscopy in a counting glass cell.

Impingers were already being used for dust sampling in the 1890s. Michaelis (1890) used an impinger, which was used in the Robert Koch laboratory in Germany, for sampling dust penetrating through filters. Nevertheless, further development and applications of impingers dates from the 1920s. An improved impinger dust sampling instrument was introduced in the U.S. in 1922 by Greenburg and Smith. Later, the Midget impinger was produced and used as standard equipment for a long period in the U.S., as well as in several other countries (Hatch et al., 1932).

1.4.4   PRECIPITATORS

The separation of dust particles in thermal and electric fields for airborne dust measurement was accomplished by thermal and electrostatic precipitators.

1.4.4.1   Thermal Precipitators

Bancroft (1920) proposed the development of a *thermal filter* and stated that thermophoresis played an important role in dust separation. Radiometric forces and methods for their measurement were investigated by Einstein (1924), Hettner (1924), and Epstein (1929). In the 1930s, very important studies were conducted using thermophoretic forces to sample dust in workplace air. Miyake (1935)

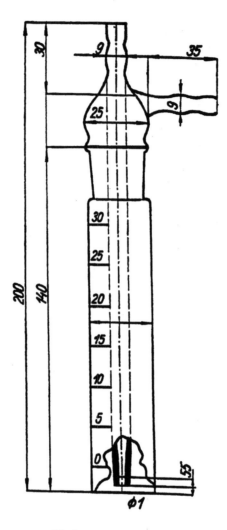

IMPINGER

FIGURE 1.5   Schematic view of a dust impinger.

experimented with a heated platinum ribbon and separated dust particles from gas flow. The most successful experiments were conducted by Green and Watson (1935) and Watson (1936). These studies enabled the construction of a thermal precipitator. The Green–Watson thermal precipitator used a nichrome wire (Ni–Cr), positioned across a slot, which was heated electrically to about 100°C. Figure 1.6 shows a schematic view of this instrument and its function.

The walls of the slot are formed by two cover slips backed by blocks of brass. The dust deposit was evaluated by light microscopy. The thermal precipitator was used for many years in several countries as a standard dust sampling instrument and underwent a number of improvements. Walkenhorst (1962) used a heated tungsten ribbon instead of a wire for improving the sampling conditions. Thermal precipitators with oscillatory or rotating collecting surfaces were designed to reduce the effects of particle overlap and to eliminate the problem of size segregation (Cember et al., 1953). Special thermal precipitators were developed and used for gravimetric sampling. The first gravimetric thermal precipitator was developed in the 1950s by Bredl and Grieve (1951). A

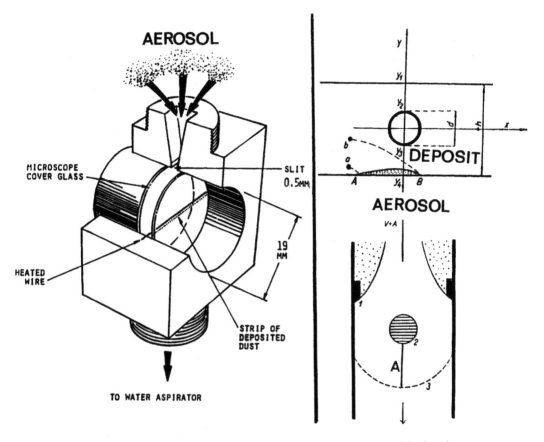

FIGURE 1.6 Schematic view of a standard thermal precipitator and its function.

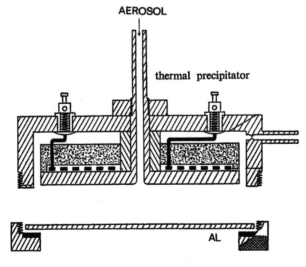

FIGURE 1.7 Schematic view of a thermal precipitator which collects dust samples large enough to weigh.

schematic view of this instrument appears in Figure 1.7. The dust was collected on the aluminum plate (AL), which could be weighed. The upper plate was heated electrically. A thermal precipitator designed for continuous aerosol sampling was developed by Orr (Orr and Martin, 1958).

1.4.4.2 Electric Precipitators

The use of electrostatic forces to remove particles from the air was demonstrated by Hohlfeld in 1824 when he applied high power to a wire suspended in a bottle filled with smoke and rapidly precipitated the smoke particles in the bottle (Mercer, 1973). The first electrofilter designed for air cleaning was developed by Cottrell in the U.S. in 1911 (Cottrell, 1911). The practical possibilities of the use of electrostatic force were recognized in 1883 by Lodge (Lodge, 1883), who attempted to use an electrostatic field to clean flue gases. Further developments of portable electric precipitators started in 1919. Tolman et al. (1919) built a small glass electric precipitator and used it to collect smoke. The first application to industrial hygiene sampling was by Bill (1919). Subsequently, a number of electric precipitators were described, e.g., by Drinker and Thompson (1925), by Drinker (1932), by Barnes and Penney (1936, 1938). The Barnes and Penney instrument was later produced commercially. The electrically-precharged dust particles were collected in a metal tube. The dust concentration was measured as mass/m^3.

1.5 PARTICLE COUNTING AND SIZING

As previously discussed, the measurement of particle number concentrations was the preferred measurement method before the 1960s. This practice was established in approximately 1916, when use of the Kotze konimeter in South Africa was the standard method. Prior to this, and after the 1960s, dust and aerosol concentrations in the workplace were measured gravimetrically as mass/m^3.

Using microscopic methods, the particles in dust samples collected by konimeters, impingers, and thermal and electric precipitators were counted, and their sizes, mainly in the range between 0.5 and 5 µm, were measured. In some cases, individual mineral particles were identified. The refractive index of transparent particles could be found by immersion methods, using a range of liquids whose indices embraced those of the particle. This procedure was applied, for example, to identify single SiO_2 (quartz) particles. Quartz has two indices of refraction, 1.544 and 1.553. Other minerals encountered in industrial dusts have higher or lower indices. By using suitable liquids (such as mononitrobenzene, tetraline, and so forth) the number of SiO_2 particles could be estimated.

If a small portion of dust is well dispersed in a liquid medium with refractive index of 1.54 (suitable oils) and examined microscopically, the quartz particles will exhibit central illumination when the objective of the light microscope is raised slightly above focus.

For the measurement of particle sizes, optical micrometers and standardized graticules were used. The eyepiece graticules consisted of a series of lines and circles of graduated size on a glass disc (Figure 1.8). The sizes of irregular particles were described in terms of arbitrary dimensions, for example, as diameters measured in one well-defined direction.

With the use of impingers, the dust particles were sampled in distilled water or ethyl alcohol. In the laboratory, samples were made up to a known volume, using, for example, a glass microcuvette. After the liquid evaporated, particles were counted and their sizes were measured.

1.6 LIMITATIONS OF THE "CLASSICAL" METHODS

Many of the imperfections and sampling and measurement errors of these methods were recognized when the methods were first used. All of the previously described instruments were plagued with problems of rebound, re-entrainment, and disaggregation of particles during sampling. The sampling times of different instruments varied between seconds and several hours. The sampling and collection

Types of graticule.

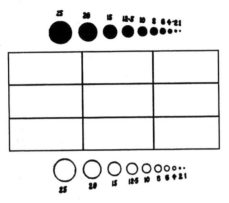

(*a*) Patterson-Cawood graticule.

$$D = \sqrt{2}^{\,n}$$

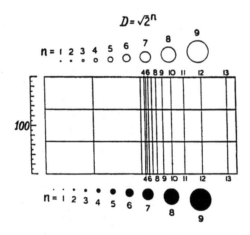

(*b*) K. R. May graticule.

FIGURE 1.8 Graticules for the estimation of particle sizes.

efficiencies of different instruments also varied substantially. For these reasons, a comparison of the concentrations measured by these instruments was practically impossible. The differences in the measured concentrations lay in the range of ±100%. Therefore, no single ratio or even approximate conversion factor was available for comparing particle counts made by two different instruments.

The importance of isokinetic sampling conditions in the measurement of dusts and aerosols was not fully recognized before the 1960s. Walton has mentioned the possible errors due to anisokinetic sampling of aerosols in 1954 (Walton, 1954): Nevertheless, the first satisfactory theory for isokinetic sampling has been described only after 1960s (Davies, 1968).

Also, the thermal precipitator, which was considered the method of choice for a long time, was later found to be of little use. Several investigations done during the 1960s and later (Mercer, 1973) show important irregularities and lack of homogeneity in particle deposits obtained by dust sampling. The theory of thermophoresis indicates that the thermophoretic force or thermophoretic velocity, at normal air pressure, depends on the particle size — the smaller particles are deposited first. Therefore, in samples obtained in a thermal precipitator, the average particle size increases continuously from the front edge (nearest the intake) to the back edge (Figure 1.9). Furthermore,

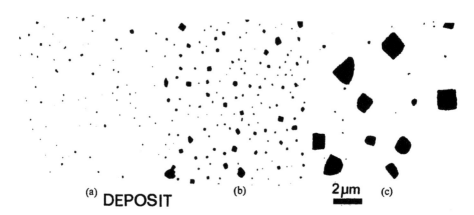

FIGURE 1.9 Segregation of aerosol particles deposited in a thermal precipitator. Front edge (a); middle (b); and rear edge (c).

nonuniform patterns of deposition exist with respect to both the number of particles per unit area and the size distributions. The collection efficiency of a thermal precipitator begins to decrease when particle sizes exceed 2 μm.

Particles of substances having large thermal conductivities are subjected to thermal forces many times greater than particles with low thermal conductivities (Schadt and Cadle, 1957).

The general conclusion, based on theoretical and experimental investigations done during the 1950s and primarily after the 1960s, is that the working principles of the "classical" methods and instruments remain useful and applicable to modern aerosol measurement techniques. However, the instruments themselves, in their original design and function, are of historical importance only.

1.7   SAMPLING BY FILTRATION

Removing dust and aerosol particles from gases by filtering them through a suitable medium provides a simple means of sample collection. The Soxhlet Filter was the standard filtration instrument used for dust sampling during the 1920s, when dust concentrations in the workplace were in the range of mg/m³. This instrument was developed by Trostel and Frevert (1923) and is shown in Figure 1.10. The instrument used a Whatman paper extraction thimble filter filled with fluffed-out cotton to reduce clogging. The gravimetric dust concentration was calculated from the change in weight of the dried thimble.

Soluble dust sampling filters were also used. A sugar-tube dust sampler was used in South Africa, and a soluble filter made of fibrous tetrachlornaphthalene crystals was used for dust sampling in France (Avy, 1956). The next development in dust-sampling filters was the paper filter disk (Silverman and Ege, 1943).

In the early 1950s, membrane filters (MFs) became the most important standard analytical and sampling filters for aerosols. These consisted of a porous membrane having a foam-like structure approximately 100 to 150 μm thick (Figure 1.11). They were prepared from one or several cellulose ester gels. The MFs had a pore volume of 75 to 80%. Pore size was controlled by the manufacturing process. The MFs were used both to weigh collected dust and to count particles with an optical microscope. Before the microscope method could be used, the MF containing the collected dust particles had to be treated with a few drops of immersion oil or an organic solvent to make it transparent.

The real history of the preparation and use of MFs began long before the 1950s. Cellulose ester MFs have been produced commercially since 1927. Production was based primarily on the research of Zsigmondy et al. in 1916 and 1918 (Spurny, 1965–1967). After World War II, the production of MFs began in the U.S., Russia, England, and Czechoslovakia.

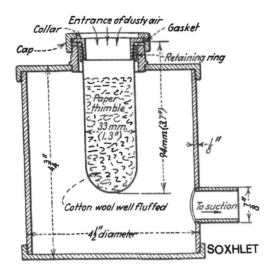

FIGURE 1.10 Schematic view of the Soxhlet gravimetric dust sampler.

FIGURE 1.11 The inner and surface structure of a membrane filter with a theoretical pore size of 0.8 μm.

The application of MFs in dust and aerosol measurement began near the end of the 1940s. Kruse had used MFs for measuring bioaerosols in Germany (1948). A Russian publication first discussed the use of MFs for the gravimetric measurement of dust concentrations in the workplace (Reznik, 1951). In the U.S., Alexander Goetz was the father of MF production and application. He used MFs for both aerosol (1953, 1956) and bioaerosol (1959) measurement. Pioneering work on the application of MFs to dust measurement was done by First and Silverman (1953), and by Fraser (1953) in the U.S. In France, the use of MFs for dust measurement was introduced by Le Bouffand (1954) and Le Bouffand and Davelu (1958). At the same time, the MF became the standard method for dust measurement in Czechoslovakia (Spurny et al., 1957). The history, production procedures, and applications of MFs to aerosol measurement had already been well documented (Spurny, 1965–67; Spurny and Gentry, 1979).

Important progress in the field of dust and aerosol measurement by pore filters was achieved after the 1960s with the invention and application of polycarbonate filters, also called Nuclepore filters or NPFs (Spurny et al., 1969).

1.8 ELUTRIATORS AND AEROSOL CENTRIFUGES

At the beginning of the 1950s, Walter (1952) had defined *air elutriation* as a process in which particles are separated on the basis of size by contrasting their settling velocity to the velocity of the current of air in which they move. Vertical elutriators were used mainly for size-fractionation or to measure the distribution of aerodynamic particle sizes, and, at that time, a very useful elutriation spectrometer was designed by Timbrell (1952, 1972). It was a portable instrument consisting of a wedge-shaped sedimentation chamber. The aerosol was drawn by a laminar air flow through the sedimentation chamber. The relationship between the particle falling speed and the distance along the particle deposit depended on the flow rate. Therefore, size/distance relationships for several flow rates could be plotted. Timbrell also used his instrument to determine the shape factors and aerodynamic diameters of fibrous particles. The deposition force for sampled particles could be substantially increased (e.g., up to a factor of 20,000) by using the aerosol centrifuges.

The operation of aerosol centrifuges was similar in principle to the horizontal elutriators, with the force of gravity replaced by centrifugal force (Mercer, 1973). The first aerosol centrifuges were also constructed in the 1950s. Sawyer and Walton (1950) designed and produced the first *conifuge*. This instrument consisted of a metal cone mounted directly on the rotor of a high-speed electric motor, and a conical metal cover that could be fastened rigidly to the cone, leaving a conical annular air space between the cone and the cover. When the unit was rotated, air was drawn into the opening at the top, pumped through the annulus, and exhausted through jet orifices at the bottom. Particles were deposited in narrow bands around the inner surface of the outer cone. The position of the center of each band was characteristic of the aerodynamic diameter of particles. The conifuge theory was formulated only after the 1960s (Stöber and Zessack, 1966).

A second, already commercially available, aerosol centrifuge was the Goetz Aerosol Spectrometer, first mentioned in 1957. This centrifuge consisted of an aluminum cone, grooved with two independent helical channels, and covered with a close-fitting conical shell. The cone rotated at speeds up to 24,000 rpm. Aerosol particles moving through the channels were subjected to a constantly increasing centrifugal acceleration that deposited them on the channel floor. The floor consisted of a thin, removable foil that covered the inner surface of the outer cone. Particles deposited on this foil formed Archimedean spirals. The length of the spiral segment was correlated to the aerodynamic particle size. The theoretical description of particle separation as performed by this centrifuge and the evaluation procedures were published only after the 1960s (Preining, 1962; Stöber and Zessack, 1966).

The next generation of aerosol centrifuges began in the 1960s, primarily with the design and development of the *spiral centrifuge aerosol spectrometer* (Kast, 1961; Stöber and Flachsbart, 1969). The Stöber Aerosol Centrifuge was later developed as a sophisticated and useful instrument (Stöber, 1972).

1.9   CONDENSATION NUCLEI COUNTING AND MEASUREMENT

Between 1880 and 1890, J. Aitken made many atmospheric observations that demonstrated the fundamental role of dust (nuclei) particles in the formation of clouds and fog. As mentioned previously, he produced the first of the Aitken condensation nuclei counters (CNCs) around 1890. The next important manually operated CNCs based on the Aitken principles were developed in Germany. Two meteorologists, Lüdeling (1903) and Scholz (1931), substantially improved the original Aitken CNC. Further improvements to the CNCs, including photographic and photoelectric sample evaluation, occurred between 1940 and 1960 (Pollak and Morgan, 1940; Verzar, 1953; Rich, 1955, Pollak and Metnieks, 1960; Podzimek, 1985).

The invention of photoelectric CNCs is dated after the 1960s. Podzimek states, in his excellent review dedicated to the 100-year evolution of Aitken counters (1985): "One of the objectives of the first International Symposium on Condensation Nuclei in Dublin (1955) was to introduce the photoelectric nuclei counter into the existing primitive methodology of atmospheric particulate counting." The instrument was introduced commercially as the Environmental One CNC (Rich, 1961; Skala, 1963). The full history of CNCs and of nuclei measurements can be found in the original Podzimek reviews (1965, 1985, 1989).

1.10   ULTRAMICROSCOPY AND OPTICAL PARTICLE COUNTERS

Tyndall's phenomenon, which was first studied in 1869 on an aerodisperse system, is caused by light scattering on particles in the air. Tyndall used his methods to demonstrate that particles well below the visible limit can be observed, counted, and measured. This was the logical basis for the later invention of ultramicroscopes, nephelometers, and optical particle counters (Gentry, 1996).

1.10.1   ULTRAMICROSCOPY

As early as 1881, ultramicroscopical observation of the motion and deposition of fine aerosol particles was described by Bodaszewsky (1881) (see Figure 1.12). The systematic development of ultramicroscopy began with the introduction of a slit ultramicroscope by Siedentopf and Zsigmondy (1903, 1904), intended specifically for the purposes of colloid chemistry.

The first design of this apparatus is also shown in Figure 1.12, and its function is schematically illustrated in Figure 1.13. Light scattered on particles was observed with an optical microscope. The individual particles were observed against a dark background. The colloid (cuvette C) was illuminated by the light source (S) and individual particles were observed in the microscope (M). Further development and application of ultramicroscopic methods continued mainly in Vienna (Ehrenhaft, 1905, 1907).

The first aerosol ultramicroscope was developed much later by Vlasenko in Russia (Deryaghin and Vlasenko, 1953). It was a flow ultramicroscope, also shown in Figure 1.13. The aerosol was aspirated through the tube (Tp) and passed through the illuminated zone in the center of this tube. Observation was made with the microscope (M) in the direction of the flow axis. The particle passing through the illuminated zone appeared as flashes of light. The counting field was defined by a diaphragm situated in the microscope eyepiece. Numerical particle counting was obtained through this eyepiece. When I visited Vlasenko in Moscow in 1962, he demonstrated this new instrument to me with pride, stating: "It is very sensitive to small particles, because the human eye is a much more sensitive apparatus than any photocell." Perhaps he was right at the time.

Visual ultramicroscopy did not permit observation of several particles while simultaneously following the movement of the particles. As early as 1919, Wells and Gehrke used a photographic recording in combination with an ultramicroscope to measure the movement of charged aerosol particles. Fuchs subsequently improved this method for simultaneously measuring size and charge of individual aerosol particles (Fuchs and Petryanov, 1933). This method is shown in Figure 1.14.

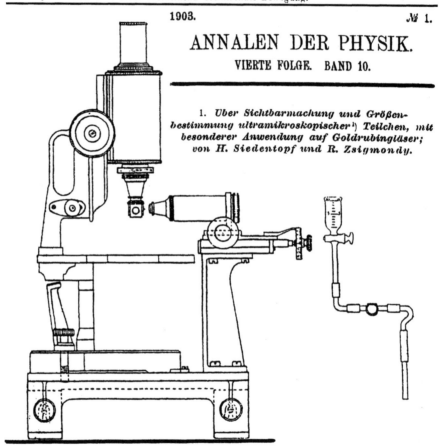

FIGURE 1.12 Ultramicroscopy article from Bodaszewsky's publication in 1881 and the ultramicroscope by Zsigmondy and Siedentopf from 1903.

The movement of individual particles in a condenser with an alternating electric field was observed and photographed. With the combination of gravity and electric force, the resulting motion of an individual particle showed a periodically changing path. By measuring the parameters L_g and L_e, an alternative evaluation of the particle size or the particle charge was possible (Kubie, 1965).

1.10.2 TYNDALLOMETRY

The measurement of the light scattered from all particles in a given volume occurred later with the development of tyndallometers and nephelometers (Berek et al., 1935; Stuke, 1955). Tyndallometers measured the light scattering through an angle of 30 degrees. In Germany, the E. Leitz Company began commercial production of tyndallometers at the end of the 1930s (Meldau, 1956). A critical evaluation of light scattering equipment existing before the 1960s was published by Hodkinson in 1965.

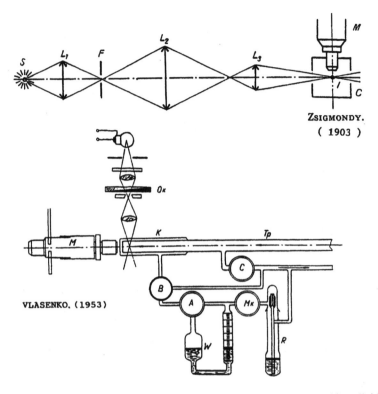

FIGURE 1.13 Schematic view of the Zsigmondy–Siedentopf ultramicroscope used in colloid chemistry, and the Vlasenko flow ultramicroscope for aerosol counting.

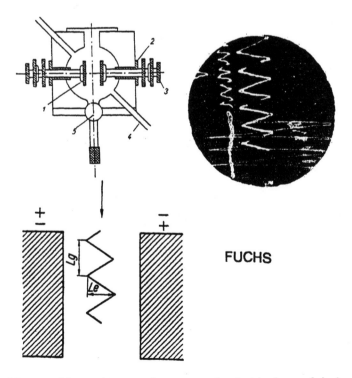

FIGURE 1.14 Diagram of the condenser used to measure the electric charge of single aerosol particles.

1.10.3 OPTICAL PARTICLE COUNTERS

A photoelectric optical particle counter, based on the theory of light scattering on individual particles, was initially developed in the second half of the 1940s (Gucker and O'Konski, 1949; Gucker, 1947; Gucker et al., 1949; Gucker and Rose, 1954). Optical particle counters measured, on line and *in situ,* the number concentration of diluted aerosols with particles larger than 0.3 μm. The particle counters were later refined to record voltage pulses of different magnitudes. These voltage pulses were correlated with light pulses coming from aerosol particles of different sizes and used to determine particle size distributions (Gucker and Rose, 1954; Fisher et al., 1955). The commercial development of optical particle counters began in the 1960s by firms such as the ROYCO company (Zinky, 1962). The full history of light scattering instrumentation for aerosol studies and measurements has been published recently by M. Kerker (1997).

1.11 MINERALOGICAL AND CHEMICAL AEROSOL ANALYSIS

As previously mentioned, quartz and other silicates, as well as some heavy metals such as lead, were the mineral dust components of interest before the 1960s. The analytical procedures and equipment available at that time, such as gravimetry, titrations, colorimetry, photometry, polarography, X-ray diffraction, and others, were used for further chemical dust and aerosol analysis.

The microscopic determination of SiO_2 (quartz) was mentioned previously. The second, more quantitative method, which was standardized internationally later, was the X-ray diffraction analysis method (Clark and Reynolds, 1936; Ballard et al., 1940; Klug et al., 1948). This method was sensitive to SiO_2 concentrations as low as 1%. Nevertheless, the method was not used until much later for routine analysis in industrial hygiene laboratories. One of the reasons mentioned by Drinker and Hatch in their book (1954) was: "The equipment (X-ray diffractometer) is expensive and the technique is not simple."

Dust samples were analyzed for heavy metals as well as SiO_2. Lead (Pb) and cadmium (Cd) were considered the most important from the viewpoint of industrial toxicology. After the sample dust was extracted and placed in a solution of strong inorganic acids, the concentration of both metals as chlorides was determined using the classical polarographic procedure.

1.12 CONCLUDING REMARKS

This short historical review is intended as a summary of my personal impressions of approximately 100 years' development in the aerosol measurement field, although it concentrates primarily on development within the period between 1920 and 1960. An exhaustive description of activities during this time is available from the published papers of the period. I have tried to remember the atmosphere and the philosophy that existed during that period. This could help younger aerosol scientists to better understand the methodological development and to evaluate the existing state of this field. (Also see Walton, W.H. and Vincent, J.H., Aerosol instrumentation in occupational hygiene: An historical perspective. *Aerosol Sci. Technol.* 28, 417-438, 1998.)

Two important realities must be considered when comparing the aerosol measurement technology available in the 1990s (Willeke and Baron, 1991) with the technology before 1960. First, development after the 1960s exploited the knowledge and the experience of the classical period. The basic sampling and measurement principles and mechanisms were developed further. Second, the development after the 1960s profited from the rapid technical and instrumentation progress in the fields of microelectronics and laser and computer techniques, as well as modern physical methods in analytical chemistry and analytical electron microscopy.

The majority of the instruments used before the 1960s were laboratory made and included no computer support and no automation. The current methodology fully utilizes all of the technical advancements made since the 1960s. Considering the most important successes in aerosol measurement techniques, I must mention, for example, the great improvements in cascade impactors (real, virtual, low pressure, and so forth) and in electric aerosol mobility analyzers, which are the logical successors of the earlier electrostatic precipitators. Enormous improvements have been made in the optical particle counters and analyzers, facilitated by modern laser and spectroscopy techniques. The same is true for the very impressive improvement of photoelectric condensation nuclei counters. New analytical aerosol filters (glass, polymer, carbon fiber, and especially the polycarbonate Nuclepore filters) have been developed and are commercially available. The developments in the field of chemical and mineralogical aerosol analysis have been very successful.

Applications of new and sensitive methods of modern analytical chemistry (chromatography, mass spectrometry, plasma and laser spectroscopy, radiation and nuclear beam methods such as PIXE, NAA, and so forth) to aerosol analysis have made it possible to determine almost all inorganic and organic components in aerosol samples. Furthermore, the applications of some of these modern analytical techniques, which are fast and sensitive enough, have made it possible to realize a new and important field of aerosol measurement and identification — single particle counting, measurement, and analysis (Spurny, 1986). The dream of Shal Friedlander (Friedlander, 1977) — the Single Particle Counter Analyzer — has almost been realized.

How about the needs and outlook for the 21st century? The research, as well as the applications of aerosol science and technology, will increase in several fields. Also, ultra-high dispersed aerosols (UHDA) will assume great importance in the near future (see also Preining, 1996), in relation to aerodispersed systems with particle sizes as small as and much smaller than 1 μm. These aerosols are of basic importance in the aerosol synthesis of nano-sized high-tech materials, and also in the field of human inhalation toxicology and its health effects. Very sensitive, on-line, *in situ* aerosol measurement and analytical techniques will be necessary for further research in the areas of both "good" and "bad" UHDA.

REFERENCES

Aitken, J., 1880, *Proc. Roy. Soc. Edinb.* 9:14-18.

Aitken, J., 1881, *Nature* (London) 23:384-385.

Aitken, J., 1888, *Nature* (London) 37:187-206.

Aitken, J., 1889, *Proc. Roy. Soc. Edinb.* 16:135-172.

Aitken, J., 1894, *Trans. Roy. Soc. Edinb.* 32:239-272.

Avy, A.P., 1956, *Les Aerosols.* Dunod, Paris.

Ballard, J.W., Oshry, H.I., and Schrenk, H.D., 1940, *U.S. Bur. Mines Rept. Invest.* 3520.

Bancroft, N.D., 1920, *J. Phys. Chem.* 24:421-433.

Barnes, E.C., 1936, *Am. J. Publ. Health* 26:274-280.

Barnes, E.C. and Penney, G.W., 1936, *J. Ind. Hyg. Toxicol.* 18:167-172.

Barnes, E.C. and Penney, G.W., 1938, *J. Ind. Toxicol.* 20:259.

Berek, M., Männchen, K., and Schäfer, W., 1935, *Zschr. für Instrumentenkunde* 56:49-56.

Bill, J.P., 1919, *J. Industr. Hyg.* 1:323-342.

Bodaszewsky, L.J., 1881, *Dinglers Polytech. J.* 5:325.

Bredl, J. and Grieve, T.W., 1951, *J. Sci. Instrum.* 28:21-29.

Cember, H., Hatch, T., and Watson, J.A., 1953, *Am. Hyg. Assoc. Quart.* 14:191-197.

Clark, G.L. and Reynolds, D.H., 1936, *J. Industr. Hyg.* 7:345-349.

Collis, E.L., 1926, *Tuberculosis-Silicosis: Occupation and Health,* International Labour Office. Brochure 32, Geneva.

Cottrell, F.G., 1911, *J. Ind. Engn. Chem.* 3:242-250.

Coulier, P.J., 1875, *J. de Pharmacie et de Chimie,* Paris, Ser. 4, 22(1):165-172 and 22(2):254-255.

Davies, C.N., 1954, *Dust Is Dangerous.* Faber and Faber Ltd., London, U.K.

Davies, C.N., 1968, *Br. J. Appl. Phys. Ser.* 2, 1:921-932.

Deryaghin, B.V. and Vlasenko, G.Y., 1953, *Priroda (Nature,* Moscow) 11.

Drinker, P. and Thompson, R.M., 1925, *J. Ind. Hyg.* 7:261-271.

Drinker, P., 1932, *J. Ind. Hyg. Toxicol.* 14:364-370.

Drinker, P. and Hatch, T., 1954, *Industrial Dust.* McGraw-Hill, Inc. New York.

Ehrenhaft, F., 1905, *Wien. Anz.* 3:213-214.

Ehrenhaft, F., 1907, *Wien. Anz.* 5:72 and 331.

Einstein, A., 1924, *Zschr. für Physik* 27:i-6.

Epstein, P., 1929, *Zschr. für Physik* 54:537-542.

First. M.N. and Silverman, L., 1953, *AMA Arch. Industr. Hyg. Occup. Med.* 7:1-14.

Fisher, M.A., Katz, S., Lieberman, A., and Alexander, N.E., 1955, *The Aeroscope: An Instrument of the Automatic Counting and Sizing of Aerosol Particles.* Proc. Natl. Air Pollution Symposium Los Angeles. p. 112.

Fraser, D.A., 1953, *AMA Arch. Industr. Hyg. Occup. Med.* 7:412-419.

Friedlander, S.K., 1977, *Smoke, Dust and Haze.* John Wiley & Sons, New York.

Fuchs, N.A. and Petryanov, J., 1933, *Kolloid Zschr.* 65:171-183.

Fuchs, N.A., 1955, *Mechanika Aerozolei.* Academy of Science Press, Moscow.

Fuchs, N.A., 1964, *The Mechanics of Aerosols.* Pergamon Press, Oxford.

Fuchs, N.A., 1971, *Collection of Aerosol Abstracts.* Karpov Institute, Moscow.

Gentry, J.W., 1996, *J. Aerosol Sci.* Suppl. 1 27:S503-S504.

Goetz, A., 1953, *A. J. Publ. Health* 43:150-163.

Goetz, A., 1956, *J. Air Pollut. Control Assoc.* 6:19-27.

Goetz, A., 1957, *Geofis. Pura Appl.* 36:49-69.

Goetz, A. and Tsuneishi, N., 1959, *AMA Industr. Health* 20:167-174.

Green, H.L. and Watson, H.H., 1935, Physical Methods for Estimation of the Dust Hazard in Industry.*Medical Res. Council Spec. Rept.* Ser. No 199.

Greenburg, L. and Smith, G.W., 1922, *U.S. Bur. Mines Rept. Invest.* 2392.

Greenburg, L., 1932, *Am. J. Publ. Health* 22:1077-1082.

Gucker, F.T., Pickard, H.B., and O'Konski, C.T., 1949, *J. Am. Chem. Soc.* 69:429-438.

Gucker, F.T., 1947, *Electronics* 20:106-110.

Gucker, F.T., and O'Konski, C.T., 1949, *Chem. Revs.* 44:373-388.

Gucker, F.T., and Rose, D.G., 1954, *Br. J. Appl. Phys.* Suppl. 3 S138-S143.

Gurney, S.W., Williams, C.R., and Meigs, R.P., 1938, *J. Industr. Hyg. Toxicol.* 20:24-31.

Hatch, T., Williams, C.E., and Dolin, B.C., 1932, *J. Industr. Hyg.* 14:301-311.

Hettner, G., 1924, *Zschr. für Physik* 27:12-22.

Hodkinson, J.R., 1965, The physical basis of dust measurement by light scattering, in Spurny, K. (Ed.) *Aerosols — Physical Chemistry and Applications.* Press of the Academy of Science, Prague. pp. 181-194.

Kast, W., 1961, *Staub* 21:215-223.

Kerker, M., 1997, *Aerosol Sci. Technol.* 27:522-540.

Klug, H.P., Alexander, L., and Kummer, E., 1948, *J. Industr. Hyg. Toxicol.* 30:166-172.

Kubie, G., 1965, On the ultramicroscopy of aerosols, in Spurny, K. (Ed.)*Aerosols — Physical Chemistry and Applications.* Press of the Academy of Science, Prague. pp. 207-219.

Le Bouffand, L., 1954, *2 eme Colloque Inter. sur las Pousieres,* Strassbourg.

Le Bouffand, L. and Davelu, M., 1958, *Staub* 18:342-351.

Lodge, J.P., 1969, *The Smoake of London.* MRC, Maxwell Reprint Comp. New York.

Lodge, O.J., 1883, *Nature* (London) 28:274-280.

Lodge, O.J., 1883, *Nature* (London) 28:297-299.

Lodge, O.J. and Clark, J.W., 1894, *Phil. Mag. Ser.* 5, 17:214-239.

Lüdeling, G., 1903, *Illustr. aeronautische Mitteilungen* 7:321-335.

May, K.R., 1945, *J. Sci. Instrum.* 22:187-193.

Meldau, U.R., 1956, *Handbuch der Staubtechnik.* VDI Verlag, Düsseldorf, Germany.

Mercer, T.T., 1973, *Aerosol Technology in Hazard Evaluation.* Academic Press, New York.

Michaelis, H., 1890, *Zschr. Hygiene* 9:389-393.

Miyake, S., 1935, *Aeronaut. Res. Inst. Tokyo,* Report 123, pp. 85-106.

Orr. C. and Martin, R.A., 1958, *Rev. Sci. Instrum.* 29:129-134.

Owens, J.S., 1922, *Proc. Royal Soc.* (London), Vol. 101:18-25.

Podzimek, J., 1965, The importance of aerosols in meteorology in Spurny, K. (Ed.) *Aerosols — Physical Chemistry and Applications.* Press of the Academy of Science, Prague. pp. 493-509.

Podzimek, J., 1985, *J. Rech. Atmos.* 19:257-274.

Podzimek, J., 1989, *Bull. Amer. Meteorological Soc.* 70:1538-1545.

Pollak, L.W. and Morgan, W.A., 1940, *Rep. Irish Meteor. Service,* Dublin.

Pollak, L.W. and Metnieks, A.L., 1960, *Geofis. Pura Appl.* 50:7-22.

Preining, O., 1962, *Staub* 22:129-133.

Preining, O., 1996, *J. Aerosol Sci. Suppl.* 1 27:s 1 S 6.

Rayleigh, Lord, 1882, *Proc. Roy. Soc.* (London), 34:414-418.

Reznik, J.B., 1951, *Gigiena i Sanitaria* (Moscow) 10:28-41.

Rich, T.A., 1955, *Geofis. Pura Appl.* 35:702-706.

Rich, T.A., 1964, *Geofis. Pura Appl.* 50:46-52.

Sawyer, K.F. and Walton, W.H., 1950, *J. Sci. Instrum.* 27:272-276.

Schadt, C.F. and Cadle, R.D., 1957, *Anal. Chem.* 29:864-868.

Scholz, J., 1931, *Zschr. Instrumentenkunde* 51:505-521.

Siedentopf, H. and Zsigmondy, R., 1903, *Ann. Phys.* 10:1-17.

Siedentopf, H. and Zsigmondy, R., 1904, *Berl. klin. Wschr.* 41:865-872.

Silverman, L. and Ege, J.F., 1943, *J. Industr. Hyg. Toxicol.* 25:185-192.

Sinclair, D., 1950, *Handbook of Aerosols.* AEC, Washington, D.C.

Singerson, G. 1870-1874, *Proc. Royal Irish Acad.* Second Series, Science. pp. 13-31.

Skala, G.F., 1963, *Anal. Chem.* 35:702-706.

Spurny, K. and Vondracek, V., 1957, *Collection Czechoslov. Chem. Cornmun.* 22:22-34.

Spurny, K. Jech, C., Sedlacek, B., and Storch, O., 1961,*Aerosoly (Aerosols)* Publ. Technical Literature, Prague.

Spurny, K., 1965-1967, *Zbl. Biol. Aerosol-Forschung* 12:369-407; 13:44-101; 13:398-451.

Spurny, K.R., Lodge, J.P., Frank, E.R., and Sheesley, D.C., 1969, *Environ. Sci. Technol.,* 3:453-458.

Spurny, K.R. and Gentry, J.W., 1979, *Powder Technol.* 24:129-142.

Spurny, K.R. (Ed.), 1986, *Physical and Chemical Characterization of Individual Airborne Particles.* Ellis Horwood. Chichester, U.K.

Spurny, K.R., 1993, *J. Aerosol Sci. Suppl.* 1. 24:S 1 — S 2.

Stöber, W. and Zessack, U., 1966, *Zschr. Biol. Aerosol-Forschung* 13:263-281.

Stöber, W. and Flachsbart, H., 1969, *Environ. Sci. Technol.,* 3:1280-1296.

Stöber, W., 1972, Dynamics of shape factors of nonspherical particles, in Mercer, T.T., Horrow, P.E., and Stöber, W. (Eds.)*Assessment of Airborne Particles.* C.C. Thomas Publ., Springfield, IL (U.S.) pp. 249-289.

Stuke, J., 1955, *Glückauf* 91:1405-1406.

Timbrell, V., 1952, *Nature* 170:318-320.

Timbrell, V., 1972, An aerosol spectrometer and its applications, in Mercer, T.T., Morrow, P.E., and Stöber, W. (Eds.) *Assessment of Airborne Particles.* C.C. Thomas Publ., Springfield, IL (U.S.) pp. 290-330.

Tolman, R.C., Reyerson, L.H., Brooks, A.P., and Smyth, D.H., 1919, *J. Am. Chem. Soc.* 41:587-589.

Trostel, L.S. and Frevert, H.W., 1923, *J. Industr. Engn. Chem.* 15:232-239.

Tyndall, J. 1870, *Proc. Roy. Inst.* 6:1-14.

Tyndall, J. 1871, *On Dust and Smoke.* RILS-PS, Vol. 2, pp. 302-313.

Union of South Africa, 1919, Miners Phthisis Prevention Committee, Johannesburg, January 10.

Verzar, F., 1953, *Arch. Meteorol. Geophys. Bioklim.* A %:372t374.

Walkenhorst, W., 1962, *Staub* 22:103-111.

Walton, W.H., 1952, *Br. J. Appl. Phys. Suppl.* 3:S 29 — S 37.

Walton, W.H., 1954, *Am. Ind. Hyg. Quat.* 15:21-25.

Watson, H.H., 1936, *Trans. Farad. Soc.* 32:1073-1078.

Wells, P.V. and Gehrke, R.H., 1919, *J. Am. Chem. Soc.* 41:312-321.

Willeke, K. and Baron, F.A., 1991, *Aerosol Measurement.* Van Nostrand Reinhold, New York.

Zeiss, 1950, Zeiss Konimeter, *Prospectus* 650.

Zinky, W.R., 1962, *Air Pollut. Control Assoc. J.* 12:578-583.

Zsigmondy, R. and Bachmann, W., 1916, German Patents 329-060 (May 9) and 329-117 (August 22).

Zsigmondy, R. and Bachmann, W., 1918, *Zschr. Anorg. Chem.* 103:119-129. *J. Soc. Chem. Ind:* 453A-458A.

2   Trends in the Chemical Analysis of Aerosols

Kvetoslav R. Spurny

CONTENTS

2.1   INTRODUCTION

Titration, colorimetry, polarography, and X-ray diffraction were the "most modern" methods for the chemical analysis of particulate and gaseous air pollutants before the 1960s (Chapter 1). Now there are 20 or more modern, sensitive analytical methods available for analyzing atmospheric and other aerosols in relatively small samples, even single particles. The first important step in aerosol measurement and analysis is to obtain a representative particulate sample. Methods have to be applied that do not disturb the dynamics of the aerosol cloud and do not change the physical and chemical state of single particles. Isokinetic sampling procedures have to be used.

The composition of atmospheric aerosols varies markedly, depending on sampler location, the proximity of significant sources of aerosols and their gaseous precursors and meteorology. Anthropogenic aerosols are concentrated in fine particles (<2.5 μm in diameter), while natural aerosols are concentrated above all in larger particles.

The existing routine analyses, for example, those that support current air quality standards in the U.S. and Europe, require only the determination of lead in total suspended particles, and the mass of

PM10 inlet   PM2,5 inlet

FIGURE 2.1   Size-separating inlets for the sampling and measurement of aerosol fractions PM 10 and PM 2.5 μm.

suspended particles less than 10 μm in aerodynamic diameter (PM 10).[1,2] Studies on visibility impairment as it relates to atmospheric particulate matter emphasize analysis of the <2.5 μm (PM 2.5) particle fraction for its major constituents, SO_4^{2-}, NO_3^-, NH_4^+, and organic and elemental carbon. Analysis in support of health effects research may determine the concentration of toxic metals, e.g., Ni, Cd, and Cr), H^+ or specific organic compounds, e.g., polyaromatic hydrocarbons, PAH, etc.[1,2]

The recent results of epidemiological and toxicological studies suggested several improvements and changes in the existing analytical philosophy and strategy (Chapter 3). More complex, extended and improved particulate analyses are required in the near future to get better and more precise information about the chemistry of atmospheric anthropogenic aerosols as well as of aerosols in several modern technological and manufacturing processes.

2.2   SAMPLING AND ARTIFACTS

Aerosol samples are most often acquired by drawing ambient or room air through filter material using a pump, with subsequent quantification of the particle mass and its chemical components by off-site laboratory analysis. As mentioned, the air sampling should be done under isokinetic conditions and by using size-selective inlets, which define the particle size fraction being sampled. Examples of such inlets for fractions of <2.5 μm and <10 μm are shown in Figure 2.1.

The sampling strategies depend on the prescribed definition of the particulate sample. The sampling in ambient air is mostly defined as PM 10 and PM 2.5. Dichotomous samplers are the most suitable equipment for such samplings.[3,4]

In workplaces, the indoor aerosol is usually sampled or directly measured in three fractions. For example, the convention of European Standard EN 481[5] requires the separate measurement of these fractions: inhalable (IN), thoracic (T), and alveolar (A) aerosol fraction. A new personal measuring system was developed recently which simultaneously separates and monitors these three fractions on-line and *in situ*. The measuring principle is a combination of inertial classification and concentration enrichment using virtual impactors, filter sampling, and aerosol photometry. The apparatus RESPICON™-3F is commercially available and has been used since 1998. Its functional schematic is illustrated in Figure 2.2.[6]

Dichotomous sampling seems to be the method of choice for size-selective fractionation by aerosol measurements at the end of the 20th century. The very useful trichotomous sampler can separate successfully three aerosol fractions by on-line sampling; e.g., the fractions PM 10, PM 2.5,

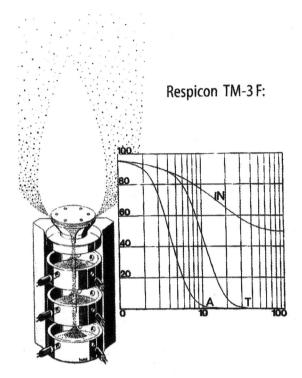

Respicon TM-3 F:

FIGURE 2.2 Pictures showing the size-selective sampling and measurement of three aerosol fractions — inhalable (IN), thoracic (T) and alveolar (A) — by the RESPICON system. (From Koch, W., Dunkhorst, W., and Lödding, H. Respicon™-3 F: A new personal measuring system for size segregated dust measurement at workplaces. *Gefahrstoffe-Reinhalt. Luft.* 57:177-184, 1997. With permission.)

and PM 1.[7] Dichotomous separators can also sample and enrich ultrafine aerosol fractions, e.g., particulates less than 0.1 μm.[8] Improved virtual impactors have been found to be very useful for fractionation of spherical and nonspherical particles.[9]

The function of virtual impactors by sampling spherical and fibrous aerosols is illustrated in Figure 2.3. The separation efficiency curves depend strongly on the Stk (Stokes number). This means that, by changing flow rate and/or inlet dimension, the particle size- and shape-selective sampling is possible.

Another fractionating system was developed for the collection of milligram quantities of condensation nuclei for chemical analysis.[10] This system can isolate three size classes of "nano-particles," having mass median diameters of 0.27, 0.12, and 0.075 μm, respectively.

The classical impinger sampler (Chapter 1) has experienced a renaissance recently.[11] A "cascade" impinger has been developed and tested. It was designated as a multistage liquid impinger (MLI) and consists of four impinger stages and a back filter. It can sample aerosols in five fractions — four in liquids and one on a filter. In Figure 2.4, the schematic of the equipment and the calibration curves are shown. The fractions in a liquid medium have cut-off sizes: 13, 6.8, 3.1, and 1.7 μm. The apparatus could have useful and special applications; e.g., sampling liquid and soluble aerosols, sampling chemically unstable substances, etc.

Artifacts. Sampling artifacts are formed by the retention and absorption of gases by the sample media and collected particles, volatilization of collected particles, gas–particles and particle–particle reactions within collected particles. Almost all sampling procedures influence the chemistry of particulate matter. Positive and negative errors, or artifacts, can be observed. The measurement of semivolatile materials, which can coexist in the aerosol and gaseous phase, can be influenced by negative artifacts. Some aerosols, such as NH_4NO_3, NH_4Cl, nitrates, PAHs, etc., are examples of

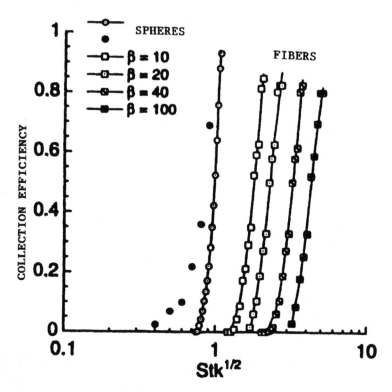

FIGURE 2.3 Calculated collection efficiencies of spherical and fibrous particles in an improved virtual impactor as a function of its Stokes numbers and the aspect ratios of fibers, β (β = fiber length/fiber diameter). (From Asgharian, B. and Godo, M.N. Transport and deposition of spherical particles and fibers in an improved virtual impactor. *Aerosol Sci. Technol.* 27:499-506, 1997. With permission.)

such matter. Positive artifacts can be produced by sorption or reaction of HNO_3, SO_2, some gaseous organics, etc., by sampling on filters. In this case, the filter material and air flow rates play important roles. All these possible artifacts must be considered when using different filters and sampling situations.[2] Some experimental results showed that dichotomous samplers with Teflon filters collected about 8% more sulfates because of adsorption and oxidation of SO_2 by the particle deposits, but about 20% less nitrates, about 30% less chlorides, and about 20% less ammonium because of volatization and reactions.[12]

It is, therefore, clear that new, artifact-free sampling methods are needed for determination of the aerosol composition as a function of particle size. One way to decrease the effects of sampling artifacts is to use different kinds of diffusion denuders and sampling trains.[13] They can be used for the on-line separating collection of vapors, volatile aerosols, and stable aerosols (see also Chapter 4).

Another solution is to use a combination of a preseparator with a steam-jet aerosol collector (SJAC).[14] Such a system is artifact-free. An SJAC consists of a wet denuder, which removes water-soluble gases from the sample air stream. Then the air is mixed with steam. The resulting high supersaturation causes aerosol particles to expand into droplets. The droplets containing dissolved aerosol species are then collected by a small cyclone. The collected solution is constantly pumped out and is on- or off-line analyzed by means of ion chromatography and flow-injection analysis.[15]

2.3 BULK ANALYSIS

The bulk analysis of aerosol samples on filters, on other organic or inorganic substrates, or in liquids is realized mainly by the use of modern analytical methods and equipment. The majority of these methods are applied as routine and standardized methodology.

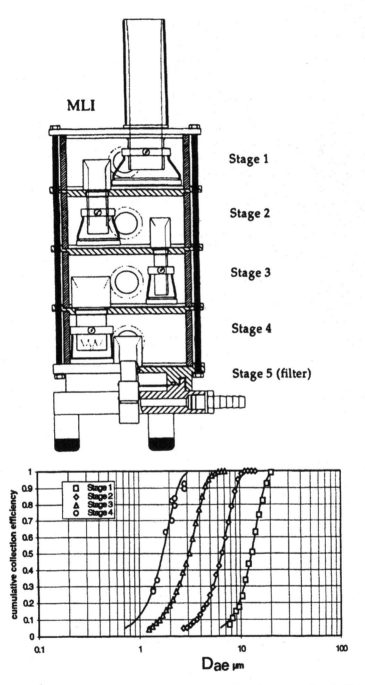

FIGURE 2.4 Schematic of the multistage liquid impinger (MLI) and an example of efficiency curves for stages 1–4 of the MLI at a flow rate of 60 L/min. (From Asking, L. and Olsson, B. Calibration at different flow rates of a multistage liquid impinger. *Aerosol Sci. Technol.* 27:39-49, 1997. With permission.)

2.3.1   ELEMENTAL ANALYSIS BY NONDESTRUCTIVE TECHNIQUES

These methods are used for direct analysis of aerosol samples on filters. The most popular methods in this category are: **XRFA** (X-ray fluorescence analysis), **PIXE** (proton-induced X-ray emission analysis), and **INAA** (instrumental neutron activation analysis).

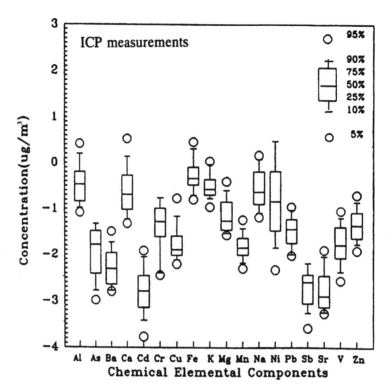

FIGURE 2.5 Box plots of elemental concentration in atmospheric aerosols of Taiwan obtained by ICP measurements. (From Wang, C.F., Yang, J.Y., and Ke, C.H. Multi-element analysis of airborne particulate matter by various spectrometric methods after microwave digestion. *Anal. Chim. Acta.* 320:207-216, 1996. With permission.)

All these methods can also be considered complementary procedures. XRFA and PIXE quantify the concentrations of elements with atomic numbers ranging from 11 (sodium) to 92 (uranium).[2,4] A combination of INAA and PIXE is preferred. Several elements are generally determined with precision in both, and the results compare well.[16] Nevertheless, for several elements the methods are actually complementary.[4] PIXE (see also Chapters 5 and 6) is more commonly applied in routine aerosol analysis than the INAA.

2.3.2 ELEMENTAL ANALYSIS BY DESTRUCTIVE TECHNIQUES

Destructive techniques for elemental analysis of aerosol samples begin with sample dissolution, usually in concentrated acid, solutions at reflux temperature, or, at about 100°C, using ultrasonic extraction. **AES** (atomic emission spectroscopy), **AAS** (atomic absorption spectroscopy), and mainly **ICP** (inductively coupled plasma emission spectroscopy) are the most commonly used methods for routine analysis of aerosol samples. For example, As, Ba, Cd, Co, Cu, Fe, Mn, Ni, Pb, Sr, V, and Zn can be satisfactorily determined by means of the mentioned techniques[2,4] (see also Chapter 7). The detection limits by improved methods can reach the ng range.[17] Important improvements in sensitivity can be achieved, e.g., by using a microwave digestion procedure in sample dissolution. Results obtained by such methods are illustrated in Figure 2.5.[18] By similar procedures, As concentrations in the range of ng/m³ were determined in samples of atmospheric aerosols.[19,20]

Several less common elements can also be satisfactorily determined in aerosol samples. Amounts of 0.1 mg/m³ of Hf and more were determined in industrial dust samples in some workplaces.[21] Be concentrations in the range of less than 0.5 ng/m³ can be measured in ambient air by AAS.[22]

2.3.3   OTHER METHODS FOR ELEMENTAL ANALYSIS

For the determination of metals, e.g., Cd, Cu, Mn, and Pb, in atmospheric aerosols, the Zeeman–AAS is a very useful and sensitive method. The aerosol can also be continuously sampled and electrostatically accumulated in a graphite furnace. In such a case the concentration detection limits are three orders of magnitude lower than those attained in a normal AAS.[23]

ICP-MS (mass spectroscopy) and **ID** (isotope dilution) **-ICP-MS** can be used successfully to determine subnanogram levels of Pt and Pd in atmospheric aerosols.[24-26] **ID-MS** makes it possible to determine Cr (III) and Cr (IV) in sampled aerosol particles. The detection limits lay at about 30 pg/m³ for Cr (III) and 8 pg/m³ for Cr (IV).[27] The **UV**-visible spectrometric method can be applied for the determination of small amounts (ca. ng/m³) of silicon in atmospheric aerosols sampled by filters.[28] The Mössbauer spectroscopy is of interest for studies of iron-containing particles of atmospheric aerosols. This technique permits not only the evaluation of the chemical compounds, in which Fe appears, but also the evaluation of the size of the particles in which the iron is contained (see also Chapter 8).

2.3.4   WATER-EXTRACTABLE ANION AND CATION ANALYSIS

Aerosols sampled on filters are extracted in bidistilled water by ultrasonic treatment, and then anions and cations are determined by different analytical techniques. The **IC** (ion chromatography) is the most applied analytical method. Several anions (NO_3^-, SO_4^{2-}, Cl^-, etc.) and cations (Cu^{2+}, Ni^{2+}, Co^{2+}, Pb^{2+}, and Fe^{2+}) can be determined by IC in airborne particulate samples. The detection limits are about 0.01 µg/ml for anions and about 1.0 to 3.0 µg/ml for cations.[29]

CZE (capillary zone electrophoresis) with an indirect UV-detector is another sensitive method for the analysis of atmospheric aerosol samples collected by cascade impactors and filters. Several anions (Cl^-, SO_4^{2-}, NO_3^-, oxalate, formiate, phosphate, acetate) and cations (NH_4^+, K^+, Ca^{2+}, Na^+, Mg^{2+}) can be determined at detection limits between 0.07 and 3.0 mg/L.[30]

SIE (selective ion electrode) is a technique that can be applied for the determination of NH_4^+. The ammonia-selective electrode uses a hydrophobic, gas-permeable membrane to separate the sample extract from an internal electrode solution of NH_4Cl. A chloride-ion selective electrode senses the fixed level of Cl^- in the internal solution, providing a reference electrode. Potentiometric measurements are made with a pH-meter. The method is applicable for analysis in the range of concentrations from 0.04 to 1700 µg/ml (as NH_3).[2]

A new collection method, the vapor condensation aerosol collection system (**VCACS**), has been shown to be very suitable for the continuous monitoring of strong acid aerosols. (See Kuzuaki, I., Coleman, C.C., Chung, H.K., et al., A continuous monitoring system for strong acidity in aerosols. *Anal. Chem.* 70, 2839-2847, 1998.) Non-H^+ cations (primarily NH^+) are conductometrically determined as the corresponding hydroxide. Total strong acid anions are then conductometrically determined by elution with carbonate/hydroxide-based eluents. Aerosol strong acidity is determined on the basis of charge balance: H^+ equivalents present = Σ anion equivalents – Σ non-H^+ cation equivalents. The system works continuously, with a detection limit of 7 to 18 nmol/m³.

2.3.5   CARBON DETERMINATION

The total carbon (TC) present as aerosol in the atmosphere can be expressed as the sum of organic carbon (OC), elemental carbon (EC), and carbonate carbon (CARBC). The contribution of CARBC to the TC in atmospheric aerosols is usually less than 5%.[2] The amount of TC is determined by combustion or decomposition of the particulate sample to CO_2. OC and EC are then determined separately by various methods, several of which have already been standardized.[2,31]

Carbon-containing material is a significant constituent of atmospheric anthropogenic aerosols. The mass fraction varies between about 10% and 40% of total mass in urban aerosols. Carbonaceous particles consist of two major components: EC, also designated as black or free carbon, which has a graphitic microstructure and is strongly light absorbing, and organic carbon (OC), which is emitted directly into the atmosphere or arises from atmospheric gas-phase photochemical processes. EC is a residue of incomplete combustion and, therefore, is an unambiguous indicator of emissions. According to the particle formation mechanism during combustion, the EC — combustion soot — is often heavily "contaminated" with organic compounds which belong to the important toxic, mutagenic, and carcinogenic substances. All emissions from incomplete combustion processes contain carcinogenic PAH (polycyclic aromatic hydrocarbons) and their nitroderivates (nitro-PAH).

The existing routine methods for measuring EC as an outdoor and indoor air pollutant, which are also used for on-line monitoring, are based on different properties of carbonaceous aerosols. The optical properties of EC are used by optical-sensitive measurement techniques. Optical absorption caused by particulate EC in the ambient aerosol is then measured, e.g., with photoacoustic spectroscopy,[32] or by attenuation of light reflectance.[31] Both methods need calibration.

A method to detect and classify combustion particles according to their PAH-content is *in situ* charging of the aerosol by UV-photoemission.[33] Commercial equipment for these methods is available, e.g., the Aethalometer and the PAS (photoelectric aerosol sensor).

Several other methods for EC determination have been developed and tested recently.[34-37] A coulometric method is able to distinguish between EC and OC and has a detection limit of about 0.16 $\mu g/m^3$.[34] A soot photometer measures the light transmission of aerosol samples collected on Nuclepore filters (NPF). The absorption optical depth and the mass of black carbon in the aerosol sample is determined from the light intensities measured for the pure NPF and for NPF with aerosol samples. This equipment also needs calibration.

The soot amount is then measured on the NPF in $\mu g/cm^2$.[35] This method was also extended for the analysis of the aerosol sample on NPF with Raman spectroscopy.[36] The Raman spectra of highly absorbing EC deposits show a strong, nonlinear rise in the Raman intensity in the high wave region (>2000 cm^{-1}). With increasing laser power, the intensity relation of the bands at 1601 cm^{-1} to 887 cm^{-1} remain unchanged. These bands result from the aromatic ring system. There is a linear relationship between EC concentration and the relative Raman intensities (I-1600/I-888) within the concentration range of 0.1 to 20 $\mu g/cm^2$. Another spectroscopic method able to measure EC concentrations is EPR (electron paramagnetic resonance) spectroscopy, described in Chapter 9.

In recent years, exhaust emissions from diesel-powered vehicles have received special attention because of their health effects. For this reason, it is important to be able to distinguish the diesel-soot aerosol in the ambient air from the soot aerosol emitted from other combustion sources. 1-nitropy-rene(1-NP) was found as a characteristic marker of diesel soot.

The soot aerosol samples on filters can be extracted by ultrasonic extraction with dichloromethane. The eluates are then analyzed by a sensitive immunoassay (ELISA) or by conventional HPLC or GC/MS. The ELISA method is the method of choice. It is useful and fast for determining 1-NP concentrations in range between about 50 and 150 pg/m^3.[37]

2.3.6 ORGANIC AEROSOL ANALYSIS

Atmospheric anthropogenic aerosols are vehicles for several hundred organic substances, of which several are toxic, mutagenic, and/or carcinogenic. Polycyclic aromatic hydrocarbons (PAHs) as well as their nitroderivates (nitro-PAHs) belong to the most important group of organic particulate air pollutants, for which concentrations are measured routinely. Their presence in the atmosphere is due principally to incomplete combustion of various materials, particularly oil fuels/petrochemicals. Their main sources are vehicles (diesel and petrol) exhaust emissions, but an important role is played by the emission from domestic oil heaters and industry.[38] The HPLC, GC/MS and similar

atmospheric concentrations (gas and particles)

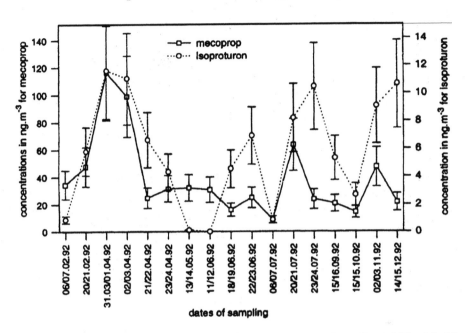

FIGURE 2.6   Annual atmospheric concentrations of two important pesticides. (From Millet, M., Wortham, H., Sanusi, A., and Mirabel, P. Atmospheric contamination by pesticides: Determination in the liquid, gaseous and particulate phases. *ESPR-Environ. Sci. Pull. Res.* 4:172-180, 1997. With permission.)

method combinations are the procedures of choice for their analysis. About 15 or more PAHs and approximately 8 nitro-PAHs are routinely measured by this methodology.[39-41]

The multiphoton ionization fast conductance (MPI–FC) technique is a new alternative method to detect combustion by-product aerosols.[42] The PAH-contaminated aerosols are sampled on-line by means of renewable water droplets. The method is extremely sensitive. Detection limits as low as 1 pg can be determined (see also Chapter 10).

Another important group of harmful organic substances which occurs in the atmosphere in gaseous as well as in particulate phase is the **agroaerosol**. The intensive use of pesticides, fungicides, etc., leads to a ubiquitous contamination of the air. Pesticides enter the atmosphere by application drift during spraying operations. Several commercially produced pesticides can be monitored in the ambient air of agricultural regions. Their presence in the atmosphere is an important problem for human health.

Several analytical procedures have been developed and used to determine the pesticide concentrations in ambient air.[43-52] Inorganic fiber filters connected to adsorbing columns (to trap vapors) are applied to collect ambient air samples of pesticides.[46] The samples on filters and adsorption columns are then extracted (e.g., with dichloromethane) by ultrasonic methods. Gas chromatography/mass spectrometry with a selected ion monitor (**GC/MS-SIM**) is a very useful procedure for quantification of about 40 different pesticides in air.[46] A recently published multiresidue method for determination of pesticides in air uses a combination of HPLC-based fractionation followed by **GC-ECD** (electron capture detector) and HPLC-UV analysis of each fraction. The detection limits vary between 0.1 and 0.01 µg/ml of the extract.[47-49] By application of this procedure for ambient air measurements on farms, relatively high concentrations of some pesticides were determined (Figure 2.6). The gaseous phase of pesticides was more concentrated than the aerosol phase in the majority of cases.

Activated carbon fiber filters (ACFF) have been used successfully for sampling gaseous and particulate pesticides and fungicides in ambient air. These filters have gas adsorbing as well as particulate collecting properties.[50-52]

The atmospheric aerosols also contain several organic compounds of natural polymers. Cellulose is a common component in atmospheric aerosols. A useful method for the detection of particulate cellulose in ambient air is based on the use of enzymes. The analytical procedure consists of three steps, in which cellulose is converted to D-glucose. Its concentration is then determined by the photometric method. Measurements in the ambient air in downtown Vienna have shown cellulose concentrations between 0.1 and 1.0 $\mu g/m^3$.[53]

2.4 ANALYTICAL AEROSOL MONITORING

In situ and on-line monitoring of aerodispersed air pollutants has several advantages. The concentration measurement results are available immediately, time-consuming and expensive laboratory procedures are not necessary, and the already mentioned sampling artifacts can be avoided. Particulate monitoring can be divided into two main categories: particulate mass measuring methods and chemical specific methods.[4] The analytical aerosol monitors belong in the second category. The existing continuous aerosol analyzers can be considered the soot monitors for measuring the concentrations of EC and OC, the heavy metal monitors, and the multicomponent aerosol monitors (see also Chapter 3).

2.4.1 Soot Monitors

A method for the measurement of EC concentrations in ambient air based on the light absorption in filter samples was published in 1984.[54] The soot monitor AETHALOMETER samples ambient air on a quartz-fiber filter tape. The difference in attenuation between exposed and blank filter is proportional to the amount of light-absorbing carbonaceous aerosol. Detection limit is about 0.01 $\mu g/m^3$ for 1 minute average.

For determination of OC and EC concentrations, an *in situ* thermal/optical analyzer was developed at the end of the 1980s.[4,55,56] The sampler provides on-line thermal/optical analysis of exposed quartz-fiber filters. OC compounds are volatilized by heating the filter to 650°C in the He-atmosphere. The OC-vapor phase compounds are passed through a MnO_2 bed heated to 1000°C. The CO_2 produced by oxidation is then reduced to CH_4 in a Ni-firebrick. CH_4 is measured in a flame ionization detector (FID). To quantify EC, the temperature is reduced to 350°C and oxygen is added to the He in order to oxidize the EC. The CO_2 is again converted to CH_4 and is measured by FID. Light transmission measurement through the filter is used to correct for the pyrolysis of OC. A commercial ambient carbon particulate monitor based on a similar principle is also available.[57,58]

Photoacoustic spectroscopy is another suitable method for the detection of EC in ambient air.[59] Ambient air is aspirated through a resonant chamber, where it is illuminated by chopped laser light. Carbonaceous particles absorb energy from the laser beam and transfer it as heat to the surrounding air. The expansion of the heated gas produces pressure pulses at the same frequency as the laser modulation. These pulses are detected by a microphone.

An improved device, the photoacoustic soot sensor (PASS), has been developed in Germany.[60,61] Its construction and function is illustrated in Figure 2.7. The equipment uses two electric microphones, which enhance the output signal. The apparatus uses an absolute calibration and has a detection limit for soot aerosol of about 0.5 $\mu g/m^3$.

An EC-based method was developed and tested for monitoring particulate concentrations in diesel exhaust. This method is also based on the use of the thermal-optical technique for the analysis of carbonaceous fractions of particulate diesel exhaust. Speciation of OC and EC is accomplished through temperature and atmosphere control and by an optical feature that corrects for pyrolytically generated carbon[62] (see also Chapter 3).

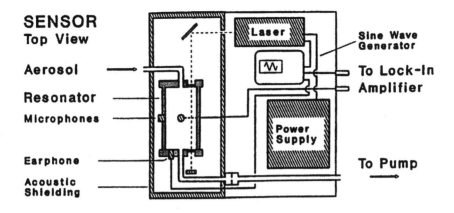

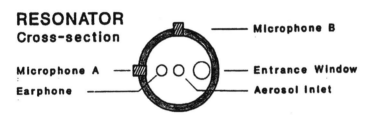

FIGURE 2.7   Schematic of the photoacoustic soot sensor and the positions of both microphones. (From Petzold, A. and Niessner, R. Photoacoustic soot sensor for *in situ* black carbon monitoring. *Appl. Phys.* B 63:191-197, 1996. With permission.)

Measurement methods for carbon black aerosol were also compared under field sampling conditions. Thermal and optical methods for concentration measurement of carbon black aerosol were operated side-by-side in an urban environment in the vicinity of a heavily traveled road. Very good correlation has been found between the Aethalometer and the ambient carbon particulate monitor.[63]

2.4.2   PARTICULATE METALS MONITORS

The monitoring of particulate metal concentrations in ambient air is another important field of the air monitoring methodology. Among the main methods for monitoring particulate metal concentrations in ambient air are the ICP techniques using AEC in argon and in air–argon plasmas, the XRF spectroscopy based on X-ray fluorescence of air samples trapped on an activated-carbon-impregnated filter tape, which is useful for metal detection over the ppm range, laser-spark spectrometry or laser induced breakdown spectroscopy (LIBS) consisting of AES detection in sparks excited by lasers.

The most well developed at the present seems to be the ICP–AES techniques.[4] A recently developed continuous monitor for measuring the metal aerosol concentrations in ambient air also uses the ICP–AES technique.[64] The ambient air is channelled by direct suction into a quartz confinement tube with ignited plasma. The metallic pollutants are identified by their atomic spectra. The lowest detection limits, e.g., for Cu, Fe, Ni, lay below 1 $\mu m/m^3$ (see also Chapter 3).

Laser-induced plasma spectroscopy (LIPS) has recently been found to be a promising method for the characterization of heavy metal aerosols.[65] The system consists of a laser unit and a detector unit, including the spectrometer (see Figure 2.8). Both units are connected to a miniaturized sensor head.

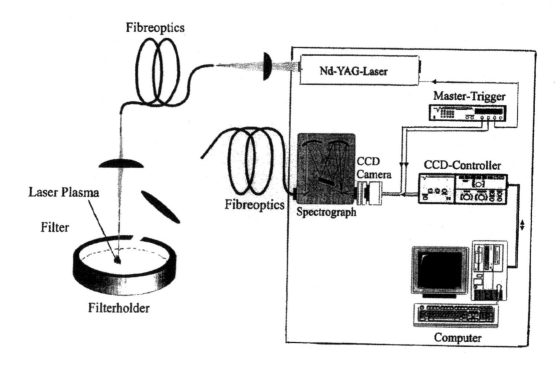

FIGURE 2.8   Schematic diagram of the setup for LIPS measurement of heavy metal aerosols. (From Neuhauser, R.E., Panne, U., Niessner, R. et al. Elemental characterization of heavy metal aerosols by Laser-Induced Plasma Spectroscopy (LIPS). *J. Aerosol Sci.* 27:Suppl. 1, S 437–S 438, 1997. With permission.)

A laser plasma is created by focusing the laser light on the surface of a glass fiber filter, which provides the necessary temperature stability for use in hot exhaust gases and a collection efficiency of nearly 100% even for ultrafine aerosol particles. The filter is placed on a filter holder which can be rotated to get spectra from different locations of the filter. The plasma emission is guided to the spectrometer.

2.4.3   ORGANIC PARTICULATE MONITORS

The methods for on-line monitoring of organic particulate pollutants are still underdeveloped. Nevertheless, we have useful monitors for the continuous measurement of PAH concentrations. The most frequently used and well tested equipment is the PAS (photoelectric aerosol sensor). It works on the principle of photoelectric ionization of aerosols induced by UV light of a mercury lamp on aerosol particles with PAH molecules adsorbed on their surfaces. The electric charge produced by ionization is collected by an electrometer and the electrical signal is proportional to the amount of PAHs adsorbed on the surface of particles. The performance of PAS depends upon particle size. Particles smaller than 1 μm are the best photon emitters.[66,67] These particles are the most important with respect to the fraction of the PAH molecules adsorbed on their surface.[68] The performance of PAS has been compared with off-line sampling techniques. This comparison showed a satisfactory linearity between the data obtained with the two methods when single PAHs were measured. The concentration of total PAHs adsorbed on particle surfaces measured by PAS techniques was an order of magnitude greater than the sum of PAHs selectively determined with off-line techniques.[69]

Among the PAH monitors are two commercially available devices, the **LIFA** (laser induced fluorescence analyzer) and the **FRFID** (fast-response flame ionization detector). Using a N_2-laser, fluorescence signals are produced by PAHs and the intensity is measured in the LIFA. The detection limits for the majority of PAHs are less than 1 μg/m³. The FRFID can also be used successfully to measure the concentration of unburned hydrocarbons in real time, e.g., in an internal combustion engine. The equipment can also continuously measure the gaseous hydrocarbons as well as the soot particulates.[70]

Another sensitive method is mass spectrometry. **EMS** (emission mass spectrometry) uses a transportable GC–MS measuring system. Aerosol is sampled on a glass fiber filter tape, and the PAH concentration is determined by GC-MS. The method is very sensitive. The detection limit for different PAHs is between approximately 1 and 100 ng/m^3.[71]

2.4.4 MULTICOMPONENT AEROSOL MONITORS

While there are monitors to determine the concentration of single elements or compounds, and only in some cases a group of compounds with very similar properties (e.g., PAHs), practically no commercial instruments exists for the rapid *in situ* measurement of the chemical composition of aerosol particles. The existing chemical analysis of aerosols has largely been conducted in an off-line manner — sample collection on site and chemical analysis in the laboratory.

Several investigations based on continuous aerosol sampling by means of a condensation process and on wet chemistry analysis have been done. The results have shown that such methods could analyze an aerosol on-line, quickly and with relatively high sensitivity.[72-76]

Two relatively well developed and tested devices have been described recently.[72-74] One of them is an automated instrument for the measurement of the chemical composition of atmospheric aerosols.[72,73] It uses a particle collection system (PCS) combined with an ion chromatograph (IC). The PCS preseparates gaseous pollutants by means of a parallel plate diffusion denuder (PPDD). Aerosols which penetrate the PPDD are introduced into a mixing chamber along with supersaturated steam. Following cooling, the steam completely condenses to liquid water. The condensed water drops contain both dissolved and insoluble particles. The microdrops are then collected using an inertial air liquid separator. The analysis of the liquid effluent is performed downstream by an IC. It is also possible to combine the liquid effluent from the denuder with the solution from the PCS before the chemical analysis. By this combination, gaseous as well as aerodispersed air components can be determined.

The collection efficiency for atmospheric aerosols has been found to be almost 100%. The detection limits for different compounds lie in the range of ng/m^3. The procedure has been applied for chemical monitoring of water soluble aerosols, but it may even be possible to measure insoluble, i.e., metallic, constituents of the atmospheric aerosol by incorporating suitable reaction steps before analysis.

A similar system for on-line chemical analysis of aerosol species combines so-called stream-jet-aerosol collector (SJAC) with an IC-analyzer.[74] Interfering gases are again removed from the sample air stream in a denuder. The air is then mixed with steam. Cooling results in a high supersaturation and causes aerosol particles to grow into large droplets. These droplets containing dissolved aerosol species are collected in a cyclone. The collected solution is continuously pumped out and is on-line analyzed by means of IC and FIA (flow-injection analysis). The detection limits lay also in the range of ng/m^3.

Condensation aerosol samplers have been used for many years. In the mid-1950s H. Cauer, a German meteorologist, used a simple device to sample and analyze atmospheric aerosols.[75] He designated it as a "condensation sphere" (see Figure 2.9). An egg-shaped metallic vessel (K) was filled with solid CO$_2$ pieces. The atmospheric water vapor condenses on the surface of the sphere and is collected in the glass vessel (B) in the form of droplets. The condensed water is then chemically analyzed in the laboratory. By simultaneous measurement of the air humidity, concentrations of soluble aerosols were determined, e.g., in ng/ml and in ng/m^3.

In another study, an automated system for chemical analysis of airborne particles, based on corona-free electrostatic collection and IC analysis, was developed and described.[76] The aerosol sample is drawn through an annular space of the electrostatic collector. Field charging occurs and the aerosol is collected on the surface of the tubular electrodes. Lack of corona discharge eliminates

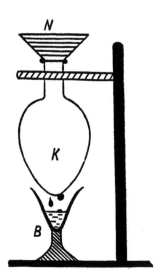

FIGURE 2.9 Schematic picture of Cauer's Condensation Sphere. (From Cauer, H. Aerosolprobenahme mit Hilfe der Wasserdapfkondensation. *Zbl. Biol. Aerosol-Forschung.* 5:6-15, 1956. With permission.)

the forming of NO_x and, consequently, the artifact forming NO_2^- and NO_3^-. The equipment is effective in collecting the fine particle range (<2.5 µm). The device is coupled in an automated fashion to an IC. After the aerosol collection period, water is pumped through the annulus and simultaneously aspirated back through the central electrode and pumped through the IC column. This is followed by analysis. The collection frequency is greater than 90%.

2.5 CHEMICAL ANALYSIS OF INDIVIDUAL AEROSOL PARTICLES

Twelve years ago, we published the first monograph dealing with the chemical analysis and identification of individual aerosol particles.[77] We prophesied that this field of chemical analysis of aerosols would expand and find broad applications in research as well as in routine practice. Our prophecy has been realized. In the meantime, the methods of analyzing individual collected particles, individual particles levitated in the gas phase, and on-line and *in situ* analysis of atmospheric aerosol particles have been further developed and improved and have become, in some cases, routine aerosol measurement procedures.

2.5.1 ANALYSIS OF INDIVIDUAL COLLECTED PARTICLES

For characterization of single collected aerosol particles, specific chemical reactions have been used since about 1950, and the analytical tool was the light microscope[77,78] (see also Chapter 11). Extensive development in this field began in the 1980s when new equipment based on physical and nuclear methods for chemical analysis became commercially available for broad use in analytical and environmental laboratories.[77,78] The most exciting development came when a combination of electron microscopical and X-ray and electron beam analysis instruments were introduced[78] (see also Chapter 12).

Electron microscopical methods, SEM (scanning electron microscopy) and TEM (transmission electron microscopy), combined with analytical procedures, like EDXA (energy dispersive X-ray analysis) and SAED (selective area electron diffraction), are frequently used methods for routine analysis and characterization of single atmospheric aerosol particles.[79-90] This methodology has also been found to be useful for the physical and chemical characterization of fine aerosol particles in the middle and upper troposphere in assessing the effect of aerosol particles on global climate, e.g., through cloud formation.[79-85] Suitable sampling devices and sample preparation methods have been developed for such applications.[80] Figure 2.10 illustrates such TEM and EDXA applications. Figure 2.10a shows the transmission of electron micrograph of particles sampled in the upper tropical troposphere. Particles labeled by arrows contain H_2SO_4. EDXA spectra characterize the single particles A(b) and B(c).

Both methods — SEM and TEM — characterize the single particles by elemental composition (EDXA) and by diffraction patterns (SEDA). In some cases, element atomic ratios and trace elements can help in finger printing special particles. Atomic ratios O/C and trace elements S and Cl were found to be specific for the identification of black carbon particles in SEM.[89,90]

In addition to SEM and TEM, new "super" microscopes are available and have been successfully used for the characterization of single collected aerosol particles. Neither SEM nor TEM is sensitive enough for observation and characterization of low nanometer-size aerosol particles. For this size range, much higher resolution is necessary. This can be achieved by STM (scanning tunneling microscope) and by AFM (atomic force microscope).[91]

Further improvements in single particle analysis means the applications of the NM (nuclear microscope). This instrument and method are now used in routine characterization and analysis of small aerosol samples as well as of single aerosol particles. NM can provide analysis with µg/g sensitivity for almost all elements of the periodic table, even in very small, micrometer size samples[92] (see also Chapter 13).

2.5.2 SINGLE PARTICLE ANALYSIS *IN SITU*, ON-LINE, AND IN REAL TIME

The physical and chemical characterization of individual aerosol particles by on-line methods can be achieved by two main methods. The single particle, solid or liquid, is observed, measured, and analyzed *in situ* in the gas phase by using different levitation methods. It means the particle can be rising into the air by several forces, e.g., radiometric and phoretic, electrostatic, electrodynamic, acoustic, etc., which act against the gravitational force.[93,94]

Another methodology characterizes the single particle in real-time in an aerosol beam, which enters the analyzing equipment, usually a special mass spectrometer.[95] Several physical and chemical properties of single aerosol particles, as well as their chemical interactions with the surrounding gaseous environment, can be observed and measured by suspending the single particle in one or more laser beams. Information about particle size and refractive index can be obtained by means of light-scattering and by using infrared (IR), Raman, and fluorescence spectroscopies, chemical compositions and their changes can be well measured.[96] Aerosol levitation in a laser beam is a useful technique for chemical characterization of single particles mainly by the application of Raman spectroscopical methods[97] (see also Chapter 14).

Electrodynamic levitation is another very useful technique which has been used in physico-chemical aerosol research since about the middle of the 1950s.[98] However, the principle was first used by Milikan at the beginning of the 20th century.[94] The electrodynamic balance or the electro-dynamic trap is now often used in investigations and studies on single levitated particles, e.g., for the measurement of electric charge, mass, size changes, drag forces, vapor condensation and evaporation, accommodation coefficients, chemical reactions, etc.

The apparatus configuration which is usually needed for such studies is shown schematically in Figure 2.11. The particle, e.g., a single microdroplet, is caught in the electrodynamic trap and irradiated by a laser beam. Its physical and chemical properties, as well as their changes in time, are observed and measured by several additional methods — microscopy, IR, Raman, and fluores-cent spectroscopy, etc.[99]

The aerosol beam technique is more or less a method for physical and chemical characterization of single particles in aerosol clouds. A large collection of single aerosol particles can be charac-terized by size (or shape) and chemical composition in a very short time, *in situ* and on-line.

When Shal Friedlander introduced in his classification of aerosol measurement instruments in the mid-1970s his nonexistent "Perfect Single Particle Counter Analyzer" as the best measurement

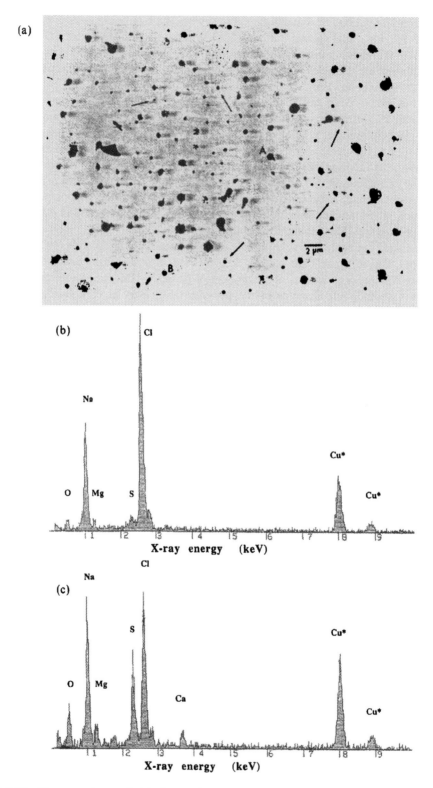

FIGURE 2.10 Electron micrograph (a) of aerosol particles sampled in the upper tropical troposphere and X-ray spectra (b, c) of single particles. (From Ikegami, M., Okada, K., Zaizen, Y., and Makino, Y. Sea-salt particles in the upper tropical troposphere. *Tellus.* 46 B, 142-151, 1994. With permission.)

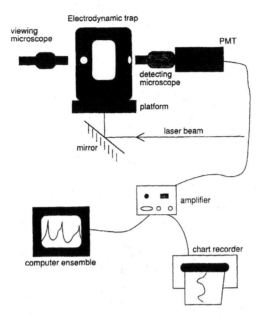

FIGURE 2.11 Schematic of a setup for the analysis of single levitated particles. (From Ngo, D. and Pinnick, R.G. Suppression of scattering resonances in inhomogeneous microdroplets. *J. Opt. Soc. Am.* A 11:1352-1359, 1994. With permission.)

instrument for aerosols in the future, he understood it as a device, which could classify the aerosol particles according to size, identify each class separately, and indicate its chemical composition in real time[100] (see Table 2.1). The research and development in this field in the last 20 years has shown that mass spectroscopy and its several modifications are the methods which are able to make Friedlander's "scientific dream" a reality.

2.5.3 MASS SPECTROSCOPIC PARTICLE COUNTERS-ANALYZERS

The *in situ* analysis of individual airborne particles by mass spectroscopy was originated by Davis, who applied for a patent on the method in 1971.[101] In Davis's method, analysis was accomplished by admitting a stream of ambient air directly into the ion source of a mass spectrometer. Particles in the air strike a hot rhenium filament, and ions of the constituent materials are produced by surface ionization.[95] The method is already 25 years old. During this time, the method has been improved and perfected, resulting in the construction of portable instruments which can be applied directly in several real situations in atmospheric environments (see also Chapters 15, 16, and 17). Several research groups in the U.S.[102-115] and in Europe[116-119] have been working in this field and have already achieved promising results.

Recent research and development investigations have contributed substantially to the understanding of atmospheric aerosol chemistry. Single particle analysis, unlike bulk analysis, allows microscopic variations in sample composition to be determined. Using this technique, we obtain information on the size and composition of individual particles in real time. Therefore, several fast chemical heterogeneous reactions, which can produce short-lived or intermediate inorganic as well as organic aerodispersed compounds, can be investigated and better understood.[103-110] Such studies contribute to understanding the production, chemistry, and ultimate fate of atmospheric aerosols. Studies in the tropospheric environment have confirmed that the secondary aerosols consist of particles which contain internal mixtures of organic and inorganic reaction products.[107-109] The improved equipment is able to analyze simultaneously both positive and negative ions from individual particles.[115-117] There does exist several already well-developed laboratory and portable instruments, differing in name,

TABLE 2.1

Characterization of Aerosol Measurement Instruments. Perfect Single Particle Counter Analyzer Is the Ideal Instrument

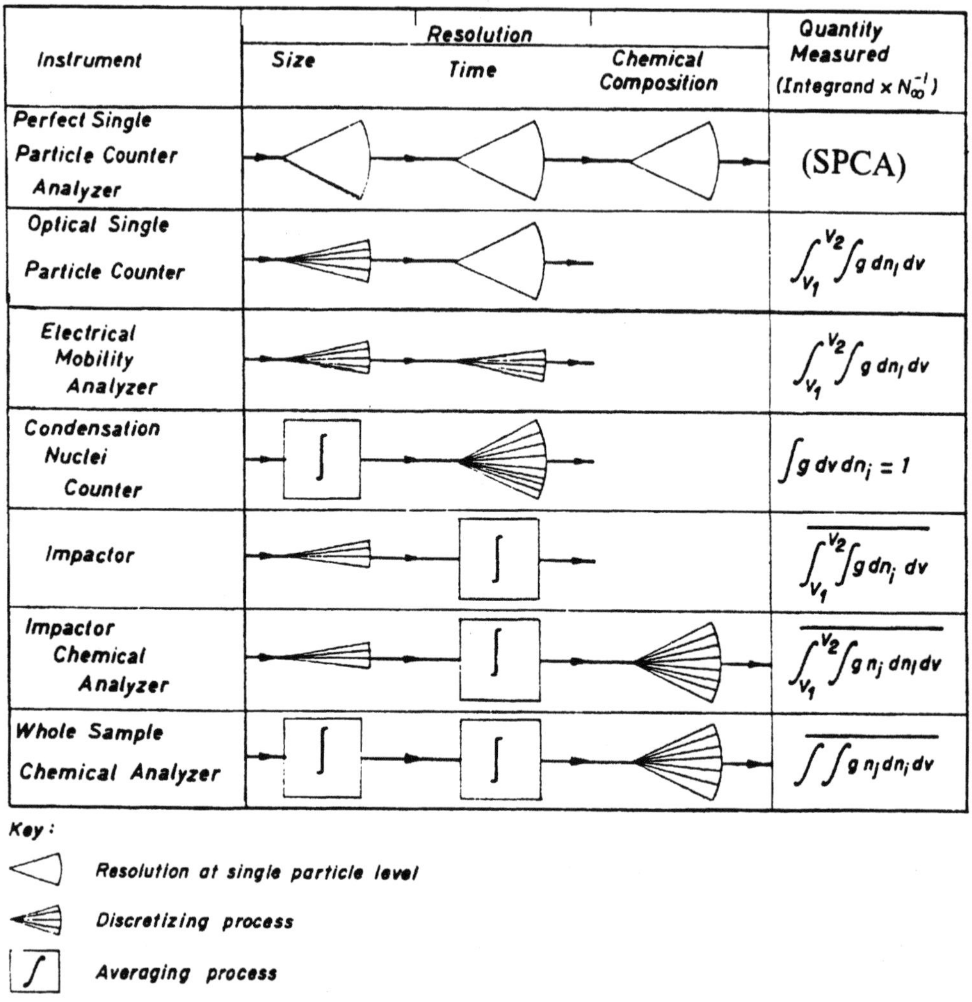

Instrument	Resolution — Size	Resolution — Time	Resolution — Chemical Composition	Quantity Measured (Integrand × N_∞^{-1})
Perfect Single Particle Counter Analyzer	(resolution at single particle level)	(resolution at single particle level)	(resolution at single particle level)	(SPCA)
Optical Single Particle Counter	(discretizing)	(resolution at single particle level)		$\int_{V_1}^{V_2}\int g\, dn_i\, dv$
Electrical Mobility Analyzer	(discretizing)	(discretizing)		$\int_{V_1}^{V_2}\int g\, dn_i\, dv$
Condensation Nuclei Counter	$\int$	(discretizing)		$\int g\, dv\, dn_i = 1$
Impactor	(discretizing)	$\int$		$\int_{V_1}^{V_2}\int g\, dn_i\, dv$
Impactor Chemical Analyzer	(discretizing)	$\int$	(discretizing)	$\int_{V_1}^{V_2}\int g\, n_j\, dn_i\, dv$
Whole Sample Chemical Analyzer	$\int$	$\int$	(discretizing)	$\int\int g\, n_j\, dn_i\, dv$

Key:

◁ Resolution at single particle level

◀ Discretizing process

$\boxed{\int}$ Averaging process

Source: From Friedlander, S.K. *Smoke, Dust and Haze.* John Wiley & Sons, New York, 1977, pp. 146-169. With permission.

design, and analytical abilities. Among the most-used instruments and methods are the **PALMS** (particle analysis by laser mass spectroscopy), the **MALDI** (matrix-assisted laser desorption/ionization) method, the **ATOFMS** (aerosol time-of-flight mass spectroscopy), and the **RTLMS** (real-time laser mass spectroscopy).

The PALMS technique uses an instrument shown schematically in Figure 2.12. Aerosols enter the instrument through a differentially pumped nozzle into a vacuum. They cross a He–Ne laser beam, which provides a trigger and rough particle size estimate from the scattered light. A light pulse from an eximer laser forms ion species in the particles. The ions are analyzed in a time-of-flight mass spectrometer. The instrument is also available in a portable version.[104]

The MALDI method performs on-line analysis of individual aerosol particles by laser desorption/ionization (LDI). Aerosol particles are drawn into a mass spectrometer and then ablated with

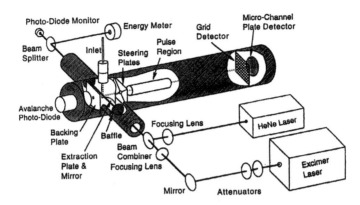

FIGURE 2.12 Diagram of the PALMS instrument. A TOF mass spectrometer is coupled to an aerosol inlet with particle detection and sizing by light scattering. (From Murphy, D.M. and Thomson, D.S. Chemical composition of single aerosol particles in Idaho Hill: Negative ion measurements. *J. Geophys. Res.* 102:6353-6368, 1997. With permission.)

a high-energy laser pulse as they pass through the source region. On-line LDI is combined with an optical particle sizing technique.[110]

The ATOFMS instrument (see also Chapter 15) is composed of three distinct regions: (1) a particle–beam interface, where particles are introduced into the instrument, (2) a light-scattering region, which uses two lasers to determine particle velocity/size information, and (3) a reflection time-of-flight mass spectrometer for particle composition analysis. It exists in two portable varieties and can, therefore, provide geographically, seasonally, and temporally resolved size and composition information of aerosol particles. Furthermore, the ATOFMS simultaneously generates positive and negative ions by the desorption and ionization of single particles.[112-115] This is particularly useful in identifying the source of atmospheric particles. Figures 2.13 and 2.14 show examples of single particle analysis of atmospheric aerosols.[114,115]

The RTLMS is an improved instrument which also uses a bipolar time-of-flight mass spectrometer. This results in increased accuracy in determining the chemical content of particles, when compared to classical off-line unipolar mass spectrometry. The simultaneous detection of a positive ion spectrum and a negative ion spectrum from the same particle for each laser shot provides a fast and precise chemical characterization of single particles.[116,117]

The application of the tandem mass spectrometry, the ion trap MS/MS techniques, to real-time aerosol particle analysis can make positive identification of components of a complex particle spectrum quite feasible and permit tracking of specific chemical species over time and space.[120]

The evaluation of single particle mass spectra is still a time-consuming procedure. Therefore, aims exist to develop on-line evaluation procedures. Such methods can evaluate the data received by real-time measurements using time-of-flight mass spectrometry. One way seems to be to perform a cluster analysis after a principal component analysis.[121]

2.6 CONCLUDING REMARKS

When we consider the recently found epidemiological and toxicological data about the harmful health effects of the fine fraction of atmospheric anthropogenic aerosols, there is no question that we need new strategies and improved methods for measuring and analyzing atmospheric particulate pollutants (see also Chapter 3). Recent research in the field of aerosol sampling and analysis show the trends in this direction. Artifact-free aerosol sampling procedures seem to be important complements to the classical filter methods. On-line bulk analysis could have, after further improvement, the ability to determine the real chemical composition of an aerosol sample.

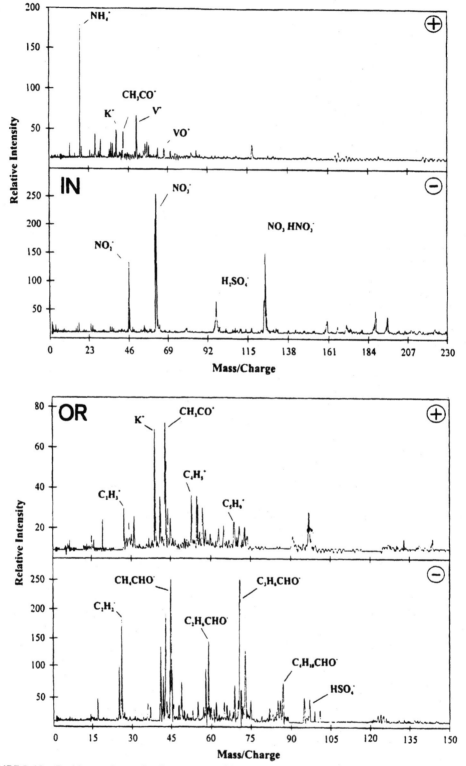

FIGURE 2.13 Positive and negative ion spectra of inorganic (IN) and organic (0) particles from ambient atmosphere. (From Gard, E., Mayer, J.E., Morrical, B.D., Dienes, T., Fergenson, D.P., and Prather, K.A. Real-time analysis of individual atmospheric aerosol particles: Design and performance of a portable ATOFMS. *Anal. Chem.* 69:4083-4091, 1997. With permission.)

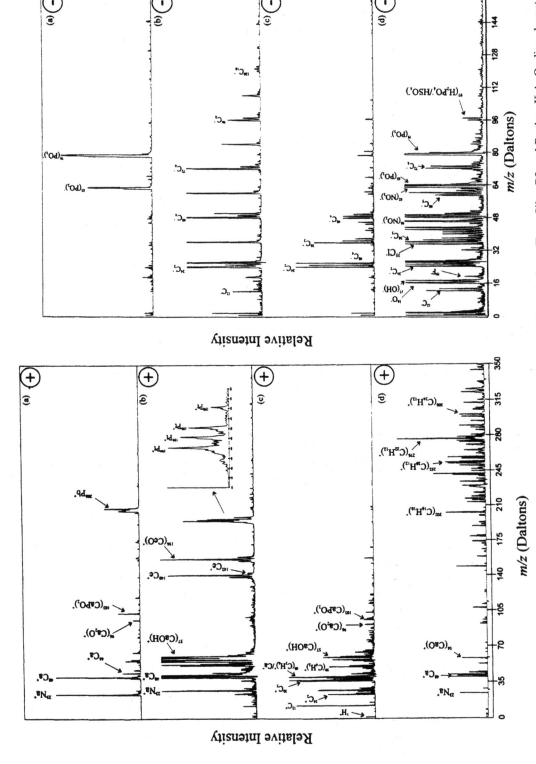

FIGURE 2.14 Positive and negative ion spectra of single particles monitoring in automobile emissions. (From Silva, P.I. and Prather, K.A. On-line characterization of individual particles from automobile emissions. *Environ. Sci. Technol.*, 31:3074-3080, 1997. With permission.)

The methods for single particle measurement and analysis, after collection on substrates or on-line and *in situ,* begin to be used for routine measurements in the troposphere as well as in the stratosphere. Their importance will increase in the future, and this methodology will essentially contribute to greater understanding of the chemical and toxicological properties of the atmospheric aerosol.

REFERENCES

1. Willeke, K. and Baron, P. *Aerosol Measurement.* Van Nostrand Reinhold, New York, 1993.
2. Appel, B.R. Atmospheric sample analysis and sampling artifacts, in K. Willeke and P. Baron (Eds.) *Aerosol Measurement,* Van Nostrand Reinhold, New York, 1993, pp. 233-259.
3. Watson, J.G. and Chow, J.C. Aerosol air sampling, in K. Willeke and P. Baron, (Eds.) *Aerosol Measurement,* Van Nostrand Reinhold, New York, 1993, pp. 622-639.
4. Chow, J.C. Measurement methods to determine compliance with ambient air quality standards for suspended particles. *Air and Waste Manage. Assoc.* 45:320-382, 1995.
5. CEN EN 481: Festlegung von Konventionen von Partikelgrößen zur Messung von Schwebstoffen am Arbeitsplatz. Beuth Verlag. Berlin, 1993.
6. Koch, W., Dunkhorst, W., and Lödding, H. Respicon™-3 F: A new personal measuring system for size segregated dust measurement at workplaces. *Gefahrstoffe-Reinhalt. Luft.* 57:177-184, 1997.
7. Lundgren, D.A., Hlaing, D.N., Rich, T.A., and Marple, V.A. PM 10/PM 2.5/PM 1 Data from a trichotomous sampler. *Aerosol Sci. Technol.* 25:353-357, 1996.
8. Sioutas, C. and Koutrakis, P. Inertial separation of ultrafine particles using condensation growth/virtual impaction system. *Aerosol Sci. Technol.* 25:424-436, 1996.
9. Asgharian, B. and Godo, M.N. Transport and deposition of spherical particles and fibers in an improved virtual impactor. *Aerosol Sci. Technol.* 27:499-506, 1997.
10. Alfos, D.J., Williams, A.L., Hagen, D.E. et al. A system for collecting milligram quantities of cloud condensation nuclei. *Aerosol Sci. Technol.* 26:415-432, 1997.
11. Asking, L. and Olsson, B. Calibration at different flow rates of a multistage liquid impinger. *Aerosol Sci. Technol.* 27:39-49, 1997.
12. Tsai, C.J. and Perng, S.N. Artifacts of aerosol particles during filter sampling. *5th AAAR Annual Conference,* Orlando, FL, Abstracts p. 80, 1996.
13. Cheng, Y.S. Condensation detection and diffusion size separation, in K. Willeke and P.A. Baron (Eds.) *Aerosol Measurement,* Van Nostrand Reinhold, New York, 1993, pp. 427-449.
14. Khlystov, A., Wyers, G.P., and Slanina, J. The steam-jet aerosol collector. *Atmos. Environ.* V 29 A:2229-2234, 1995.
15. Mikuska, P., Khlystov, A., and ten Brink, H.M., Wyers, G.P., and Slanina, J. Artifact-free method for size-resolved chemical analysis of ambient aerosols. *J. Aerosol. Sci.* 28, Suppl. 1, S 443-S 444, 1997.
16. Salma, I., Maenhaut, W., Annegarn, H.J. et al. Combined application of INAA and PIXE for studying the regional aerosol composition in Southern Africa. *J. Radioanal. Nuclear Chem.* 216:143-148, 1997.
17. Thomas, S., Morawska, L., Hofinger, N., and Selby, M. Investigation of the source of blank problems in the measurement of lead by ICP. *J. Anal. Atom. Spectr.* 12:553-556, 1997.
18. Wang, C.F., Yang, J.Y., and Ke, C.H. Multi-element analysis of airborne particulate matter by various spectrometric methods after microwave digestion. *Anal. Chim. Acta.* 320:207-216, 1996.
19. Wang, C.F., Huang, M.F., Chang, E.E., and Chiang, P.C. Assessment of closed vessel digestion methods for elemental determination of airborne particulate matter by ICP–AES. *Anal. Sci.* 12:201-207, 1996.
20. Wang, C.F., Jeng, S.L., and Shieh, F.J. Determination of As in airborne particulate matter by ICP. *J. Anal. Atom. Spectr.* 12:61-67, 1997.
21. Hulanicki, A., Surgiewicz, J., and Jaron, I. Determination of Hf air dust filters by ICP. *Talanta.* 44:1159-1162, 1997.
22. Bhat, P.N. and Pillai, K.C. Be in environment air, water and soil. *Water, Air, and Soil Pollution.* 95:133-146, 1997.
23. Altman, E.L., Lvov, B.V., Paniches, N.A., and Fedorov, P.N. Determination of metals in atmospheric aerosols by Zeeman–AAS with absorption pulse restoration. *J. Anal. Chem.* 51:824-828, 1996.

24. Maynard, A.D., Northage, C., Hemingway, M., and Bradley, S.D. Measurement of short-term exposure to airborne soluble Pt in the platinum industry. *Ann. Occup. Hyg.* 41:77-94, 1997.

25. Barefoot, R.R. Determination of Pt at trace levels in environmental and biological materials. *Environ. Sci. Technol.,* 31:309-314, 1997.

26. Balcerzak, M. Analytical methods for the determination of Pt in biological and environmental materials. A review. *Analyst.* 122:67R-74R, 1997.

27. Nusko, R. and Heumann, K.G. Cr(III)/Cr(IV) speciation in aerosol particles by extractive separation and thermal ionization isotope dilution mass spectrometry. *Fresenius J. Anal. Chem.* 357:1050-1055, 1997.

28. Wang, C.F., Tu, F.H., and Jeng, S.L., Determination of Si in airborne particulate matter by UV-visible spectrometry. *Anal. Chim. Acta.* 342:239-245, 1997.

29. Rahmalan, A., Abdullah, M.Z., Sanagi, M.M., and Rashid, M. Determination of heavy metals in air particulate matter by ion chromatography. *J. Chromatogr.* 739:233-239, 1996.

30. Mihaly, B., Molnar, A., and Meszaros, E. Application of capillary zone electrophoresis for atmospheric aerosol research. *Annali di Chimica* (Italy). 86:439-447, 1996.

31. Petzold, A. and Niessner, R. Method comparison study on soot-selective techniques. *Microchimica Acta.* 117:215-237, 1995.

32. Petzold, A. and Niessner, R. Photoacoustic soot sensor for *in situ* black carbon monitoring. *Appl. Phys.* B 63:191-197, 1996.

33. Burtscher, H. Measurement and characteristics of combustion aerosols with special consideration of photoelectric charging and charging by flame ions. *J. Aerosol Sci.* 23:549-595, 1992.

34. Petzold. A. and Niessner, R. Coulometric determination of soot in ambient air. *Gefahrstoffe-Reinhalt. Luft.* 56:173-177, 1996.

35. Bussemer, M. and Heintzenberg, J. Laboratory calibration of a soot photometer for the spectral light absorption of atmospheric aerosol samples. *J. Aerosol Sci.* 27, Suppl. 1, S 693-S 694, 1996.

36. Keller, S. and Heintzenberg, J. Quantification of graphitic carbon on polycarbonate filters by Raman spectroscopy. *J. Aerosol Sci.* 28:Suppl. 1, S 609-S 610, 1997.

37. Zühlke, J., Fröschl, B., and Niessner, R. Immunological detection of 1-nitropyrene. A possible marker compound for diesel soot. *J. Aerosol Sci.* 27:Suppl. 1, S 707-S 708, 1996.

38. Baek, S.O., Field, R.A., Goldstone, M.E., Kirk, P.W., Lester, J.N., and Perry, R. A review of atmospheric PAH sources, fate and behavior. *Water, Air, Soil Pollution.* 60:279-300, 1991.

39. Librano, V. and Fazzino, S.D. Quantification of PAH and nitro-PAH in atmospheric particulate matter of Augusta city. *Chemosphere.* 27:1649-1656, 1993.

40. Goelen, E., Lambrechts, M., and Geyskens, F. European sampling intercomparison for aromatic and chlorinated hydrocarbons. *Ann. Occup. Hyg.* 41:527-554, 1997.

41. Hoja, H., Marquet, B., Lofti, H., Penicault, B., and Lachatre, G. Application of liquid chromatography-mass spectrometry in analytical toxicology. A review. *J. Anal. Toxicol.* 21:116-126, 1997.

42. Gridin, V.V., Litani-Barilai, I., Kadosh, M., and Schlechter, I. A renewable liquid droplet method for on-line pollution analysis by multi-photon ionization. *Anal. Chem.* 69:2098-2102, 1997.

43. Sherma, J. and Shafic, T.M. A multiclass, multiresidue analytical method for determining pesticide residues in air. *Arch. Environ. Contamin. Toxicol.* 3:57-71, 1975.

44. Wehner, T.A., Woodrow, J.E., Kim, Y.H., and Seiber, J.E. Multiresidue analysis of trace organic pesticides in air. In L.H. Keith (Ed.) *Identification and Analysis of Organic Compounds in Air,* Butterworths Publ., Woburn, MA, 1985, pp. 273-290.

45. Seiber, J.N. Glotfelty, D.E., Lucas, A.D., McChesney, M.D., Sagebiel, J.C., and Wehner, T.A. A multiresidue method by HPLC-based fractionation and GC determination of trace levels of pesticides in air and water. *Arch. Environ. Contamin. Toxicol.* 19:583-592, 1990.

46. Haraguchi, K., Kitamury, E., Yamashita, T., and Kido, A. Simultaneous determination of trace pesticides in urban air. *Atm. Environ.* 28:1319-1325, 1994.

47. Millet, M., Wortham, H., Sanusi, A., and Mirabel, P. A multiresidue method for determination of trace levels of pesticides in air and in water. *Arch. Environ. Contamin. Toxicol.* 31:543-556, 1996.

48. Millet, M., Wortham, H., Sanusi, A., and Mirabel, P. Atmospheric contamination by pesticides: Determination in the liquid, gaseous and particulate phases. *ESPR-Environ. Sci. Pull. Res.* 4:172-180, 1997.

49. Hargrave, B.T., Barrie, L.A., Bidleman, T.F., and Welch, H.E. Seasonality in exchange of organochlorines between arctic air and seawater. *Environ. Sci. Technol.* 31:3258-3266, 1997.

50. Moriyama, N., Murayama, H., Kitajima, E., and Urushiyama, Y. Sampling of airborne pesticides using a quartz fiber filter and an activated carbon fiber filter. *Eisie Kagaku.* 36:299-303, 1990.

51. Kawata, K., Noriyama, N., Kasahara, M., and Urushiyama, Y. GC determination of deposited pesticides in aerial application using activated carbon fiber paper for sample collection. *Bunseki Kagaku.* 39:423-425, 1990.

52. Kawata, K., Moriyama, N., and Urushiyama, Y. Simple GC determination of fungicides and chlorothalonil in air using activated carbon fiber paper for sampling. *Bunseki Kagaku.* 39:601-604, 1990.

53. Kunit, M. and Puxbaum, H. Enzymatic determination of the cellulose content of atmospheric aerosols. *Atm. Environ.* 30:1233-1236, 1996.

54. Hansen, A.D.A., Rosen, H., and Novakov, T. The Aethalometer — an instrument for the real-time measurement of optical absorption by aerosol particles. *Sci. Total Environ.* 36:191-196, 1984.

55. Turpin, B.J., Cary, R.A., and Huntzicker, J.J. An *in situ,* time resolved analyzer for aerosol organic and elemental carbon. *Aerosol. Sci. Technol.* 12:161-171, 1990.

56. Turpin, B.J. and Huntzicker, J.J. A descriptive analysis of organic and elemental carbon concentrations. *Atm. Environ.* 25 A, 207-215, 1991.

57. Patashnick, H. and Rupprecht, E. A new network-ready instrument for the measurement of particulate carbon in the atmosphere. *J. Aerosol Sci.* 27:Suppl. 1, S 691-S 692, 1996.

58. Rupprecht and Patashnick Co. Inc. Ambient carbon particulate monitor. Prospectus, 1997.

59. Adams, K.M. Real time, *in situ* measurements of atmospheric optical absorption in the visible via photoacoustic spectroscopy. *Appl. Opt.* 27:4052-4056, 1988.

60. Petzold, A. and Niessner, R. Novel design of a resonant photoacoustic spectrophone for elemental carbon mass monitoring. *Appl. Phys. Lett.* 66:1284-1287, 1995.

61. Petzold, A. and Niessner, R. Photoacoustic soot sensor for *in situ* black carbon monitoring. *Appl. Phys.* B 63:191-197, 1996.

62. Birch, M.E. and Cary, R.A. Elemental carbon-based method for monitoring occupational exposures to particulate diesel exhaust. *Aerosol. Sci. Technol.* 25:221-241, 1996.

63. Berghmans, P., Pauwels, J., Roekens, E., and Bogaert, R. Comparison of methods for the concentration measurement of black aerosol in ambient air. *J. Aerosol Sci.* 27:Suppl. 1, S 689-S 690, 1996.

64. Gomes, A.M., Sarrette, J.P., Madon, L., and Almi, A. Continuous emission monitoring of metal aerosol concentrations in atmospheric air. *Spectrochim. Acta.* B 51, 1695-1705, 1996.

65. Neuhauser, R.E., Panne, U., Niessner, R. et al. Elemental characterization of heavy metal aerosols by Laser-Induced Plasma Spectroscopy (LIPS. *J. Aerosol Sci.* 27:Suppl. 1, S 437-S 438, 1997.

66. Hart, K.M., McDow, S.R., Giger, W., Steiner, D., and Burtscher, H. The correlation between *in situ* real time aerosol photoemission intensity and PAH concentration in combustion aerosols. *Water, Air, Soil Pollut.* 68:75-90, 1993.

67. McDow, J. PAH and combustion aerosol photoemission. *Atm. Environ.* 24 A:1911-1916, 1990.

68. Baek, S.O., Goldstone, M.E., Kirk, P.W., Lester, J.N., and Perry, R. Phase distribution and particle size dependency of PAH in the urban atmosphere. *Chemosphere.* 22:503-520, 1991.

69. Agnesod, G., DeMaria, R., Fontana, M., and Zublena, M. Determination of PAH in airborne particulate: comparison between off-line sampling techniques and automatic analyzer based on photoelectric aerosol sensor. *Sci. Total Environ.* 189/190:443-449, 1996.

70. Sun, J.H. and Chan, S.H. A time-resolved measurement technique for particulate number density in diesel exhaust using a fast-response flame ionization detector. *Meas. Sci. Technol.* 8:1-8, 1997.

71. Beckmann, A., Bröker, G., Gärtner, A. et al. State of the art of continuously monitoring measuring techniques for the determination of PAH. *Gefahrstoffe-Reinhalt. Luft.* 56:405-410, 1996.

72. Poruthoor, S.K. and Dasgupta, P.K. An automated instrument for the measurement of atmospheric aerosol composition. *Am. Lab.* February, 1997, pp. 51-56.

73. Zellweger, C., Baltensperger, U., Ammann, M. et al. Continuous automated measurement of the soluble fraction of atmospheric aerosols. *J. Aerosol Sci.* 28:Suppl. 1, S 155-S 156, 1997.

74. Mikuska, P., Khlystov, A., Ten Brink, H.M., Wyers, G.P., and Slania, J. A system for on-line chemical analysis of aerosol species. *J. Aerosol Sci.* 28:Suppl. 1, S 445-S 446, 1997.

75. Cauer, H. Aerosolprobenahme mit Hilfe der Wasserdapfkondensation. *Zbl. Biol. Aerosol-Forschung.* 5:6-15, 1956.

76. Liu, S. and Dasgupta, K.P. Automated system for chemical analysis of airborne particles based on corona-free electrostatic collection. *Anal. Chem.* 68:3638-3644, 1996.

77. Spurny, K.R. Physical and chemical characterization of individual airborne particles. Ellis Horwood, Chichester, U.K., 1986.
78. Fletcher, R.A. and Small, J.A. Analysis of individual collected particles, in K. Willeke and P.A. Baron (Eds.) *Aerosol Measurements,* Van Nostrand Reinhold, New York, 1993, pp. 260-295.
79. Mouri, H. and Okada, K. Stratospheric solid particles collected at 24 km altitude. *J. Meteorol. Soc. Japan,* February, 67-75, 1992.
80. Ikegami, M., Okada, K., Zaizen, Y., and Makino, Y. Sea-salt particles in the upper tropical troposphere. *Tellus.* 46 B, 142-151, 1994.
81. Niimura, N., Okada, K., Kai, K. et al. A method for the identification of Asian dust-storm particles mixed internally with sea salt. *J. Meteorol. Soc. Japan.* 72, 777-784, 1994.
82. Zaizen, Y., Ikegami, M., Okada, K. and Makino, Y. Aerosol concentration observed at Zhankye in China. *J. Meteorol. Soc. Japan.* 73:891-897, 1995.
83. Zaizen, Y., Ikegami, M., Tsutsumi, Y., Makino, Y., and Okada, K. Number concentration and size distribution of aerosol particles in the middle troposphere over the western Pacific Ocean. Atm. Environ. 30:1755-1762, 1996.
84. Okada, K., Ikegami, M., Zaizen, Y. et al. Soot particles in the middle troposphere over Australia. *Int. Symposium on Atm. Chemistry,* Nagoya, 1997, pp. 1-4.
85. Okada, K. Wu, P.M., and Tanaka, T. A light balloon-borne sampler collecting stratospheric aerosol particles for electron microscopy. *J. Meteorol. Soc. Japan.* 75:753-760, 1997.
86. Turpin, B. and Huang, P.F. Reduction of sampling and analytical errors for electron microscopic analysis of atmospheric aerosols. *Atm. Environ.* 30:4137-4148, 1996.
87. Ebert, M., Dahmen, J., Hoffmann, P., and Ortner, H.M. Examination of clean room aerosol composition by total reflection X-ray analysis and electron probe microanalysis. *Spectrochimica Acta.* B 52:967-975, 1997.
88. Weinbruch, S., Wentzel, M., Kluckner, M., Hoffmann, P., and Ortner, H. Characterization of individual atmospheric particles by element mapping in electron probe microanalysis. *Microchimica Acta.* 125:137-141, 1997.
89. Ormstad, H., Gaarder, P.I., and Johansen, B.C. Quantification and characterization of suspended particulate matter in indoor air. *Sci. Total Environ.* 193:185-196, 1997.
90. Stoffyn-Egli, P., Potter, T.M., Leonard, J.D., and Pocklington, R. The identification of black carbon particles with the analytical SEM. *Sci. Total Environ.* 198:211-223, 1997.
91. Köllensperger, G., Friedbacher, G., Grasserbauer, M., and Dorffner, L. Investigation of aerosol particles by atomic force microscopy. *Fresenius J. Anal. Chem.* 358:268-272, 1997.
92. Orlic, I., Watt, F., and Tang, S.M. Nuclear microscopy of individual atmospheric aerosol particles. *J. Aerosol Sci.* 27: Suppl 1, S 661-S 662, 1996.
93. Davis, E.J. Electrodynamic levitation of particles, in K. Willeke and P.A. Baron (Eds.) *Aerosol Measurement,* Van Nostrand Reinhold, New York, 1993, pp. 452-470.
94. Davis, J.D. A history of single aerosol particle levitation. *Aerosol Sci. Technol.* 26:212-254, 1997.
95. Stoffels, J.J. and Allen, J. Mass spectrometry of single particles *in situ,* in K.R. Spurny (Ed.) *Physical and Chemical Characterization of Individual Airborne Particles.* Ellis Horwood, Chichester, U.K., 1986, pp. 380-399.
96. Brandt, E.H. Levitation in physics. *Science* 243:349-355, 1989.
97. Thurn, R. and Kiefer, W. Raman-microsampling technique applying optical levitation by radiation pressure. *Appl. Spect.* 38:78-83, 1984.
98. Straubel, E. and Straubel, H. Electro-optical measurement of chemical and physical changes taking place in an individual aerosol particle, in K.R. Spurny *Physical and Chemical Characterization of Individual Airborne Particles,* Ellis Horwood, Chichester, U.K., 1986, pp. 127-160.
99. Ngo, D. and Pinnick, R.G. Suppression of scattering resonances in inhomogeneous microdroplets. *J. Opt. Soc. Am.* A 11:1352-1359, 1994.
100. Friedlander, S.K. *Smoke, Dust and Haze.* John Wiley & Sons, New York, 1977, pp. 146-169.
101. Davis. W.D. U.S. Patent 3770954, 1973.
102. McKeown, P.J., Johnston, M.V., and Murphy, D.M. On-line single particle analysis by laser desorption mass spectroscopy. *Anal. Chem.* 63:2069-2073, 1991.
103. Thomson, D.S. and Murphy, D.M. Laser-induced ion formation thresholds of aerosol particles in a vacuum. *Appl. Optics.* 33:6818-6826, 1993.

104. Thomson, D.S. and Murphy, D.M. Analyzing single aerosol particles in real time. *CHEMTECH.* 24:30-35, 1994.
105. Carson, P.G., Johnston, M.V., and Wexler, A.S. Laser desorption/ionization of ultrafine aerosol particles. *Rapid. Commun. in Mass Spectr.* 11:993-996, 1997.
106. Middlebrook, A.M., Thomson, D.S., and Murphy, D.M. On the purity of laboratory generated sulfuric acid droplets and ambient particle studied by laser mass spectroscopy. *Aerosol Sci. Technol.* 27:293-307, 1997.
107. Murphy, D.M., Thomson, D.S., and Kauzhny, M. Aerosol characteristics at Idaho Hill during the OH photochemistry experiment. *J. Geophys. Res.* 102:6325-6330, 1997.
108. Murphy, D.M. and Thomson, D.S. Chemical composition of single aerosol particles at Idaho Hill: Positive ion measurement. *J. Geophys. Res.* 102:6341-6352, 1997.
109. Murphy, D.M. and Thomson, D.S. Chemical composition of single aerosol particles in Idaho Hill: Negative ion measurements. *J. Geophys. Res.* 102:6353-6368, 1997.
110. Mansoori, B.A, Johnston, M.V., and Wexler, A.S. Matrix-assisted laser desorption/ionization of size- and composition-selected aerosol particles. *Anal. Chem.* 68:3595-3601, 1996.
111. Carson, P.B., Johnson, M.V., and Wexler, A.S. Real-time monitoring of the surface and total composition of aerosol particles. *Aerosol Sci. Technol.* 26:291-300, 1997.
112. Noble, C.A. and Prather, K.A. Real-time measurement of correlated size and composition profiles of individual atmospheric aerosol particles. *Environ. Sci. Technol.,* 30:2667-2680, 1996.
113. Liu, D.Y., Rutherford, D. Kinsey, M., and Prather, K.A. Real-time monitoring of pyrotechnically derived aerosol particles in the troposphere. *Anal. Chem.* 69:1808-1814, 1997.
114. Silva, P.I. and Prather, K.A. On-line characterization of individual particles from automobile emissions. *Environ. Sci. Technol.,* 31:3074-3080, 1997.
115. Gard, E., Mayer, J.E., Morrical, B.D., Dienes, T., Fergenson, D.P., and Prather, K.A. Real-time analysis of individual atmospheric aerosol particles: Design and performance of a portable ATOFMS. *Anal. Chem.* 69:4083-4091, 1997.
116. Hinz, K.P., Kaufmann, R., and Spengler, B. Simultaneous detection of positive and negative ions from single airborne particles by real-time mass spectrometry. *Aerosol Sci. Technol.* 24:233-242, 1996.
117. Hinz, K.P., Kaufmann, R., Jung, R., Greweling, M., Rews, F., and Spengler, B. Characterization of atmospheric particles using bipolar on-line laser mass spectrometer and multivariate classification. *J. Aerosol Sci.* 28: S 305-S 306, 1997.
118. Weiss, M., Verheijen, P.J.T., Marijnissen, J.C.M., and Scarlett, B. On the performance of an on-line time-off-flight mass spectrometer for aerosols. *J. Aerosol Sci.* 28:159-171, 1997.
119. Grootveld, C.J., Weiss, M., Marijnissen, J.C.M., and Scarlett, B. Particle size-dependent triggering of the on-line aerosol TOF mass spectrometers. *J. Aerosol Sci.* 28:S 605-S 606, 1997.
120. Reily, P.T.A., Gieray, R.A., Yang, M., Whitten, W.B., and Ramsey, J.M., Tandem mass spectrometry of individual airborne microparticles. *Anal. Chem.* 69:36-39, 1997.
121. Maeder, M. and Ebel, S. Data evaluation for the characterization of individual aerosol particles using time-of-flight mass spectrometry. *Chemometrics and Intelligent Laboratory Systems.* 37:205-207, 1997.

3 New Concepts for Sampling, Measurement, and Analysis of Atmospheric Anthropogenic Aerosols

Kvetoslav R. Spurny

CONTENTS

3.1 INTRODUCTION

The development of atmospheric pollution by anthropogenic aerosols since the 1950s can be divided into two periods. In the first, between about 1950 until 1980, the most important particulate pollutants were emitted from heavy industries, coal burning, and vehicles. Aerosols with predominantly coarse particles lying in the thoracic particle size range (particle diameter less than 10 μm, designated as PM 10 fraction) were considered representative indicators in evaluating the health risk for the general population.

While the concentrations of PM 10 in urban ambient air were as high as 1 mg/m³ or more during the 1950s and 1960s, later, because of the application of efficient air cleaning systems (electro- and fiber mat filters), the particulate industrial emissions, and therefore also their atmospheric concentrations, were substantially decreased. They lay mostly in the range less than 100 μg/m³. Blue sky over the cities was an important visible result of these air cleaning strategies.

Since 1980 and mainly at the beginning of the 1990s, measurements done in several urban atmospheric environments have shown an increasing trend in the concentration of highly dispersed aerosols; i.e., of aerosols with particle diameters less than 2 μm. There are several reasons for this development and for qualitative changes in the physical and chemical properties of the atmospheric anthropogenic aerosols. The total emissions of fine particles from transportation, fuel combustion, etc., now consist of practically all fine aerosols (size under 2 μm) only. Particulate automobile exhaust emissions as well as emissions of gas and oil combustion are in the fine particle size range, and the coarse particulates in the industrial emissions are very efficiently separated by air cleaning equipment. The fine aerosol can be dispersed in the atmosphere very homogeneously, can be transported for long distances, and because of its relatively high residence time in the atmosphere, accumulation with a consequently increasing concentration can also occur. Furthermore, as the chemical composition of the atmospheric anthropogenic aerosols depends strongly on the particle size, we have a new set of atmospheric particulate air pollution problems.

One consequence is probably the change in health effects on the general population, especially on children, the elderly, and other vulnerable persons. Several epidemiological investigations were done during the last decade showing that correlations do exist between ambient air concentrations of aerodispersed (particulate) pollutants and the health risk for the general population. These observed health effects lay at air particulate mass concentrations below the existing air quality standards (AQS). These studies indicate that, for example, increases in human mortality and morbidity have been associated with levels of air particulate pollutions significantly lower than those previously thought to affect human health. Despite some uncertainties, these new epidemiological data are considered sufficient to be considered as a serious basis for the reevaluation of existing air particulate pollution philosophy and strategy. These new results and observations call for several basic consequences and changes in definitions, physical and chemical characterization of air particulate pollutants, including their toxicology, sampling, measurement, and analytical procedures, and last but not least, new air quality standards.

3.2   EXISTING PARTICULATE AIR POLLUTION

As already mentioned, since 1980 important changes in physical and chemical characterization of air particulate pollutants has been observed and quantitatively determined. The previous heavy air pollution by coarse dust and dark smokes disappeared from the skies of cities in industrial countries and remained in environmentally undeveloped countries, e.g., Eastern Europe, Russia, China, etc.

Formerly, in the industrialized world, ambient particulates were dominated by coal smoke due to incomplete burning (Figure 3.1). The coarse particulate fraction was responsible for the relatively high ambient air anthropogenic aerosol concentrations. Mean annual particulate concentration values lay, for example, in the U.S. in the range of 100 μg/m³ and more during the 1960s (Table 3.1).[1] Because of high SO₂ emissions, the sulfate concentrations were also elevated.

Several disastrous particulate and gas pollution periods have been reported since the 1930s — the Meuse Valley in Belgium (1930), Donora, Pennsylvania (1948), Poza Rica, Mexico (1950), London (1952), New York (1953), Minneapolis (1956), and several worldwide episodes in 1962 (eastern U.S., London, Rotterdam, Hamburg, Osaka). These have made it obvious that the air quality

FIGURE 3.1   Stack plumes of dust and smoke which characterized the particulate air pollution situation during the 1960s.

TABLE 3.1
Particulate Analyses from Selected Urban Locations:
Arithmetic Mean Values for 1964 Expressed
as Micrograms per Cubic Meter

	Atlanta	Chicago	Boston	Pittsburgh
Suspended particulates	97	176	144	174
Benzene-soluble organics	7.6	9.5	9.1	10.0
Sulfates	4.8	19.0	18.1	17.5
Nitrates	2.0	2.5	2.3	3.0
Chromium		0.014	0.007	0.021
Cadmium	Most samples are below minimum detectable quantity			
Iron		1.6	0.9	2.8
Lead		0.9	0.9	0.9
Manganese		0.07	0.03	0.21
Nickel		0.042	0.076	0.026
Tin		0.02	0.01	0.01
Titanium		0.02	0.02	0.05
Zinc		0.95	0.33	1.06

Source: From Stern, A.C. (Ed.) *Air Pollution*, Academic Press, New York, 1968.
With permission.

of a community may deteriorate enough to damage the health of its citizens. Unfortunately, the toll of excess mortality and morbidity in these disasters was not appreciated at the time and protective measures were not taken.[1]

Similar as well as higher particulate air pollution still exists in highly industrialized but environmentally backward countries. For example, in China coal burning is the dominant source of outdoor air pollution. More than 75% of China's primary energy needs are supplied by domestic coal. Heavy air particulate and SO_2 pollution dominates in urban areas. Existing Chinese air quality standards for particulates are as high as 200 $\mu g/m^3$ (annual mean). Nevertheless, 1995 particulate outdoor concentrations lay between 250 and 410 $\mu g/m^3$ as the annual mean and median of all larger cities.[2] Similar situations still exist in the former communist countries in Eastern Europe. Figure 3.2

demonstrates air particulate and SO_2 concentrations measured in the northern part of Bohemia (Czech Republic). Coal-fired power plants in this region produce 35% of the electricity for the whole country. The brown coal of this region is of low quality and contains between 1 and 5% sulfur. Examples of measurements done in 1992/1993 are illustrated in Figure 3.2. They are comparable with concentrations measured during the London smog period in 1952.[3,4]

The emission situation has changed considerably in the highly industrialized countries since about the 1980s, mainly because of improved burning and abatement technologies. The changes in the quality and quantity of particulate air pollutants have been well documented in the literature since about the 1980s. For example, in the U.S. approximately 2.6 million tons of particulates emitted in 1995 into the ambient air had approximately the following source distribution: 27% from transportation, 35% from fuel combustion, 35% from nonchemical industrial processes, and 3% from chemical industrial processes.[5]

The ambient air concentrations of particulates in the U.S., measured as PM 10, had been decreasing since the 1970s and are still decreasing somewhat. They lay, for example, in 1988 between 25 (rural) and 34 (urban) µg/m³ as annual means and as U.S. weighted average. They were as low as about 20 (rural) and 26 (urban) in 1995. This means they are considerably lower than the existing air quality standard of 50 µg/m³ (annual mean).

3.2.1 CHANGES IN PARTICLE SIZES

Because of qualitative changes in the emission situation since about the 1980s, changes have been observed in the particle size distributions of the atmospheric anthropogenic aerosol. While the concentration of coarse particles (in the approximate size range of 2 to 10 µm) has been considerably decreased in the past and still has a decreasing trend, an increase in the concentration of fine (≤ 2.5 µm) and ultrafine (< 1 µm) particles can be confirmed in several new observations and measurements.[6-9] Results of measurements in Hamburg (Germany) have shown that the mass concentration (c) of the fine fraction of atmospheric anthropogenic aerosols continuously increased between 1976 and 1986 (Δc/year + 8%).[6]

Very new measurements in the eastern part of Germany have confirmed this trend. Concentrations of ultrafine particulates doubled between 1991 and 1996, while the particulate mass concentrations of fine particulate fractions (≤ 2.5 µm) decreased by about 50% during this same time.[7] Similar results have been obtained from measurements in the urban area of Vienna.[8] Concentrations of particulate black carbon (BC) (the fine dispersed carbonaceous aerosols) were measured in this area in 1985 and in 1994. Comparison of this data has shown that the BC content of the atmospheric anthropogenic aerosol has increased 27% (outside the heating period by 106%) over the past 10 years, while the total mass concentration of airborne particulates decreased 44% (36% outside the heating period). If the current trends continue, the urban aerosol may one day contain BC — the very fine aerosol fraction — as a major component. Similarly, very high concentrations of ultrafine particulates were recently measured in Santiago, Chile, during the period between 1992 and 1994.[9]

3.2.2 CHANGES IN FINE PARTICLE CHEMISTRY

The qualitative changes in the chemistry of emissions as well as the distribution of emission sources influence the chemical composition of air particulates. Fuel, oil, and gasoline combustion as well as the combustion of wastes, etc., produce fine and ultrafine particulates containing many organic and inorganic toxic substances. The majority of emissions from vehicles and gas and fuel combustion consist mainly of particles smaller than 2 or 3 µm. However, fine and ultrafine particles are also emitted from several industrial sources. The modern separation and filtration methods used in the industrial air cleaning technology are efficient for fine and coarse particles. The ultrafine particles penetrate and are emitted into the atmosphere. This particle fraction contains important inorganic as well as organic toxic substances. Such as example is illustrated in Figure 3.3.[10] The fine and ultrafine atmospheric anthropogenic aerosols have relatively long residence time in the troposphere;

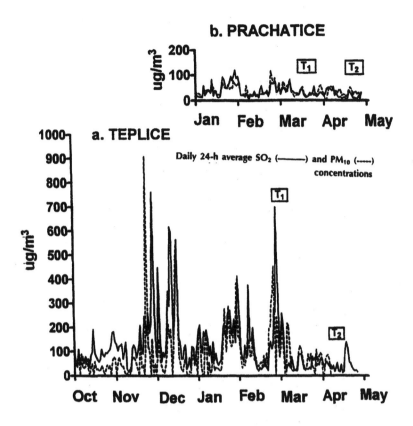

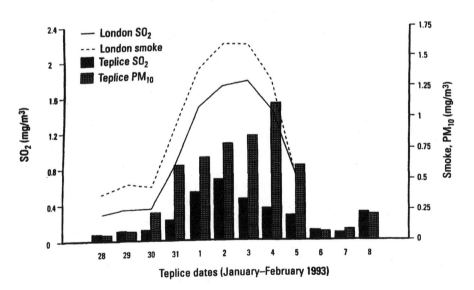

FIGURE 3.2 Daily 24-hr. average concentrations of PM 10 and SO$_2$ measured in the highly polluted region of North Bohemia (Teplice) and in a rural region of South Bohemia (Prachatice). (From Sram, R.J., Benes, I., Binkova, B. et al. Teplice program — The impact of air pollution on human health. *Environ. Health Perspectives* 104:Suppl. 4, 699-714, 1996. With permission.)

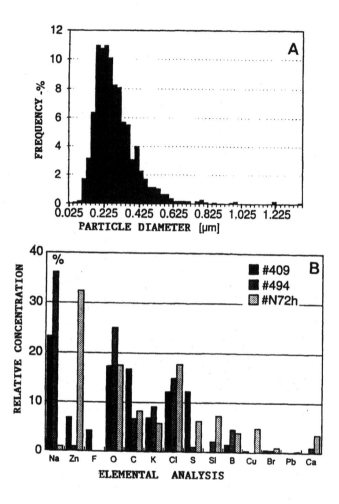

FIGURE 3.3 Distribution of ultrafine particles penetrated through high-efficiency industrial gas filters (A) and their chemical composition (B). (From Nachtweyh, K., Wetzel, A., Dikov, Y., and Rammnesse, W. Emission control in hazardous waste incineration. *Gefahrstoffe-Reinhalt. Luft.* 56:285–289, 1996. With permission.)

its concentration can be enhanced by meteorological events, such as inversion; and it can be transported for long distances. Therefore, it can be detrimental to the rural atmosphere. Figure 3.4a, b, c show such examples of recent measurements in the Shenandoah and Great Smoky Mountains National Parks.[11]

As already mentioned, fine carbonaceous particles are becoming the major component of the existing atmospheric anthropogenic aerosol.[12-22] They are working as vehicles for several organic substances, mainly for the PAH, oxy-PAH, etc.[12,20–22] Another group of fine and ultrafine anthropogenic aerosols characterizing air pollution situations since the end of 1980s deals with the application of agrochemicals and with the "modern" waste combustion technologies which produce highly toxic dioxins.[23-25]

3.2.3 INORGANIC SUBSTANCES IN FINE AND ULTRAFINE PARTICLES

Fine aerosol particles are formed from the condensation of hot vapors during the combustion process and from gas-to-particles conversion in the atmosphere. Some important toxic components, such as sulfates, nitrates, carbon-containing species, and metals (including transition metals) contribute to the increased health risk for the general population.[26-34]

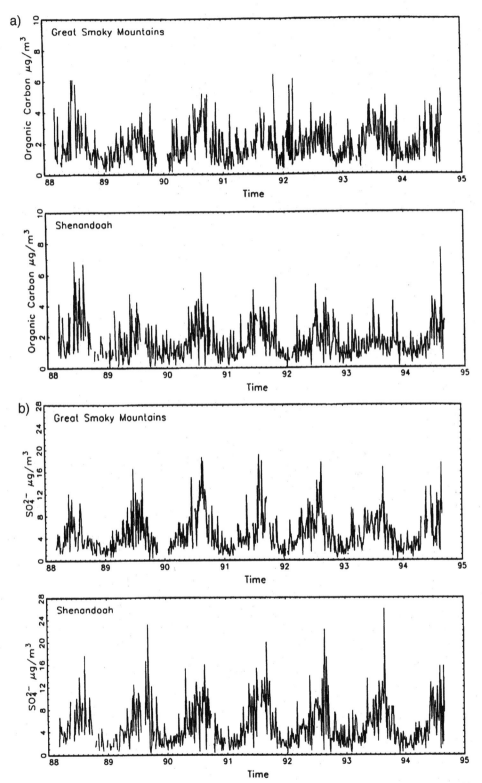

FIGURE 3.4 Concentrations of organic carbon (a), particulate sulfate (b), and nitrate (c) in the ambient air in mountain regions. (From Day, E.D., Malm, W.C., and Kreideweis, S.M. Seasonal variations in aerosol composition and acidity at Shenandoah and Great Smoky Mountains National Parks. *J. Air Waste Manage. Assoc.* 47:411-418, 1997. With permission.)

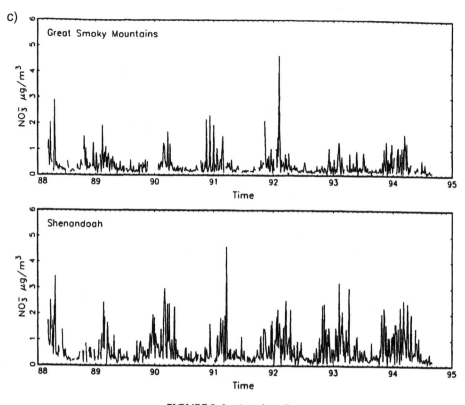

FIGURE 3.4   (continued)

On the other hand, the coarse particles are formed by mechanical processes and consist mainly of Fe, Al, and Si from the crustal material, and Na and Cl from sea salt.

Atmospheric anthropogenic aerosols originate from various sources. In urban areas, the sources of anthropogenic aerosols contribute about 70% of the total particulate air pollution. The heavy and transition metals form 10 to 20% of the atmospheric anthropogenic aerosols. Table 3.2 indicates elements that can often be correlated to specific emission sources.[26] Using the cluster analysis group of elements can characterize different groups of emission sources,[27] see Table 3.3. There are, of course, differences in urban and suburban atmospheres. Measurements done recently in the large industrial cities of Taegu and Pohang in South Korea are shown in Table 3.4.[33] The data demonstrate the influence of vehicle traffic and industry. Personal exposure to toxic and transition metals in highly polluted cities can be very serious, but depends also, of course, on the indoor and outdoor residence time of the urban population (Figure 3.5).[34]

In summary, that the most important inorganic air pollutants, which are bonded mainly on fine and ultrafine particulates, are sulfates, nitrates, carbon black, and toxic heavy and transition metals.

3.3   AIR PARTICULATE EPIDEMIOLOGY

Over the past decade a number of epidemiological studies have concluded that ambient air particulate exposure is associated with increasing mortality and morbidity among the general population. These studies have been conducted in several geographical locations and have involved a range of populations. The health effects were more aggravated among children, the elderly, and other vulnerable segments of populations.[35-42]

TABLE 3.2
Indicator Elements of Each Source Type

Source Type	Indicator elements
Oil combustion	V, S, As
Coal combustion	Al, Sc, S
Steel industry	Mn, Fe, Cr
Nonferrous metal	Cu, Zn, As, Sm, Sb
Cement, Limestone	Ca
Refuse combustion	Zn, K, Pb, Sb
Automobile exhaust	Pb, Br
Diesel exhaust	C
Soil dust	Al, Si, Ti, Sc, Mn
Marine (sea salt)	Na, Cl

Source: From Kasahara, M. Chemical composition of atmospheric aerosol particles, in N. Fukuta and P.E. Wagber (Eds.) *Nucleation and Atmospheric Aerosols.* A. Deepak Publ., 1992, pp. 385-393. With permission.

TABLE 3.3
Possible Sources of Elements as Obtained with Cluster Analysis

Soil	Industrial	Traffic	Fuel oil	Sulfate	Unidentified
Si–Fe–Ca	S–Mn–Br–Ni	Sr–Zn–Pb			Al–P
	Cu–V–Ti–Cr–K				
Si–Fe–Ca–K–Na–Zn–Mn	Cr–Al–Se	C–Pb–Ti–Br	V–Ni–Cu	H–S	
Si–Fe–K–Ca–S–Ti–Zr	Cl–Cu–Zn–Zr–V–Cr	Br–Pb			Al–Mn
Ca–Fe–Mn–Al–Cu–Ti	Na–Cl–Zn	C–Pb–Br–V–Ni–Si		H–S–K	

Source: From Miranda, J., Cahill, T.A., Morales, J.R., et al. Determination of elemental concentrations in atmospheric aerosols in Mexico City. *Atmospheric Environ.* 28:2299-2306, 1994. With permission.

Schwartz et al.[43,44] investigated correlations between TSP (total suspended particles) concentrations and acute respiratory illness (relative risk of croup disease) in five German communities. They found a positive correlation between TSP-concentrations and relative health risk. The health effects could be well established at TSP-concentrations above 10 $\mu m/m^3$ and were relatively high at TSP-concentrations at the air quality standard (AQS) levels.

Similar results have been obtained by Pope et al.[45-49] by investigating mortality and PM 10 particulate air concentrations in the Utah Valley. A significant positive association between nonaccidental mortality and PM 10 pollution was observed. The association between mortality and PM 10 was largest for respiratory disease death; next largest was cardiovascular death; and it was smallest for all other deaths. The relative risk of death increased monotonically with PM 10 and was well determined below the AQS concentrations (e.g., below 50 $\mu g/m^3$).

Positive correlations between particulate air pollution and daily mortalities in Detroit were confirmed by Schwartz in 1991.[44] A significant correlation was found between TSP concentrations and daily mortality. Concentrations of each increase of 100 $\mu g/m^3$ (TSP) resulted in a 6% increase in mortality. Cancer incidence and mortality in a cohort of 6000 Seventh Day Adventists, non-smokers,

TABLE 3.4
Ambient Concentrations (mean + SD)
of Elemental Inorganic Ion Fraction

Variable	Unit	Taegu urban	Taegu suburban	Pohang urban	Pohang suburban
As	ng m^{-3}	25.6 ± 16.5	13.1 ± 8.4	21.0 ± 13.8	30.5 ± 17.4
Fe	ng m^{-3}	1193 ± 534	1071 ± 643	1847 ± 806	5719 ± 2329
K	ng m^{-3}	708 ± 281	698 ± 344	1009 ± 383	1131 ± 399
Mg	ng m^{-3}	708 ± 454	477 ± 300	847 ± 409	1056 ± 369
Mn	ng m^{-3}	54.0 ± 28.2	50.2 ± 26.5	67.0 ± 42.8	819 ± 445
Ni	ng m^{-3}	25.6 ± 14.7	16.2 ± 10.7	17.6 ± 10.1	29.6 ± 14.4
Pb	ng m^{-3}	190 ± 109	134 ± 87.0	120 ± 47.8	165 ± 64.8
Se	ng m^{-3}	17.7 ± 10.5	8.3 ± 6.3	16.9 ± 7.5	18.3 ± 8.6
Ti	ng m^{-3}	30.6 ± 14.8	20.1 ± 11.5	27.8 ± 12.5	79.6 ± 35.4
V	ng m^{-3}	38.6 ± 25.4	26.2 ± 19.3	23.7 ± 13.9	38.1 ± 29.3
Zn	ng m^{-3}	316 ± 196	286 ± 135	228 ± 116	460 ± 213
Cl$^-$	µg m^{-3}	3.4 ± 2.9	2.7 ± 1.6	5.3 ± 2.7	6.4 ± 3.4
SO$_4^{2-}$	µg m^{-3}	22.8 ± 11.4	17.3 ± 12.5	18.0 ± 5.7	26.5 ± 6.2
NO$_3^-$	µg m^{-3}	5.1 ± 3.1	3.7 ± 2.9	3.0 ± 2.3	3.1 ± 1.7
TSP	µg m^{-3}	113 ± 49.4	83.6 ± 42.2	94.6 ± 34.7	172 ± 55.4
Number of data		96	73	68	69

Source: From Baek, S.O., Choi, J.S., and Hwang, S.M. A quantitative estimation of source contributions to the concentrations of atmospheric suspended particulate matter in urban, suburban, and industrial areas of Korea. *Environ. International* 23:205-213, 1997. With permission.

who were residents of California for a 6-year period, and the relationship to long-term ambient concentrations of TSPs were studied by Mills et al. in 1991[50] and by Abbey et al.[51,52] Risk of malignant neoplasms in females increased with exceedance frequencies for TSPs. Statistically significant increased risk of airway obstructive diseases, chronic bronchitis and asthma could also show positive correlations to TSP mass concentrations. The association between total daily mortality and particulate air pollution was also investigated in St. Louis and in counties in eastern Tennessee by Dockery et al. in 1992.[53] Concentrations of PM 10, PM 2.5, the elemental composition of these particulates, and aerosol acidity were measured daily during the period of study. Total mortalities were found to increase in 16% from each 100 µg/m^3.

Schwartz and Dockery found increased mortalities associated with TSPs in Philadelphia in 1992.[54] Total mortality was estimated to increase by 7% with each 100 µg/m^3 increase in TSPs. Positive correlations between TSP concentration and mortality was also found by Ostro in 1993. Each 10 µg/m^3 increase in TSPs was correlated with a 0.3 to 1.5% increase in mortality. Ostro considered it very important to determine the specific constituents of particulate matter that may be responsible for the health effect. In critical evaluations of several studies done in 1994 and 1995,[55-57] correlations of TSP exposure with respiratory diseases in children, the elderly, and other vulnerable persons and the general population are considered real and significant. The consistency of these findings across multiple reinforcing end points argues strongly for causality. While no epidemiologic study can prove this causality, all these results support the conclusion that current airborne particle standards are unlike to protect public health.[53] Last but not least, in several recent publications and critical reviews, correlations between increased particulate concentrations and health effects on the general population were repeatedly found.[54-57] Schwartz and Morris, for example, confirm in 1995 that increasing PM 10 concentrations were associated with increasing daily admissions for ischemic heart diseases in the Detroit urban area. Abbey et al. confirmed in a

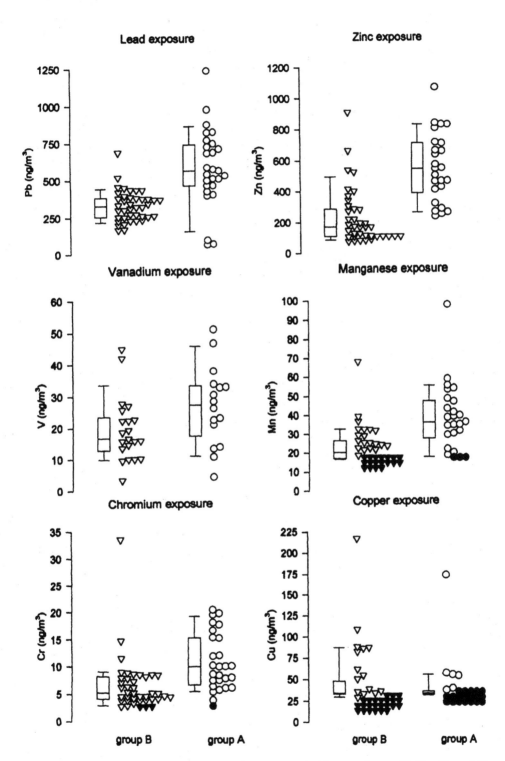

FIGURE 3.5 Box plots and distribution of individual personal airborne values to V, Mn, Cr, and Cu as a function of two population groups: (A) a group of residents spending most of their time outdoors, and (B) a group of residents spending most of their time indoors. (From Riveros-Rosas, H., Pfeifer, G.D., Lynam, D.T., et al. Personal exposure to elements in Mexico City air. *Sci. Total Environ.* 198:79-96, 1997. With permission.)

study in 1995[57] a continuously increased risk in respiratory diseases of nonsmoking population. A good relationship between daily mortality of elderly (>65 y) persons and elevated PM 10 concentrations was found during a study in Sao Paulo, Brazil, in 1995.[58] An increase in PM 10 to 100 $\mu g/m^3$ was associated with an increase in overall mortality equal to approximately 13%.

Pope reviewed and summarized his epidemiological studies in the Utah Valley in 1996.[49] He confirmed that Utah Valley has provided an interesting and unique opportunity to evaluate the health effects of respirable particulate air pollution (PM 10) for several reasons: (1) It has moderately high average PM 10 levels, and during low-level temperature inversion episodes, local emissions may become trapped in a stagnant air mass near the valley floor, resulting in highly elevated PM 10 concentrations; (2) the valley experienced the intermittent operation of a local integrated steel mill, the largest single particulate pollution source; (3) valley residents have very low smoking rates; (4) levels of SO_2, O_3, and aerosols with strong acidity are relatively low.

Several studies specific to Utah Valley have evaluated associations between various indicators of health and PM 10 pollution. Each of these individual studies had limitations imposed by data and analytic constraints. Taken together, however, they suggest a coherence or cascade of associations across various health end points for a specific location and population. Apparent health effects of elevated PM 10 pollution observed in Utah Valley include: (1) decreased lung functions; (2) increased incidence of respiratory symptoms; (3) increased school absenteeism; (4) increased respiratory hospital admissions; (5) increased mortality, especially respiratory and cardiovascular mortality; and (6) possibly increased lung cancer.

Correlations between asthma and the pollution including particulates has also been investigated.[59,60] Positive results were obtained by a study in Santa Clara County, California.[60] PM 10 concentrations were measured during winters between 1986 and 1992. The results demonstrated an association between ambient wintertime PM 10 concentration and exacerbation of asthma in areas where the principal source of PM 10 was residential wood combustion. High breast cancer incidence and mortality rates, relative to other parts of New York State, have been experienced for many years in Nassau and Suffolk counties, Long Island (NY). A case control study was used to evaluate the relationship between breast cancer risk and residential proximity to industrial facilities and traffic for pre- and postmenopausal women. A significantly elevated risk of breast cancer was observed among postmenopausal women. Unfortunately, no quantitative air pollution data and species were reported. Nevertheless, there is no question that particulate air pollution could play an important role.[61]

Results similar to those in the U.S. were also obtained in some other industrial countries. In a Canadian study, associations were reported between mortality and particulate air pollution.[62] Particulate air pollution was associated with cardiopulmonary and lung cancer mortality but not with mortality of other causes. The air particulate concentrations were in the range usually measured in North American states adjoining Canada. The effect of long term exposure to air pollutants in a cross-sectional population was investigated in Switzerland.[63,64] Among others, significant effects on lung function were found for particulates (PM 10) in the total population. PM 10 showed a very consistent effect of 3.4% change in lung function per 10 $\mu g/m^3$. A correlation between pulmonary functions of school children was found in highly polluted northern Bohemia (Figure 3.2).[65] Significant effects on lung function were found and, furthermore, the depressed lung function of children were found to be already chronic.

3.3.1 COARSE AND FINE PARTICULATES

In several later investigations, the air particulate concentrations were measured in two modes — PM 10 and PM 2.5.[66,67] The studied health effects could, therefore, be correlated to both particle size fractions. The correlations fit much better for the PM 2.5 than for the PM 10 modes.

In a positive cohort study, Dockery et al.[66] estimated the effects of air pollution and mortality, while controlling for individual risk factors, including SO_2, O_3, aerosol acidity, and coarse and fine

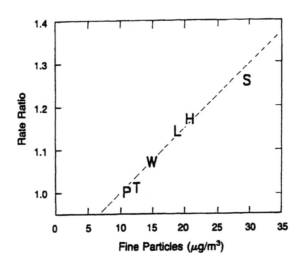

FIGURE 3.6 Estimated adjusted mortality-rate ratios and pollution levels (PM 2.5) in six cities. (From Dockery, D.W., Pope, C.A., Xiping, Xu, et al. An association between air pollution and mortality in six U.S. cities. *N. Engl. J. Med.* 329:1753-1759, 1993. With permission.)

particulate concentrations. Air pollution was positively associated with death from lung cancer and cardiopulmonary disease but not with death from other causes considered together. City-specific mortality rates, adjusted for a variety of health risk factors, were associated with the average levels of air pollutants in the cities. Mortality was more strongly associated with the levels of inhalable, fine particulates (PM 2.5) (Figure 3.6). The fine-particulate air pollution, or a more complex pollution mixture associated with fine particulate matter, is, therefore, considered a contributor to excess mortality in certain U.S. cities (e.g., Portage, Wisconsin [P]; Topeka, Kansas [T]; Watertown, MA [W]; St. Louis, MO [L]; Harrison, TN [H]; and Steubenville, OH [S]).

Similar conclusions were reported in a later publication by Schwartz et al. in 1996.[67] The epidemiologic analysis of the majority of American studies suggests that fine particle mass (PM 2.5) is associated with increased daily mortality. It suggests that these associations are not attributable to the sulfate or acidic composition of these particles. Nevertheless, the knowledge of the source of these fine particles points to the importance of combustion processes.

3.4   AIR PARTICULATE TOXICOLOGY

The already mentioned results of a number of epidemiological studies have concluded that ambient air particulate exposure is associated with increasing mortality and morbidity. The mechanisms of these adverse health effects are, nevertheless, practically unknown.

It is the aim of inhalation toxicology to investigate and describe such mechanisms. There is toxicological evidence for adverse effects from polluted air proven on an animal model.[68] In order to assess the adverse effects of urban levels of air pollution (Sao Paulo), rats were used as biological indicators in chronic exposure experiment. The results obtained suggest that chronic exposure to urban levels of air pollution (particulates, gases, and vapors) may cause respiratory lesions in rats.

Similar investigations are now available for testing the toxicity and/or carcinogenicity of real ambient air particulates by means of inhalation exposure on animal models. For such animal inhalation exposure studies, the atmospheric anthropogenic aerosol has to be fractionated and preconcentrated. Techniques have already been developed and tested which enable us to increase the original concentration of ambient air particulates to levels of 25 to 30 times greater. A suitable fractionater and concentrator with two or more virtual impactors in series has been described.[69,70]

This concentrator consists of a high-volume impactor with a 2.5 or 1 μm cutoff and two virtual impactors with a 0.15 μm cut-off size.

Laboratory investigations have proved that the process of particle enhancement, fractionation, and water condensation does not change the chemical composition of particulates. Such equipment combined with the rat inhalation exposure method could be applied as a transportable test procedure for health risk evaluation of different polluted sites, e.g., in cities, etc.

Numerous controlled toxicological investigations of individual chemical species have clearly shown that specific constituents of ambient air particulate matter (e.g., DEP (diesel exhaust particles), PAH, oxy- and nitro-PAH, acidic aerosols, metallic aerosols, etc.) are associated with adverse biological effects, including carcinogenicity.[71-73] Furthermore, insoluble, isometric, as well as fibrous dust particles, which are also components of air particulates, may exert a carcinogenic or cocarcinogenic effect in the respiratory tract.[74] But the PM 10, PM 2.5, and PMI 1 samples from ambient air are, in reality, a heterogeneous mixture that varies in constituents, particle sizes, and chemical composition, depending on geographical location, meteorology, and source emissions (see Air Particulate Chemistry). Complex mixtures present difficult problems for toxicological studies and risk assessment.[75] Schlesinger evaluated the abilities for toxicological evidence of health effects from ambient air particulates in a critical review.[76] He underlines the difficulties connected with toxicological studies of chemical mixtures.[77,78] It is likely not possible *a priori* to predict the nature of any interaction based merely on stated exposure conditions.

It is clear that specific components of air particulates alone, or as components of mixtures with other pollutants, can produce adverse biological responses consistent with human morbidity findings.

In most animal studies, the exposure concentrations of total exposures needed to produce any effect were well above those experienced by the general human population. For this reason, in terms of plausibility for increased human mortality, there are no toxicological data to allow any conclusion. On the other hand, toxicological studies have not yet indicated that specific components of atmospheric particulates *cannot* produce increased mortality/morbidity.[76] The responsible chemical constituents or markers and/or chemical reactions of ambient air particulates are to be determined and specified.

3.4.1 CYTOTOXICITY

The *in vitro* cyto- and genotoxicity tests have been used for evaluating the inhalation hazards from airborne particulates for a long time.[79,80] Furthermore, the application of the *in vitro* testing procedures makes it possible to correlate the genotoxicity, mainly the mutagenicity, to the physical (e.g., particle size) and chemical (e.g., PAH, oxy-, and nitro-PAH, etc.) properties of air particulate samples. For example, a predominance of the direct-acting mutagenicity found in the particulate matter of Barcelona (Spain) could well be attributed to a major contribution of the PAH transformation processes.[81]

Human cell mutagenicity was measured in Los Angeles air.[82] The mutagenic potency obtained from sampled particulates was found to be due to ubiquitous emission sources, e.g., motor vehicle traffic or stationary sources of fuel combustion. Furthermore, the human cell mutagen concentration in Los Angeles air was found to be one order of magnitude greater than at the background site. The city aerosol is, therefore, a source of human cell mutagens.

Investigations were conducted on the urban air of Bologna (Italy).[83] The mutagenicity of total (PM 10) and of size fractions (particle diameter in μm: <0.4, 0.4 to 0.7, 0.7 to 1.1, 1.1 to 3.3, > 3.3) of urban particulates was identified using tests on *Salmonella typhimurium*. There was no correlation between total and/or coarse particle concentration in air and mutagenic activity. The correlations increased as particle size decreased; moreover, the finer the particles, the greater the mutagenicity was. Therefore, the PM 10 concentration does not seem to be representative of air quality, at least with regard to the mutagenicity. The PM 2.5 seems to combine a better air quality concept with effective health risk (Figure 3.7).

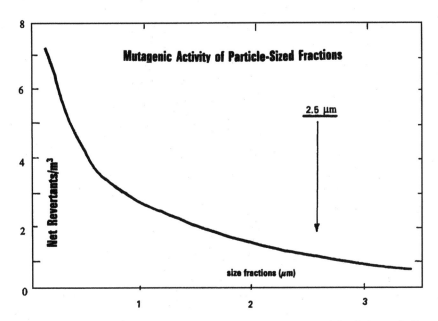

FIGURE 3.7   Mutagenic activity of air particulates as a function of particle diameter, indicated in net revertants per cubic meter. (From Pagano, P., De Zaiacomo, T., Scarcella, E., Bruni, S., and Calamosca, M. Mutagenic activity of total and particle-sized fractions of urban particulate matter. *Environ. Sci. Technol.,* 30:3512-3516, 1996. With permission.)

3.4.2 TOXIC MECHANISMS

The results of some recently published epidemiological studies, as well as animal toxicology and *in vitro* experiments, support the hypothesis that physical and chemical properties of single particles are involved in toxic, genotoxic, and carcinogenic health effect mechanisms of the inhaled particulates. Particle size, shape, and electrical charge, and particulate surfaces and solubilities are the most important physical parameters that can be correlated to observed toxic effects. These parameters also substantially influence the particle lung deposition and clearance rates.[86-96] Fine and ultrafine particles penetrate the deep lung compartments, and their deposition fraction depends strongly on the particle diameter and breathing rates. Figure 3.8 shows computed as well as measured deposition fractions of ultrafine particles (diameters less than 0.1 μm) in human extrathoracic and intrathoracic regions. it can be seen that the intrathoracic deposition efficiencies of 20 nm-particles can be as high as 50%. Particles smaller than 10 nm already have a very high nasal deposition rate.[92] Furthermore, electric charge also enhances airway deposition of fine and ultrafine particles.[96]

Similarly, particle clearance depends on particle size. Experimental data obtained on rat studies are summarized in Figure 3.9. Fine and ultrafine, insoluble particles (e.g., carbon black, diesel exhaust, TiO_2 ultrafine and silica) have long and very long clearance half times.[95]

The hypothesis that fine and ultrafine particles, when inhaled, can be very toxic to the lung is, therefore, supported by their high deposition efficiency in the lower respiratory tract, by their slow clearance rates, by their large numbers per unit mass, and by their increased surface areas available for interactions with cells.[85-87] Fine and ultrafine particles have been shown to be highly toxic to rats.[97] They are taken up poorly by lung macrophages[98] and are capable of penetrating the pulmonary epithelium into the interstitium.[99]

Preconcentrated ambient particles drawn from the outside air in Boston were inhaled by rats at concentrations less than 300 μg/m³. After three days of exposure, the mortality was 37%.[100] Additionally, in several measurements and experiments, the dependency on chemical composition of air particulates was also proved.

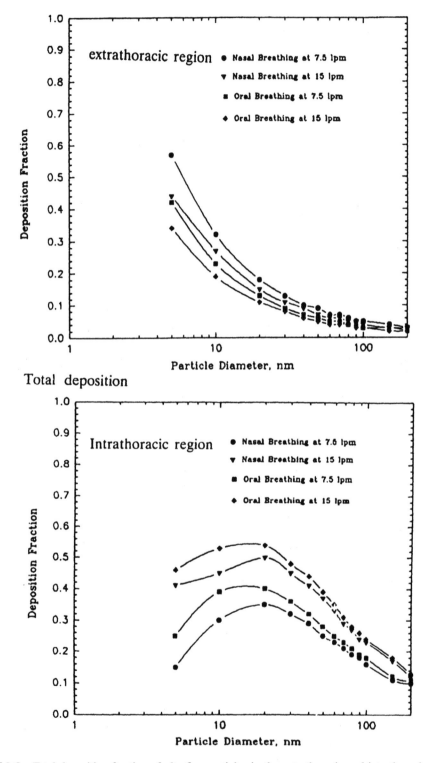

Total deposition

FIGURE 3.8 Total deposition fraction of ultrafine particles in the extrathoracic and intrathoracic region of the human airways. (From Cheng, K.H. and Swift, D.L. Calculation of total deposition fraction of ultrafine aerosols in human extrathoracic and intrathoracic regions. *Aerosol Sci. Technol.* 22:194-201, 1995. With permission.)

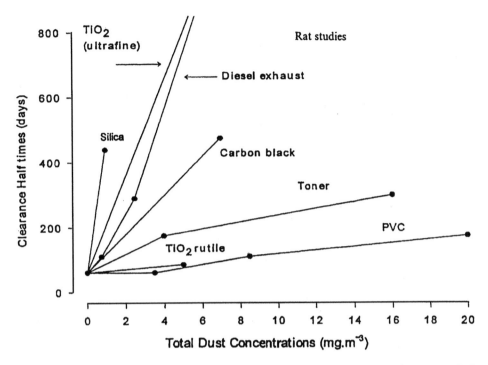

FIGURE 3.9   Clearance half-times after chronic inhalation of different dusts, including fine and ultrafine dusts, in rats. (From Kaufman, J.W., Scherer, P.W., and Yang, C.C.G. Predicted combustion product deposition in the human airway. *Toxicology.* 115:123-128, 1996. With permission.)

Particle-cell interactions could be potentially amplified by the presence of radicals on their surfaces. Transition metals, such as Fe, V, Ni, Ti, etc., which are the most important inorganic constituents of air particulates (Figure 3.10), release, in interactions with cells, metal ions such as ferric ions, which catalyze the production of hydroxyl free radicals via the Fenton reaction.[101-103] The free radical activity of particle surfaces could have several "damaging" effects on cells with which they make contact: lipid peroxidation, protein oxidation, DNA strand breaks, or antioxidant depletion.[103]

There also exists the direct production of short-lived free radicals (OH, HO_2, H_2O_2) in particulates in the air by several chemical reactions. Their concentrations are approximately an order of magnitude higher in polluted urban regions than in rural and remote areas.[105]

Particle toxicity and carcinogenicity is also enhanced by numerous organic constituents — e.g., PAH, oxy- and nitro-PAH, etc. — which are adsorbed on or absorbed in fine and ultrafine insoluble particles, such as soot particulates.[17,106]

The majority of such particles with organic toxicants are produced by different combustion processes. The relative contents of such organic toxic pollutants can differ substantially as a function of location. In a study in Germany, relative contents of some organic pollutants were determined in the air particulate mass at seven sampling sites (Figure 3.11). It can be seen that the particulates were loaded with different amounts of toxic substances (PAH, PCB, PCPh). The relative amount of toxic or carcinogenic substance in a mass unit of the particulates (e.g., ng or pg per 100 μg of particulates) can be designated as a "specific particulate toxicity/carcinogenicity."[17]

Further toxicological research (studies *in vitro* and *in vivo*) is needed for a better understanding of the toxic mechanisms of inhaled fine and ultrafine particulates. Nevertheless, the existing knowledge suggests that fine and ultrafine particulates, depending on their chemical composition, are associated with the observed serious health effects.[87]

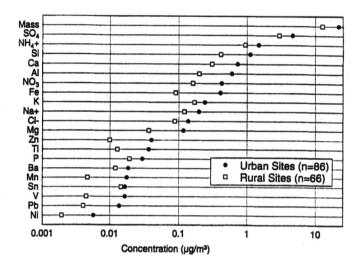

FIGURE 3.10 Concentrations of metals and other components in particulates, sampled in urban and rural sites in Canada. (From Brook, J.R., Dann, T.F., and Burnett, R.T. The relationship among TSP, PM 10, PM 2.5 and inorganic constituents of atmospheric particulate matter at multiple Canadian locations. *J. Air Waste Manage. Assoc.* 47:2-19, 1997. With permission.)

3.4.3 CARCINOGENIC HEALTH RISK FROM PARTICULATES IN THE AMBIENT AIR

Based on results published in epidemiological and animal studies, carcinogenic risk to classical air pollutants, including particulates, can be estimated. A governmental commission in Germany has estimated unit carcinogenic risk per a unit concentration (1 $\mu g/m^3$) for arsenic, benzene, Cd, diesel soot, PAH, (B[a]p), and dioxin.[107,108]

Based on concentrations measured in different areas and sites in Germany, the total carcinogenic risk for the general population (for an exposure of 70 years) was calculated. A risk of 1:2500 (4.10^{-4}) was recommended as a limit. Results illustrated in Table 3.5 and in Figure 3.12 show that, in practically all regions of Germany, the calculated risk is higher or considerably higher than the recommended limit. Furthermore, as seen in Table 3.5, elemental carbon (EC) and PAH (benzo[a]pyrene, B[a]P) are the most important contributors to the total carcinogenic risk. Both these particulate pollutants are in the fine and ultrafine particle size mode. Urban residents are additionally exposed to high soot and PAH concentrations in public transport systems. In public busses in Munich (Germany), the measured soot concentrations are in the range between 10 and 20 $\mu g/m^3$.[109]

3.5 AIR PARTICULATE FRACTIONS AND CHEMISTRY

Several modern methods of analytical chemistry have made it possible to obtain a detailed picture of the chemical composition of atmospheric particulates — atmospheric anthropogenic aerosol (AAA) — in different countries and regions.[110-112] The AAA is an aerodispersed system of solid and liquid particles with different sizes, shapes, and chemical composition. Also, the chemical composition of single particles can be anisotropic (i.e., there can be different compositions on the surface and inside the particles). The AAA is a product of several emission sources. It is also produced directly in the atmosphere by several physicochemical mechanisms, and during its residency there, it undergoes several further physical, physicochemical, and chemical processes. The resulting product is a polydispersed system of chemically heterogeneous particles, with complex toxic and carcinogenic potential. Biochemically active components may be present at the particle surface or inside the particles. The AAA can contain short-lived chemical species, including free radicals and other metastable components.

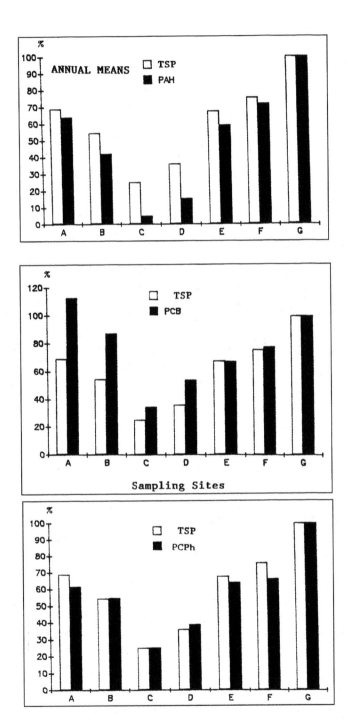

FIGURE 3.11   Relative mass of PAH (polycyclic aromatic hydrocarbons), PCB (polychlorinated biphenyls), and PCPh (polychlorinated phenols) bonded on TSP (total suspended particles). Results of measurements at 7 sites (industrial — A, F, G, residential — B, D, E, and remote — C) in Germany. (From Spurny, K.R. Atmospheric anthropogenic aerosol and its toxic and carcinogenic components. *Wissenschaft und Umwelt*, ISU, Aachen, Germany. 2:139-151, 1993. With permission.)

TABLE 3.5
Mean Total Cancer Risk

Area	Benzene μg/m³	Benzene risk (×10⁻⁵)	Soot as EC μg/m³	Soot as EC risk (×10⁻⁵)	PAH as B(a)P ng/m³	PAH as B(a)P risk (×10⁻⁵)	Σ risk (×10⁻⁵)
Rural	1.4	1.8	3.2	25.9	0.8	5.6	33
Residential	3.1	2.8	7.2	37.5	1.7	11.9	52
Traffic	9.9	9.0	8.7	43.5	1.8	11.9	64.4
Industrial	2.7	2.4	4.5	22.5	2.1	15.4	40

Source: From Tesseraux, I. and Kappos, A.D. Actual German air pollution levels. Risk estimation for respiratory diseases and mortality. *Experimental and Toxicologic Pathology,* 1997, in press. With permission.

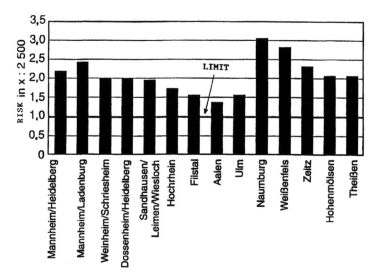

FIGURE 3.12 Evaluated carcinogenic risks in 15 German urban areas. (From Hanss, A., Herzig, S., and Lutz-Holzhauer. Assessment of risk of cancer caused by air pollution. *Gefahrstoffe-Reinhalt. Luft.* 57:71-74, 1997. With permission.)

The AAA is formed by two basic mechanisms: dispersion and condensation (including chemical reactions). The resulting AAA is a mixture of primary and secondary aerosols, with polymodal particle size distribution. Given that the components are of different origins, they have different chemical compositions (i.e., chemical composition is particle-size dependent). The fine fraction is associated with the most toxic inorganic and organic air pollutants. The fine AAA fraction could be designated as "lung toxic aerosol mode" and could serve as a possible marker for dosage of toxic and carcinogenic components of particulate air pollutants (Figure 3.13).

3.5.1 PARTICLE SIZE FRACTIONS

Figure 3.14 shows the major features of the mass distribution of particle sizes found in the atmosphere. The "nucleation" range, with particle diameters less than 0.1 μm, is termed the "ultrafine particles" as in a "highly dispersed aerosol." This fraction includes the fraction of "nanometer particles," defined as particles with diameters below 50 nm.[110] Since the particles less than 0.1 μm coagulate relatively fast (having lifetimes of less than 1 hour), from the standpoint of sampling and measurement as well as from a toxicological standpoint, the "ultrafine fraction" is defined as the

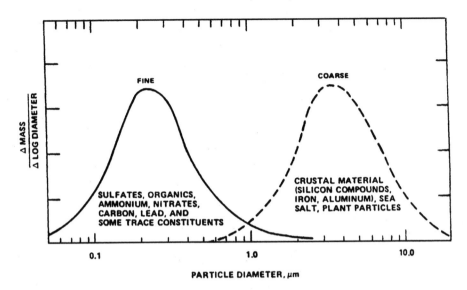

FIGURE 3.13 Idealized mass/size distribution and chemical composition for urban aerosol. (From Hidy, G.M. Summary of the California aerosol characterization experiment. *J. Air Poll. Control Assoc.* 25:1106-1114, 1975. With permission.)

fraction with particle diameters less than 1 μm. It includes the nucleation and a great part of the "accumulation" mode. The particle fraction designated "fine particles" is the fraction with particle diameters less than 2.5 μm. This fraction is usually called the PM 2.5 fraction. It includes the nucleation and accumulation modes. All particles with diameters below 10 μm are designated as the "respirable" or "thoracic" fraction and also as the PM 10 fraction. This includes the nucleation, accumulation, and part of the "coarse" modes. The TSP (total suspended particles) is then designated the fraction including the coarse, accumulation, and nucleation modes.

TSP and PM 10 were first introduced and standardized in 1971.[114] The PM 2.5 (fine fraction) and PM 1 (ultrafine fraction) are considered to be standardized in the proposed revision of Air Quality Standards in the U.S.[115]

3.5.2 Relationships and Evaluations

Data exist on relationships among the measured parameters TSP, PM 10, and PM 2.5, e.g., from measurements from Canada.[104] On average across all sites, PM 2.5 accounted for 49% of the PM 10, and PM 10 accounted for 44% of the TSP. However, there was considerable variability among sites, with the mean ratio ranging from 0.36 to 0.65 (Figure 3.15). Correlations of TSP/PM 10 and PM 10/PM 2.5 show similar variation (Figure 3.16). PM 2.5 concentrations tended to increase from summer to winter. Coarse particles (2.5 μm < diameter < 10 μm) were found to exhibit the opposite seasonal patterns. In some measurements in the U.S., fine particles (PM 2.2) accounted for 52 to 61% of total PM 10 mass.[105] TSP (particles with aerodynamic diameters less than 50 μm) represent a very broad size range (Figure 3.17). They include fractions of PM 10, PM 2.5, PM 1, and PM 0.1. Furthermore, the particle size fractions between 10 and 50 μm represent very coarse particles, which differ substantially in their origin and chemistry from all other fractions. The TSP standard, therefore, has a very poor toxicological definition.

PM 10 has been defined from a physiological standpoint only. The mechanisms by which inhaled particulates induce adverse health effects depends, initially, on their depth of penetration and deposition and retention in the lung. The response to deposited material has been shown to be markedly greater when the tracheobronchial and alveolar regions of the respiratory tract (the intrathoracic region as opposed to the extrathoracic regions) were involved.[35] Since it has been

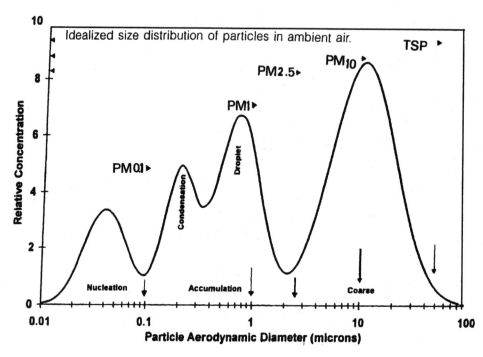

FIGURE 3.14 Three mode size distributions of AAA and the existing (TSP, PM 10) and proposed standard-ized size fraction definitions. (From Chow, J.C. Measurement methods to determine compliance with ambient air quality standards for suspended particles. *J. Air Waste Manage. Assoc.* 45:320-383, 1995. With permission.)

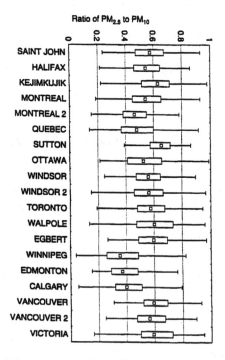

FIGURE 3.15 Distributions of the mass ratio of PM 2.5 to PM 10 measured at different sites in Canada. (From Brook, J.R., Dann, T.F., and Burnett, R.T. The relationship among TSP, PM 10, PM 2.5 and inorganic constituents of atmospheric particulate matter at multiple Canadian locations. *J. Air Waste Manage. Assoc.* 47:2-19, 1997. With permission.)

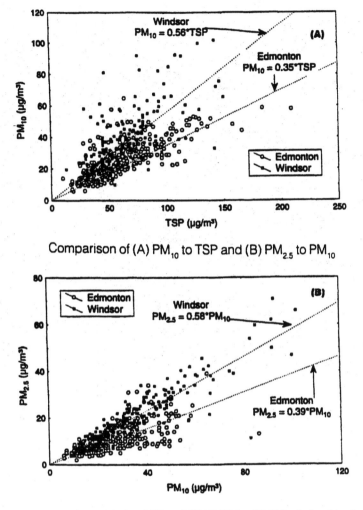

Comparison of (A) PM_{10} to TSP and (B) $PM_{2.5}$ to PM_{10}

FIGURE 3.16 Comparison of (A) PM 10 to TSP and (B) PM 2.5 to PM 10 relationship observed at two sites in Canada. (From Brook, J.R., Dann, T.F., and Burnett, R.T. The relationship among TSP, PM 10, PM 2.5 and inorganic constituents of atmospheric particulate matter at multiple Canadian locations. *J. Air Waste Manage. Assoc.* 47:2-19, 1997. With permission.)

demonstrated that particles with an aerodynamic diameter of 10 μm and less enter the thoracic region (the bronchus and bronchioles), the U.S. Environmental Protection Agency has proposed the term "thoracic particles" to describe such particles, and the criterion mass levels, relating to thoracic particles, was designated PM 10.

Nevertheless, based on the results of recent investigations, the parameters PM 10 and TSP do not correctly and sufficiently represent the health risk potential of the particulate air pollutants. The existing PM 10 definition, based on the particle size only, neglects the role of "particle chemistry and toxicity" (Table 3.6). The PM 2.5 fraction, recently proposed and already being used in routine measurements, seems to be a much better standard with respect to the toxic and carcinogenic potency of particulate air pollutants.

There does exist considerable variations in the size distribution of ambient air particles. However, there is generally a clear separation into fine and coarse modes, with a dividing point between 1.0 and 3.0 μm, where the mass of the two modes is at minimum (Figure 3.13).[116] Condensation aerosols do not normally grow above 2 μm, and significant concentrations of dispersion aerosols are not normally found below 2 μm. Therefore, the cut-off point for fine AAA at 2.5 μm seems

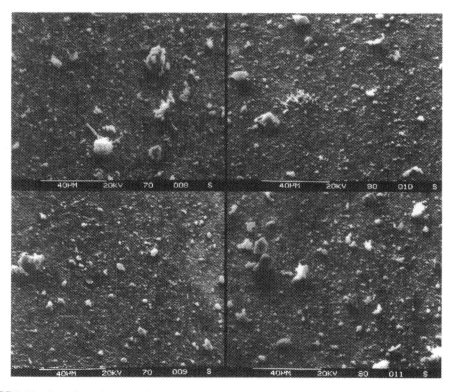

FIGURE 3.17 Scanning electron micrographs illustrating the large particle size differences of an urban particulate sample.

acceptable and reasonable. PM 1 and PM 0.1, with particles below 1 μm and 0.1 μm, respectively, could probably represent the "most toxic modes" of the AAA. Both of these fractions are characterized by high particle number and particle surface concentrations, while their mass concentrations are low or very low and not easy to measure (see Figures 3.18 and 3.19).[117]

3.5.3 PARTICLE SIZE AND CHEMICAL COMPOSITION

Numerous publications deal with the composition of the AAA.[118-135] The ambient air particulate concentrations, as well as their inorganic and organic compositions, undergo broad variations depending on the sampling site, day and season, meteorology, etc. Examples of such variations obtained by measurements in Germany are illustrated in Figures 3.20 through 3.23.[17] The particle size distribution, as well as chemical composition of the AAA, is determined mainly by the emission sources. In several studies, correlations between emission sources and ambient air particle characterizations were well confirmed.[120-125]

From a toxicological standpoint, the particle-size dependent chemical composition of the AAA is of basic importance. Temporal and spatial variations of the chemical compositions of PM 2.5 and PM 10 fractions were also established. For example, from measurements in California, nitrates, sulfates, ammonia, and inorganic and organic carbon were the most abundant species in the PM 2.5 fraction. The coarse particle fraction (2.5 μm < diameter < 10 μm) was enriched with soil-related elements, mainly with Al, Si, Ca, and Fe.[122] Similar results have been obtained from measurements of trace metals in the urban air in Illinois.[118] Metals, such as V, Cr, Mn, Zn, Se, and Pb were incorporated predominantly into the fine particulate fraction. Particle size distributions of Fe, Pb, Mn, Cu, and Ni were measured in the ambient air of Los Angeles.[126] Zn, Cu, and Ni were found mainly in particles with diameters of less than 1 μm. The noble metals, such as Pt, Pd, and Rd,

---

TABLE 3.6
The Most Important Components of the
Ambient Air Particulate Pollutants

What Is In Pm 10?

Trace metals & metal ions
Carbon particles
Products of incomplete combustion
Photochemical reaction products
Nitrates
Sulfates
Road dust & fine soil
Biological materials

Plus copollutants

Ozone
Sulfur dioxide
Nitrogen oxides
Organic vapors
Carbon monoxide
Acids
Free radicals
Formaldehyde

Source: From Phalen, R.F. The association of air pollution and mortality. *Fourth Int. Aerosol Conference.*
Los Angeles, pp. 964-965, 1994. With permission.

---

which are released in particulate emissions from autocatalysts, are also present in the fine and very fine fractions of the AAA.[127]

The organic toxic and carcinogenic particulate compounds are highly dispersed aerosols. The polycyclic aromatic hydrocarbons (PAHs) belong among the most common and ubiquitous organic air pollutants. They are present in the atmosphere as volatile and semivolatile fractions, but mainly as fine and very fine particulates. The importance of PAHs to air pollution chemists and public health professionals was recognized in the 1940s.[106] PAHs, as their oxy- and nitro- derivatives, are currently measured in the atmospheric environment. PAHs undergo chemical and photochemical reactions producing 2- and 4-ring PAHs with higher mutagenic activity than their precursors. The hydroxyl (OH) and nitrate (NO_3) radical-initiated reactions lead to the formation of mutagenic nitro-PAHs, including nitro-dibenzopyranones.[128-131] It has been demonstrated in several publications that the PAHs are bounded practically on fine and very fine particulates.[132-135] From measurements in Germany, more than 90% of PAHs were found in the particle fraction with particle diameters less than 1 μm.[133,134]

From measurements of the ambient air at traffic intersections in Taiwan, the content of PAHs in the fraction with particle diameters of less than 1 μm was in the range of 50% (Table 3.7).[132] Measurements of PAHs and elemental carbon in streets and tunnels were taken in California.[119] The maximal contents of PAHs were found in the very fine particulate fraction (Figure 3.24).[119] The elemental carbon (EC) size distributions were mono- and bimodal. The peaks of bimodal EC size distributions were in the size ranges 0.05 to 0.12 μm and 0.5 to 1.0 μm. The first peaks were attributed to primary emissions from combustion sources, while the second peaks were attributed to the accumulation of secondary reaction products on primary aerosol particles. Distributions of polychlorinated dibenzo-p-dioxins and dibenzo-furans correlated to the particle sizes of the AAA

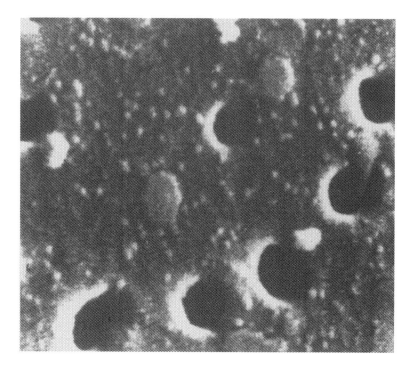

0,1 μm

FIGURE 3.18 A Nuclepore filter sample of nanosized particulates from a rural site.

were investigated in Germany.[135] By means of cascade impactor fractions, the size-dependent concentrations of both dioxins were established. The dioxins were bonded to particles with aerodynamic diameters below 1.5 μm. Nevertheless, the majority of the dioxin mass was connected with particles smaller than 0.5 μm.

3.6 SAMPLING, ANALYTICS, MARKERS, AND MONITORING

Better physical and chemical particulate characterization is needed for a more sensitive evaluation of the health risk of the AAA. This task begins with the use of the new sampling strategy and sampling equipment, as well as with suitable and, from a toxicological standpoint, recommended chemical characterization.

3.6.1 SAMPLING

The sampling procedure of the PM 10 mass fraction is well developed and standardized, and equipment is commercially available for disposition.[110,136] Cascade and virtual impactors are the methods of choice for the fractionation of particles within the size range less than 10 μm. Virtual impactors classify particles according to their aerodynamic diameter (Figure 3.25). In such a device, an acceleration nozzle (jet) directs particle-laden air toward a collection probe, which is slightly larger than the jet. Particles larger than a certain size cross the air streamlines due to their inertia and enter the collection probe, whereas smaller particles follow the deflected streamlines. A dichotomous sampler is already in use for separately measuring the PM 2.5 and the coarse fraction, with particle diameters between 2.5 and 10 μm.

Recently, a trichotomous sampler was developed and used for sampling the particulate mass fractions PM 10, PM 2.5, and PM 1. The equipment (Figure 3.26) uses a standard PM 10 size-selective

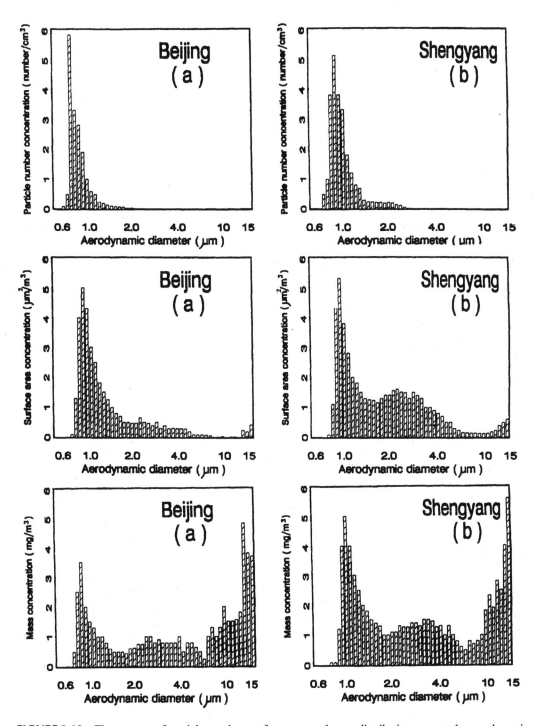

FIGURE 3.19   The patterns of particle number, surface area, and mass distributions versus the aerodynamic size of particles. (From Da-Tong Ning, Zhong, L.X. and Chung, Y.S. Aerosol size distribution and elemental composition in urban areas of Northern China. *Atm. Environ.* 30:2355-2362, 1996. With permission.)

inlet. In addition, it contains two high-volume virtual impactors in series, one for the 2.5 μm classification and the other for the 1 μm classification.[137] The multistage low pressure impactors have been found suitable for the sampling of the fine (less than 2.5 μm) and ultrafine less than

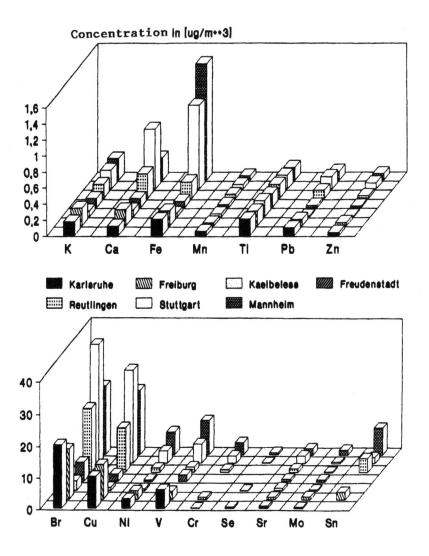

FIGURE 3.20 Distributions of element concentrations (weekly means) measured at seven sampling sites (industrial cities: Mannheim, Karlsruhe, and Stuttgart; residential sites: Feiburg, Freudenstadt, Reutlingen, and remote site — Kälbeles scheuer — elevation 1000 m in the Black Forest mountains) in western Germany. (From Spurny, K.R. Atmospheric anthropogenic aerosol and its toxic and carcinogenic components. *Wissenschaft und Umwelt,* ISU, Aachen, Germany. 2:139-151, 1993. With permission.)

1 μm) particulate fractions.[138] For example, the Berner low pressure cascade impactor makes it possible to select particles even in the nanosize range. This cascade impactor was recently used in combination with AFM (atomic force microscopy) to measure ultrafine particles within the AAA in Vienna.[139] By means of the AFM procedure, size and morphology of nanometer-sized particles can be easily studied (Figure 3.27). Sioutas et al. have developed and used a low pressure impactor which samples particles in the size range between 0.1 and 2.5 μm.[140]

Electrostatic collectors can also be used as suitable alternatives for collecting fine aerosol fractions. A new annular electrostatic collector was designed and used for sampling particles in the size range of less than 2.5 μm. The AAA is drawn through the annular space, field charging occurs, and the aerosol is collected on the surface of the tubular electrodes. This sampling procedure has an advantage compared with sampling on filters in that no chemical artifacts occur.[141]

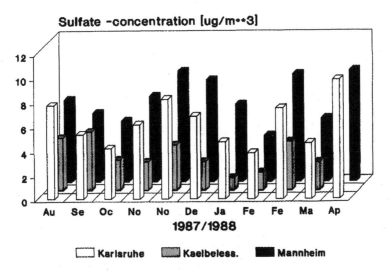

FIGURE 3.21   Sulfate concentrations measured at three sites (see Figure 3.20) in Germany. (From Spurny, K.R. Atmospheric anthropogenic aerosol and its toxic and carcinogenic components. *Wissenschaft und Umwelt*, ISU, Aachen, Germany. 2:139-151, 1993. With permission.)

3.6.2   ANALYTICS

The composition of AAA varies markedly, depending upon sampler location, the proximity of significant sources of aerosols and their gaseous precursors, and meteorology. As mentioned, the most important particle size fraction from a health effect standpoint is the PM 2.5 mode. Filters are the most frequently used sampling media. The amount of sample available for chemical analysis is small and, therefore, sensitive analytical methods have to be applied. These are already well developed, and their precision and accuracy are well known.[110,136,142]

For the determination of elements and inorganic compounds, techniques of physical analytical methods are most useful. X-ray fluorescence analysis (XRF) is commonly used for nondestructive element analysis of ambient aerosols. The proton-induced X-ray emission (PIXE) is also a nonde-structive, multielement procedure in which protons excite the atoms of a sample. Another nonde-structive method is neutron activation analysis (NAA). The particulate sample is bombarded with neutrons, and the radioactivity involved is subsequently measured. A standard destructive method for trace metal analysis is atomic absorption spectroscopy (AAS) and inductively coupled plasma emission spectroscopy (ICP). For ion analysis in soluble sample extractions, ion chromatography (IC) is the most useful analytical tool.

Studies of organic compounds in AAA samples are necessary for identifying important toxic and carcinogenic components. PAHs belong to the most frequently measured and studied organic compounds. High performance liquid chromatography (HPLC) as well as gas chromatographic mass spectrometry (GC-MS) are useful tools for identifying PAHs, their derivatives, and several old organic substances in the AAA samples.[110,142]

In addition to the inorganic and organic bulk analysis of the AAA, several modern physical methods make it possible to analyze and identify individual aerosol particles on filters[143] or *in situ* and on-line in an aerosol beam.[144] Their applications are useful and increasingly popular.

3.6.3   MARKERS AND MONITORING

Complete and accurate chemical analysis of the AAA samples is possible and useful for several basic studies of the atmospheric environment and for toxicological research. Nevertheless, practical applications of such results in health-risk assessment is difficult. Furthermore, our understanding

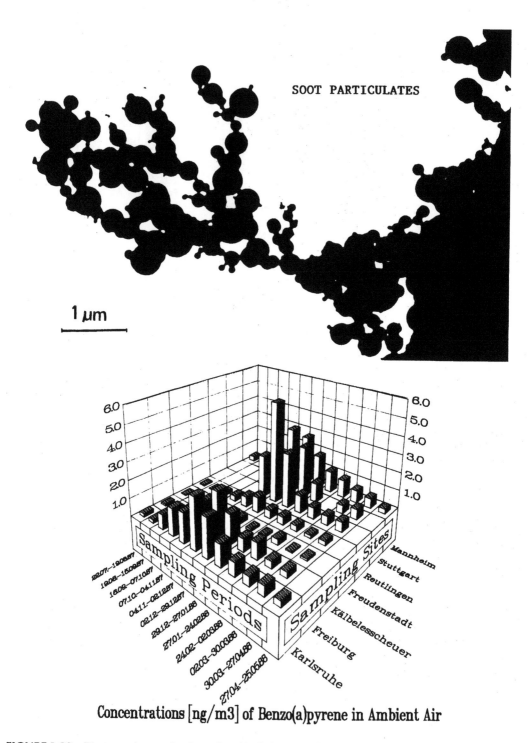

FIGURE 3.22   Electron micrograph of a carbon black (soot) particle agglomerate as a vehicle for PAH and the distribution of benzo-a-pyrene (BaP) concentrations at seven sampling sites in western Germany measured between July and April. (From Spurny, K.R. Atmospheric anthropogenic aerosol and its toxic and carcinogenic components. *Wissenschaft und Umwelt*, ISU, Aachen, Germany. 2:139-151, 1993. With permission.)

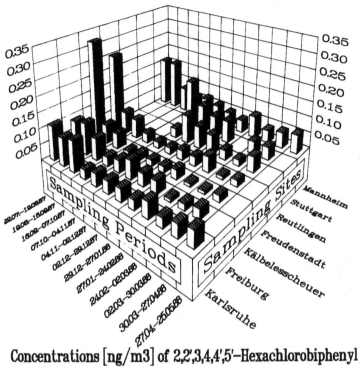

Concentrations [ng/m3] of 2,2',3,4,4',5'-Hexachlorobiphenyl

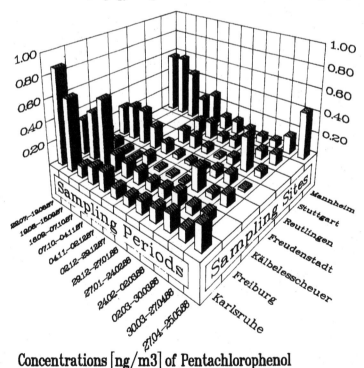

Concentrations [ng/m3] of Pentachlorophenol

FIGURE 3.23   Example of the distributions of PCB (polychlorinated biphenyls, top), and PCPh (polychlorinated phenols, bottom) at seven sampling sites in western Germany between July and April. (From Spurny, K.R. Atmospheric anthropogenic aerosol and its toxic and carcinogenic components. *Wissenschaft und Umwelt*, ISU, Aachen, Germany. 2:139-151, 1993. With permission.)

TABLE 3.7
Particle Size Distribution of Cumulative Fraction (%) for Individual PAHs in the Ambient Air of Traffic Intersection

PAHs	<1.0 µm (%)	<2.5 µm (%)	<10 µm (%)	<25 µm (%)
Nap	61.3	76.8	92.0	94.9
Acpy	55.9	70.6	86.2	95.6
Acp	56.7	71.7	87.3	94.5
Flu	57.9	74.7	88.8	94.2
PA	60.9	78.0	92.6	96.9
ANT	55.2	71.4	87.8	94.4
FL	53.2	76.3	92.7	97.2
Pyr	55.3	82.1	95.0	97.9
CYC	57.6	74.4	88.5	94.6
BaA	51.5	78.7	92.0	95.9
CHR	51.8	73.8	88.2	94.8
BbF	53.5	76.7	92.4	96.6
BkF	53.8	73.3	90.5	96.3
BeP	53.3	75.5	88.3	97.9
BaP	51.8	73.4	90.0	95.7
PER	45.7	70.6	88.0	94.9
IND	50.4	76.0	93.7	97.5
DBA	48.4	70.0	88.9	92.4
BbC	48.6	69.5	85.1	92.6
Bghip	43.5	66.0	86.0	94.4
COR	37.8	64.0	85.4	94.9

Source: From Sheu, H.L., Lee, W.J., Tsai, J.H., et al. Particle size distribution of PAH in the ambient air of a traffic intersection. *J. Environ. Sci. Health.* A 31:1293-1316, 1996. With permission.

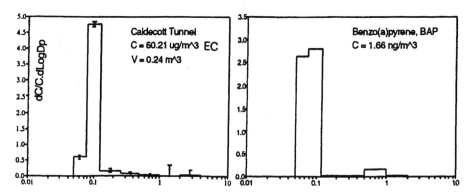

FIGURE 3.24 Size distribution of elemental carbon (EC) and benzo(a)pyrene (BAP) measured in California. (From Venkataraman, C. and Friedlander, S.K. Size distribution of PAH and elemental carbon 1,2. *Environmental Sci. Technol.* 28:555-572, 1994. With permission.)

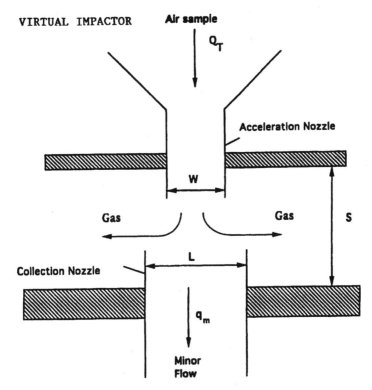

FIGURE 3.25   Schematic diagram showing the principle of a virtual impactor. (From Koutrakis, P. and Sioutas, C. Physico-chemical properties and measurement of ambient particles, in R. Wilson and J. Spengler (Eds.)*Particles in Our Air.* Harvard School of Public Health. Boston, U.S.A., 1996, pp. 15-40. With permission.)

of the health effects of fine particulate matter is far from complete. Results of already published epidemiological and toxicological studies suggest that fine particulates and their several chemical components are involved in the mechanisms leading to observed health effects. With respect to this fact and to protect the general population, mainly children, the elderly, and other vulnerable people, analysis and identification of some special markers, whose toxic and carcinogenic effects are known, should be a useful approach for the near future. Furthermore, it could take 30 years or more to identify precise toxic and carcinogenic mechanisms. AAA sample analysis, as well as real-time and on-line monitoring of, for example, carbon black, polycyclic aromatic hydrocarbons, and transition metals, should be recommended.

3.6.3.1   The Measurement of Elemental Carbon-Carbon Black Aerosol

Carbon is one of the most abundant constituents of ambient particulates. It can be present as elemental carbon (EC), which is nonvolatile, and as organic carbon (OC), which is volatile. The EC is a very important constituent of the fine and very fine particulate fraction and a vehicle for many organic and several inorganic compounds.

The observed EC mass fraction varies between 10 and 40% of the total AAA. EC is a residue from incomplete combustion of fuels. In urban atmospheres heavily impacted by automotive emissions, diesel engine exhaust has been recognized as the main source of EC. According to the particle formation mechanisms during combustion, the graphitic nucleus of a combustion particle is covered with organic compounds, whereby the ratio of OC to the total carbon strongly depends on the emission source.[136]

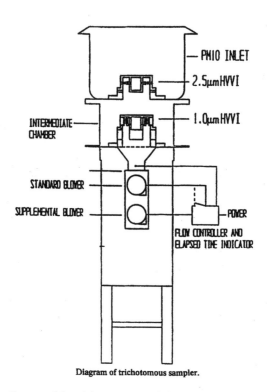

Diagram of trichotomous sampler.

FIGURE 3.26   Schematic diagram of the trichotomous particle sampler. (From Lundgren, D.A., Hlaing, D.N., Rich, T.A., and Marple, V.A. PM 10/PM 2.5/PM1 Data from a trichotomous sampler. *Aerosol Sci. Technol.* 25:353-357, 1996. With permission.)

All emissions from incomplete combustion processes contain PAHs as the most prominent representative of the organic fraction. There are several methods that can be used for concentration measurements of EC, OC, and TC (total carbon).[145-148] Some of them were standardized and compared in interlaboratory and field measurements.[146] Some of these procedures can be used as on-line and direct-reading monitoring methods. They can be used easily at sampling sites for getting analytical information complementary to the measurement of fine and/or very fine particulate concentrations.

3.6.3.1.1   Aethalometer

The working principle of the aethalometer is schematically illustrated in Figure 3.28. Black carbon aerosol is collected on a filter, and the optical transmission is continuously measured. The rate of optical attenuation is proportional to the rate of increase of black carbon surface loading; i.e., proportional to black carbon concentration and the air flow rate.[149] The aethalometer is commercially available and has been recognized as a part of the air quality monitoring network in some countries.

3.6.3.1.2   Photoacoustic Soot Measurement

The photoacoustic method is another procedure that can be successfully used for the measurement of EC concentrations *in situ*. The existing equipment was further developed and improved to be able to detect low EC concentration in the atmosphere. The detection limits lie in the area of $0.5 \ \mu m/m^3$.[147]

3.6.3.1.3   Thermal–Optical Methods

In the thermal–optical method, speciation of organic, carbonate, and elemental carbon is accomplished through temperature and atmosphere control. A schematic of the instrument is shown in

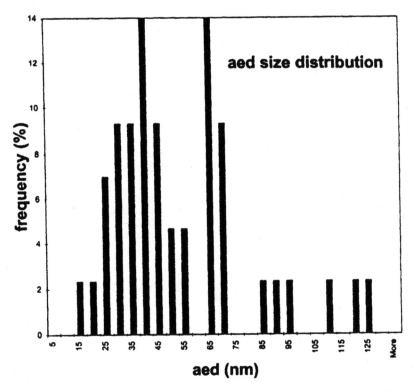

FIGURE 3.27 An example of the aerodynamic (aed) distribution of AAA in the nanometer size region. (From Köllensperger, G., Friedbacher, G., Grasserbauer, M., and Dorffner, L. Investigation of aerosol particles by atomic force microscopy. *Fresenius J. Anal. Chem.* 358:268-173, 1997. With permission.)

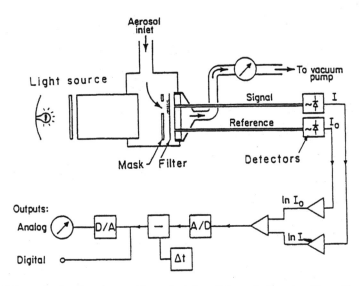

FIGURE 3.28 Schematic showing the working principles of the aethalometer. (From Hansen, A.D.H. The development of the aethalometer. Fifth *GIV Colloquium*, Frankfurt (Germany), pp. 1-7, 1994. With permission.)

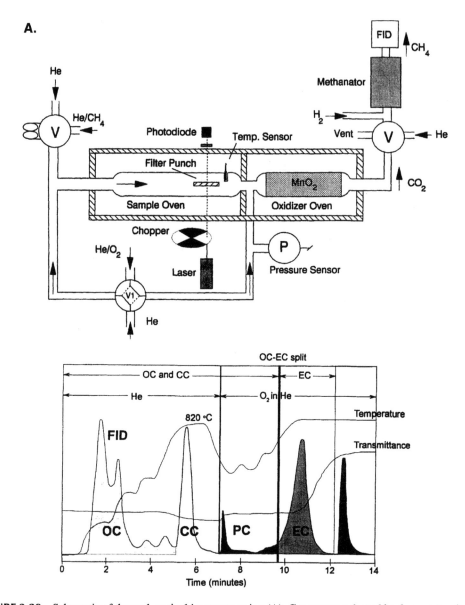

FIGURE 3.29 Schematic of thermal–optical instrumentation (A). Gas stream selected by four-port switching valve (V_1). Pure He used during first stage of analysis, and O_2(2%)–He mix used during second stage. Example of a thermogram (L) for sample containing rock dust (carbonate source) and diesel exhaust. Peaks correspond to organic (OC), carbonate (C), and elemental (EC) carbon. (From Birch, M.E. and Cary, R.A. Elemental carbon-based method for monitoring occupational exposures to particulate diesel exhaust.*Aerosol Sci. Technol.* 25:221-141, 1996. With permission.)

Figure 3.29. The AAA are sampled on quartz-fiber filters. A part of the filter (filter punch) is placed into the sample oven. He–Ne laser light passed through the filter allowing continuous monitoring of the filter transmittance. OC and EC are reported as $\mu m/cm^2$. Total EC and OC on the filter are calculated by multiplying reported values by the sample deposit area. A flame ionization detector (FID) is used for quantification (as CH_4) of evolved carbon.[150] An example of the instrument output, called "thermogram," is shown in Figure 3.29.

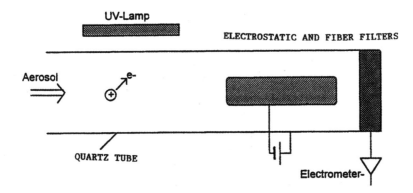

FIGURE 3.30 Schematic illustrating the working principles of the PAS (photoelectric aerosol sensor) for the detection of aerosols loaded with PAHs. (From Burtscher, H. Aerosol characterization by PAS (photoelectric aerosol sensor. *Fifth GIV Colloquium,* Frankfurt (Germany), pp. 1-5, 1994. With permission.)

The three traces appearing in the thermogram correspond to temperature, filter transmittance, and detector (FID). The analysis proceeds essentially in two stages. In the first, organic and carbonate carbon are volatilized from the sample in a pure He atmosphere as the temperature is stepped to about 820°C. Evolved carbon is catalytically oxidized to CO_2 in a bed of granular MnO_2 (at 900°C), reduced to CH_4 in a Ni/firebrick methanator (450°C) and quantified as CH_4 by a FID. During the second stage of the analysis, pyrolysis corrections and EC measurements are made.

3.6.3.1.4   Analytical Scanning Electron Microscopy

Carbon black aerosol can also be detected by means of the methods for the analysis of individual particles. The analytical scanning electron microscope (ASEM), which is able to identify light elements such as O, C, etc., has been proved to be able to differentiate between carbonaceous and noncarbonaceous particles. The first were identified in the ASEM as particles with an O/C atomic ratio of less than 0.15. Morphology (shape and surface texture) and trace element content (S, Cl, metals, etc.) were used to classify the black carbon according to source.[151]

3.6.3.2   PAH Measurement and Monitoring

Because of their carcinogenic nature, PAHs have been studied more extensively than any other compound. Their sampling and analysis are captious, because they are partionated in both gas and particulate phases, and, therefore, their collection is susceptible to sampling artifacts. Analysis of PAHs is typically carried out with HPLC or after thermal separation with MS. The first technique separates the different PAHs, while the second determines their mass. Well standardized sampling and analysis methods for PAHs are available.[152]

Fast and *in situ* and on-line detection methods for PAHs, including commercially available equipment, are also available for use.[146,147] The photoelectric aerosol and PAHs sensor (PAS) is an example of such modern devices.

3.6.3.2.1   PAS Technique for the Detection of the Total PAHs

The photoelectric aerosol sensor measures the near-UV photoemission produced by aerosols contaminated with PAHs.[153,154] The working principle of this commercially produced device is schematically shown in Figure 3.30. The aerosol (e.g., AAA) flows through a quartz tube and is irradiated by the light of a UV lamp. Then, by the emission of photoelectric mechanisms, electrons are separated in an electrostatic "filter" with lower potential. Positively charged aerosol particles are then collected in a fiber filter with high efficiency. These are measured by a sensitive electrometer. The indicated electric charge is an indirect measure of the emitted photoelectrons and can be correlated to the total PAHs bound on the AAA.

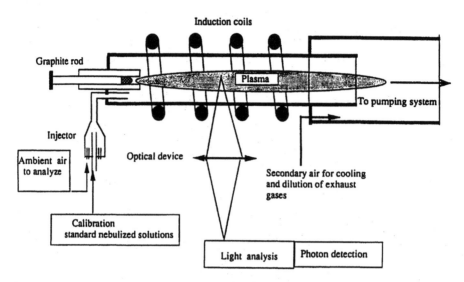

FIGURE 3.31 Schematic showing the principle of the air plasma torch for the *in situ* detection of metal aerosols. (From Gomes, A.M., Sarrette, J.P., Madon, L., and Almi, A. Continuous emission monitoring of metal aerosol concentrations in atmospheric air. *Spectrochimica Acta B.* 51:1695-1705, 1996. With permission.)

Furthermore, the PAS method can be combined in a single apparatus with the aethalometer. Both carbon black aerosol and total PAHs concentrations are then detected at the same time and on the same site.

3.6.3.3 Metallic and Ionic Components

As already mentioned, the analytics of metals and ions in the AAA samples are well developed and sensitive. Several procedures, such as XRF, PIXE, NAA, AAS, ICP, etc., can be successfully used for the analysis of metals in AAA samples. THe ion chromatography (IC) is the method of choice for the determination of salts and acids, sulfates, nitrates, H_2SO_4, etc.[136,142]

The transition metals, such as Fe, V, Ni, Mn, etc., as well as carcinogenic metals, such as As, Cd, Cr, Ni, etc., belong in the inorganic markers of the AAA, whose concentrations should be measured routinely. Such measurements provide important complementary information to the PM 2.5 data.

There also exist methods for continuous monitoring of metal aerosols in ambient air.[155,156] For example, the inductively coupled plasma atomic emission spectroscopy (IPC–AES) was recently modified for continuous measurement of ambient air metal concentrations. The principle of the method is illustrated in Figure 3.31. The AAA to be analyzed is introduced into the air plasma torch. The plasma is ignited and excited directly in air by 64 MHz tuned-line oscillator. The metals are identified by their atomic spectra. Metal concentrations are calibrated by nebulizing standard aqueous solutions. Concentrations of several metals, such as Cd, Co, Cr, Cu, Fe, Ni, Pb, and Zn, can be continuously detected with detection limits between 0.01 and 60 $\mu g/m^3$. The detection limits for Cu, Ni, Fe, Co, and Cr are less than 2 $\mu g/m^3$.[157]

In some cases, the valency of the measured metal has to be specified. Cr(III) and Cr(VI) are good examples. Cr(O), Cr(III), and Cr(VI) are the most common states of chromium in aerosols and dusts. Whereas the Cr(III) species is essential for plants and animals in trace concentrations, Cr(VI) is toxic and carcinogenic even at very low levels. An isotope dilution mass spectrometric method (IDMS) was recently developed for Cr(O), Cr(III), and Cr(VI) speciation in aerosol particles.[158] The aerosol particles collected on a glass filter can be analyzed by this procedure (Figure 3.32). The method is highly sensitive. Detection limits of 30 pg/m^3 for Cr(III) and of 8 pg/m^3 for Cr(VI) can be achieved.

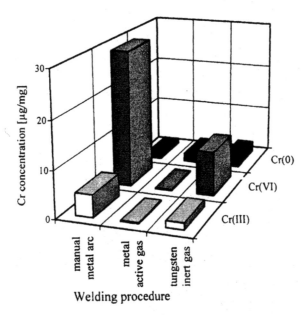

FIGURE 3.32   The speciation of Cr(O), Cr(III), and Cr(VI) in aerosols from different welding procedures for stainless steel. (From Nusko, R. and Heumann, K.G. Cr(III)/Cr(VI) speciation in aerosol particles by extractive separation and thermal ionization isotope dilution mass spectrometry. *Fresenius J. Anal. Chem.* 367:1050-1055, 1997. With permission.)

3.6.3.4   Cellular Genotoxic Markers

The *in vitro* cellular test for determining the genotoxic as well as the nongenotoxic effects of AAA extracts are already well developed and their applications also have several advantages.[79,80] From a practical point of view, one important aspect of such tests is that the results represent effects, e.g., the mutagenicity, produced by the total mixture of toxic substances which were extracted from the AAA sample. These effects are produced by chemical and not physical properties of sampled particulates. For example, the results of such tests correlate very well with the PAHs concentrations in ambient air as well as with the concentrations of the fine (PM 2.5) particulate fraction. The reason for the latter correlation is the fact that the PAHs and other toxic and carcinogenic substances are bound on such fine and very fine particles.

The cellular tests are useful mainly for measurements in cities with heavy automobile traffic. In such urban atmospheres, automobiles are the main source of AAA (Figure 3.33). The carbonaceous aerosols exhausted by automobiles, mainly by diesel engines, are carriers of several highly mutagenic organics.[159]

3.7   RECENT STUDIES

The trend demonstrating the importance of the fine fraction of particulate pollutants in the atmospheric environment is continuing in several recently published epidemiological, as well as measurement, studies.[160-165] Several studies have found associations between particulate air pollution and total and adult mortality. The relationship between postneonatal infant mortality and airborne particulate matter (PM 10) was investigated and evaluated in the U.S. in a recent study.[160] This study involved analysis of cohorts consisting of approximately 4 million infants born between 1989 and 1991 in states that report relevant covariants. This included 86 metropolitan statistical areas in the U.S. Overall postneonatal mortality rates (per 1000 births) were 3.1 among infants with low particulate exposure (less than 28 $\mu g/m^3$), 3.5 among infants with medium particulate exposure

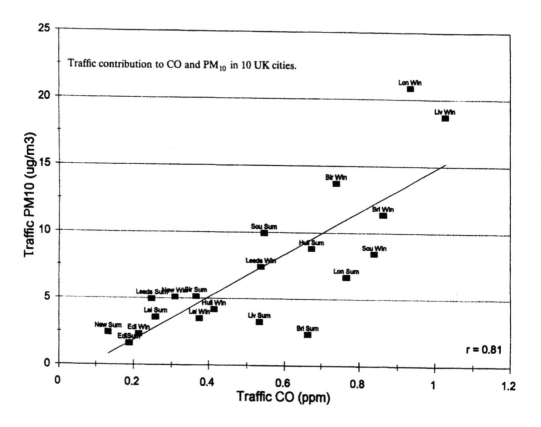

FIGURE 3.33   The correlation between AAA (PM 10) concentration and the concentration of CO. Traffic-generated CO is well correlated (cor. coeff. 0.81) with traffic-derived PM 10. (From Deacon, A.R., Derwent, R.G., Harrison, R.M., et al. Analysis and interpretation of measurement of suspended particulate matter at urban background sites in the U.K. *Sci. Total Environ.* 203:17-36, 1997. With permission.)

(between 28.1 and 40.0 $\mu g/m^3$), and 3.7 among highly exposed (more than 40.1 $\mu g/m^3$) infants. This suggests that airborne particulate matter is associated with a risk of postneonatal morbidity, and that the particulate matter may also influence an infant's chance of survival.

When looking for possible heavy toxic components of fine atmospheric particulates, a very important source is diesel vehicles. Submicron diesel particle mass size distributions display three log-normal modes that are centered at 0.09, 0.2, and 0.7 to 1.0 μm of particle aerodynamic diameter (Figure 3.34). The size distributions of elemental carbon (EC) and organic carbon (OC) are quite different: EC peaks at 0.1 μm, and OC peaks between 0.1 and 0.3 μm.[161] The carcinogenic PAHs and nitro-PAHs exist in the particulate matter emitted from gasoline and, mainly, from diesel engines. Remarkably high mutagenic and carcinogenic aromatic nitroketones, e.g., 3-nitrobenzanthrone (3-nitro-7H-benz(d,e)anthracene-7-one) were found in the polar extracts of diesel exhaust and airborne particulates.[162]

The importance of the fine atmospheric particulate fraction has been confirmed and documented by recent measurements in eastern Germany.[163] The most important findings were: (1) Particles larger than 2.5 μm at the end of the 1990s are rare in the European urban atmosphere, so that it is open to debate whether the inhalation of these particles is relevant for human health. This is most likely true for particles larger than 0.5 μm. (2) Since particle number and mass concentrations can be considered poorly correlated variables, more insight into the health related aspects of particulate air pollution will be obtained by correlating respiratory responses with mass and number concentration of ambient particles. Both statements are supported by measurement data (Figure 3.35).

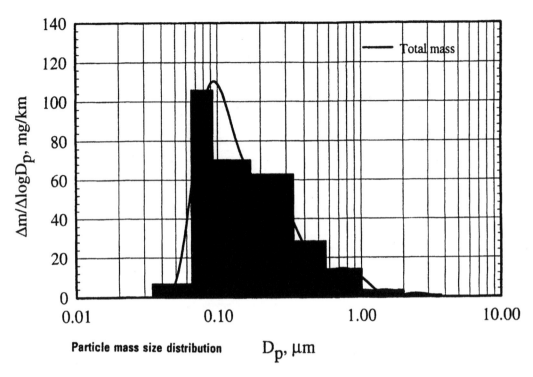

FIGURE 3.34   The size distribution (particle diameter D_p) of particles emitted in diesel engine exhaust. (From Kerminen, V.M., Mäkela, T.E., Ojanen, C.H., et al. Characterization of particulate phase in the exhaust from a diesel car. *Environ. Sci. Technol.*, 31:1883-1889, 1997. With permission.)

The importance of the PM 2.5 air particulate fraction was further documented by recent measurements in industrialized regions in Poland.[164] This is illustrated in Figure 3.36. The monthly mean of atmospheric aerosol concentrations and their distribution between fine (aerodynamic particle diameters less than 2.5 µm) and coarse (aerodynamic particle diameters between 2.5 and 10 µm) particulate fractions showed similar pictures in all measured regions. The concentrations of the PM 2.5 fraction were much more representative of the polluted atmospheres.

Important roles of the fine (PM 2.5) and coarse (PM 10) atmospheric fractions during different meteorological situations was documented in recent extensive measurements in the U.K.[165] In winter, when the combustion sources, including traffic, are the most important, and the wind speeds are low, the PM 10 concentrations correlate well to the PM 2.5 values (Figure 3.37). In contrast, when the wind plays an important role, no such correlations exist (Figure 3.38A). Furthermore, because of the dilution effect, the soot (EC) concentrations decrease substantially with the increasing windspeed (Figure 3.38C).

Chemical analysis of PM 10 showed that analyzed components from primary and secondary pollutants are better correlated with PM 2.5 than PM 10. The mass of the fine fraction material can be accounted for by vehicle emissions and secondary aerosol. This supports the view that PM 2.5 is predominantly form polluting sources, while a significant part of the coarse material is natural dusts. Vehicles are shown to be major contributors to levels of fine particle matter during winter and to make a higher contribution to PM 2.5 than PM 10.

3.8   DISCUSSION AND CONCLUSIONS

Recent epidemiologic studies have reported consistent associations between inhalable particle (PM 10) concentrations and increased daily mortality. Nevertheless, PM 10 consists of two size

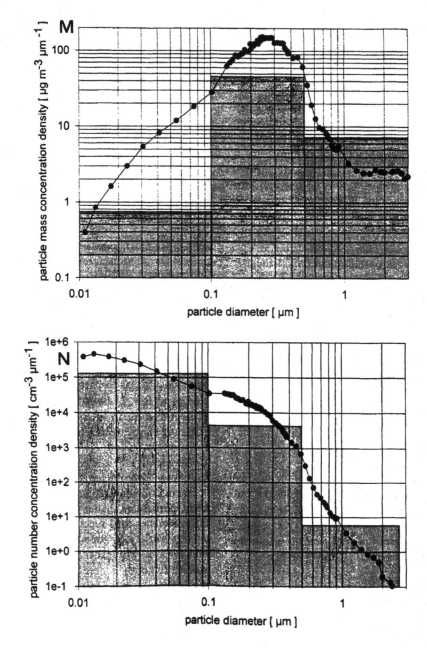

FIGURE 3.35 Average mass density function of ambient particles measured during the winter (M) and average number density function (N). (From Tuch, Th., Brand, P., Wichmann, H.E., and Heyder, J. Variation of particle number and mass concentration in various size ranges of ambient aerosols in Eastern Germany. *Atm. Environ.* 31(24)4193-4197, 1997. With permission.)

fractions, fine and coarse, which have different physiological and source characteristics. This combined analysis of the association of fine and coarse particles measured in six U.S. cities suggests that ambient fine particle exposures, and not coarse particle exposures, are specifically responsible for the observed associations with daily mortality. The implication is that toxicological studies of mechanisms and control strategies to promote public health should focus on fine particles which are produced by direct emissions and secondary reactions of combustion-related pollutants.[67]

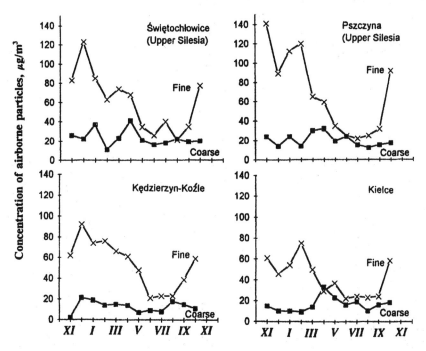

FIGURE 3.36 Monthly mean concentrations in $\mu m/m^3$, of fine (D_p 2.5 μm and coarse (D_p = 2.5 to 10 μm) measured at sites in Poland. (From Pastuszka, J. Study of PM 10 and PM 2.5 concentrations in Southern Poland. *J. Aerosol Sci.* 28, Suppl. 1, S 227- S 228, 1997. With permission.)

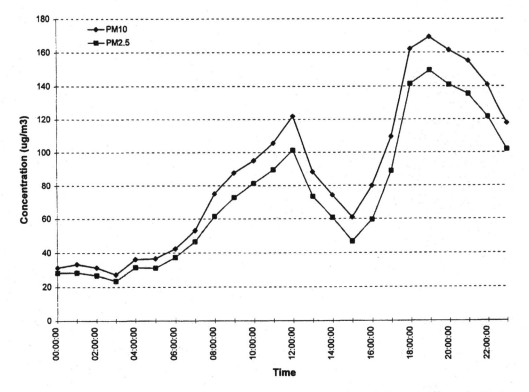

FIGURE 3.37 Diurnal variations in PM 10 and PM 2.5 measurements in U.K. in December. (From Harrison, R.M., Deacon, A.R., Jones, M.R., and Appleby, R.S. Sources and processes affecting concentrations of PM 10 and PM 2.5 particulate matter in Birmingham (U.K.). *Atm. Environ.* 31(24)4103-4117, 1997. With permission.)

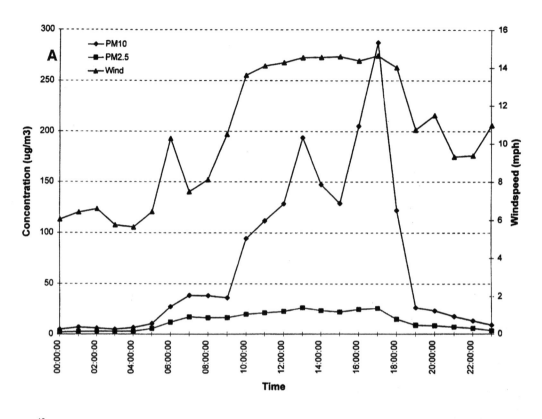

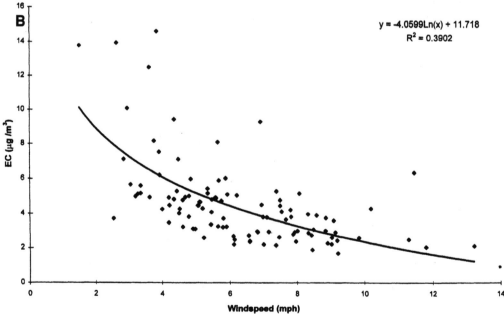

FIGURE 3.38 Diurnal variations in PM 2.5 and PM 10 and local windspeed (A), and relationship between daily mean elemental carbon (EC) concentration and windspeed (C). Measurements taken in the U.K. in spring. (From Harrison, R.M., Deacon, A.R., Jones, M.R., and Appleby, R.S. Sources and processes affecting concentrations of PM 10 and PM 2.5 particulate matter in Birmingham (U.K.).*Atm. Environ.* 31(24)4103-4117, 1997. With permission.)

The increased production and ambient air concentration of fine and very fine particulates has been a characteristic of the atmosphere since about the 1980s. The particulate emissions of automobiles, mainly diesel engines, are the most probable reason for the increasing toxic and carcinogenic potency of the AAA and for the observed health effects. Our understanding of these health effects of fine particulates is far from complete, and it could take 30 or more years to identify a precise mechanism. Nevertheless, the results of the existing toxicological studies (*in vivo* and *in vitro*) suggest that both physical (particle size, shape, surface, and biopersistence) as well as chemical (solved and leached toxic chemicals and surface catalytic reactions) properties of the fine (PM 2.5) particulate fraction are involved in the observed health effects. The fractions PM 2.5 and PM 1 are, therefore, more representative for the evaluation of the health risk of the general population than the existing PM 10 fraction.

Standardized reference measurement methods in the near future should be based on the PM 2.5 fraction. Additionally, the very fine fraction (PM 1) and the coarse particulate fraction ($2.5 \, \mu m < PM < 10 \, \mu m$) could be measured. Chemical analysis for some markers, such as carbon black (EC), PAHs, metals, etc., could serve as important complementary information about the toxic potential of the AAA.

New AQS for the PM 2.5 particulate fraction and, of course, an important decrease of fine particulate emissions (combustion processes and automobile exhausts) should be the immediate consequence of the changed situation in the atmosphere. There also exists first proposals for the AQS of PM 2.5, which should lie approximately in the range of $10 \, \mu m/m^3$ (annual mean). The U.S. EPA proposal is $15 \, \mu m/m^3$.

REFERENCES

1. Stern, A.C. (Ed.) *Air Pollution,* Academic Press, New York, 1968.
2. Florig, H.K. China's air pollution risks. *Environ. Sci. Technol.,* 31:274 A-279 A, 1997.
3. Sram, R.J., Benes, I., Binkova, B. et al. Teplice program — The impact of air pollution on human health. *Environ. Health Perspectives* 104:Suppl. 4, 699-714, 1996.
4. Horstman, D., Kotesovec, F., Vitnerova, N., and Leixner, M.P. Pulmonary functions of school children in highly polluted Northern Bohemia. *Arch. Environ. Health.* 52:56-62, 1997.
5. Raber, L.R. EPA's air standards. *Chemical and Engn. News.* April 14, 10-18, 1997.
6. Winkler, P. and Kaminski, U. Increasing submicron particle mass concentration in Hamburg. *Atmospheric Environ.* 22:2871-2883, 1988.
7. Tuch, Th., Heyder, J., Heinrich, J., and Wichmann, H.E. Changes of the particle size distribution in an eastern Germany city, in U. Mohr (Ed.) *Relationships Between Respiratory Disease and Exposure to Air Pollution.* Sixth Int. Inhalation Symposium. Hannover (Germany) Paper A19, 1997.
8. Hitzenberger, R., Fohler-Norek, C., Dusek, U., Galambos, Z., and Sidla, S. Comparison of recent, 1994, black carbon data with those obtained in 1985 and 1986 in the urban area of Vienna, Austria. *Sci. Total Environ.* 189/190, 275-280, 1996.
9. Trier, A. Submicron particles in an urban atmosphere. *Atmospheric Environ.* 31:909-919, 1997.
10. Nachtweyh, K., Wetzel, A., Dikov, Y., and Rammnesse, W. Emission control in hazardous waste incineration. *Gefahrstoffe-Reinhalt. Luft.* 56:285-289, 1996.
11. Day, E.D., Malm, W.C., and Kreideweis, S.M. Seasonal variations in aerosol composition and acidity at Shenandoah and Great Smoky Mountains National Parks. *J. Air Waste Manage. Assoc.* 47:411-418, 1997.
12. Gray, H.A. and Cass, G.R. Characteristics of atmospheric organic and elemental carbon particle concentration in Los Angeles. *Environ. Sci. Technol.,* 20:580-589, 1986.
13. Hildemann, L.M., Mazurek, M.A., and Cass, G.R. Quantitative characterization of urban sources of organic aerosol by high-resolution gas chromatography. *Environ. Sci. Technol.,* 25:1311-1325, 1991.
14. Rogge, W.F., Hildemann, L.M., Mazurek, M.A., Cass, G.R., and Simoneit, B.R.T. Sources of fine organic aerosol 1. *Environ. Sci. Technol.,* 25:1112-1125, 1991.
15. Rogge, W.F., Hildemann, L.M., Mazurek, M.A., Cass, G.R., Simoneit, B.R.T. Sources of fine organic aerosols 2. *Environ. Sci. Technol.,* 27:636-651, 1993.

16. Rogge, W.F., Hildemann, L.M., Mazurek, M.A., Cass, G.R., and Simoneit, B.R.T. Source of fine organic aerosol 4. *Environ. Sci. Technol.*, 27:2700-2711, 1993.

17. Spurny, K.R. Atmospheric anthropogenic aerosol and its toxic and carcinogenic components. *Wissenschaft und Umwelt*, ISU, Aachen, Germany. 2:139-151, 1993.

18. Rogge, W.F., Hildemann, L.M., Mazurek, M.A., Cass, G.R., and Simoneit, B.R.T. Sources of fine organic aerosol 5. *Environ. Sci. Technol.*, 27:2736-2744, 1993.

19. Rogge, W.F., Hildemann, L.M., Mazurek, M.A., Cass, G.R., and Simoneit, B.R.T. Sources of fine organic aerosol 6. *Environ. Sci. Technol.*, 28:1375-1388, 1994.

20. Smith, D.J.T., Harrison, R.M., Luhana, L., et al. Concentrations of particulate airborne polycyclic aromatic hydrocarbons and metals collected in Lahore, Pakistan. *Atmospheric Environ.* 30:4031-4040, 1996.

21. Tremolada, P., Burnett, V., Calamari, D., and Jones, K.C. Spatial distribution of PAHs in the U.K. atmosphere using pine needles. *Environ. Sci. Technol.*, 30:3570-3577, 1996.

22. Cui, W., Machir, J., Lewis, L., Eatough, D.J., and Eatough, N.L. Fine particulate organic material at Meadview during the project MOHAVE summer intensive study. *J. Air Waste Manage. Assoc.* 47:357-369, 1997.

23. Bolt, A. and DeJong, A.P.J.M. Ambient air dioxin measurement in the Netherlands. *Chemosphere.* 27:73-81, 1993.

24. König, J., Theisen, J., Günther, W.J., Liebl, K.H., and Büchen, M. Ambient air levels of polychlorinated dibenzofurans and dibenzo(p)dioxins at different sites in Hessen(Germany). *Chemosphere.* 26:851-861, 1996.

25. Dyke, P., Coleman, P., and James, R. Dioxins in ambient air. *Chemosphere.* 34:1191-1201, 1997.

26. Kasahara, M. Chemical composition of atmospheric aerosol particles, in N. Fukuta and P.E. Wagber (Eds.) *Nucleation and Atmospheric Aerosols.* A. Deepak Publ., Hampton, VA, 1992, pp. 385-393.

27. Miranda, J., Cahill, T.A., Morales, J.R., et al. Determination of elemental concentrations in atmospheric aerosols in Mexico City. *Atmospheric Environ.* 28:2299-2306, 1994.

28. Chow, J.C., Fujita, E.M., Watson, J.G., et al. Evaluation of filter based aerosol measurement during the 1987 in Southern California Air Quality Study. *Environ. Monitoring and Assessment* 30:49-80, 1994.

29. Cahill, T.A., Morales, R., and Miranda, J. Comparative aerosol studies of Pacific rim cities. *Atmospheric Environ.* 30:747-749, 1996.

30. Beceiro-Gonzales, E., Andrade-Garda, J.M., Serrano-Velasco, E., and Lopez-Mahia. Metals in airborne particulate matter in La Coruna (Spain). *Sci. Total Environ.* 196:131-139, 1997.

31. Janssen, N.A.H., Van Manson, D.F.M., Van Der Jagt, K. et al. Mass concentration and elemental composition of airborne particulate matter at street and background locations. *Atmospheric Environ.* 31:1185-1193, 1997.

32. Sandroni, V. and Migon, C. Significance of trace metal medium-range transport in the western Mediterranean. *Sci. Total Environ.* 196:83-89, 1997.

33. Baek, S.O., Choi, J.S., and Hwang, S.M. A quantitative estimation of source contributions to the concentrations of atmospheric suspended particulate matter in urban, suburban, and industrial areas of Korea. *Environ. International* 23:205-213, 1997.

34. Riveros-Rosas, H., Pfeifer, G.D., Lynam, D.T., et al. Personal exposure to elements in Mexico City air. *Sci. Total Environ.* 198:79-96, 1997.

35. Withey, J.R. A critical review of the health effects of atmospheric particulates. *Toxicol. Industrial Health.* 5:519-554, 1989.

36. Lebret, E., Houthuijs, D., and Dusseldorp, A. Acute and chronic studies in air pollution epidemiology. *Int. Symposium on Environmental Health Hazards in Central and Eastern Europe.* WHO, Sosnowiec, Poland, 1994.

37. Phalen, R.F. The association of air pollution and mortality. *Fourth Int. Aerosol Conference.* Los Angeles, pp. 964-965, 1994.

38. Perkins, M.S., Phalen, R.F., Kleinman, M.T. et al. A bibliography on particulate air pollution and human health. Colloquium on Particulate Air Pollution and Human Mortality. University of California, Irvine, p. 1-46, 1994.

39. Gamble, J.F. and Lewis, R.J. Health and respirable particulate (PM 10) air pollution. *Environ. Health Perspectives.* 104:838-850, 1996.

40. Wilson, R. and Spengler, J.P. *Particulates in Our Air.* Harvard University Press, Cambridge, MA, 1996.
41. Spurny, K.R. Atmospheric particulate pollutants and environmental health. *Arch. Environ. Health.* 51:415-416, 1996.
42. Vedal, S. Ambient particles and health. *J. Air Waste Manage. Assoc.* 47:551-581, 1997.
43. Schwartz, J., Spix, C., Wichman, H.E., and Malin, E. Air pollution and acute respiratory illness in five German communities. *Environ. Res.* 56:1-14, 1991.
44. Schwartz, J. Particulate air pollution and daily mortality in Detroit. *Environ. Res.* 56:204-213, 1991.
45. Pope, C.A., Schwartz, J., and Ransom, M.R. Daily mortality and PM 10 pollution in Utah. *Arch. Environ. Health.* 47:211-217, 1992.
46. Pope, C.A., Dockery, D.W., and Schwartz, J. Review of epidemiological evidence of health effects of particulate air pollution. *Inhalation Toxicology.* 7:1-18, 1995.
47. Pope, C.A., Thun, M.J., Mohan, M. et al. Particulate air pollution as a predictor in mortality in a prospective study of U.S. adults. *Am. J. Respir. Crit. Care Med.* 151:669-674, 1995.
48. Pope, C.A. Adverse health effects of air pollutants in a nonsmoking population. *Toxicology.* 111:149-155, 1996.
49. Pope, C.A. Particulate pollution and health. A review of the Utah Valley experience. *J. Exposure Analysis and Environ. Epidemiology.* 6:23-34, 1996.
50. Mills, P.K., Abbey, D., Beeson, W.L., and Petersen, F. Ambient air pollution and cancer in California Seventh-day Adventists. *Arch. Environ. Health.* 45:271-280, 1991.
51. Abbey, D.E., Petersen, F.P., Mills, P.K., and Beeson, W.L. Long term ambient air concentration of TSP, O_3, SO_2 and respiratory symptoms in a nonsmoking population. *Arch. Environ. Health.* 48:33-46, 1993.
52. Abbey, D.E., Hwand, B.L., Burchette, R.J., Vancuren, T., and Mills, P.K. Estimated long-term ambient concentrations of PM 10 and development of respiratory symptoms in a nonsmoking population. *Arch. Environ. Health.* 50:139-152, 1995.
53. Dockery, D.W., Schwartz, J., and Spengler, J.D. Air pollution and daily mortality: Associations with particulate and acid aerosols. *Environ. Res.* 59:362-373, 1992.
54. Schwartz, J. and Dockery, D.W. Increased mortality in Philadelphia associated with daily air pollution concentrations. *Am. Rev. Respir. Dis.* 145:600-604, 1992.
55. Schwartz, J. Air pollution and daily mortality: A review and meta analysis. *Environ. Res.* 64:36-52, 1994.
56. Schwartz, J. Air pollution and hospital admission for the elderly in Birmingham, Alabama. *Am. J. Epidemiol.* 139:589-598, 1994.
57. Schwartz, J. and Morris, R. Air pollution and hospital admissions for cardiovascular disease in Detroit, Michigan. *Am. J. Epidemiol.* 142:23-35, 1995.
58. Saldiva, P.H.N., Pope, G.A., Schwartz, J., et al. Air pollution and mortality in elderly people. *Arch. Environ. Health.* 50:159-164, 1995.
59. Ostro, B.D., Lipsett, M.J., Mann, J.K., et al. Indoor air pollution and asthma. *J. Respir. Crit. Care Med.* 149:1400-1406, 1994.
60. Lipsett, M., Hurley, S., and Ostro, B. Air pollution and emergency room visits for asthma in Santa Clara County. *Environ. Health Perspectives.* 105:216-222, 1997.
61. Lewis-Michl, E.L., Melius, J.M., Kallenbach, L.R., et al. Breast cancer risk and residence near industry or traffic in Nassau and Suffolk Counties, Long Island, New York. *Arch. Environ. Health.* 51:255-265, 1996.
62. Bates, D.V. Particulate air pollution. *Thorax* 51:S 3- S 8, 1996.
63. Schwela, D. Exposure to environmental chemicals relevant for respiratory hypersensitivity. *Toxicol. Letters.* 86:131-142, 1996.
64. Ackermann-Liebrich, U., Leuenberger, P., Schwartz, J., et al. Lung function and long term exposure to air pollutants in Switzerland. *Am. J. Respir. Crit. Care Med.* 155:122-129, 1997.
65. Horstman, D., Kotesovec, F., Vitnerova, N., and Leixner, M. Pulmonary functions of school children in highly polluted Northern Bohemia. *Arch. Environ. Health.* 52:56-62, 1997.
66. Dockery, D.W., Pope, C.A., Xiping, Xu, et al. An association between air pollution and mortality in six U.S. cities. *N. Engl. J. Med.* 329:1753-1759, 1993.
67. Schwartz, J., Dockery, D.W., and Neas, L.M. Is daily mortality associated specially with fine particles? *J. Air and Waste Manage. Assoc.* 46:927-939, 1996.
68. Saldiva, P.N.H., King, M., Delmonte, V.L.C., et al. Respiratory alterations due to urban air pollution. *Environ. Res.* 57:19-33, 1992.

69. Sioutas, C., Koutrakis, P., and Ferguson, S.T. A technique to expose animals to concentrated fine ambient aerosols. *Environ. Health Perspectives.* 103:172-177, 1995.
70. Sioutas, C. and Koutrakis, P. Inertial separation of ultrafine particles using a condensation growth/virtual impaction system. *Aerosol Sci. Technol.* 25:424-436, 1996.
71. Mauderly, J.L. Diesel exhaust, in M. Lippmann (Ed.) *Environmental Toxicants,* Van Nostrand Reinhold, New York, pp. 119-162, 1992.
72. Heinrich, U., Fuhst, R., Rittinghausen, S., et al. Chronic inhalation exposure of Wistar rats and two different strains of mice to diesel engine exhaust, carbon black, and titanium dioxide. *Inhal. Toxicol.* 7:533-556, 1995.
73. Ichinose, T., Furuyama, A., and Sagai, M. Biological effects of diesel particulates (DEP). *Toxicology.* 99:1106-1114, 1995.
74. Heinrich, U. Carcinogenic effect of solid particles, in U. Mohr (Ed.) *Toxic and Carcinogenic Effects of Solid Particles in the Respiratory Tract,* ILSI Press, Int. Life Sci. Inst., Washington, D.C., 57-73, 1994.
75. Lewtas, J. Complex mixtures of air pollutants. *Environ. Health Perspectives.* 100:211-218, 1993.
76. Schlesinger, R.B. Toxicological evidence for health effects from inhaled particulate pollution. *Inhal. Toxicol.* 7:99-109, 1995.
77. Schlesinger R.B. Nitrogen oxides, in M. Lippmann (Ed.) *Environmental Toxicants,* Van Nostrand Reinhold, New York, 412-453, 1992.
78. Lang, L. Assessing risk of chemical mixtures. *Environ. Health Perspectives.* 103:142-145, 1995.
79. Seemayer, N.H., Hadnagy, W., Behrend, H., and Tomingas, R. Indicators of potential health risk by airborne particulates: cytotoxic, mutagenic and carcinogenic effects on mammalian cells *in vitro,* in *Environmental Hygiene,* Springer Verlag, 54-59, 1988.
80. Seemayer, N.H., Hadnagy, W., Behrend, H., and Tomingas, R. Inhalation hazards from airborne particulates evaluated by *in vitro* cyto- and genotoxicity testing. *Exp. Pathology.* 37:228-230, 1989.
81. Bayona, J.M., Casellas, M., Fernandez, P., Solanas, A.M., and Albaiges, J. Sources and seasonal variability of mutagenic agents in the Barcelona city aerosol. *Chemosphere.* 29:441-450, 1994.
82. Hanningan, B.W., Cass, C.R., Penman, B.W. et al. Human cell mutagens in Los Angeles air. *Environ. Sci. Technol.,* 31:438-447, 1997.
83. Pagano, P., De Zaiacomo, T., Scarcella, E., Bruni, S., and Calamosca, M. Mutagenic activity of total and particle-sized fractions of urban particulate matter. *Environ. Sci. Technol.,* 30:3512-3516, 1996.
84. Seaton, A., MacNee, W., Donaldson, K., and Godden, D. Particulate air pollution and acute health effects. *Lancet.* 345:176-178, 1995.
85. Ferin, J., Oberdörster, G., Penney, D.P., Soderholm, S.C., et al. Increased pulmonary toxicity of ultrafine particles. *J. Aerosol Sci.* 21:381-384, 1990.
86. Oberdörster, G., Ferin, J., and Lehnert, B.E. Correlation between particle size, *in vivo* particle persistence and lung injury. *Environ. Health Perspectives.* 102:Suppl. 5, 173-179, 1995.
87. Oberdörster, G. Effects of ultrafine particles in the lung and potential relevance to environmental particles, in J.M.C. Marijnissen and L. Gradon (Eds.) *Aerosol Inhalation.* Kluwer Academic Publ., Dordrecht, 1996, pp. 165-174.
88. Spurny, K.R. Atmospheric particulate pollutants and environmental health, physicochemical standpoint, in J.M.C. Marijnissen and L. Gradon (Eds.) *Aerosol Inhalation.* Kluwer Academic Publ., Dordrecht, 1996, pp. 175-186.
89. Pui, D.Y.H. and Da-Ren, Chen. Nanometer particles. *J. Aerosol Sci.* 28:539-544, 1997.
90. Brown, R.C., Hoskins, J.A., and Johnson, N.F. (Eds.) Mechanisms in fibre carcinogenesis. NATO ASI Ser. 223, Plenum Press, London, 1991.
91. Cheng, Y.S., Yeh, H.C., and Swift, D. Aerosol deposition in human nasal airway for particles 1 nm to 20 nm. *Radiation Protection Dosimetry.* 38:41-47, 1991.
92. Cheng, K.H. and Swift, D.L. Calculation of total deposition fraction of ultrafine aerosols in human extrathoracic and intrathoracic regions. *Aerosol Sci. Technol.* 22:194-201, 1995.
93. Kaufman, J.W., Scherer, P.W., and Yang, C.C.G. Predicted combustion product deposition in the human airway. *Toxicology.* 115:123-128, 1996.
94. Cheng, Y.S., Yeh, H.C., Guilmatte, R.A., et al. Nasal deposition of ultrafine particles in human volunteers and its relationship to airway geometry. *Aerosol. Sci. Technol.* 25:274-291, 1996.

95. Soutar, C.A., Miller, B.G., Gregg, N., et al. Assessment of human risks from exposure to low toxicity occupational dusts. *Ann. Occup. Hyg.* 41:123-133, 1997.

96. Cohen, B., Xiong, J., and Li, W. The influence of charge on the deposition behavior of aerosol particles with emphasis of singly charged nanometer sized particles, in J.M.C. Marijnissen and L. Gradon (Eds.) *Aerosol Inhalation.* Kluwer Academic Publ., Dordrecht, 1996, pp. 127-142.

97. Oberdörster, G., Gelein, R.M., Ferin, J., and Weiss, B. Association of particulate air pollution and acute mortality: Involvement of ultrafine particles. *Inhalation Toxicology.* 7:111-124, 1995.

98. Godleski, J.J., Hatch, V., Hauser, R., et al. Ultrafine particles in lung macrophages of healthy people. *Am. J. Respir. Crit. Care Med.* 151:A 254-A 266, 1995.

99. Ferin, J., Oberdörster, G., Sonderholm, S.C., and Gelein, R. Pulmonary tissue access of ultrafine particles. *J. Aerosol Med.* 4:57-68, 1991.

100. Godleski, J.J., Sioutas, C., Katler, M., et al. Death from inhalation of concentrated ambient air particles in animal models of pulmonary disease. *Am. J. Respir. Crit. Care Med.* 153:A 15-A 18, 1996.

101. Gilmour, P.S., Brown, D.M., Lindsay, T.G., et al. Adverse health effects of particles: Involvement of iron in generation of hydroxyl radical. *Occup. Environ. Med.* 53:817-822, 1996.

102. Li, X.Y., Gilmour, P.S., Donaldson, K., and MacNee, W. Free radical activity and pro-inflammatory effects of particulate air pollution *in vivo* and *in vitro. Thorax.* 51:1216-1222, 1996.

103. Donaldson, K. Beswick, P.H., and Gilmour, P.S. Free radical activity associated with the surface of particles. *Toxicology Letters.* 88:293-298, 1996.

104. Brook, J.R., Dann, T.F., and Burnett, R.T. The relationship among TSP, PM-10, PM-2.5 and inorganic constituents of atmospheric particulate matter at multiple Canadian locations. *J. Air Waste Manage. Assoc.* 47:2-19, 1997.

105. Kao, A.S. and Friedlander, S.K. Temporal variations of particulate air pollution: A marker for free radical dosage and adverse health effects. *Inhalation Toxicology.* 7:149-156, 1995.

106. Finlayson-Pitts, B.J. and Pitts, J.N. Tropospheric air pollution: Ozone, airborne toxins, polycyclic aromatic hydrocarbons, and particles. *Science.* 276:1045-1052, 1997.

107. Hanss, A., Herzig, S., and Lutz-Holzhauer, C. Assessment of risk of cancer caused by air pollution. *Gefahrstoffe-Reinhalt. Luft.* 57:71-74, 1997.

108. Tesseraux, I. and Kappos, A.D. Actual German air pollution levels. Risk estimation for respiratory diseases and mortality in U. Mohr, Ed., *Relationships Between Respiratory Disease and Exposure to Air Pollution,* ILSI (Int. Life Sci. Inst.) Washington, D.C., 1998, pp. 238-241.

109. Zielinski, M., Römmelt, H., and Fruhmann, G. Ambient air soot concentrations in Munich public transportation systems. *Sci. Total Environ.* 196:107-110, 1997.

110. Chow, J.C. Measurement methods to determine compliance with ambient air quality standards for suspended particles. *J. Air Waste Manage. Assoc.* 45:320-383, 1995.

111. Spurny, K.R. (Ed.) *Physical and Chemical Characterization of Individual Airborne Particles.* Ellis Horwood, Chichester, U.K., 1986.

112. Willeke, K. and Baron, P.A. (Eds.) *Aerosol Measurement.* Van Nostrand Reinhold, New York, 1993.

113. Hidy, G.M. Summary of the California aerosol characterization experiment. *J. Air Poll. Control Assoc.* 25:1106-1114, 1975.

114. *Federal Register,* Environmental Protection Agency national primary and secondary ambient air quality standards. *Federal Register.* 36:8186, 1971.

115. Raber, L.R. EPA's air standards. *Chemical Engineering News.* April 14, 10-18, 1997.

116. Lundgren, D.A. and Burton, R.M. The effect of particle size distribution on the cut point between fine and coarse ambient mass fractions. *Fourth International Aerosol Conference,* Los Angeles, U.S.A., 1994, pp. 330-331.

117. Da-Tong Ning, Zhong, L.X. and Chung, Y.S. Aerosol size distribution and elemental composition in urban areas of Northern China. *Atm. Environ.* 30:2355-2362, 1996.

118. Sweet, C.W. and Vermette, S.J. Sources of toxic trace elements in urban air in Illinois. *Environ. Sci. Technol.,* 27:2502-2510, 1993.

119. Venkataraman, C. and Friedlander, S.K. Size distribution of PAH and elemental carbon 1,2. *Environmental Sci. Technol.* 28:555-572, 1994.

120. Kaplan, I.R. and Gordon, R.J. Non-fossil-fuel fine particle carbon aerosol in Southern California. *Aerosol Sci. Technol.* 21:343-359, 1994.

121. Watson, J.G., Chow, J.C., Lowenthzal, D.H., et al. Differences in carbon composition of source profiles for diesel and gasoline-powered vehicles. *Atm. Environ.* 28:2493-2505, 1994.

122. Chow, J.C., Watson, J.G., Fujita, E.M., Lu, Z., and Lawson, D.R. Temporal and spatial variations on PM 2.5 and PM 10 aerosol in Southern California air quality study. *Atm. Environ.* 28:2061-2080, 1994.

123. Kao, A.S. and Friedlander, S.K. Chemical signatures of the Los Angeles aerosol. *Aerosol Sic. Technol.* 21:283-293, 1994.

124. Kao, A.S. and Friedlander, S.K. Frequency distribution of PM 10 chemical components and their sources. *Environ. Sci. Technol.,* 29:19-28, 1995.

125. Jones, J.C., Duarte-Davidson, R., and Cawse, P.A. Changes in the PCB concentrations of U.K. air between 1972 and 1992. *Environ. Sci. Technol.,* 29:272-275, 1995.

126. Lyons, J.M., Venkataran, C., Main, H.H., and Friedlander, S.K. Size distribution of trace metals in the Los Angeles atmosphere. *Atm. Environ.* 27:237-249, 1993.

127. Helmers, E. Platinum emission rate of automobiles with catalytic converters. *Environ. Sci. and Pollution Res.* 4:100-103, 1997.

128. Atkinson, R. and Arey, J. Atmospheric chemistry of gas-phase polycyclic aromatic hydrocarbons: Formation of atmospheric mutagens. *Environ. Health Perspectives.* 102:Suppl. 4, 117-126, 1994.

129. Fan, Z., Chen, D., Birla, P., and Kamens, R.M. Modeling of nitro-polycyclic aromatic hydrocarbons formation and decay in the atmosphere. *Atm. Environ.* 29:1171-1181, 1995.

130. Sasaki, J., Arey, J., and Harger, W.P. Formation of mutagens from photooxidation of 2-4 ring PAH. *Environ. Sci. Technol.,* 29:1324-1334, 1995.

131. Scheepers, P.T. J., Martens, M.H.J., Velders, D.D., et al. 1-Nitropyrene mutagenicity of diesel exhaust-derived particulate matter. *Environmental and Molecular Mutagenesis.* 25:134-147, 1995.

132. Sheu, H.L., Lee, W.J., Tsai, J.H., et al. Particle size distribution of PAH in the ambient air of a traffic intersection. *J. Environ. Sci. Health.* A 31:1293-1316, 1996.

133. Schnelle, J., Wolf, K., Frank, G., et al. Particle size-dependent concentrations of PAH. *Analyst.* 121:1301-1304, 1996.

134. Schnelle, J., Wolf, K., Frank, B., et al. Size-dependent concentrations of PAH in ambient air. *Gefahrstoffe-Reinhalt. Luft.* 57:23-25, 1997.

135. Kaupp, H., Towara, J., and McLachlan, M.S. Distribution of polychlorinated dibenzo-p-dioxins and dibenzofurans in atmospheric particulate matter with respect to particle size. *Atm. Environ.* 28:585-593, 1994.

136. Koutrakis, P. and Sioutas, C. Physico-chemical properties and measurement of ambient particles, in R. Wilson and J. Spengler (Eds.) *Particles in Our Air.* Harvard School of Public Health. Boston, U.S.A., 1996, pp. 15-40.

137. Lundgren, D.A., Hlaing, D.N., Rich, T.A., and Marple, V.A. PM 10/PM 2.5/PM1 Data from a trichotomous sampler. *Aerosol Sci. Technol.* 25:353-357, 1996.

138. Marple, V.A., Rubow, K.L., and Olson, B.A. Inertial, gravitational and thermal collection techniques, in K. Willeke and P. Baron (Eds.) *Aerosol Measurement.* Van Nostrand Reinhold, New York, 1993, pp. 206-232.

139. Köllensperger, G., Friedbacher, G., Grasserbauer, M., and Dorffner, L. Investigation of aerosol particles by atomic force microscopy. *Fresenius J. Anal. Chem.* 358:268-173, 1997.

140. Sioutas, C., Ferguson, S.T., Wolfson, J.M., et al. Inertial collection of fine particles using a high-volume rectangular geometry conventional impactor. *J. Aerosol Sci.* 28:1015-1028, 1997.

141. Liu, S. and Dasgupta, P.K. Automated system for chemical analysis of airborne particles based on corona-free electrostatic collection. *Anal. Chem.* 68:3638-3644, 1996.

142. Appel, B.R. Atmospheric sample analysis and sampling artifacts, in K. Willeke and P. Baron (Eds.) *Aerosol Measurement,* Van Nostrand Reinhold, New York, 1993, pp. 233-259.

143. Fletcher, R.A. and Small, J.A. Analysis of individual collected particles, in K. Willeke and P. Baron (Eds.) *Aerosol Measurement,* Van Nostrand Reinhold, New York, 1993, pp. 260-295.

144. Spurny, K.R. (Ed.) Physical and chemical characterization of individual airborne particles. Ellis Horwood, Chichester, U.K., 1986.

145. Petzold, A. and Niessner R. Novel design of a resonant photoacoustic spectrophone for elemental carbon mass monitoring. *Appl. Phys. Lett.* 66:1285-1287, 1995.

146. Petzold, A. and Niessner, R. Method comparison study of soot-selective techniques. *Mikrochim. Acta.* 117:215-237, 1995.

147. Petzold, A. and Niessner, R. Photoacoustic soot sensor for *in situ* black carbon monitoring. *Appl. Phys. B.* 63:191-197, 1996.
148. Petzold, A. and Niessner, R. Coulometric determination of soot in ambient air. *Gefahrstoffe-Reinhalt. Luft.* 56:173-177, 1996.
149. Hansen, A.D.H. The development of the aethalometer. Fifth *GIV Colloquium,* Frankfurt (Germany), pp. 1-7, 1994.
150. Birch, M.E. and Cary, R.A. Elemental carbon-based method for monitoring occupational exposures to particulate diesel exhaust. *Aerosol Sci. Technol.* 25:221-141, 1996.
151. Stoffy-Egli, P., Potter, T.M., Leonard, J.D., and Pocklinger, R. The identification of black carbon particles with the analytical electron microscope. *Sci. Total. Environ.* 198:211-223, 1997.
152. Lodge, J.P. (Ed.) *Methods of Air Sampling and Analysis.* Lewis Publ. Inc. Chelsea, MI, U.S.A., 1989.
153. Burtscher, H. and Schmidt-Ott, A. *In situ* measurement and condensation of polyaromatic hydrocarbons on ultrafine particles by means of photoemission. *J. Aerosol:Sci.* 17:699-703, 1986.
154. Burtscher, H. Aerosol characterization by PAS (photoelectric aerosol sensor. *Fifth GIV Colloquium.* Frankfurt (Germany), pp. 1-5, 1994.
155. Altman, E.L., Lvov, B.V., Panichev, N.A., and Fedorov, P.N. Determination of metals in atmospheric aerosols by Zeeman atomic absorption spectrometry with absorption pulse restoration. *J. Anal. Chem.* 51:824-828, 1996.
156. Balcerzak, M. Analytical methods for the determination of platinum in biological and environmental materials. *Analyst.* 122:67 R- 74 R, 1999.
157. Gomes, A.M., Sarrette, J.P., Madon, L., and Almi, A. Continuous emission monitoring of metal aerosol concentrations in atmospheric air. *Spectrochimica Acta B.* 51:1695-1705, 1996.
158. Nusko, R. and Heumann, K.G. Cr(III)/Cr(VI) speciation in aerosol particles by extractive separation and thermal ionization isotope dilution mass spectrometry. *Fresenius J. Anal. Chem.* 367:1050-1055, 1997.
159. Deacon, A.R., Derwent, R.G., Harrison, R.M., et al. Analysis and interpretation of measurement of suspended particulate matter at urban background sites in the U.K. *Sci. Total Environ.* 203:17-36, 1997.
160. Woodruff, T.J., Grillo, J., and Schoendorf, K.C. The relationship between selected causes of postneonatal infant mortality and particulate air pollution in the United States. *Environ. Health Perspectives.* 105(6)608-612, 1997.
161. Kerminen, V.M., Mäkela, T.E., Ojanen, C.H., et al. Characterization of particulate phase in the exhaust from a diesel car. *Environ. Sci. Technol.,* 31:1883-1889, 1997.
162. Takeji, E., Suzuki, H., Watanabe, T., et al. 3-nitrobenzanthrone, a powerful bacteria mutagen and suspected human carcinogen found in diesel exhaust and airborne particulates. *Environ. Sci. Technol.,* 31:2772-2776, 1997.
163. Tuch, Th., Brand, P., Wichmann, H.E., and Heyder, J. Variation of particle number and mass concentration in various size ranges of ambient aerosols in Eastern Germany. *Atm. Environ.* 31(24) 4193-4197, 1997.
164. Pastuszka, J. Study of PM 10 and PM 2.5 concentrations in Southern Poland. *J. Aerosol Sci.* 28, Suppl. 1, S 227- S 228, 1997.
165. Harrison, R.M., Deacon, A.R., Jones, M.R., and Appleby, R.S. Sources and processes affecting concentrations of PM 10 and PM 2.5 particulate matter in Birmingham (U.K.). *Atm. Environ.* 31(24) 4103-4117, 1997.

Section II

Bulk Analysis

4 Filtration and Denuder Sampling Techniques

A-M. N. Kitto and I. Colbeck

CONTENTS

4.1 INTRODUCTION

Accurate measurements of atmospheric gases and particulate species are essential to facilitate an understanding of the important chemical and physical processes which govern the fate of these components in the troposphere. Most of these species are present in the atmosphere in trace levels, which creates a difficult task for the analytical chemist. An additional complication is that some of the atmospheric constituents exist in the atmosphere partitioned between gaseous and particulate forms. In some cases, this involves adsorption/desorption processes, as in the case of polycyclic aromatic hydrocarbons (Yamasaki et al., 1982).

$$PAH \rightarrow PAH$$

Determination of gaseous species at trace concentrations (ppt levels) as well as discrimination from particulate forms is a challenge to scientists, especially analytical chemists.

A variety of methods have been used for sampling atmospheric components. These include: filtration (Okita et al., 1976; Appel et al., 1980, 1981; Spicer et al., 1979, 1982; Allen et al., 1989), diffusion denuder (Ferm, 1979, 1986; Forrest et al., 1982; Possanzini et al., 1983; Koutrakis et al., 1988; Harrison and Kitto, 1990), and spectroscopic methods such as tunable diode laser spectroscopy (TDLS) (Schiff et al., 1988), differential optical adsorption spectroscopy (DOAS) (Platt et al., 1980), Fourier transform infrared spectroscopy (FTIR) (Hanst, 1971; Tuazon et al., 1978). This chapter discusses the integrative methods in detail.

4.2   FILTRATION TECHNIQUES

This is the most widely used technique for sampling aerosols and gaseous components, because of its low cost and simplicity. Different types of filters have been used for atmospheric sampling. The most common filters used for particulate sampling are glass fiber, quartz fiber filters, and Teflon (PTFE) membrane filters. Teflon (PTFE) membrane is an excellent collection media for atmospheric particles, because it is inert toward chemical species. Gaseous compounds are collected on subsequent filters. Nylon filters have been used frequently in measurements of nitric and hydrochloric acids (Spicer et al., 1982; Shaw et al., 1982; Grosjean, 1982; Goldan et al., 1983; Hildemann et al., 1984; Hering et al., 1988; Dasch et al., 1989; Sturges and Harrison, 1989; Harrison and Allen, 1990). Variable background nitrate and chloride levels on the nylon filters are reduced to acceptable levels by routine pretreatment of filters through a repeated extraction with deionized water. Additionally, impregnated filters are also used for measurement of atmospheric gases:

NaCl-impregnated filters have been used for nitric acid collection (Forrest et al., 1980; Appel et al., 1980)

Tetrabutyl ammonium hydroxide-impregnated filters for nitric acid (Huebert and Lazrus, 1980)

KOH-impregnated filters for SO_2 (Nodop and Hanssen, 1986)

Acid-impregnated (phosphoric, citric, and oxalic acids) filters have been widely used for ammonia collection (Ferm, 1979; Harrison and Kitto, 1990).

4.2.1   FILTER PACK

Filter packs are assemblages of filters in series, on which particulate matter is collected on the first filter (usually Teflon or quartz fiber) and gases on subsequent filters (e.g., HNO_3 on nylon; SO_2 on carbonate-impregnated; and NH_3 on acid-impregnated filters). Significant errors are inherent in the sample enrichment process involved in the filtration technique. Both negative and positive interferences are associated with filter pack samplers. These interferences are described in the following sections.

4.2.2   ARTIFACTS IN THE FILTER PACK TECHNIQUE

1. Volatilization of ammonium particles from the first filter.
   Ammonium particulates (ammonium nitrate and ammonium chloride) are very sensitive to temperature, relative humidity, and acidity of aerosols (Stelson and Seinfeld, 1982; Appel et al., 1980; Pio and Harrison 1987). Any change in these parameters during the sampling period influences the position of the equilibrium.

$$NH_4NO_3 \rightleftarrows NH_3 + HNO_3$$

$$NH_4Cl \rightleftarrows NH_3 + HCl$$

An increase in temperature will shift the equilibrium toward the gaseous components, resulting in an overestimation of HNO_3 and NH_3 (in the case of NH_4NO_3) and NH_3 and HCl (in the case of NH_4Cl). Appel et al. (1981) reported a loss of 50% of ammonium nitrate at temperatures of 29 to 35°C and relative humidity ~30%. This is the main disadvantage of the filter pack technique.

2. Gas–particle reactions.

A typical example of gas–particle artifact formation is the reaction of gaseous ammonia with acidic aerosols precollected on the first filter (Klockow et al., 1979; Ferm, 1979)

$$2NH_3 + H_2SO_4 \rightarrow (NH_4)_2SO_4$$

This process will lead to an underestimation of NH_3 and acidity (H^+) and an over-estimation of ammonium particulate concentrations. The reaction of hydrochloric acid with particulate nitrate is also possible (Appel and Haik, 1981)

$$NH_4NO_3 + HCl \rightarrow NH_4Cl + HNO_3$$

which leads to an overestimation of nitric acid measurements and underestimation of HCl and NH_4NO_3. The reaction of acid gases with basic particles, which may exist in the atmosphere, is another example:

$$2HNO_3 + CaCO_3 \rightarrow Ca(NO_3)_2 + H_2O + CO_2$$

The outcome of this reaction will be an underestimation of nitric acid. Appel et al. (1980) reported that 6–22% of nitric acid was retained by the particles on the Teflon filters and the retention increased with particle load.

3. Particle–particle interactions.

A reaction between particles collected on the first filter may occur. An example of this type is the reaction of particulate nitrate with sulfuric acid (Appel and Haik, 1981; Appel et al., 1984)

$$NH_4NO_3 + H_2SO_4 \rightarrow NH_4HSO_4 + HNO_3$$

This will overestimate HNO_3 and underestimate both particulate nitrate and aerosol acidity. Additionally, samples influenced by marine aerosols are subject to interferences due to the reaction of sea salt with atmospheric strong acids (Hitchcock et al., 1980)

$$2NaCl + H_2SO_4 \rightarrow 2HCl + Na_2SO_4$$

This process will lead to an underestimation of acidic aerosols and overestimation of HCl. In the eastern United States where high concentrations of acidic aerosol are observed (Koutrakis et al., 1988; Waldman et al., 1990), the concentrations of particulate nitrate were very low, and nitric acid represented a large fraction of total inorganic nitrate (TIN). The reaction of acidic aerosols with particulate nitrate in the first filter was conjectured for this artifact.

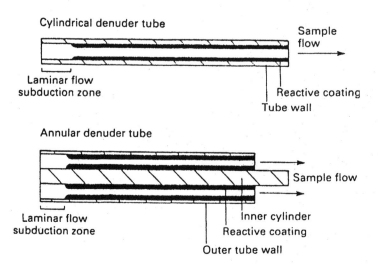

FIGURE 4.1   Schematic diagram of a tubular and annular denuder.

4. Loss of reactive species to inlet surfaces and inside the filter pack.
   Reactive species such as nitric, nitrous, and hydrochloric acids may be adsorbed/absorbed
   to the walls of the filter pack and/or by the filter (Appel et al., 1980, 1988b). Most filters
   retain a certain degree of reactive gaseous species which increases with high relative
   humidity. Quartz and glass fiber filters retain acidic gases to some degree. Interferences
   in measurements of acidic gases will occur if these filters are used (without pretreatment)
   for precollection of particles (Appel et al., 1980). Teflon filters were found to retain the
   lowest amount of HNO_3, and thus are widely used for the collection of particulate matter
   (Appel et al., 1984). However, Goldan et al., (1983) reported losses of up to 55% of
   nitric acid collected by a nylon filter preceded by a Teflon filter. The loss was not
   retrievable from the prefilter. Loss of nitric acid to inlet devices such as cyclones or
   impactors has been observed (Appel et al., 1988b). The loss increased with relative
   humidity and residence time.

4.3   DENUDER TECHNIQUES

The concept of a denuder sampler is that air is drawn through a conduit (cylindrical or annular),
the gaseous components are removed by diffusion to the walls of the denuder while the particulate
matter passes through the tube unaffected (Figure 4.1). The denuder walls are coated with a specific
reagent which selectively reacts with the gas of interest. An efficient denuder must meet the
following conditions: the gas flow must be stable and laminar; the collection capacity of the surface
is infinitely large and acts as a perfect sink; and analyte (gas of interest) is not generated nor
destroyed inside the denuder. The fundamental principle of the denuder sampler relies on the fact
that the diffusion of gaseous species are (3 to 6) orders of magnitude higher than those of particles
(Forrest et al., 1982).

Denuder tubes were first considered on a theoretical basis in the 1890s (Townsend, 1900). In
the 1930s, Nolan and Guerrini (1935) considered diffusional removal of particles in both rectangular
and cylindrical tubes. The earliest application of a diffusion denuder in atmospheric sampling was
for scrubbing undesirable gases from a stream of air, and as a tool for measurement of the diffusion
coefficients of gaseous species. Cider et al. (1969) used a diffusion denuder coated with potassium
hydroxide and magnesium perchlorate for the removal of water vapor from a stream of a sulfate
containing aerosols which were detected by a hydrogen flame chemiluminescence detector. Denuders coated with lead oxide were used for the removal of sulfur dioxide gas from ambient air in

measurements of atmospheric sulfate aerosols (Durham et al., 1978; Cobourn et al., 1980). The removal of sulfur dioxide by the denuder was found to be quantitative, and the diffusion coefficient of sulfur dioxide was determined using the Gormley–Kennedy treatment for diffusion of gases from a stream flowing through a cylindrical tube (Gormley and Kennedy, 1949).

Mathematical expressions relating to collection or penetration of particles and gases through channels of various geometrics have been derived. For cylindrical tubes, with laminar flow, several investigators have attempted to improve on the original work of Gormley and Kennedy (1949) (Davis and Parkins, 1970; Tan and Hsu, 1971; Bowen et al., 1976). They have all expressed penetration, for particles, at distance L along the tube for particles, flow rate Q and diffusion coefficient D as a function of the parameter μ where

$$\mu = \frac{\pi DL}{Q}$$

The numerical solution obtained by Bowen et al. (1976) for μ between 1×10^{-7} and 1 is most accurate. The following analytical solutions have the accuracy of four significant figures as compared to Bowen et al.'s result:

$$C/C_o = 0.81905e^{-3.6568\mu} + 0.09753e^{-22.305\mu} + 0.0325e^{-56.961\mu} + 0.01544e^{-107.62\mu}$$

for $\mu > 0.02$

$$C/C_o = 1.0 - 2.5638\,\mu^{2/3} + 1.2\,\mu + 0.17767\,\mu^{4/3}$$

for $\mu \leq 0.02$.

C is the mean radial concentration and C_o is the ambient concentration. These expressions are only valid if the flow rate is sufficiently low to ensure laminar flow, i.e., the Reynolds number is less than 2000. The Reynolds number (R_e) is given by:

$$R_e = \frac{dV\rho}{\eta} = \frac{4Q\rho}{\pi d\eta}$$

where V is the velocity of gas in the tube, d is the internal diameter of the tube, ρ is the gas density, and η is the gas viscosity. Laminar flow is achieved a short distance from the inlet. The minimum inlet length, L_{min}, necessary to fully develop laminar flow is given by:

$$L_{min} = 0.05R_e d$$

Hence, L_{min} is solely dependent upon the sampling rate Q.

The Gormley–Kennedy solution does not apply to annular tubes. Possanzini et al. (1983) proposed the following empirical equation:

$$C/C_o = A\exp(-\alpha\mu_a)$$

where A and α are experimentally determined constants. μ_a is defined as $\pi DL(d_1 + d_2)/4Q(d_2 - d_1)$, and d_2 and d_1 are the outer and inner diameters, respectively. Possanzini et al. (1983) determined A and α with sulfur dioxide concentrations in the range of 0.29 to 1.45 mg m^{-3} in air and sampling rates between 0.072 and 2.4 m^3 hr^{-1}. The following equation, reported by Possanzini et al. (1983), is only valid for annular tubes where μ is large.

$$C/C_o = (0.82 \pm 0.10)\exp(-22.53 \pm 1.22 \ \mu_a)$$

Inlet length considerations for annular denuders are not fundamentally different from those for cylindrical denuders. If the annular gap is small relative to the radius of curvature, then

$$L_{min} = 0.04R_e x$$

where $x = (d_1 + d_2)/2$.

4.3.1 TUBULAR DENUDERS

Some of the undesirable processes encountered in the filtration sampler can be hindered or minimized when gaseous pollutants are removed prior to collection of aerosols by means of a diffusion denuder. Originally, cylindrical denuders were used for sampling atmospheric species. Ferm (1979) used an oxalic acid denuder for the collection of ammonia to avoid an ammonia-acidic aerosols reaction. Stevens et al. (1978) used a battery of 16 parallel tubes arranged in a circle for the separation of ammonia from acidic sulfate particulate. Braman et al. (1982) measured NH_3 and HNO_3 using a tungstic acid (H_2WO_4)-coated tube. Forrest et al. (1982) used 48 tubes coated with an Na_2CO_3 film for HNO_3 measurements; the NO_3 particles were collected on the filter following the tube. Slanina et al. (1981) used a cylindrical thermodenuder system for speciation of atmospheric strong acid. HNO_3 and NH_3 gases were collected on NaF, H_3PO_4-coated denuders, respectively. Strong acid aerosols (H_2SO_4) were thermally evaporated in a heated glass tube, and the small fragments formed were collected on the walls of the tube. The system consists of seven tubes connected in series. NH_4NO_3 was also volatilized and collected as NH_3 and HNO_3 gases. Shaw et al. (1982) measured nitric acid and particulate nitrate in a system termed the *denuder difference method* (DDM), in which two samplers are run in parallel. In one sampler, both particulate and gaseous nitrate are collected on a nylon filter, while in the parallel sampler, HNO_3 is removed by an MgO denuder before collection of particulate nitrate on the nylon; the difference between the nitrate collected on the two nylon filters gives the nitric acid concentration. Lewin et al. (1986) used a denuder tube system consisting of 15 parallel tubes for the collection of ammonia prior to the collection of the particulates on the Teflon filter. Ferm (1986) used a Na_2CO_3-coated denuder for collection of nitric acid and a Na_2CO_3-impregnated filter for the collection of particulate nitrate. This set-up takes no account of possible artifacts from the dissociation of NH_4NO_3 and NH_4Cl, and the interferences caused by nonselective removal of other nitrogen compounds. The connection of two denuders in series of this type permits the determination of HNO_3 and HONO, and the effect of other nitrogen interferences (Ferm and Sjodin, 1985).

4.3.2 ANNULAR DENUDERS

The development of cylindrical denuders has successfully overcome some of the inherent artifacts in the filtration technique. However, the efficiency of diffusion tubes depends on maintaining full laminar flow. This considerably limits the amount of material collected in the tube. To collect a sufficient amount of mass above the detection limit of the analytical instruments, an extreme length of denuder and/or longer sampling periods are required. Both of these are undesirable options. One reason is that atmospheric conditions do not persist for long periods. Any change in temperature, pressure, or relative humidity during sampling will affect the phase of the atmospheric species. Short time sampling becomes necessary to avoid perturbation of the atmospheric species, especially for pollutants that are partitioned between the gas and particulate phase. Additionally, short sampling is important in size distribution studies of atmospheric aerosols. Using a parallel multitube system which can operate at a high flow rate and for short periods (Forrest et al., 1982; Lewin and Hansen, 1984) is an option. However, the disadvantage of using such a multi assembly is that a large volume

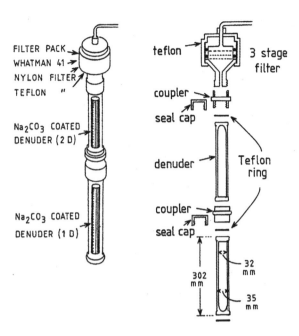

FIGURE 4.2   Annular denuder system.

of solvent is required for extraction. Furthermore, the coating and extraction procedure for such a system is cumbersome.

The development of the annular denuder sampler (Possanzini et al., 1983) overcame the limitations of cylindrical diffusion tubes. In the annular denuder, air flows between two coaxial glass tubes (Figure 4.2). The surface area is increased and a single denuder can operate at large flow rates and short sampling periods with a high collection efficiency and a sufficient amount of mass for analysis. Comparing the performance of the cylindrical denuder with an annular denuder (dimensions: $d_2 = 3.3$ cm, $d_1 = 3.0$ cm [d_2 is the internal diameter of the external tube; d_1 is the external diameter of the internal tube]) at a given collection efficiency shows that for a given flow rate, the annular denuder can achieve equivalent collection efficiency at 1/30 of the length of the cylindrical denuder, or for a given denuder length, the annular denuder can be operated at 30 times the flow rate of the cylindrical denuder. Working with high flow rates (10 to 30 1 min^{-1}) enables scientists to collect samples over short time periods. The numerical basis of computing collection efficiencies in annular denuders has been established (Winiwarter, 1989). Multiple annular denuders bearing as many as 12 concentric tubes have been described, and computer programs are available for calculating the collection efficiency of such denuders (Coutant et al., 1989).

4.3.3   POSSIBLE INTERFERENCES IN THE DENUDER SAMPLER

The effectiveness of the denuder diffusion sampler depends on a complete discrimination between the gas species and particulate matter. Complete separation of gases and particulate is not an easy task. Particles may deposit inside the denuder due to gravitation and diffusion processes, which leads to an error in the measurements. Gravitation losses afflict large particles (aerodynamic diameter >1 μm), which are predominantly of crustal origin. These particles have a large capacity to neutralize acid particles, and if acidic aerosols are measured without prior separation of these particles, significant error may occur.

Losses due to diffusion become important for small particles (<0.1 μm aerodynamic diameter). Deposition of particles due to gravitation is eliminated by setting the denuder in the vertical position. Losses in the denuder have been calculated from the Gormley–Kennedy equation to be 5% for particles with diameter of 0.01 μm (diffusion coefficient $2.4*10$ cm²s⁻¹). Experimental measurements by Forrest et al. (1982) indicate that losses ranged from 0.2 to 2.2% for aerosol particles between 0.3 μm and 0.6 μm to about 4 to 5% for 1 to 2 μm particles. These levels are considered acceptable as a fraction of the total aerosol collected.

Experimental evidence has shown that negligible amounts of particles are deposited inside annular denuders. Possanzini et al. (1983) demonstrated this by passing aerosols larger than 0.3 μm in diameter through an annular denuder. Aerosols were counted upstream and downstream of the denuder, and no significant changes were observed when the denuder was inserted. Ferm (1986) investigated the artifact due to diffusion, electrostatic effects, and evaporation of particles inside the denuder and found that these artifacts were within the experimental error. Particle loss inside an annular denuder equipped with a cyclone or impactor was less than 3% of the total concentration for particles in the range of 1.5 to 2.77 μm aerodynamic diameter (Koutrakis et al., 1988). Kitto (1991) investigated the transit of monodisperse aerosols of 0.5 μm aerodynamic diameter inside an annular denuder and observed a 2.3% loss. More recently, Ye et al., (1991) investigated particle loss inside both coated and uncoated annular denuders over a wide range of particle sizes and degree of charge: neutral, Boltzman charge, and single charge. They reported that particle loss inside the uncoated denuder increased as particles became charged.

Because the deposition problem primarily affects large particles, some workers have adopted the use of a cyclone or preimpactor (Fox et al., 1988; Koutrakis et al. 1988). The use of cyclone or preimpactor should overcome problems associated with the transit of large particles through the denuder, but opens up possible problems of gas–particle and particle–particle reactions in the impactor (Appel et al., 1988a), and losses of reactive gases to impactor walls (Grosjean, 1985).

Denuder samplers may also be subject to evaporation of particles during transition. Larson and Tayler (1983) studied the evaporation of ammonium salts inside denuders. Both theoretical treatment and experimental measurements demonstrated that these interferences were insignificant as long as the HNO_3 to particulate nitrate ratio is not too small or too large. Laboratory experiments conducted by Forrest et al. (1982) showed a slight loss of ammonium nitrate (1.9%) inside the tube; however, in their experiment, the denuder was heated above the ambient temperature (temperature difference between entering and exiting air was 6°C) which may have an effect on the evaporation of ammonium salts. However, based on field measurements, Eatough et al. (1985) reported that these artifacts were very small. The rate of evaporation is roughly proportional to the diffusion coefficient of the gas, the total surface area of the volatile NO_3 particles, and the difference between the equilibrium concentration and the actual concentration of HNO_3 surrounding the particles (Ferm, 1986). Since the equilibrium concentration is very small (in ppb levels) and Aitken particles are a minor part of particulate matter, the evaporation will be negligible compared to the ratio between the actual concentration of particulate nitrate HNO_3 and the mean residence time in the denuder, i.e., the time required to achieve the equilibria is greater than the particle residence time inside the denuder. Kinetic measurements on the evaporation of ammonium salts have shown a constraint on the achievement of the equilibrium (Harrison et al., 1990).

The second problem with annular denuders is that of possible artifact formation. This has been noted principally with a Na_2CO_3/glycerol-coated denuder (Ferm and Sjodin, 1985; Febo et al., 1986; Allegrini et al., 1987). These interferences, originate from nitrogen dioxide (NO_2), peroxyacetyl nitrate (PAN), organic nitrates, and organic nitrate species in the atmosphere. The removal of NO_2 by carbonate denuders is not quantitative; only 0.5 to 2% of NO_2 is removed by denuders (Allegrini et al., 1987; Perrino et al., 1988). However, NO_2 exhibited a fairly constant mass distribution along the denuders. Thus, the quantity of the interferant can be taken into account by using two carbonate denuders in series (Koutrakis et al., 1988; Perrino et al., 1990). More recently, Appel et al. (1990) used carbonate denuders for HONO measurements and found only 0.2% of NO_2 was retained as NO_2^-. Also it has been reported that HONO collected on the first Na_2CO_3-coated denuder may be subsequently released from the denuder when a purified air stream is passed through the denuder (Perrino et al., 1990).

Another possible interferant is peroxyacetyl nitrate (PAN) which could be hydrolyzed to nitrite in alkaline solution (Penkett et al. 1977). Ferm and Sjodin (1985) reported up to 10% retention of PAN by a carbonate-coated cylindrical tube. Allegrini et al. (1987) estimated that less than 5% of PAN was retained on the denuder walls. PAN concentrations in semirural areas are in the range of 0.05 to 0.4 ppbv (Davies et al., 1989). Interferences from PAN (in such locations) would be very small and acceptable within the limit of experimental error. The use of tandem denuders in series, each coated with the same material, should solve the problem of retention of interferant gaseous components. The use of an NaCl-coated denuder followed by two Na_2CO_3/glycerol-coated denuders has been suggested (Febo et al., 1986; Koutrakis et al., 1988; Perrino et al., 1990), which permits unbiased determination of HNO_3 and HONO. However, in the urban environment, NO_2 concentrations are much higher than those of HNO_3 and retention of NO_2 could be significant. Febo et al. (1993) demonstrated that the application of a three denuder system coated with ($NaCl$–Na_2CO_3–Na_2CO_3) for nitric and nitrous measurements in polluted area were subject to interferences from NO_2 and SO_2. Nitrous acid concentrations obtained by $NaCl$–Na_2CO_3–Na_2CO_3 denuders were sixfold higher than those of tetrachlormercurate (TCM)–Na_2CO_3–Na_2CO_3.

Negative artifacts might occur when carbonate-coated denuders are used for the determination of HONO. These arise from conversion of nitrite to nitrate in the presence of ozone and other atmospheric oxidants (Febo et al., 1986a; Perrino et al., 1990). In this case, HONO concentrations will be underestimated. The addition of glycerol to the coating material (Na_2CO_3) has shown to reduce, substantially, the oxidation of NO_2 to NO_3 (Febo et al., 1986a) and increase the capacity of the denuder toward SO_2 and HNO_3 (Allegrini et al., 1987). Appel et al. (1988b) reported that nitrate was observed in the second denuder and averaged 14% of the front denuder. It was inferred that this was due to the oxidation of the retained NO_2-yielding. To eliminate such interferences, it has been suggested that an additional denuder be used to collect HNO_3 separately without the HONO. Perrino et al. (1990) used an NaCl-coated denuder for the HNO_3 followed by two carbonate denuders. However, it is possible to use carbonate denuders alone for measurements of both HONO and HNO_3 in an unpolluted area. Since HNO_3 exhibits a diurnal distribution with a maximum during the afternoon, and HONO during nighttime hours, a relatively insignificant loss of HONO due to oxidation may be possible if sampling starts and ends in the morning (Appel et al., 1990; Kitto and Harrison, 1992a).

Brauer et al. (1990) reported that NO_3^- was detected on the second denuder, in some cases at levels higher than those found on the first denuder. However, their measurements were conducted in an indoor environment at high temperatures and low relative humidity. De Santis et al. (1988) compared nitric acid measurements obtained by an annular denuder preceded by a cyclone made from Teflon with the same system without a cyclone and manifold. Their results were lower than those measured without the Teflon cyclone, and from laboratory experiments, machined Teflon parts such as a manifold and cyclones showed a noticeable absorption toward nitric acid. A triple path denuder, each with a different coating, has been used to investigate artifact formation (Davis et al., 1988, 1994). Results show that particulate loss may result both from surface scavenging through the system conduit parts and control losses from incomplete gas-to-particle conversion. They concluded that oxalic acid is not a suitable coating for NH_3 removal when using X-ray diffraction analysis because of likely losses from the denuder walls.

The annular denuder is usually followed by a filter pack for the collection of particulate species. The annular denuder/filter pack system (ADS), illustrated in Figure 4.2, has been widely used for measuring atmospheric gases and particulate species (Koutrakis et al., 1988, 1989; Sickles et al., 1988, 1989, 1990; Fox et al., 1988; Eatough et al., 1988a; Harrison and Kitto, 1990). In some cases, cyclones or impactors were used prior to the denuder to remove large particles (Sickles et al., 1988, 1989; Koutrakis et al., 1988; Cox et al., 1988).

4.4 DENUDER APPLICATIONS

4.4.1 INORGANIC COMPOUNDS

4.4.1.1 Ammonia

Ammonia is the dominant neutralizing agent in the lower troposphere and plays an important role in the chemistry of the atmosphere. Emissions of ammonia arise predominantly from decomposition of animal manure and the use of fertilizers (Asman and Janssen, 1987; Erisman et al., 1989). Ammonia reacts with acidic constituents of the atmosphere. The reactions of ammonia with both nitric and hydrochloric acids are reversible and are temperature and relative humidity dependent.

Measurements obtained by filter packs overestimated the concentration of atmospheric ammonia as mentioned earlier. Diffusion denuders which are able to distinguish gaseous ammonia from particulate ammonium overcome this problem. The first application of a diffusion denuder for measurements of atmospheric ammonia was reported by Ferm (1979). A cylindrical tube was coated with 1.5% oxalic acid in methanol. After sampling, the denuder was extracted with $0.1M$ NaOH and analyzed for ammonium ion using an ion-selective electrode. The experimental efficiency of the denuder was 90.6%, which was determined by connecting two denuders in series. A diffusion coefficient of 0.247 cm^{2}sec^{-1} was calculated from the removal of ammonia at different segments of the denuder. Bos et al. (1980) demonstrated that diffusion denuders can be automated. The entire procedure was automated from coating to extraction and calorimetric analysis. The limitation of this method is that long sampling periods are required, because cylindrical denuders are operated at low rates (0.2 to 2 1 min^{-1}). Different coating materials have been used for ammonia collection. These include: oxalic acid (Ferm 1979; Lewin and Hansen, 1984; Cadle 1985; Allegrini et al., 1987; Andersen and Hovmand, 1994), tungstic acid (Braman et al., 1982; McCleany et al., 1982), citric acid (Koutrakis et al., 1988; Brauer et al., 1991; Waldman et al., 1990; Kitto, 1993), phosphoric acid (Rosenberg et al., 1988), and phosphorous acid (Rapsomanikis et al., 1988; Harrison and Kitto, 1990). The amount of acid material used for coating varies from 1 to 5% in alcohol (methanol or ethanol). Oxalic acid was found to evaporate from denuder walls and to be collected on the filter which would cause error if strong acid aerosols are measured. Lewin and Hansen (1984) reported that oxalic acid in ethanol resulted in a more uniform layer. Extreme care should be taken when coating/extracting the ammonia denuder. A clean environment is recommended to prevent contamination from indoor ammonia, which is much greater in concentration than atmospheric ammonia (Li and Harrison, 1990; Brauer et al., 1991). A denuder sampler for ammonia and particulate ammonium would consist of an annular denuder coated with one of the acid materials mentioned above, followed by a filter pack comprising a Teflon filter and acid-impregnated filter (cellulose). Particulate ammonium concentration is the sum of the ammonium ion found on both filters.

4.4.1.2 Nitric and Nitrous Acids

Nitric acid is one of the major acidifying components in the atmosphere and may be neutralized if sufficient ammonia is available and the meteorological parameters favor the formation of particulate nitrate. Nitric acid is formed in the atmosphere through oxidation of NO and NO$_2$. The oxidation of nitrogen oxides by OH is the main mechanism of daytime formation. Nitric acid measurements using filtration techniques are fraught with artifacts, which primarily arise due to dissociation of particulate ammonium nitrate. The application of a denuder sampler provided a solution for such artifacts. Various types of denuder samplers have been applied for the determination of inorganic nitrates, such as the denuder difference method (Shaw et al., 1982), in which particulate nitrate is collected on a nylon filter with or without prior removal of nitric acid by the denuder. The difference between the nitrate on the nylon with or without the denuder is the nitric acid concentration. In the same category, where the denuder extract is not analyzed, MgO has been used as a coating material for the removal of HNO_3 (Shaw et al., 1982). When the denuder extract is analyzed after sampling, different coating materials have been used. These include: tungstic acid (Braman et al., 1982), NaF (Rosenberg et al., 1988), NaCl (Allegrini et al., 1987; Febo et al., 1993), Na_2CO_3 (Febo et al., 1986; De Santis et al., 1985; Ferm, 1986; Allegrini et al. 1987, Koutrakis et al., 1988, Kitto and Harrison, 1992b). Collection of nitric acid by a carbonate/glycerol-coated denuder followed by ion chromatography has become the most popular among workers in the area of trace gas analyses. This approach is relatively simple and easy to construct but labor intensive. The advantage of this procedure is that close to theoretical collection efficiency is achieved and simultaneous measurements of acidic gases are accomplished. Analysis of the denuder extract by ion chromatography provides multicomponent (such as Cl^-, NO_2^-, NO_3^-, and SO_4^{2-}) determination in one single injection. Carbonate-coated denuders are efficient for the collection of nitrous acid (Ferm and Sjodin, 1985; Brauer et al., 1990; Koutrakis et al., 1988; Kitto and Harrison, 1992a), sulfur dioxide (Allegrini et al., 1987), and hydrochloric acid (Allegrini et al., 1987; Harrison and Kitto, 1990), although some compounds may interfere in nitrous acid measurements using carbonate-coated denuders. The use of multiple denuders in series would solve this problem, and Perrino et al. (1990), suggested the use of three denuders in series. The first denuder was coated with NaCl, and the other two were coated with carbonate (Na_2CO_3/glycerine). Febo et al. (1993) reported that such a system is susceptible to interferences, especially in urban environments. The authors suggested the use of tetrachloromercurate (TCM) instead of NaCl before two carbonate denuders and have recently reported the use of such a coating (Allegrini et al., 1994).

A combined impactor/honeycomb denuder/filter pack system has been reported (Koutrakis et al., 1993). The advantage of such a system is that a length of only 3.8 cm yields the same collection capability as a length of 21 cm for the annular denuder. The complete system is shown in Figure 4.3. At a sampling rate of 10 l/minute the circular array of inlet nozzles causes particles larger than 2.1 μm to impact on the impactor plate. The glass honeycomb denuder is a cylinder with a height of 3.8 cm and a diameter of 4.7 cm, containing 212 hexagonal glass tubes. Experimental results for the collection of HNO_3, and NH_3 indicate that this design functions better than the annular denuder.

4.4.2 ORGANIC COMPOUNDS

4.4.2.1 Environmental Tobacco Smoke (ETS)

Cylindrical denuders were first used to study the distribution of nicotine between the gas and particulate phase (Eudy et al., 1986). Nicotine in the gas phase was removed by an acid-coated denuder which was connected to the inlet of a chemical ionization mass spectrometer. The study showed that acid-coated denuders were efficient in removing the gas phase nicotine. The application of denuder samplers for the collection of gas phase nicotine in indoor environments has been widely adopted by many workers (Eatough et al., 1986c, 1987b; Caka et al., 1990). Both cylindrical and annular denuders have been used for monitoring ETS and are generally coated with benzenesulfonic acid (BSA). The particulate nicotine was collected by a quartz filter followed by a glass fiber filter impregnated with benzenesulfonic acid. Eatough et al. (1987b; 1989) calculated a diffusion coefficient of 0.063 ± 0.007 cm sec^{-1} for the gas phase nicotine and reported a detection limit of 0.02 mMole m^{-3} for a 6 h sample. Results obtained by an annular denuder sampler (Koutrakis et al., 1989) were compared to those of a filter pack sampler (Caka et al., 1990). The filter pack results were approximately 25% less than those measured by the annular denuder sampler. The authors inferred that this was due to evaporation of particulate nicotine from the filter pack sampler. Indeed, this was confirmed by the denuder sampler wherein most of the particulate nicotine evaporated from the quartz filter and was collected by a benzenesulfonic acid impregnated filter. It was concluded that the filter pack sampler would underestimate particulate nicotine if used alone (Caka et al., 1990; Koutrakis et al., 1989).

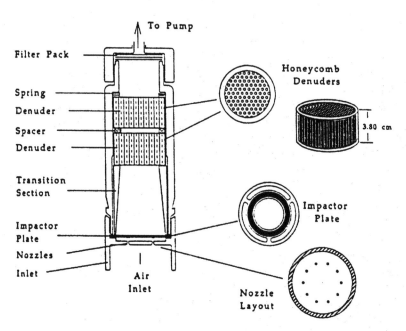

FIGURE 4.3 Honeycomb denuder sampler. Reproduced with permission from Environmental Science and Technology (1993, Vol. 27, p 2497) ©American Chemical Society. (From Koutrakis P., Sioutas C., Ferguson S.T., Wolfson, J.M., Mulik J.D., and Burton R.M., 1993, Development and evaluation of a glass honeycomb denuder/filter pack system to collect atmospheric gases and particles. *Environ. Sci. Technol.*, 27, 2497-2501. With permission.)

A cylindrical denuder has been used to investigate the effects of the dilution of mainstream tobacco smoke upon the evaporation and diffusion of nicotine from smoke particles (Lewis et al., 1994) in both a continuous and puffing sampling method. Figure 4.4 shows the experimental layout for the puffing (puffwise) method. A 1.5 m tube, of internal radius 0.78 cm, was coated with oxalic acid. The maximum percentage of nicotine collected by the denuder due to particle diffusion, rather than nicotine vapor diffusion, was determined to be 0.49% by comparing the amount of solanesol collected by the denuder and the filter.

4.4.2.2 Alkyl Sulfate

Diffusion denuder samplers have been used for measurements of dimethyl sulfate (DMS) and monomethyl sulfuric acid (MSA) in air (Eatough et al., 1986a, 1986b) and in power plant plumes (Eatough et al., 1986b; Hansen et al., 1987). Both nylon denuders and carbonate-coated annular denuders have been used (Eatough et al., 1986a, b). In the nylon denuder, both dimethyl sulfate and monosulphonic acid were measured indirectly as a sulfate and monomethyl sulfate (Hansen et al., 1986; Eatough et al., 1987a) which were identified by ion chromatography. A carbonate-coated denuder has been used to measure monomethyl sulfuric acid in the Los Angeles basin (Eatough et al., 1986a, b). Interferences from sulfur dioxide were not significant. However, it was assumed that monomethyl sulfate was not hydrolyzed to sulfate on a carbonate-coated denuder during sampling or storage. This assumption has not been proven experimentally.

4.4.2.3 Chlorinated Organic Compounds

Chlorinated organic compounds in ambient air have been determined using a multisection annular denuder sampler (Lane et al., 1988; Johnson et al., 1986). The sampler consisted of an annular denuder connected to a dichotomous sampler. A filter pack sampler was used in parallel for comparison with the annular denuder sampler. In both samplers, Tenax was used after the filter as

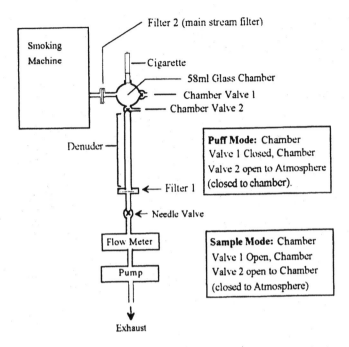

FIGURE 4.4 Schematic diagram of puffwise sampling method.

a backup absorber. The collection efficiencies of annular denuders for lindane, a-hexachlorohexane, and hexachlorobenzene were determined as a function of temperature and relative humidity. Field measurements in the vicinity of Lake Ontario showed that the majority of chlorinated organic compounds exited in the gas phase, and chlorinated aerosol concentrations increased as the temperature decreased. This study demonstrated that annular denuder samplers can be used to investigate the phase distribution of chlorinated organic compounds.

4.4.2.4 Carboxylic Acids

Carboxylic acids, such as formic and acetic acids, are formed in the atmosphere during photochemical smog episodes. Although present in low and sub ppb range, they are important contributors to atmospheric acidity. These compounds are distributed between the gas and particulate phase. Annular denuders coated with bicarbonate or KOH have been used for the collection of gas phase organic acids (e.g., Rosenberg et al., 1988; Lawrence and Koutrakis, 1994). Using an annular denuder sampler, organic acid has been identified in fog (Winiwarter et al., 1988). An intercomparison of methods to measure carboxylic acids showed that the annular denuder sampler was in good agreement with other methods (Kenne et al., 1989).

4.4.2.5 Formaldehyde and Other Ketones

Formaldehyde contributes to the formation of photochemical pollutants such as ozone and peroxy acetyl nitrate (PAN) (National Research Council, 1991). Concentrations of formaldehyde in indoor environments are much greater than those outdoors (Hines et al., 1993). The reaction of aldehydes with 2,4-dinitrophenyl hydrazine (2,4-DNP) has been utilized for measurements of atmospheric aldehydes. Possanzini et al. (1987) used in annular denuder sampler for measurements of aldehydes such as formaldehyde, acetaldehyde, and propaldelhyde. The denuder was coated with 2,4-DNP and the hydrazone derivatives were identified by high performance liquid chromatography (HPLC) coupled with a UV detector. Dasgupta et al. (1988) developed a continuous diffusion denuder for real-time measurements of gaseous components in the atmosphere. More recently, Fan and Dasgupta

(1994) described a continuous method for determination of formaldehyde at the parts per trillion level. The method utilized a Nafion 811 hydrophilic membrane tube-based diffusion scrubber with water as the scrubbing liquid. The authors reported a detection limit of 9 pptv and a lag time of 4.6 min.

4.4.2.6 Total Organic Compounds

Appel et al. (1983) pioneered the use of diffusion denuders for the determination of total organic compounds in air. Organic compounds in the gas phase were collected by a cylindrical denuder. The denuder surface was etched with HF then coated with a silicone deposited from pentane solution. After evaporation of the pentane, the denuder was coated with chromatographic grade 400–200 mesh Al_2O_3 beads. Particulate organic compounds were collected on a filter followed by a fluidized bed of Al_2O_3. Eatough et al. (1988b) used an activated carbon-impregnated filter as a surface for the denuder for the collection of gas phase organic compounds. Particulate organic compounds were collected on a quartz filter followed by an activated carbon sorbent. Results obtained by annular denuder samplers compared to those of filter packs showed that filter packs underestimate concentrations of particulate organic compounds (Aghdair et al., 1988; Eatough et al., 1988a). The loss of particulate organic compounds collected on a quartz filter ranged from 40 to 70% of the total organic matter (Eatough et al., 1989).

4.4.3 OTHER APPLICATIONS

Diffusion denuders have been used to investigate the formation of artifacts during sampling. The presence of oxidants such as ozone and nitrous acid (HONO) may alter the chemical composition of reactive species. Williams and Grosjean, (1988) suggested the use of a carbonate-coated denuder to remove acidic gases and KI-coated denuders for the removal of ozone prior to collection of particulate organic compounds. Aldehydes are converted to organic acids in the presence of the oxidants. Thus, it is recommended that these oxidant components be removed prior to the collection of organic compounds (Williams and Grosjean, 1990).

4.5 AUTOMATED DENUDERS

4.5.1 Thermodenuders

The first thermodenuder was introduced with the aim of speciation of sulfate-containing aerosols, and, in particular, the separation of sulfuric acid aerosols from ammonium hydrogen sulfate and ammonium sulfate aerosols. In their initial application, cylindrical denuders were heated to 137°C. At this temperature, sulfuric acid aerosols dissociate and are subsequently collected on uncoated denuders, while the other sulfate-containing aerosols are not affected and may be collected on Teflon filters. At the end of the sampling period, the uncoated denuder may be leached with deionized water and analyzed for sulfate. Nissener and Kloclow (1980) reported that 90% of sulfuric acid aerosols were collected by this method. Later, NaCl was used for collection and showed 99% collection of acidic aerosols. Based on previous work, Slanina (1981) described a method for the separation of sulfate and nitrate aerosols, in addition to sulfuric acid. The system consisted of multiple denuders connected in series with or without heating. The first set of denuders consisted of denuders coated with NaF, H_3PO_4, and KOH. The sodium fluoride denuder was heated to 115 to 135°C to collect H_2SO_4 aerosols and HNO_3 gas decomposed from NH_4NO_3 aerosols; the phosphoric acid denuder collected NH_3 from dissociated NH_4NO_3 particulates, while the NaOH denuder collected SO_2 gas. Ammonium hydrogen sulfate NH_4NSO_4 aerosols were dissociated in an NaF-coated denuder heated to 215 to 235°C. The resulting H_2SO_4 was collected by the same denuder (NaF), while the evolved NH_3 was collected on a third denuder coated with H_3PO_4. Complete separation of the species was not attained. Later, Slanina et al. (1985) reported a thermodenuder for the separation and collection of sulfuric acid aerosols and ammonium sulfate. This method relies on the reaction of sulfuric acid with copper–copper(II) oxide and decomposition of ammonium sulfate aerosols at 220°C. Copper sulfate formed in the denuder was decomposed to SO_2, which was analyzed by a flame photometric detector. A set of denuders were used prior to the copper–copper(II) oxide denuder to eliminate the interferant compounds: a carbonate-coated denuder was used to remove SO_2, and an $NaHSO_4/Ag_2NO_3$ denuder was used for H_2S. Concentrations of sulfuric acid have also been determined using denuders coated with manganese and palladium oxide. After sampling, the denuders were heated to 800°C and the trapped sulfur compounds were thermally desorbed as H_2S, which was then analyzed by gas chromatography (GC) with a flame photometric detector (Lindqvist, 1985). The interferant compounds were removed from the air stream using a set of denuders.

Thermodenuder techniques have also been used in the determination of inorganic nitrogen species. Denuders coated with $Al_2(SO_4)_3$ have been used for nitric acid determination (Lindqvist, 1985). The concentration of nitric acid was determined from thermal desorption of nitrate trapped inside the denuder as NO_x, which was subsequently converted to NO and analyzed by a denuder GC equipped with a photoionization detector. Braman et al. (1982) used a tungstic acid-coated denuder for nitric acid collection. After sampling, the nitric acid was converted to NO_2 by heating the denuder to 350°C and analyzed by a chemiluminescence technique. Roberts et al. (1987), using the same method, reported difficulties due to interferences from other nitrogen compounds. Intercomparison studies of tungsten oxide (WO_x) denuders with other denuders and filter techniques have shown that measurements with tungsten oxide were artificially higher than other methods (Goldan et al., 1983; Eatough et al., 1985; Anlauf et al., 1985; Appel et al., 1988b).

Keuken et al. (1989) described a very sophisticated thermodenuder method for the determination of NH_3. An annular denuder was coated with vanadium pentoxide, and heated rapidly at the end of the sampling period to 700°C using a detachable preheated oven. Interferences from other nitrogen compounds were observed. Simultaneous measurements of HNO_3 and NH_4NO_3, using a thermodenuder, have been reported by Klockow et al. (1989). His method consisted of two sets of denuders in parallel. Each set contained three denuders in series: the first two were at ambient temperature, while the last was heated to 110°C. $MgSO_4$ was used as the coating for one set of denuders, while in another set $BaSO_4$ was used as the coating for the first denuder, activated charcoal for the second, and NaF for the third. Nitric acid was collected by the first denuder from each set. Nitrogen interferant compounds were collected by all three $MgSO_4$ denuders, but only the first two of the other system. After sampling, the $MgSO_4$-coated denuders were heated to 700°C consecutively and analyzed for NO_x. NO_x released from the second denuder was used to correct for HNO_3 and NH_4NO_3. Intercomparison measurements of thermodenuders with other types of denuders have shown that higher HNO_3 concentrations are collected by the thermodenuder technique. A tungstic acid–thermodenuder-NO_x analyzer has been developed for the continuous measurement of ammonia (Romer et al., 1994). The method is based on preconcentration and conversion of NH_3 to NO over a gold catalyst and detection of NO by a chemiluminescence analyzer (Figure 4.5). As air is sampled through the denuder, NH_3 diffuses to the tungsten trioxide-coated walls and is converted to ammonium tungstate. Because two denuders are alternately sampled and analyzed, measurements may be made continuously. After sampling for 15 minutes, the denuder is heated to 350°C, upon which decomposition of the ammonium salt and desorption of NH_3 occurs. Transportation to the gold catalyst is effected in an He/O_2 carrier gas. Although interfering substances will also be absorbed, the main interferant (HNO_3) can be eliminated by applying temperature programmed desorption. Linearity and good accuracy were reported for NH_3 concentrations up to 500 ppb.

4.5.2 Wet Denuders

In this technique, the denuder walls are continuously kept wet, and the effluent is removed and analyzed continuously. This is achieved by introducing the scrubber liquid at one end of the denuder and collecting it at the other end. Keuken et al. (1988) were first to report the application of a wet

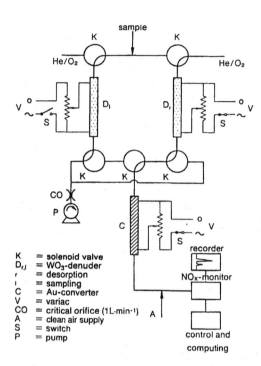

FIGURE 4.5   Automated instrument for continuous measurement of ammonia. (Adapted from Romer F.G., van den Beld L., van Elzakker B.G., and Memmen M.G., 1994, Automated thermodenuder system for continuous measurement of atmospheric ammonia concentrations, in *Physico-chemical Behaviour of Atmospheric Pollutants*, Angeletti G. and Restelli G. (Eds.), Report EUR 15609/2 EN, European Commission. With permission.)

denuder for measurements of atmospheric gaseous species. The denuders were continuously rotated electrically at a rate of 40 rpm. The scrubber liquid used formed a thin layer (0.5 mm) on the surface of the annulus. Two denuders were operated in parallel. One for the HCl, HNO₃, and NH₃ and the second for SO₂ and H₂O₂ collection. The scrubber liquids were 0.5 mM HCOOH–HCOONa buffer (pH = 3.7), and the other was a phosphate-buffered solution (pH = 7) containing 4-hydroxyphenylacetate, peroxidase, and formaldehyde. After completion of the sampling cycle (40 min.), the denuders were tilted and the scrubber liquids were pumped out of the denuders into an autosampler tray collection. The denuders were then washed with 2 ml of the scrubber liquids before introducing a new liquid scrubber (26 ml). To avoid interference, the first denuder was operated with a cyclone to remove coarse particles, and in the second denuder, the air was premixed with NO to remove O₃ (which interferes with the H₂O₂ measurements) from the air stream prior to the inlet of the denuder. The authors reported 90% collection efficiency for the measured species. The limit of detections were 290, 80, 130, 190, and 7 ppt for NH₃, HCl, HNO₃, SO₂ and H₂O₂, respectively. A continuous flow denuder for the measurement of NH₃ has been developed in the Netherlands (Wyers et al., 1991). Air is sampled through a rotating annular denuder, which is coated with an absorption solution. This solution is continuously withdrawn from the denuder and passes through a denuder, where NH₃ diffuses out of the solution through the membrane into very pure water, where the concentration is measured by conductivity (Figure 4.6).

Different versions of wet denuders have been investigated by other research groups (Simon et al., 1991; Vecera and Dasgupta, 1991). Various problems are experienced with wet denuders, such as maintaining a uniform layer (Dasgupta, 1993). Taira and Kanda (1993) have proposed a wet effluent denuder which may be coupled to a gas analyzer so that continuous measurements of HNO₃ and HNO₂ may be made. The design is shown in Figure 4.7. The denuder is a 0.16 cm i.d. glass tube of 50 cm length. Using a 0.05 M NaOH absorption solution with a flow rate of 0.12

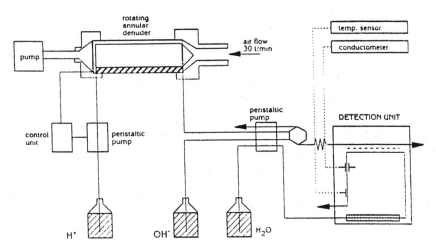

FIGURE 4.6 On-line wet denuder system to measure ammonia. (Adapted from Slanina J., Wyer G.P., Kieskamp W.M. and ten Brink H.M., 1994, Analytical chemistry in the troposphere, in *Physico-chemical Behaviour of Atmospheric Pollutants*. Proceedings of the sixth European Symposium. Report EUR 15609/1 EN. With permission.)

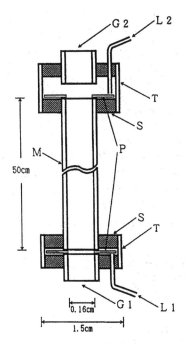

FIGURE 4.7 Schematic diagram of wet effluent diffusion denuder: G1 and G2 — air inlet/outlet glass tube; L1 and L2 — liquid inlet/outlet PTFE tube; M — main denuder tube; P — filter paper; S — silicon rubber support; T — glass tube. (From Taira M. and Kanda Y., 1993, Wet effluent diffusion denuder for sampling of atmospheric gaseous nitric acid. *Anal. Chem.*, 65, 3171-3173. ©American Chemical Society. With permission.)

ml/minute, the collection efficiency for HNO_3 was in excess of 94% for a sampling rate of 4 l/minute. Higher efficiencies were obtained at lower flow rates.

It has been proposed that a wet parallel plate denuder offers better collection efficiency and lower particle losses than the annular design (Simon and Dasgupta, 1993). Collection efficiency for SO_2 for two parallel plates of 50 × 300 mm active area and separation 3 mm (Figure 4.8) is essentially quantitative at a sampling rate of 10 1 minute^{-1}. Using this method, an automated

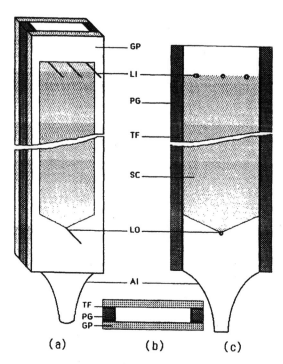

FIGURE 4.8 Wet parallel denuder: (a) overall appearance, (b) view of one plate, and (c) top cross section. GP — glass plate; LI — hypodermic needle liquid inlets; PG — Plexiglas spacer; TF — Teflon film; SC — silica coating; LO — liquid outlet; AI — air inlet. (From Simon P.K. and Dasgupta P.K., 1993, Wet effluent denuder coupled liquid/ion chromatography systems: annular and parallel plate denuders. *Anal Chem.*, 65, 1134-1139. ©American Chemical Society. With permission.)

technique for the continuous analysis of aerosols has been reported (Simon and Dasgupta, 1995). Particles are continuously transferred to a liquid stream via condensation of supersaturated vapor to promote particle growth followed by collection using impaction and thermophoresis. Combined with a wet wall parallel plate diffusion denuder, detection limits for sulfate and nitrate of 2.2 and 5.1 ng m^{-3}, respectively, for an eight-minute sample are reported.

4.5.3 Denuder Scrubbers

Dasgupta et al. (1986) pioneered and developed the denuder scrubber system for continuous measurements of atmospheric components. In the beginning, a cation-exchange membrane tube (Nafion) of wet internal diameter 700 μm, wall thickness 75 μm, and length 30 cm long was used for the collection of NH$_3$. Dilute H$_2$SO$_4$ was pumped through the outer tube. Later, Dasgupta used a PTFE membrane tube of 25 cm length, 3000 μm i.d., and 100 μm wall thickness for measurements of nitric acid. The PTFE served as an anion-exchange membrane. The relaxation time for a typical analyte species was of the order of 0.75 to 3 minutes for a 50- to 100 μm-thick membrane. Thus, the response time for a continuous analyzer based on membrane-based collectors may be restricted by this requirement. Furthermore, interferences are experienced with anion exchange membranes, such as oxidation of sulfite to sulfate and nitrite to nitrate, due to the strong affinity of the anion exchange membrane for the higher oxidized ions. The problems encountered in the early use of PTFE membrane were overcome by using a porous hydrophobic PTFE membrane (Dasgupta et al., 1986) which was examined for SO$_2$ measurements. The relaxation time was far faster than the hydrophilic membrane. The overall efficiency of the hydrophobic membrane depends on the pore size, wall thickness, and the fractional surface porosity (Dasgupta, 1993). A number of approaches based on diffusion scrubbers have been developed for the measurement of trace gases (Tanner et al., 1986; Genfa et al., 1989, 1991; Lindgren and Dasgupta, 1989).

Tanner et al. (1986) employed a denuder scrubber of annular geometry for H_2O_2 measurements. In this design, the air flows through the annular space of the outer tube, while the scrubbing liquid flows through the concentrically placed membrane tube. The membrane tube was filled with solid PTFE to reduce the interior hold-up volume. Ozone was removed from the air stream prior to the denuder by premixing the air with NO. Based on the earlier work of Tanner et al. (1986), Dasgupta et al. (1988) developed a denuder scrubber of annular geometry for measurements of HCHO, H_2O_2, and SO_2 using a porous polypropylene membrane (CelgardX-20, surface porosity 0.4, pore size 0.02 μm, wall thickness 25 μm). The scrubbing liquids were $0.1M$ H_2SO_4, water, and 5 μM HCHO for HCHO, H_2O_2, and SO_2, respectively. A fluorescent detector was used and specific reagents were added to produce fluorescent products. The instrument switched between sample and zero periodically, and the full cycle ranged from 5 to 10 minutes. The detection limits were 100, 30 and 175 pptv for HCHO, H_2O_2, and SO_2, respectively. Continuous measurements of atmospheric ammonia have been made utilizing the porous membrane denuder scrubber of annular geometry connected to a fluorometric detector (Dasgupta et al., 1988). The method is based on the reaction of o-phthalaldehyde with primary amine (ammonia), which forms a fluorescent compound. Water was used as the scrubber liquid, and o-phthalaldehyde and sulfite solution were added to form a fluorescent product. The full cycle (sample to zero mode) was 5 minutes, and the detection limit was 45 pptv. Attempts have been made to measure acidic gases (HCl, HO_3, HONO, and SO_2) using the denuder scrubber with ion chromatography (Lindgren and Dasgupta, 1989; Dasgupta and Lindgren, 1989).

Generally, the collection efficiency of the above devices is relatively low and is also subject to change due to deposition of aerosol particles.

4.6 INTERCOMPARISON STUDIES

Several field intercomparison studies for measurement of atmospheric species have been reported in the literature. The majority of these studies were conducted in the United States and Canada (Spicer et al., 1982; Anlauf et al., 1985; Roberts et al., 1987; Hering et al., 1988; Solomon et al., 1988; Tanner et al., 1989; Dasch et al., 1989; Sickles et al., 1990b; Benner et al., 1991), and a few were carried out in continental Europe (Ferm et al., 1988; Harrison and Kitto, 1990). Most of these investigations focused on inorganic nitrate components. Results from these studies suggest that measurements of HNO_3 by the filter pack technique are overestimated, and the coated denuder thermal desorption approach may not produce credible measurements of NH_3 and HNO_3.

Spicer et al. (1982) reported results of HNO_3 employing four different methods. These were: filter pack (FP), denuder difference (DD), chemiluminescence, and Fourier transform infrared (FTIR) spectroscopy. All four methods showed general agreement. Anlauf et al. (1988) compared measurements of HNO_3 using FP, tungstic acid (TA), and tunable diode laser spectroscopy (TDLAS). Concentrations of HNO_3 obtained by the FP and TA techniques were 16% lower than those of TDLAS for daytime measurements. Nighttime measurements by the TA method were twofold higher than those of FP. In a separate study, nitric acid concentrations measured by the TA method were threefold higher than those of the FP (Roberts et al., 1987).

In September 1985, extensive Intercomparison methods for measurements of nitrogen species was performed in southern California. The results of these studies were published in 1988 (Hering et al., 1988). A wide range of techniques were employed. These included the FD, DD, AD, TDLAS, TA, and FTIR methods. Comparison of the means of the methods showed the FP to be 36% high, the DD to be 1% low, the AD to be 21% low, and TDLAS to be 13% low. On the few occasions when the concentrations of HNO_3 were within the detection limit of FTIR, general agreement between the FTIR and the FP, DD, AD, and TDLAS was obtained. It was inferred that decomposition of ammonium nitrate in the filter pack technique was the major source of the artifact. For the same campaign, Wiebe et al. (1990) compared the ammonia measurements. They found that most of the methods correlated highly with the FTIR method. They postulated that differences in the NH_3 results could be due to NH_3 absorption by inlet components, sampler location or any analytical measurement bias. Solomon et al. (1986) reported results of HNO_3 collected in Los Angeles over a period of one year using the DD and FP methods. Twenty-four hour samples were collected at 8 sites once every 6 days. The annual average concentration of HNO_3 measured by the FP were 80% higher than those of DD technique. The high temperatures and low relative humidities which prevail most of the year in southern California had a greater impact on the volatilization of ammonium nitrate in the filter pack, which contributed to the high HNO_3. Benner et al. (1991) compared HNO_3 measurements made by the FP and AD methods during January and February 1986 near Page, Arizona. The FP measurements exceeded those of AD by 11%. During September and October 1986, Sickles et al. (1990b) conducted an intercomparison study of measurements of HNO_3 employing the AD, FP, transition flow reactor (TFR), and TDLAS in Research Triangle Park, North Carolina. Results obtained by the TDLAS diverged from those obtained by FTIR techniques, while the other two methods (AD and FP) did not show a significant difference. The means of daily ratios of AD, FP, and TFR to TDLAS showed that the AD were 5% low, the FP 8% high and the TFR 36 to 76% high. During the winter of 1987/1988, Dasch et al. (1989) carried out an intercomparison study of measurements of HNO_3 in Warren, Michigan, using the FP and AD methods. The average concentrations obtained by the two methods were in good accord. This is expected, since the measurements were made in winter when the temperatures are low and relative humidities are high in Michigan.

Few intercomparison studies were conducted in continental Europe. Ferm et al. (1988) conducted an intercomparison study of total nitrate and total ammonium, using the denuder, filter pack, and total filter (gaseous and particulate collected on one filter) methods. Various institutions participated in the study. The results from all the participants were compared, and the relative standard deviations for the same technique were 15% and 20% for total nitrate and total ammonium, respectively. The relative standard deviations remained the same when different techniques were compared. It was inferred that all the methods employed provided comparable results. Two separate field Intercomparison studies of the filter pack and denuder methods for measurements of reactive gaseous (HCl, HNO_3, and NH_3) and particulate (Cl^-, NO_3^-, SO_4^{2-}, and NH_4^+) pollutants were conducted in southern England. In the first study, a tubular denuder (TD) was compared to the filter pack (Harrison and Kitto, 1990). The results are summarized in Table 4.1. Volatilization of ammonium nitrate and ammonium chloride was observed in the filter pack. This resulted in overestimation of HNO_3, NH_3, and HCl typically of 12%, 22%, and 13%, respectively, relative to the denuder method. In the second study, measurements obtained by the annular denuder were compared to those of the filter pack (Kitto, 1992). Results obtained from the second study were comparable to those of the first study. In this study, the filter pack results exceeded those of the annular denuder by 7%, 10%, and 17% for HNO_3, HCl, and NH_3, respectively. A comprehensive intercomparison study for measurements of inorganic nitrogen components was conducted near Rome, Italy, in September 1988. The participants were from countries of the European Economic Community (EEC). The samples from all the participants were analyzed by the hosting institution. Different sampling methods were employed, including FP, AD, TD, total filter (TF), thermal desorption denuder (TDD), and wet denuder (WD). Various types of filter pack and denuder configurations were used. Results of this study have been compiled by Allegrini et al. (1990). However, the results obtained from different methods were not compared. We have compared the results from the different methods employed. General agreement was found between various types of filter pack methods. The denuder methods showed comparable results. However, when the results of the filter pack methods were compared to those of the denuder techniques, nitric acid concentrations measured by the filter pack methods were higher than those of the denuder techniques, and consequently, particulate nitrate measured by the filter pack was lower than that obtained by the denuder methods.

TABLE 4.1
Summary of Airborne Concentrations and Ratio of Methods

Species	Range of Concentrations Denuder Sampler (neq m^{-3})	Mean and s.d of Ratios: Denuder/Filter Pack
SO_4^{2-} (particulate)	28–666	0.99 ± 0.06
NO_3^- (total)	20–630	0.97 ± 0.10
NO_3^- (particulate)	3–495	1.05 ± 0.11
HNO_3 (gas)	2–111	0.89 ± 0.22
Cl^- (total)	13–155	0.99 ± 0.13
Cl^- (particulate)	10–147	1.08 ± 0.13
HCl (gas)	3–45	0.85 ± 0.24
NH_{4+} (particulate)[*]	33–802	1.06 ± 0.08
NH_3 (gas)[$]	4–226	0.88 ± 0.21
NH_4^+ (particulate)[$]	22–735	1.00 ± 0.12
$NH_3 + NH_4^+$ (total)[$]	89–775	0.98 ± 0.12
$NH_3 + NH_4^+$ (total)[*]	75–875	0.99 ± 0.07

[*] Using Na_2CO_3 denuder
[$] Using H_3PO_3 denuder

Source: From Harrison R.M. and Kitto A.-M.N., 1990, Field inter-comparison of filter pack and denuder sampling methods for reactive gaseous and particulate pollutants. *Atmos. Environ.,* 24A, 2633-2640. With permission.

Measurements of NH_3/NH_4^+, SO_2/SO_4^{2-} and HNO_3/NO_3^- by tubular denuder and filter pack have been reported for five different sites in Denmark (Andersen and Hilbert, 1993; Andersen and Hovmand, 1994). In general, good agreement was found for the determination of the sum of NH_3 plus NH_4^+ by the two methods. Even though sampling artifacts were seen for the NH_3 determination by the filter pack, the concentrations were too low to affect the correlation with particulate NH_4^+. The filter pack typically underestimated HNO_3 concentrations compared to the denuder (Figures 4.9 and 4.10). This was probably due to adsorption on already collected particulate matter on the particle filter. However, an overestimation by the denuder due to interferences or particle deposition could not be excluded. Denuder measurements underestimated the total amount of SO_2 and SO_4^{2-} by about 10 to 15% compared to the filter pack.

It is clear that both filter pack and denuder sampling systems are potentially subject to systematic errors. While it is feasible to predict such errors from a theoretical basis, experimental measurements provide interesting insights into the behavior of sampling systems. For many species, and ammonia in particular, the usability of the filter pack determination will depend on the specific location, since measurements are influenced by climate, pollution, and emission characteristics.

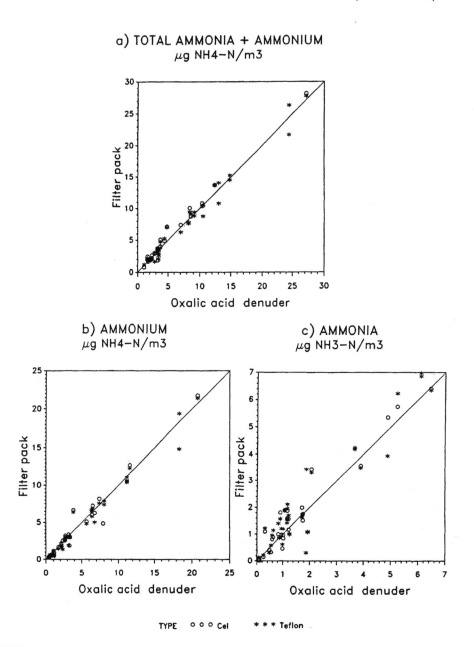

FIGURE 4.9 Filter pack with cellulose particle filter (Cel) and Teflon particle filter (Teflon) compared to denuder measurements of (a) total NH_3 and NH_4^+, (b) particulate NH_4^+ and (c) gaseous NH_3. (From Andersen H.V. and Hilbert G. (1993) Measurements of NH_3/NH_4^+ and HNO_3/NO_3^- by denuder and filter pack. National Environmental Research Institute, Denmark, *NERI Technical Report No. 73.*)

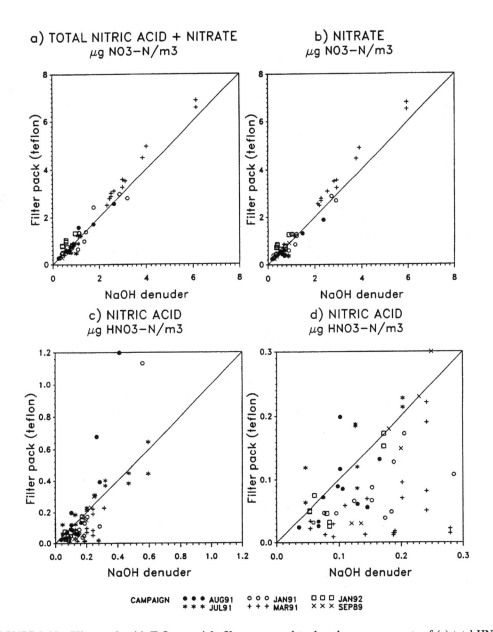

FIGURE 4.10 Filter pack with Teflon particle filter compared to denuder measurements of (a) total HNO_3 and NO_3^-, (b) particulate NO_3^-, (c) gaseous HNO_3 and (d) gaseous HNO_3 for low concentrations. (From Andersen H.V. and Hilbert G. (1993) Measurements of NH_3/NH_4^+ and HNO_3/NO_3^- by denuder and filter pack. National Environmental Research Institute, Denmark, *NERI Technical Report No. 73*.)

REFERENCES

Aghdair N., Sutton S., Hansen L.D., Eatough D.J., Lewis E.A., and Farber R.J., 1988, Diffuse denuders for the identification of artifacts in the sampling of semi-volatile particulate organic compounds. *Proceedings of the 81st APCA Annual Meeting,* Paper 88-53.3.

Allegrini I., Febo A., and Perrino C., 1988, Field intercomparison exercise on nitric acid and nitrate measurement: Method and data. *Air Pollution Research Report No. 22,* Commission of the European Communities, Rome, 18-24 September 1988.

Allegrini I., De Santis F., Di Palo V., Febo A., Perrino C., and Possanzini M., 1987, Annular denuder method for sampling reactive gases and aerosols in the atmosphere. *Sci. Tot. Environ.,* 67, 1-16.

Allegrini I., Febo A., and Perrino C., 1994, Role of nitrous acid in the oxidation processes in the Mediterranean area, in *Physico-chemical Behaviour of Atmospheric Pollutants,* Angeletti G. and Restelli G. (Eds.), Report EUR 15609/2 EN, European Commission.

Allen A.G., Harrison R. M., and Erisman J.W., 1989, Field measurements of the dissociation of ammonium nitrate and ammonium chloride aerosols. *Atmos. Environ.,* 23, 1591-1599.

Andersen H.V. and Hovmand M.F., 1994, Measurements of ammonia and ammonium by denuder and filter pack. *Atmos. Environ.,* 28, 3495-3512.

Andersen H.V. and Hilbert G., 1993, Measurements of NH_3/NH_4^+ and HNO_3/NO_3^- by denuder and filter pack. National Environmental Research Institute, Denmark, *NERI Technical Report No. 73.*

Anlauf K.G., Fellin P., Wiebe H.A., Schiff H.I., MacKay G.I., Braman R.S., and Gilbert R., 1985, A comparison of three methods for measurements of atmospheric nitric acid and aerosol nitrate and ammonium. *Atmos. Environ.,* 19, 325-333.

Anlauf K.G., MacTavish D.C., Wiebe H.A., Schiff H.J., and MacKay G.I., 1988, Measurements of atmospheric nitric by filter method and comparisons with tunable diode laser and other methods. *Atmos. Environ.,* 22, 1579-1586.

Appel B.R., Wall S.M., Tokiwa Y., and Haik H., 1980, Simultaneous nitric acid, particulate nitrate and acidity measurements in ambient air. *Atmos. Environ.,* 14, 549-554.

Appel B.R. and Haik H., 1981, Atmospheric particulate nitrate sampling errors due to reaction with particulate and gaseous strong acids. *Atmos. Environ.,* 15, 1087-1088.

Appel B.R., Tokiwa Y., and HaiK H., 1981, Sampling of nitrate in ambient air. *Atmos. Environ.,* 15, 283-289.

Appel B.R., Tokiwa Y., and Kothny E.L., 1983, Sampling of carbonaceous particles in the atmosphere. *Atmos. Environ.,* 17, 1787-1796.

Appel B.R., Povard V., and Kothny E.L., 1988a, Loss of nitric acid within devices intended to exclude coarse particles during atmospheric sampling. *Atmos. Environ.,* 22, 2535-2540.

Appel B.R., Tokiwa Y., Kothny E.L., Wu R., and Povard V., 1988b, Evaluation of procedures for measuring atmospheric nitric acid and ammonia. *Atmos. Environ.,* 22, 1565-1573.

Appel B.R., Tokiwa Y., Haik H., and Kothny E.L., 1984, Artifacts particulate sulfate and nitrate formation on filter media. *Atmos. Environ.,* 18, 409-416.

Appel B.R., 1990, Discussion on the fate of nitrous acid on selected collection surfaces. *Atmos. Environ.,* 24A, 717-718.

Appel B.R., Winer A.M., Tokiwa Y., and Biermann H.W., 1990, Comparison of atmospheric nitrous acid measurements by annular denuder and differential optical absorption systems. *Atmos. Environ.,* 24A, 611-616.

Asman W.A.H. and Janssen A.J., 1987, A long-range transport model for ammonia and ammonium ion for Europe. *Atmos. Environ.,* 21, 2099-2119.

Benner C.L., Eatough D.J., Eatough N.L., and Bhardwaja P., 1991, Comparison of annular denuder and filter pack collection of HNO_3, HNO_2, SO_2, and particulate-phase nitrate, nitrite and sulfate in the south-west desert. *Atmos. Environ.,* 25A, 1537-1545.

Bos R.J., 1980, Automatic measurement of atmospheric ammonia. *J. Air Poll. Control Assoc.,* 30, 1222-1224.

Bowen B.D., Levine S., and Epstein N., 1976, Fine particle deposition in laminar flow through parallel plate and cylindrical channels. *J. Colloid Interface Sci.,* 54, 375-390.

Braman R.S., Shelley T.J., and McClenny W.A., 1982, Tungstic acid for preconcentration and determination of gaseous and particulate ammonia and nitric acid in ambient air. *Annal Chem.,* 54, 358-364.

Brauer M., Koutrakis P., Wolfson J.M., and Spengler J.D., 1989, Evaluation of the gas collection of an annular denuder system under simulated atmospheric conditions. *Atmos. Environ.,* 23, 1981-1989.

Brauer M., Ryan P.B., Suh H.H., Koutrakis P., and Spengler J.D., 1990, Measurement of nitrous acid inside two research houses. *Environ. Sci. Technol.*, 24, 1521-1527.

Brauer M., Koutrakis P., Keeler G.J., and Spengler J.D., 1991, Indoor and outdoor concentrations of inorganic acidic aerosols and gases. *J. Air Waste Manage. Assoc.*, 41, 171-181.

Browning W.E. Jr., and Ackley R.D., 1962, Characterization of millimicron radioactive aerosols and their removal from gases. U.S. Atomic Energy Commission, TID-7641, pp. 130-147.

Cadle S.H., 1985, Seasonal variation in nitric acid, nitrate, strong aerosol acidity and ammonia in an urban area. *Atmos. Environ.*, 19, 181-188.

Caka F.M., Eatough D.J., Lewis E.A., Tang H., Hammond S.K., Leaderer B.P., Koutrakis P., Spengler J.D., Fasano A., McCarthy J., Ogden M.W., and Lewtas J., 1990, Intercomparison of sampling techniques for nicotine in indoor environments. *Environ. Sci. Technol.*, 24, 1196-1203.

Cobourn W.G., Hjukic-Husar J., and Husar R.B., 1980, Monitoring of sulphuric acid episodes in St. Louis, Missouri. *J. Geophys. Res.*, 85, 4487-4494.

Coutant R.W., Callahan P.J., Kuhlman M.R., and Lewis R.G., 1989, Design and performance of a high-volume compound annular denuder. *Atmos. Environ.* 23, 2205-2211.

Dasch J.M., Cadle S.E., Kennedy K.G., and Mulawa P.A., 1989, Comparison of annular denuders and filter packs for atmospheric sampling. *Atmos. Environ.*, 23, 2775-2782.

Dasgupta P.K., McDowell W.L., and Rhee J.-S, 1986, Porous membrane-based diffusion scrubber for sampling of atmospheric gases. *Analyst*, 111, 87-90.

Dasgupta P.K., Dong S., Hwang H., Yang H.C., and Genfa Z., 1988, Continuous liquid-phase fluorometry coupled to a diffusion scrubber for the real-time determination of atmospheric formaldehyde, hydrogen peroxide and sulfur dioxide. *Atmos. Environ.*, 22, 949-964.

Dasgupta P.K. and Lindgren P.F., 1989, Inlet pressure effects on the collection efficiency of diffusion scrubbers. *Environ. Sci. Technol.*, 23, 895-897.

Dasgupta P.K., 1993, Automated measurement of atmospheric trace gases, in *Measurement Challenges in Atmospheric Chemistry*. Edited by Newman L., ACS publication, Advances in Chemistry Series 232, Washington, D.C.

Davies T.J., Jones B.M.R. and Dollard G.J., 1989, Report on the Halvergate dry deposition study. September 1989, AERE Report, M 3778, Harwell.

Davis B.L., Johnson L.R., Johnson B.J., and Hammer R.J., 1988, A triple-path denuder instrument for ambient particulate sampling and analysis. *J. Atmos. Oceanic Tech.*, 5, 5-15.

Davis B.L., Deng Y., Anderson D.J., Johnson L.R., Detwiler A.G., Hodson L.L., and Sickles J.E., 1994, Limits of detection and artifact formation of sulfates and nitrates collected with a triple-path denuder. *Atmos. Environ.*, 28, 2485-2491.

Davis H.R. and Parkins G.V. (1970) Mass transfer from small capillaries with wall resistance in the laminar flow regime. *Appl. Sci. Res.*, 22, 20-30.

De Santis F., Febo A., Perrino C., Possanzini M., and Liberti A., 1985, Simultaneous measurements of HNO_3, HONO, HCl and SO_2 in air by means of high-efficiency annular denuder. *Proceedings of ECE workshop on Advancement in Air Pollution Monitoring Equipment and Procedures*. June 3-6, Freiberg, F.R.G.

De Santis F., Febo A., and Perrino C., 1988, Negative interference of teflon sampling devices in the determination of nitric acid and particulate nitrate. *Sci. Tot. Environ.*, 76, 93-99.

Durham J.L., Wilson W.E., and Bailey E.B., 1978, Application of an SO_2 denuder for continuous measurements of sulfur in submicrometric aerosols. *Atmos. Environ.*, 12, 883-886.

Durham J.L., Spiller L.L., and Ellestad T.G., 1987, Nitric acid-nitrate aerosol measurements by a diffusion denuder: A performance evaluation. *Atmos. Environ.*, 21, 589.

Eatough D.J., White V.F., Hansen L.D., Eatough N.L., and Ellis E.C., 1985, Hydration of nitric acid and its collection in the atmosphere by diffusion denuders. *Anal. Chem.*, 57, 743-748.

Eatough D.J., White V.F., Hansen, L.D., Eatough N.L., and Cheney J.L., 1986a, Identification of the gas phase dimethyl sulfate and mono methyl sulfuric acid in the Los Angeles atmosphere. *Environ. Sci. Technol.*, 9, 867-872.

Eatough D.J., Peterson M., Mooney R.L., Bartholomew D., Hansen L.D., Cheney J.L., and Eatough N.L., 1986b, The identification and chemistry of dimethyl sulfate in the atmosphere. *Proceedings of 7th World Clean Air Congress* August 25-29, Sydney, Australia.

Eatough D.J., Benner C.L., Mooney R.L., Bartholomew D., Steiner D.S., Hansen, L.D., Cheney J.L., and Lewis E.A., 1986c, Gas and particle nicotine in environmental tobacco smoke. *Proceedings, 79th Annual Meeting of the Air Pollution Control Assoc.*, June 22-27, Minneapolis.

Eatough D.J., Brutsch M., Lewi L., Hansen L.D., Lewis E.A., Eatough N.L., and Farber R.J., 1987a, Diffusion denuder sampling system for the collection of gas and particulate phase organic compounds. *Transaction of APCA specialty conference, Visibility Protection-Research and Policy Aspects.* Sept. 1986, Grand Teton National Park, WY, pp. 397-406.

Eatough D.J., Benner C.L., Bayona J.M., Caka F.M., Tang H., Lewis L., Lamb J.D., Lee M.L., Lewis E.A., and Hansen L.D., 1987b, Sampling the gas phase nicotine in environmental tobacco smoke with a diffusion denuder and a passive sampler. *Proceedings, EPA/APCA Symposium on Measurements of Toxic and Related Air Pollutants.* Air Pollution Control Assoc.

Eatough D.J., Lewis E.A, Lewi L., and Hansen L.D., 1988a, Identification of positive and negative artifacts in the sampling of particulate organic compounds. *Aerosol Sci. Technol.,* 10, 438-449.

Eatough N.L., McGregor S., Lewis E., Eatough D.J., Huang A.A., and Ellis E.C., 1988b, Comparison of six denuder methods and a filter pack for the collection of ambient $HNO_3(g)$, HONO(g) and SO_2 in 1985 NSMC study. *Atmos. Environ.,* 22, 1601-1618.

Eatough D.J., Benner C.L., Bayona J.M., Caka F.M., Richards G., Lamb, J.D., Lewis E.A., and Hansen L.D., 1989, The chemical composition of environmental tobacco smoke. I. Gas phase acids and bases. *Environ. Sci. Technol.,* 23, 679-687.

Eudy L.W., Thome F.A., Heavner D.L., Green C.R., and Ingebrethsen B.J., 1986, Studies on the vapor-particulate phase distribution environmental nicotine by selective trapping and detection methods. Proceedings, 79th Annual Meeting of the Air Pollution Control Assoc., June 22-27, Minneapolis.

Fan Q. and Dasgupta P.K., 1994, Continuous automated determination of atmospheric formaldehyde at the parts per trillion level. *Anal. Chem.,* 66, 551-556.

Febo A., De Santis F., and Perrino C., 1986, Measurements of atmospheric nitrous acid and nitric acid by means of annular denuder, in *Physico-chemical Behaviour of Atmospheric Pollutants,* Angeletti G. and Restelli G. (Eds.). D. Reidel Publishing Company, Dordrecht, Holland.

Febo A., Perrino C., and Cortilello M., 1993, A denuder technique for the measurement of nitrous acid in urban atmospheres. *Atmos. Environ.,* 27A, 1721-1728.

Ferm M., 1979, Method for determination of atmospheric ammonia. *Atmos. Environ.,* 13, 1385-1393.

Ferm M and Sjodin A., 1985, A sodium carbonate coated denuder for determination of nitrous acid in the atmosphere. *Atmos. Environ.,* 19, 979-983.

Ferm M., 1986, A Na_2CO_3-coated denuder and filter for determination of HNO_3 and particulate nitrate in the atmosphere. *Atmos. Environ.,* 20, 1193-1201.

Ferm M., Areskoug H., Hanssen J.E., Hilbert G., and Lattila H., 1988, Field intercomparison of measurement techniques for total NH_4 and total NO_3 in ambient air. *Atmos. Environ.,* 22, 2275-2281.

Forrest J., Tanner R.L., Spandau D.J., D'Ottavio T., and Newman L., 1980, Determination of total inorganic nitrate utilizing collection of nitric acid on NaCl-impregnated filters. *Atmos. Environ.,* 14, 137-144.

Forrest J., Spandau D.J., Tanner R.L., and Newman L., 1982, Determination of atmospheric nitrate and nitric acid employing a diffusion denuder with a filter pack. *Atmos. Environ.,* 16, 1473-1485.

Fox D.L., Stockburger L., Weathers W., Spicer C.W., Mackay G.I., Schiff H.I., Eatough D.J., Mortenson F., Hansen L.D., Shepson P.B., Kleindienst T.E., and Edney E.O., 1988, Intercomparison of nitric acid diffusion denuder methods with tunable diode laser absorption spectroscopy. *Atmos. Environ.,* 22, 575-585.

Genfa Z., Dasgupta P.K., and Dong S., 1989, Measurement of atmospheric ammonia. *Environ. Sci. Technol.,* 23, 1467-1474.

Genfa Z., Dasgupta P.K., and Cheng Y.S., 1991, Design of a straight inlet diffusion scrubber. Comparison of particle transmission with other collection devices and characterization for the measurement of hydrogen peroxide and formaldehyde. *Atmos. Environ.,* 25A, 2717-2729.

Goldan P.D., Kuster W.C., Albritton D.L., Fehsenfield F.C., Connell P.S., Norton R.B., and Huebert B.J., 1983, Calibration and tests of the filter- collection method for measuring clean-air, ambient levels of nitric acids. *Atmos. Environ.,* 17, 1355-1364.

Gormley P.G. and Kennedy M., 1948, Diffusion from a stream flowing through a cylindrical tube. *Proc. R. Irish Ac.,* 52, 163-169.

Grosjean D., 1982, Quantitative collection of total inorganic atmospheric nitrate on nylon filters. *Anal. Lett.,* 15, 785-796.

Grosjean D., 1985, Wall loss of gaseous pollutants in outdoor teflon chambers. *Environ. Sci. Technol.,* 19, 1059-1065.

Hansen, L.D., White V.F., and Eatough D.J., 1986, Determination of the gas phase dimethyl sulfate and monomethyl sulfuric acid. *Environ. Sci. Technol.*, 20, 872-878.

Hansen L.D., Eatough D.J., Cheney J.L., and Eatough N.L., 1987, The formation of dimethyl sulfate in power plants plumes. Proceedings of 80th Annual Meeting of the Air Pollution Control Assoc., June 21-26, New York.

Hanst P.L., 1971, Spectroscopic methods of air pollution measurements, *Advan. Environ. Sci. Technol.*, 2, 91-213.

Harrison R.M. and Kitto A.-M.N., 1990, Field inter-comparison of filter pack and denuder sampling methods for reactive gaseous and particulate pollutants. *Atmos. Environ.*, 24A, 2633-2640.

Harrison R.M. and Allen A.G., 1990, Measurements of atmospheric HNO_3, HCl, and associated species on a small network in Eastern England. *Atmos. Environ.*, 24A, 369-376.

Harrison R.M., Sturges W.T., Kitto A.-M.N., and Li Y., 1990, Kinetics of evaporation of ammonium chloride and ammonium nitrate aerosol. *Atmos. Environ.*, 24A, 1883-1888.

Hering S.V., Lawson D.R., Allegrini I., Febo A., Perrino, C., Possanzini M., Sickles J.E. II, Anlauf K.G., Wiebe A., Appel B.R., John W., Ondo J., Wall S., Braman R.S., Sutton R., Cass G.R., Solomon P.A., Eatough D.J., Eatough N.L., Ellis E.C., Grosjean D., Hicks B.B., Litomack J.D., Horrocks J., Knapp K.T., Ellestad T.G., Paur R.J., Mitchell W.J., Pleasant M., Peake E., MacLean A., Pierson W.R., Brachaczek W., Schiff H.I., MacKay G.I., Spicer C.W., Stedman D.H., Winer A.M., Biermann H.W., and Tuazon E.C., 1988, The nitric acid shootout: field comparison of measurement methods. *Atmos. Environ.*, 22, 1519-1540.

Hildemann L.M., Russell A.G., and Cass G., 1984, Ammonia and nitric acid concentrations in equilibrium with atmospheric aerosols: Experimental vs. theory. *Atmos. Environ.*, 18, 1737-1750.

Hitchcock D.R., Spiller L.L., and Wilson W.E., 1980, Sulphuric acid aerosols and HCl release in coastal atmospheres: Evidence of rapid formation of sulphuric acid particles. *Atmos. Environ.*, 14, 165-182.

Huebert B.J. and Lazrus A.L., 1980, Tropospheric gas-phase and particulate nitrate measurements. *J. Geophys. Res.*, 85, 7322-7328.

Johnson N.D., Barton S.C., Lane D.A., and Schroeder W.H., 1985, Development of gas/particle fractionating sampler for chlorinated organics. *Proceedings of the 78th Annual Meeting of the Air Pollution Control Assoc.*, June 16-21, Detroit.

Johnson N.D., Barton S.C., Thomas G.H.S., Lane D.A., and Schroeder W.H., 1986, Field evaluation of a diffusion denuder based gas/particle sampler for chlorinated organic compounds. *Proceedings of the 79th Annual Meeting of the Air Pollution Control Assoc.*, June 24-29, San Francisco.

Kenne W.C., Talbot R.W., Andreae M.O., Beecher K., Berresheim H., Castro M., Farmer C., Galloway J.N., Hoffmann M.R., Li S.-M., Maben J.R., Munger J.W., Norton R.B., Pszenny A.A.P., Puxbaum H., Westberg H., and Winiwarter W., 1989, An intercomparison of measurement systems for vapor and particulate phase concentrations of formic and acetic acids. *J. Geophys. Res.*, 94, 6457-6471.

Keuken M.P., Schoonebeek C.A.M., Wensveen-Louter A., and Slanina J., 1988, Simultaneous sampling of NH_3, HNO_3, HCl, SO_2, and H_2O_2 in ambient air by a wet annular denuder system. *Atmos. Environ.*, 22, 2541-2548.

Keuken M.P., Wayers-Ijpelaan A., Mols J.J., Otjes R.P., and Slanina J., 1989, The determination of ammonia in ambient air by an automated thermodenuder system. *Atmos. Environ.*, 23, 2177-2185.

Kitto, A.M., 1991, Ph.D. Thesis, Essex University, U.K.

Kitto A.M. and Harrison R.M., 1992a, Nitrous and nitric acid measurements at sites in south-east England. *Atmos. Environ.*, 26A, 235-241.

Kitto A.M. and Harrison R.M., 1992b, Processes affecting concentrations of aerosol strong acidity at sites in eastern England. *Atmos. Environ.*, 26A, 2389-2399.

Kitto A.M., 1992, Intercomparison of annular denuder and filter pack sampling methods for collection of acidifying species in south east England. *J. Aerosol Sci.*, 23, S699-S702.

Klockow D., Jablonski B., Nießner R., 1979, Possible artifacts in filter sampling of atmospheric sulphuric acid and acidic sulfates. *Atmos. Environ.*, 13, 1665-1676.

Klockow D., Nießner R., Malejczyk M., Kiendl H., vom Berg B., Keuken M.P., Wayers-Ypellan A., and Slanina J., 1989, Determination of nitric acid and ammonium nitrate by means of a computer-controlled thermodenuder system. *Atmos. Environ.*, 23, 1131-1138.

Koutrakis P. Wolfson J.M., Slater J.L., Brauer M., Spengler J.D., Stevens R.K., and Stone C.L., 1988, Evaluation of an annular denuder-filter pack system to collect acidic aerosols and gases. *Environ. Sci. Technol.*, 22, 1463-1468.

Koutrakis P., Wolfson J.M., and Spengler J.D., 1988, An improved method for measuring aerosol strong acidity: Results from nine-month study in St. Louis, Missouri and Kingston, Tennessee. *Atmos. Environ.*, 22, 157-162.

Koutrakis P., Fasano A.M., Slater J.L., Spengler J.D., McCarly, and Leader B.P., 1989, Design of personal annular denuder sampler to measure atmospheric aerosols and gases. *Atmos. Environ.*, 23, 2767-2773.

Koutrakis P., Sioutas C., Ferguson S.T., Wolfson, J.M., Mulik J.D., and Burton R.M., 1993, Development and evaluation of a glass honeycomb denuder/filter pack system to collect atmospheric gases and particles. *Environ. Sci. Technol.*, 27, 2497-2501.

Lane D.A., Johnson N.D., Barton S.C., Thomas G.H.S., and Shroeder W.H., 1988, Development and evaluation of a novel gas and particle sampler for semivolatile chlorinated organic compounds in ambient air. *Environ. Sci. Technol.*, 23, 2767-2773.

Larson T.V. and Tayler G.S., 1983, On the evaporation of ammonium nitrate aerosol. *Atmos. Environ.*, 17, 2489-2495.

Lawrence J.E. and Koutrakis P., 1994, Measurements of atmospheric formic and acetic acids: methods evaluation and results from field studies. *Environ. Sci. Technol.*, 28, 957-964.

Lewin E.F. and Hansen K.A., 1984, Diffusion denuder assembly for collection and determination gases in air. *Anal. Chem.*, 56, 842-847.

Lewin E.E., De Pena R.G., and Shimshock J.P., 1986, Atmospheric gas and particle measurement at a rural northeastern U.S. site. *Atmos. Environ.*, 20, 59-70.

Lewis D.A., Colbeck I. and Mariner D.C., 1994, Diffusion denuder method for sampling vapor-phase nicotine in mainstream tobacco smoke. *Anal. Chem.*, 66, 3525-3527.

Li Y. and Harrison R.M., 1990, Comparison of indoor and outdoor concentrations of acid gases, ammonia and their associated salts. *Environ. Technol.*, 11, 315-320.

Lindgren P.F. and Dasgupta P.K., 1989, Measurement of atmospheric sulfur dioxide by diffusion scrubber coupled ion chromatography. *Anal. Chem.*, 61, 19-24.

Lindqvist F., 1985, Determination of nitric acid in ambient air by gas chromatography/photoionization detector after collection in a denuder. *J. Air Poll. Control Assoc.*, 35, 19-23.

McClenny W.A., Kaneda K., Yanaka T., and Sugai R., 1982, Tungstic acid technique for monitoring nitric acid and ammonia in ambient air. *Anal. Chem.*, 54, 365-369.

Mulawa P. and Cadle S.H., 1985, A comparison of nitric acid and particulate nitrate measurements by the penetration and denuder difference methods. *Atmos. Environ.*, 19, 1317-1324.

Niessner R. and Klockow D., 1980, A thermoanalytical approach to speciation of atmospheric strong acids. *Int. J. Environ. Anal. Chem.*, 8, 163-175.

Nodop K. and Hanssen J.E., 1986, Field intercomparison of sampling and analytical methods for SO_2 and SO_4 in ambient air, in *Physico-chemical Behaviour of Atmospheric Pollutants*, Angeletti G. and Restelli G. (Eds.). D. Reidel Publishing Company, Dordrecht, Holland.

Nolan J.J. and Guerrini V.H., 1935, Diffusion coefficients and velocities of fall in air of atmospheric condensation nuclei. *Pro. R. Ir. Acad.*, 43, 5-24.

Okita T. Morimoto S., Izawa S., and Konno W., 1976, Measurement of gaseous and particulate nitrates in the atmosphere. *Atmos. Environ.*, 10, 1085-1089.

Penkett S.A. Sandalls F.J., and James B.M.R., 1977, PAN measurements in England; analytical methods and results. VDI-Ber 270, 47-54.

Perrino C., De Santis F., and Febo A., 1988, Uptake of nitrous acid and nitrogen oxides by nylon surfaces: Implication of nitric acid measurements. *Atmos. Environ.*, 22, 1925-1930.

Perrino C., De Santis F., and Febo A., 1990, Criteria for the choice of a denuder sampling technique devoted to the measurements of atmospheric nitrous and nitric acids. *Atmos. Environ.*, 24A, 617-626.

Platt U., Perner D., Harris G.W., Winer A.M., and Pitts J.N., 1980, Observation of HONO in an urban atmosphere by differential optical absorption. *Nature*, 285, 312-314.

Pio C.A. and Harrison R.M., 1987, Vapor pressure of ammonium chloride aerosols: Effect of temperature and humidity. *Atmos. Environ.*, 21, 2711-2715.

Possanzini M., Febo A., and Liberti A., 1983, New design of high-performance denuder for sampling of atmospheric pollutants. *Atmos. Environ.*, 17, 2065-2610.

Possanzini M., Cicciolo P., di Palo V., and Draisci R., 1987, Determination of low boiling aldehydes in air and exhaust gases by using annular denuder combined with HPLC techniques. *Chromatographia*, 23, 829-834.

Possanzini M., Palo V.Di, and Liberti A., 1988, Annular denuder method for determination of H₂O in the ambient atmosphere. *Sci. Tot. Environ.*, 77, 203-214.

Rapsomanikis S., Wake M., Kitto A.-M.N., and Harrison R.M., 1988, Analysis of atmospheric ammonia and particulate ammonium by a sensitive fluorescence method. *Environ. Sci. Technol.*, 22, 948-952.

Roberts J.M., Norton R.B., Goldan P.D., and Fehsenfeld F.C., 1987, Evaluation of the tungsten oxide denuder tube technique as a method for the measurement of low concentrations of nitric acid in the troposphere. *J. Atmos. Chem.*, 5, 217-238.

Romer F.G., van den Beld L., van Elzakker B.G., and Memmen M.G., 1994, Automated thermodenuder system for continuous measurement of atmospheric ammonia concentrations, in *Physico-chemical Behaviour of Atmospheric Pollutants*, Angeletti G. and Restelli G. (Eds.), Report EUR 15609/2 EN, European Commission.

Rosenberg C., Winiwarter W., Gregori M., Pech G., Casensky V., and Puxbaum H., 1988, Determination of inorganic and organic volatile acids, NH₃, particulate SO₄, NO₃ and Cl in ambient air with an annular diffusion denuder system. *Fresenius Z. Anal. Chem.*, 331, 1-7.

Schiff H.C., Harris G.W., and Mackay G.Z., 1988, in *The Chemistry of Acid Rain: Sources and Atmospheric Processes*. Johnson R.W. and Gordon G.E. (Eds.), ACS Symposium Series, 349, American Chemical Society, Washington, D.C., pp. 274-288.

Shaw R.W. Jr., Stevens R.K., and Bowermaster J.W., 1982, Measurements of atmospheric nitrate and nitric acid: The denuder difference experiment. *Atmos. Environ.*, 16, 845-853.

Sickles J.E., II and Hodson L.L., 1989, Fate of nitrous acid on selected collection surfaces. *Atmos. Environ.*, 22, 2321-2324.

Sickles J.E., II, Bach W.D., and Spiller L.L., 1984, Comparison of several techniques for determining dry deposition flux, in *Precipitation Scavenging, Dry Deposition and Resuspension*. Pruppacher H.R., Semonin R.G., and Slinn W.G.N. (Eds.), Elsevier, New York.

Sickles, J.E., II, Perrino C., Allegrini I., Febo A., Possanzini M., and Paur R.J., 1988, Sampling and analysis of ambient air near Los Angeles using an annular denuder system. *Atmos. Environ.*, 22, 1619-1625.

Sickles J.E., II, Hodson L.L., Rickman E.E., Seager Jr. M.L., Hardison D.L., Turner A.R., Sokol A., Estes E.D., and Paur R.J., 1989, Comparison of the annular denuder system and the transition flow for measurements of selected dry deposition species. *J. Air Poll. Control Assoc.*, 39, 1218-1224.

Sickles J.E., II, Hodson L.L., McClenny W.A., Paur R.J., Ellestad T.G., Mulik J.D., Anlauf K.G., Wiebe H.A., Mackay G.I., Schiff H.I., and Bubacz D.K., 1990, Field Comparison of methods for the measurements of gaseous and particulate contributors to acidic dry deposition. *Atmos. Environ.*, 24A, 155-165.

Sickles J.E., II, and Hodson L.L., 1990, Discussion on the fate of nitrous acid on selected collection surfaces. *Atmos. Environ.*, 24A, 717-718.

Simon P.K., Dasgupta P.K., and Vecera Z., 1991, Wet effluent denuder coupled liquid/ion chromatography systems. *Anal Chem.*, 63, 1237-1242.

Simon P.K. and Dasgupta P.K., 1993, Wet effluent denuder coupled liquid/ion chromatography systems: annular and parallel plate denuders. *Anal Chem.*, 65, 1134-1139.

Simon P.K. and Dasgupta P.K., 1995, Continuous automated measurement of the soluble fraction of atmospheric particulate matter. *Anal Chem.*, 67, 71-78.

Sjodin A. and Ferm M., 1985, Measurements of nitrous acid in an urban area. *Atmos. Environ.*, 19, 985-992.

Sjodin A., 1988, Studies of the diurnal variation of nitrous acid in urban air. *Atmos. Environ.*, 22, 1086-1089,

Slanina J., Lamoen-Doornenbal L.V., Lingerak W.A., and Meilof W., 1981, Application of a thermodenuder analyser to the determination of H₂SO₄, HNO₃ and NH₃ in air. *Int. J. Environ. Anal. Chem.*, 9, 59-70.

Slanina J., Schoonebeek C.A. Klockow D., and Niessner R., 1985, Determination of sulfuric acid and ammonium sulfate by means of a computer-controlled thermodenuder system. *Anal. Chem.*, 57, 1955-1960.

Slanina J., Wyer G.P., Kieskamp W.M. and ten Brink H.M., 1994, Analytical chemistry in the troposphere, in *Physico-Chemical Behaviour of Atmospheric Pollutants*. Proceedings of the sixth European Symposium. Report EUR 15609/1 EN.

Smith B., Watman J., and Fisher B., 1969, Interactions of airborne particles with gases. *Environ. Sci. Technol.*, 3, 558-562.

Solomon P.A., Larson S.M., Fall T., and Cass G., 1988, Basinwide nitric acid and related species concentrations observed during the Claremont Nitrogen Comparison Study. *Atmos. Environ.*, 22, 1587-1595.

Spicer C.W., Howes Jr., J.E., Bishop T.A., Arnold L.H. and Stevens R.K., 1982, Nitric acid measurement methods: An intercomparison. *Atmos. Environ.*, 16, 1487-1500.

Stelson A.W. and Seinfeld J.H., 1982, Relative humidity and temperature dependence of the ammonium nitrate dissociation constant. *Atmos. Environ.*, 16, 983-992.

Stevens R.K., Dzubay T.G., Russwurm G.M., and Rickel D., 1978, Sampling and analysis of atmospheric sulfates and related species. *Atmos. Environ.*, 12, 55-68.

Sturges W.T. and Harrison R.M., 1989, The use of nylon filters to collect HCl: Efficiencies, interferences and ambient concentrations. *Atmos. Environ.*, 23, 1987-1996.

Taira M. and Kanda Y., 1993, Wet effluent diffusion denuder for sampling of atmospheric gaseous nitric acid. *Anal. Chem.*, 65, 3171-3173.

Talbot R.W., Vigen A.S. and Harriss R.C., 1990, Measuring tropospheric HNO_2, problems and prospects for nylon filter and mist chamber techniques. *J. Geophys. Res.*, 95, 7553-7561.

Tan C.W. and Hsu C.J., 1971, Diffusion of aerosols in laminar flow in a cylindrical tube. *J. Aerosol Sci.*, 21, 117-124.

Tanner R.L., Markovits G.Y., Ferreri E.M., and Kelly T.J., 1986, Sampling and determination of gas-phase hydrogen peroxide following removal of ozone by gas-phase reaction with nitric oxide. *Anal Chem.* 58, 1857-1865.

Tanner R.L., Kelly T.J., Dezaro D.A., and Forrest J., 1989, A comparison of filter, denuder, and real-time chemiluminescence techniques for nitric acid determination in ambient air. *Atmos. Environ.*, 23, 2213-2222.

Townsend J.A., 1900, The diffusion of ions into gases. *Phils. Trans. R. Soc. Lond., Ser. A*, 193, 129-158.

Tuazon E.C., Graham R.A., Winer A.M., Easton R.R., Pitts Jr., J.N., and Hanst P.L., 1978, A kilometre pathlength Fourier-transform infrared system for the study of trace pollutants in ambient and synthetic atmosphere. *Atmos. Environ.*, 12, 865-875.

Vecera Z. and Dasgupta P.K. 1991, Measurements of atmospheric nitrous acid with a wet effluent diffusion denuder and low pressure ion chromatography-post column reaction detector. *Anal Chem.*, 63, 2210-2216.

Vermetten A.W.M., Asman W.A.H., Buijsman E., Mulder W., Slanina J., and Waijers- IJpelaan A., 1985, Concentration of NH_3 and NH_4 over The Netherlands. *VDI Berichte* 560, 241-251.

Waldman J.M., Lioy P.J., Thurston G.D., and Lippmann M., 1990, Spatial and temporal patterns in summertime sulfate aerosol acidity and neutralization with metropolitan area. *Atmos. Environ.*, 24A, 115-126.

Wall S.M., John W., and Ondo J.L., 1988, Measurements of aerosol size distribution for nitrate and major ionic species. *Atmos. Environ.*, 22, 1649-1656.

Wiebe H.A., Anlauf K.G., Tuazon E.C., Winer A.M., Biermann H.W., Appel B.R., Solomon P.A., Cass G.R., Ellestad T.G., Knapp K.T., Peake E., Spicer C.W., and Lawson D.R., 1990. A comparison of measurements of atmospheric ammonia by filter packs, transition-flow reactors, simple and annular denuders and fourier transform infrared spectroscopy. *Atmos. Environ.*, 24A, 1019-1028.

Williams E.L. and Grosjean D., 1988, Annular denuder for atmospheric oxidants. Presented before the division of environmental chemistry, American Chemical Society, Sept. 25-30, Los Angeles.

Williams E.L. and Grosjean D., 1990, Removal of atmospheric oxidants with annular denuder. *Environ. Sci. Technol.*, 24, 811-814.

Winiwarter W., 1989, A calculation procedure for the determination of the collection efficiency in annular denuders. *Atmos. Environ.*, 23, 1997-2002.

Winiwarter W., Puxbaum H., Fuzzi, S., Facchinin M.C., Orsi G., Beltz N., Enderle K., and Jaeschke W., 1988, Organic acid gas and liquid-phase measurements in Po Valley fall–winter conditions in the presence of fog. *Tellus*, 40B, 348-357.

Wyers G.P., Otjes R.P., Vermeulen A.T., de Wild P.J., and Slanina J., 1991, Measurements of vertical concentration gradients of ammonia by continuous flow denuders. EC Air Pollution Research Report 39, 173-178.

Yamasaki J., Kuwata, and Miyamoto H., 1982, Effects of ambient temperature on aspects of airborne polycyclic aromatic hydrocarbons. *Environ. Sci. Technol.*, 16, 189-194.

Ye Y., Tsai, C.-J., Pui D.Y.H., and Lewis C.W., 1991, Particle transmission characteristics for an annular denuder ambient sampling system. *Aerosol Sci. Technol.*, 14, 102-111.

5 Aerosol Analysis by a PIXE System

Václav Potoček

CONTENTS

5.1 INTRODUCTION

The aim of this chapter is to summarize an experience with PIXE in analyses of aerosol samples. There are, of course, many papers on this topic, including notable reviews, because the analysis of aerosols is one of the most frequent applications of the PIXE technique. Most of the pertinent articles are listed in Chapter 12 of Reference 1, in which the state of the art in aerosol applications of PIXE is well described. Many papers appear in the regular conferences on PIXE and its applications. They have been held every third year since 1977, and their proceedings are published as special issues of *Nuclear Instruments and Methods in Physics Research B*. Nevertheless, some aspects of PIXE seem to be disregarded even though they are important in the interpretation of the data. Some of these aspects are discussed here.

The studies of ambient aerosol play — or should play — an important role in many east European countries, and data and results from each of them may contribute to the study, or even to the prevention, of air pollution and its transport over national borders. Considering the urgent need for comparable data, some specific features of the research work in eastern Europe are not often mentioned in published articles and probably are not fully appreciated by readers. Some of these are mentioned here also.

Following is a short summary of the PIXE method as a tool in the analysis of ambient aerosol, followed by a more detailed description of one particular laboratory and its activities. It should serve as an illustration and, perhaps, give some inspiration. The last paragraph points out some aspects of a specific feature of PIXE — the possibility of remote and shared data processing.

5.2 THE PIXE METHOD

PIXE (particle-induced X-ray emission) as an analytical method was first reported at 1970 from Lund, Sweden.[2] PIXE has grown rapidly since that time and within a few years it acquired great popularity as a powerful tool in multielemental analyses.

The technique is based on the spectral analysis of the X-rays emitted from matter under bombardment by energetic ions. Most frequently they are charged particles, i.e., ions with no remaining electrons, as protons or alpha particles. Since discrete X-ray energies are strictly determined by the atomic properties of the matter under study, they are called *characteristic X-rays*. It is on these characteristic X-rays that the element analysis is based. PIXE is related to two well-known X-ray analytical methods that are frequently used with commercially available devices: X-ray fluorescence (XRF) and electron microprobe (EMP). The characteristic X-rays are emitted under X- or gamma-photons in the former, and electrons are the incident particles in the latter.

The incident electron beam that provides object imaging in electron microscopy is effectively employed in EMP, provided the microscope is equipped with X-ray detection and signal procession devices and with an appropriate software to process the data. The great advantage is the complementarity of the visual and the analytical information. Nevertheless, the intensive bremsstrahlung of the light electrons, which lose much more incident energy in collisions with the target compared with protons, makes the detection limits in EMP (i.e., the ratio of the minimum informative signals to the background) three orders of magnitude higher than those of PIXE. Therefore, far smaller concentrations of the elements under study can be detected with PIXE than with EMP.

To compare PIXE and XRF, both methods are fast, nondestructive, multielement, and sensitive analytical tools that are almost ideal for the elemental analysis of an aerosol. XRF seems to be even less destructive because there are no thermal effects in the targets due to the loss of energy of the incident radiation. The advantage of PIXE is that extremely small targets can be effectively analyzed, compared with XRF, because of the well-defined dimension of the incident proton beam spot on the target. The spot can include all the available current of the incident beam in PIXE. On the other hand, the smaller the analyzed spot in XRF, the greater is the fraction of the incident radiation that is lost at the collimator. The detection limits are similar in both methods, but in XRF, they are more dependent on the relation between the energy of the incident radiation and the energy to be detected.

A basic PIXE set-up consists of a suitable accelerator, such as Van de Graaff or cyclotron, ion beam optics, a target holder, an energy-dispersive X-ray detector, such as Si(Li), Ge(Li), or HPGe crystals, and signal processing electronic devices with a multichannel analyzer at the end of the chain. It is completed with the vacuum target chamber and/or the beam exit window. The collected X-ray spectra are processed using suitable software. In principle, it is necessary to state net spectral intensities in a so-called fitting procedure and derive either relative or absolute element concentrations from them. There are several approaches in how to do this, but the results should be independent of the method. Unfortunately, this cannot be taken as a rule because some software packages are not universal, and are merely well-suited only to certain types of samples. If another type of spectra was processed with such a package, a notable error could arise because of an improper matrix correction.

According to characteristic X-ray energies, the elements with atomic numbers $Z > 15$ can be analyzed using PIXE with no major difficulty. The elements with atomic numbers $10 < Z < 16$ must be analyzed with special care; the proper description of the X-ray attenuation in both the analyzed matter and the analytical set-up is of vital importance for reliable results that are very sensitive to it within the corresponding region of X-ray energies. The elements with lower atomic numbers cannot be analyzed with standard PIXE equipment.

The matter to be analyzed should be named the *sample,* and the object prepared for PIXE analysis should be referred to as the *target.* They are almost identical in ambient aerosol analyses because no other target preparation is necessary after the sampling except to put the filter or the backing foil with the aerosol on the target holder. If some other techniques, such as fixation, metallization, or even mineralization, are used, the possibility of some contamination and/or loss of elements must be taken into account. The results obtained from PIXE, after some target preparation that dissolves the sample, may differ from the results obtained with unchanged matter if there are elements concentrated in the surface layers, i.e., if the average element concentrations in the whole matter differ from the concentrations on its surface. The X-rays originate in the layer with a thickness of some 20 micrometers in the proton-induced X-ray emission, and even less in the alpha-induced X-ray emission, depending on the target composition that affects both the loss energy of the incident particles (i.e., their ability to cause the characteristic X-rays) and the attenuation of the X-radiation.

PIXE can be processed either in a vacuum or in an atmosphere. In the latter case, we speak of external beam PIXE, because the incident particles are carried out from the vacuum through an exit window. It must be a foil that can withstand the great difference in pressure on each side and under the thermal and radiation influence of the incident beam, and at the same time it must be thin enough to remove a minimum of the incident particle energy. There are several problems in external beam PIXE, especially the safety of the accelerator if the exit window is disrupted. The results are more uncertain because of a loss and a straggling of the incident particle energy in the exit foil and in the atmosphere, as well as the spread of the beam. External beam PIXE is used for wet samples, for instance. Moreover, it can be useful if the targets are insulators, because the atmosphere can compensate for both the locally accumulated electric charge and the local heating at the incident beam spot. It is also necessary to use external beam PIXE if the analyzed objects are too large to be put into the vacuum chamber. Aerosol targets are, for the most part, very compatible with a vacuum.

The other criterion is the size of the incident particle beam. Broad-beam PIXE uses the diameter of several millimeters collimated from a broader beam that repeatedly passes a diffuser to acquire the uniform distribution of the beam current over the spot at the target. On the other hand, micro-PIXE works with a beam that is focused to micrometers. While broad-beam PIXE enables one to obtain averaged element concentrations, micro-PIXE is a method for studying single aerosol particles. One important difference between the two methods is that current densities are much higher in micro-PIXE. Therefore, the targets must be protected against thermal and radiative damage, and local charging must be effectively suppressed in a micro-PIXE analysis. Since micro-PIXE cannot be performed with an external beam that cannot be focused to a defined dimension, the targets must be metallized and the beam current must be much lower than in broad-beam PIXE. The spectra with the same signal-to-noise ratio are acquired in long time exposures in micro-PIXE, and, therefore, less data can be obtained in the same time. There is another mode in micro-PIXE that is also well known in EMP. It is the scanning analysis when the incident beam passes an area at the target surface and maps of element distributions are created. Since there is no direct connection to the visual information in PIXE, the identification of an area in the map with a separate particle is not so easy as in EMP.

The last criterion in PIXE is the target thickness. According to the definition, a target is thin if it does not change the energy of the incident beam and if it does not absorb the X-rays. Otherwise, it must be considered to be of an intermediate thickness, or even an infinitely thick target. The results are quite reliable for thin targets because no prediction concerning the composition of the matrix must be done before the data are processed. The concentrations may be expressed in absolute units in the thin targets with no standards because the inevitable parameter — the integrated beam current — can be measured directly using the Faraday cup. This cannot be done with a thick target nor with the external incident particle beam. Aerosol samples can be considered thin if the thin backing foils or filters can be used and if the particles are limited in size to micrometers. If the particles are thicker (even 5 micrometers or more) and/or if thicker filters are used, which can incorporate small particles under their surfaces (such as nitrocellulose filters), the targets should not be supposed to be thin. Data processing is notably more complex and time consuming if thick targets are analyzed. The drop in energy of the incident particles in the target must be taken into account as well as the X-ray attenuation and the secondary X-ray fluorescence induced by the absorbed characteristic X-rays. All the effects depend on the element composition of the target, i.e., on the element concentrations that are to be determined in the analysis and on the low-Z-element concentrations that cannot be determined with PIXE. Therefore, an iterative data analysis should be used, and an input estimation of matrix element concentrations is necessary.

PIXE can be performed together with other ion beam techniques, such as PIGE (particle-induced gamma-ray emission) and RBS (Rutherford backscattering). Provided the analytical set-up is equipped with the appropriate detectors and signal procession devices, the spectra can be collected at the same time, and the complementary information can be employed in the interpretation of the results and even in data analysis.

The arrangement of a PIXE laboratory, i.e., its analytical set-up complete with the software and methodology, can be referred to as the PIXE system. To obtain absolute concentrations, the set-up must be calibrated using correct standards. Nevertheless, the entire system should be validated in some interlaboratory calibration to confirm the reliability of the output data.

5.2.1 PIXE ANALYSES OF THE ATMOSPHERIC AEROSOL

It should be said that PIXE is, in general, a very suitable tool for aerosol studies. It is a nondestructive, fast, and simultaneously multielemental analytical technique that requires extremely low amounts of the material to be analyzed and extremely low target areas. Element concentrations can be found for city aerosol samples from the air volume of some hundreds of liters and the sample area can be as small as 1 cm in diameter, or even less. Using such small amounts of the material to be analyzed can make other analytical techniques less effective. Consequently, the samples can be collected in short intervals, and variations in element concentrations within one day can be measured. Since extremely small spots of the material are acceptable in PIXE, the aerosol particles can be divided into size fractions by a cascade impactor, and coarse, respirable, and fine aerosols can be analyzed separately. Typically, the concentrations of 10 to 20 elements, concentration ratios, and enrichment factors can be determined from a single PIXE spectrum. Moreover, each sample can be stored to be reanalyzed later.

Aerosol samples are very suitable for PIXE analysis because all the techniques priorities can be employed. Since there is no other target preparation necessary for PIXE analysis except the collection of aerosol onto a filter or a backing foil, the method enables us to avoid or to considerably reduce the possibility of contamination or the loss of elements. Aerosol samples can often be considered as thin targets, i.e., the determination of absolute element concentrations is possible and can even be performed with no standards. This feature is very important because, in fact, there are no universal standards of aerosol samples because of the great variety in physical and chemical properties of aerosols from different sources.

The use of PIXE in aerosol studies has always depended on the particular conditions in laboratories that were, for the most part, established in nuclear research departments. Therefore, researchers were at first interested in the development of the methodologies and other problems related to the technique rather than the application. On other hand, since the very beginning, the filters or backing foils covered with aerosol particles have been used as very suitable targets because they could be easily prepared and analyzed. Moreover, the impressive possibilities of PIXE could easily be demonstrated with these alone. Therefore, laboratories soon began to cooperate with institutes specializing in and responsible for air pollution control and research.

There are several different approaches to the study of aerosols in the ambient air by PIXE. Many papers have appeared that include results and interpretations. Nevertheless, experience shows that they must be read carefully and critically if they are to be used in a comparative study. They often concentrated on the description of the device and methodology of the analysis, while information about the sampling itself was poor or nonexistent. There are different needs and strategies according to the aim of the study. The summaries of three types of ambient aerosol research follow.

5.2.1.1   A Study of Long-Range Transport of the Dispersed Matter

To perform this, only fine particles (up to 2 micrometers) are collected onto filters or backing foils. Different sampling conditions may cause serious mistakes in the interpretation of the results and/or in the comparison of data from different sources. In general, if larger particles are accepted, more local influences should be expected because of the smaller distances that these particles can travel before sedimentation. Nevertheless, even if the sampling devices and methodologies are the same, other information should be taken into account, namely actual meteorological conditions and air mass histories. This implicitly includes information about the size distribution of the collected aerosol particles at the sampling point, as well as the information about the change it undergoes between its source and the receptor areas. Computer models are used to describe transport conditions and to correlate them with the elemental compositions of aerosol samples. PIXE analysis with a broad beam can be simple and very accurate because of the homogeneity and small thicknesses of the targets that include fine particles only. It can be very useful to follow it with a microanalysis of single particles, or a scan picture over the target area to state element distribution, especially if it can be identified with single particles.

While analysis with the broad beam gives data for statistical evaluation, the microanalysis serves for detailed study of the morphological, physical, and chemical properties of the dispersed matter. Therefore, microanalysis can contribute to a better understanding of the processes in the transport, but it cannot serve as a basis for a general overview. No conclusions concerning long-range transport of air pollution should be done, for instance, if only some tens of particles or even fewer were analyzed.

5.2.1.2   A Study of Air Pollution Sources that Affect a Sampling Point

To take all possible sources into account, both natural and anthropogenic, no particles should be avoided because of their size. On the other hand, only relevant particle sizes must be taken into account. For example, if the aim is to study health effects of air pollution, the respirable particles up to 10 micrometers in the mean dimension are important. Broad-beam PIXE analysis includes an uncertainty as to the surface heterogeneity in target thickness, and microanalysis is either nonrepresentative or too time-consuming. The complementarity of the two approaches is evident, but they can hardly be performed without the cooperation of specialized laboratories.

Provided a sufficient number of samples is collected, sources can be discovered and their element profiles can be described if a suitable receptor model is used. The more elements that are taken into account, the better is the resolution that enables us to state the number of the main sources and their relative contributions to the air pollution at the locality, and to describe and even identify them.

5.2.1.3   A Study of Health Effects of the Aerosol

This type of aerosol research is probably most sensitive to the proper and correct interpretation of the results, which must correspond well with the methodology. It is very important to separate aerosol particles according to their sizes in correspondence to sectors of the respiratory tract. Moreover, physical and chemical characteristics of particles such as morphology, stability or instability in body fluids and tissues etc., must be considered. Therefore, neither the statement of the sampling strategy nor the sampling itself should be the business of a PIXE laboratory but of a medical one. The effectiveness of PIXE analysis is affected by target preparation because the direct analysis of the aerosol on filters or backing foils cannot give complete information about the real behavior of the matter in the body, about release of elements, etc. Nevertheless, even if the targets are prepared using mineralization, ashing, or other specific techniques, PIXE can be very useful because it is multielemental and far less material consuming compared with other techniques. Moreover, rare samples can be stored after a PIXE analysis to be analyzed repeatedly. It has a great advantage if samples are taken from animals or humans and must be obtained noninvasively.

5.3 THE PRAGUE INVESTIGATIONS

5.3.1 THE PIXE SYSTEM IN PRAGUE

At the late 1970s, the PIXE program was introduced at the ion beam laboratory of the Department of Physical Electronics, Faculty of Nuclear Sciences and Physical Engineering of the Czech Technical University in Prague. The equipment for the analyses with energetic light ions was constructed there, using the central ion beam from the 2.5 MV Van de Graaff accelerator of the Nuclear Centre, Charles University.

The equipment includes the analytical chamber, which is devoted to the analyses using PIXE. The chamber has a removable 7.5-μm-thick kapton window at its top and a horizontal carousel that can hold 20 standard slide frames and/or 25 mm rings. The target holder is, therefore, compatible with two frequently used target shapes, but it can also hold targets with irregular shapes if their dimensions do not exceed the available space. Moreover, the bottom of the chamber can be removed if large objects are to be analyzed with the external proton beam. Therefore, PIXE analysis of various types of samples may be conducted both in high vacuum and in an atmosphere with defined pressure, or even in open air.[3] It should be pointed out that the unique feature of the equipment — the vertical proton beam — allows the analysis of liquid samples with no drying.

The proton beam with energies from about 0.6 MeV to about 2.4 MeV and with currents from 0.1 nA to about 100 nA is used for the analysis. The central part of the broad beam is selected by a diaphragm. Its diameter is usually either 6 mm in vacuum or 2 mm if the external beam is used. The homogeneity of the beam current is checked semiautomatically and includes the beam profile at any time within an analytical run. Otherwise, the focused beam may be used with a diameter of about 1 mm in the target surface. The set-up is completed with the Si(Li) detector PGT and signal processing electronics from Canberra.

Collected PIXE spectra are transported from the multichannel analyzer to a PC and stored on floppy disks or forwarded to the workstation using a local area network. Since the LAN is connected to the Internet, the spectra can be sent to remote computers, and spectra from other laboratories can be received. The spectra are processed by nonlinear least-squares fitting to find the intensities of separate spectral lines corrected for background, overlapping, pile-up effect, and escape peaks. On the basis of its output data and PIXE-relevant library parameters, the amounts of separate elements are derived either in nanograms per unit area for thin targets or in ppm for thick targets. The software is described below in more details.

The Prague PIXE system is compatible with the PIXE system in Ghent, Belgium, because it includes some common features of software as the format of input/output data. The set-up is also compatible with the PIXE device in Lund, Sweden, because targets with the same shape can be analyzed. This facilitates interlaboratory comparative analyses and data exchange and makes direct cooperation easier. The calibration procedure was described in Reference 4, and the intercalibration experiments are mentioned below.

The system has been proved to be ready for analysis in environmental research. Hundreds of samples per year can be analyzed, with the average number of 10 to 15 detected elements. The reproducibility is up to 10%, typically 3 to 5% depending on the concentrations. At present, some routine analyses are performed with it, especially of solid aerosols. At the same time, further development of the equipment and methodology for different kinds of samples is taking place to improve the effectiveness and accuracy of the analyses and to make it more user friendly.

5.3.2 THE DATA PROCESSING SOFTWARE PACKAGE IN PRAGUE

Two main programs are used in Prague to process PIXE data[5]; their names, TROJAX and TROCON, include the name Troja which is the site of the laboratory. Both programs were written in Fortran 77 with no compiler-dependent enhancements. Therefore, they can be compiled for any computer. It must be taken into account, of course, that TROJAX may need a long time to process a complex spectrum if a PC AT or IBM 386 equivalent is used with no coprocessor, while a UNIX workstation enables it to be processed in 1 to 2 minutes at most. TROCON is very fast if thin targets are expected and can be used in such a case even with a PC XT. On the other hand, if data from thick targets are to be processed, the time consumption may be comparable with TROJAX.

TROJAX was developed from HEX-83,[6] a gift from the Lund PIXE laboratory. Some features were modified, especially the background description and user interface, according to AXIL, which is used in the Ghent laboratory.[7,8] It can process any spectrum, provided it is imported in a one-colon ASCII file where the numbers are interpreted as contents of up to 8192 channels of a multichannel analyzer. TROJAX generates the basic output (direct access file) for each spectrum. Other files may also be created that include detailed information about the fitting process and its results in text format. Another option is one-colon ASCII files of the fitted region of the experimental spectrum, the corresponding fitted spectrum, and also the fitted background. They can be imported to a commercial software package to be presented with ease. For example, Sigma Plot, Quattro Pro, and MS Excel have been used for processing the multitasking job of MS Windows together with TROJAX. Each of these optional files can be generated later from the basic output file that contains complete output information. Therefore, no graphics features have been incorporated into TROJAX, as has been done in other software. Experience with a number of software packages used in analytical techniques confirms that the quality of home-make graphics can hardly be compared with the flexibility and ease of graphics software that is obtainable at reasonable prices.

TROJAX was compared with several other programs in the analysis of some spectra used in a published comparative work.[9] The coincidence in peak areas seemed to be reasonably good because they did not differ more than 5% from the peak areas found in Ghent, where the spectra came from. Also, an experiment in intercalibration with the Lund PIXE system was carried out. The set of aerosol samples collected in the Czech Republic was analyzed to compare with the element concentrations with those that had been analyzed in Lund. The samples were poor, and almost all concentrations were found close to their detection limits. Nevertheless, no systematic errors were observed, and the differences did not exceed acceptable values.[4]

The basic output files of TROJAX are readable by TROCON provided both the programs are used at the same computer. TROCON can also read one of the optional text files as the input. Therefore, TROJAX and TROCON may be used with different operational systems, such as UNIX and DOS and, moreover, TROCON can even process data from any other fitting software if they are imported in the format of a TROJAX text file.

TROCON was completely developed in Prague. If it is used in thick target data processing, it works in the iterative loop that includes not only the influence of the matrix but also the contribution of analyzed trace elements to the proton energy degradation and X-ray attenuation in the target. At the same time, it states the analyzed depth, i.e., the maximum depth that the registered X-photons originated from, for each particular element. No estimations of the trace element concentrations nor the analyzed depth must be stated in advance by the user as in some other thick target data processing programs. The output of TROCON consists of tables that include lists of elements, concentrations, standard deviations, and detection limits in one table per sample. The tables can be converted to one-colon ASCII files that include the list of sample names and corresponding lists of concentrations, standard deviations, and detection limits for particular elements. These files can be directly imported to a commercial spreadsheet such as Quattro Pro, Lotus 1-2-3, or MS Excel.

TROCON, in its thick target mode, has been compared with the program GUPIX (see Reference 10) in analyses of glass targets. If identical input conditions are stated by the user, the results differ by less than 5%.

5.3.3 THE COMPARATIVE ANALYSES

In principle, PIXE enables interlaboratory cooperation in shared projects. Moreover, repeated analyses of aerosol samples should be possible, as well as the division of the analytical steps both in time and in space, i.e., they can be performed independently in different laboratories. To test this, an experiment has been prepared with a set of samples collected at Klet in the southern Czech Republic. The samples could be taken as thin as possible and homogeneous. Each filter was cut into four parts: A,B,C,D. Part A was analyzed with the PIXE set-up in Prague, and part B was later analyzed with the PIXE set-up in Ghent. Parts C and D were archived. About one year later, part A was again analyzed in Prague, together with part C. Moreover, the PIXE spectra of part B were transferred from Ghent to Prague via electronic mail and processed with the Prague software. According to the element list from Ghent, 15 element concentrations were received from each filter.

Therefore, five data sets were available:

A1 — the filters analyzed in Prague just after the sampling; the spectra were stored
Ba — the filters analyzed in Ghent several months after the sampling; the spectra were processed with the Ghent software immediately
A2 — the filters analyzed in Prague once more, about 1 year after the sampling
C — the filters analyzed in Prague about 1 year after the sampling
Bb — the spectra from Ba processed with the Prague software

Provided the filters were homogeneously covered with the aerosol, the analyses were error-free, the PIXE systems were comparable in quality, and no contamination nor loss of elements occurred during the transport and storage of the targets, the results should be equal, i.e., independent of their origin from A1, A2, Ba, Bb, or C. Such a presumption was, of course, not fulfilled; some of the corresponding trace concentrations differed even more than 100%. Nevertheless, the results had similar concentration ratios in all elements, and they were, of course, of the same order of magnitude. It cannot be determined what caused the differences: sample inhomogeneity, transport, storage, or differences in analysis and data processing. Since the results received with Prague software tended clearly to lower concentrations depending on the time after sampling and since the results from Ba and Bb, as well as those from A1 and Ba, differed up to some 30% only, it seemed to be evident that it was rather the influence of manipulation of the targets than the analysis and data processing which caused the differences. Nevertheless, since even the differences of some tens of percents are acceptable in the case of trace elements, the results compared rather well. The possibility of independently performed sampling, spectra collection, and data processing seemed to be well confirmed.

5.3.4 THE INTERCALIBRATION EXPERIMENTS

The first experiment with an intercomparison of the Prague PIXE set-up with other analytical devices was done at the very beginning of its routine operation.[11] The other experiment was prepared within the Swedish–Czech project, Source Characterization of the Central European Aerosol. The same targets were analyzed in the latter, i.e., there were no different cuts. They were filters with fine aerosol particles from 24-hour sampling. The first PIXE spectra were collected in Prague, then the targets were transported to Lund and analyzed once more.

It must be said that the preliminary results were encouraging but not fully satisfying because some systematic shift in concentrations appeared as well as a spread in separate elements: the absolute differences tended to be greater for some of them and lower for others. The spread could be reduced in the intercalibration as well as the shift itself, but the latter had first to be explained because it could have such causes as, for example, inaccuracies in software or changes in the samples between the analyses. Since the software was found to bring no systematic differences in stated peak areas nor in their conversions to element concentrations, it seemed to be improbable that there was a difference in data processing. On the other hand, the interval between the analyses was several months, and most of the targets were irradiated repeatedly at the same spots with proton beams. Both these facts could cause some loss of elements. Therefore, the experiments indicated some limits in the real possibility of repeated analyzes, even though they were nondestructive. Nevertheless, even those differences did nothing to change the presumption that there is no inherent problem in interlaboratory cooperation in the analyses of identical sets of samples.

5.3.5   THE ANALYSES OF AEROSOL SAMPLES USING THE PRAGUE PIXE SYSTEM

There have been several problems in Czech data from outdoor solid aerosol analyses. A great part of the published data originated in the samples collected on nitrocellulose filters (Synpor) that were widely used there in air pollution monitoring. Moreover, according to the valid local norm, there is no upper limit of particle size in collection to date, i.e., the sampling devices do not remove nonrespirable particles. The special problem is a relatively small number of sampling points that, in many cases, are even criticized for being nonrepresentative. Therefore, the data are still practically incomparable with others because of the incompatibility in the filters and in the sampling devices.

Some attempts to change this can be observed in research projects. They are, of course, performed in single localities instead of monitoring networks, they last for limited time intervals only, and do not strictly respect Czech normative methodologies. Nevertheless, they bring new experience and inspiration, and also some data that can be included in a system. A part of the research activity is reported here.

The first attempts to use the PIXE method in the Czech Republic appeared in the late 1970s. Except for the idea to use it in an investigation of semiconductors and other industrial materials, the possibility of using PIXE in environmental research was considered from the very beginning. Because of its unique features, a PIXE program was also created in the Institute of Landscape Ecology of the Czechoslovak Academy of Sciences. It was strongly influenced by the activities of PIXE groups in Lund and Ghent. The first experiments were performed with the old Van de Graaff accelerator at the Institute of Nuclear Research of the Czechoslovak Academy of Sciences in Rez near Prague, which was the only available accelerator at that time. Since the accelerator was slated to be abolished, consent was obtained to build the experimental external-beam PIXE set-up despite the danger of exit foil damage. The set-up became the first one in Eastern Europe in which the proton beam was carried out into the open air.

The permanent PIXE set-up described above was put into regular operation in 1987, and the analyses of aerosol samples from the Department of Aeroecology of the Institute of Landscape Ecology have been performed there since that time. Since the activities of the department were terminated in 1990, a part of the program was adopted by the owner of the PIXE system.

Several series of samples were collected in the sampling devices of the Department of Aeroecology from 1986 to 1990 to be analyzed by PIXE. Four of the sampling points were built in Prague and later included in the network of Automated Emission Monitoring. The other stations were built in the south of the Czech Republic, close to the frontiers with Germany and Austria. The latter were used in studies of air pollution transport.

Some experiments were performed with nonstandard sampling devices. The station at Mount Klet, located near the astronomical observatory, is one example. The sampling device included three pipes equipped with pneumatic valves that were controlled by an electrical signal derived from the measurement of wind direction. One of the pipes was open to north wind only, another to south wind only, and the remaining pipe served for ordinary continuous sampling. Very small flow rates were used in extremely long sampling intervals of 1 to 2 weeks. The aerosol particles were collected on Nuclepores (100 mm in diameter, 0.2 μm hole sizes) and were assumed to represent what was transported from the Czech Republic to Austria in the north wind and vice versa, including the integration in the time intervals. The experiment was terminated unfinished in 1990 when not enough data were received to make conclusions. The samples were employed in the interlaboratory experiment mentioned above.

The activities of the Prague ion beam group in this field has been connected with the transportable sampling station. Four separate time series of solid aerosol samples were collected in

1989/90 and in 1992/93 at two places in Prague. The sampling station was first placed in the southwest town outskirts facing the dominant wind direction and later it was moved to a place north of the town center. Both locations were traffic-free but not too far from highways.

The first and second series of samples were collected in the two places using the usual Czech methodology, with no preseparation of aerosol particles. The samples were collected on Czech nitrocellulose filters (Synpor) and analyzed by PIXE. The aerosol masses per air volume, element masses per air volume and per total aerosol mass, concentration ratios, and enrichment factors were stated in the data processing. All these parameters were compared for the two locations, and some differences were observed that corresponded to general expectations concerning the influence of the town area but they could hardly be compared with published data from other countries to evaluate the absolute concentration levels because of the reasons listed above.

The Synpor filters were found to be not fully suitable for PIXE because of their thicknesses with a great span of particular values. Moreover, they had rather high levels of intrinsic contents of Ca, Fe, Zn, and some other elements. The concentrations had no typical values but depended on dates of production.

To overcome the difficulties, some improvements were made in the sampling devices and in the development of the analytical methodology and data interpretation to make them more compatible with written and unwritten international standards. Therefore, the third sampling series was a comparison of methodologies. The sampling probe was equipped with the single plate impactor on its entry as a preseparator of the aerosol particles, and Nuclepore filters with hole sizes of 0.4 μm and 0.2 μm were used instead of Synpors. The impactor was constructed with the parameters used in the Swedish–Czech project, Source Characterization of the Central European Aerosol, i.e., only particles less than 2 μm were collected at an air flow rate of 10 l/s. Synpor filters were used in the parallel sampling to compare them with the Nuclepore filters. It was found that the sampling in about 1 hour was sufficient for the PIXE analyses with the Nuclepore filters. On the other hand, the uncertainty caused by the properties of the Synpor filters made it necessary to use the minimum sampling time of about 6 hours with them. Some elements, such as Pb, Zn, K, were found in notably higher concentrations in fine particle fraction of the aerosol. The data clearly showed incompatibility caused by different sampling techniques and thus indicated some uncertainties in other published Czech data.

The last series was an attempt to receive data about Prague solid aerosol that could be internationally comparable from technical and methodological viewpoints. Three sampling devices were in parallel operation in the 24-hour experiment in September 1993. Device #1 included only Nuclepore filters (0.4 μm hole diameters), while the equivalent device #2 was equipped with a one-stage impactor that removed particles above 2 μm. The samples were collected in one-hour intervals in both devices. Device #3 was the standard sampler used for 24-hour aerosol sampling onto Synpor filters as in the usual air pollution monitoring in the Czech Republic. A cold front with wind and rain went across the day before, but all the sampling passed on a rather warm, sunny day with no wind. The night temperature was close to zero, the sky was clear, and a temperature inversion occurred in the morning. The next day was warm (up to 20 degrees Celsius) and calm. Two days later, all the samples were analyzed by PIXE, and the results were evaluated. The element masses in ng per unit area were normalized to air volume unit, and also the fractions of fine particles were stated for each sample.

The final step consisted of the statement of time series for single element amounts, element mass ratios, and enrichment factors. The short intervals of the sampling made it possible to evaluate the changes in element concentrations throughout the day. Notable increases in the concentrations during the morning temperature inversion were found for K, Ca, Zn, Mn, Cr, and other elements but far less for Fe. The concentrations of S remained unchanged except for statistical fluctuations. Comparing the devices, it was confirmed that the concentrations of some elements, such as K and Cl, were relatively higher in the fine aerosol particles. The results obtained from the Synpor filter, using PIXE spectra of several blank filters from the same packet to correct the intrinsic element concentrations, showed good coincidence with the averaged results from Nuclepores that were used in the device with no preseparation, but the information about the variations within 24 hours was lost.

5.4 CONCLUDING REMARKS

In spite of the differences in the intercalibration experiments, the tests clearly indicated that there was a possibility of interlaboratory cooperation in analyses and data processing. In principle, some specialization can take place to make it possible to use the most suitable analytical set-ups and software packages for particular types of samples or to employ the accelerator times more effectively. This is, of course, not limited to the analysis of aerosols.

It must be noted also that there is some uncertainty as a consequence of the lack of standards for real aerosol samples. Unfortunately, there is principally no evidence that an analytical set-up or a software program gives results that could be considered certain, but only a more or less justified trust in the experience of some laboratories. It is perhaps one reason published results of interlaboratory analyses or comparisons are so rare.

Provided the spectra processing software package is independent of the analytical device and even of the operational system of the computer, it can be used to process the data from other laboratories. X-ray spectra and the information about analytical conditions can easily be transported on floppy disks or by e-mail and then processed together. This could be a great advantage in the following events:

- PIXE spectra for a shared project can be taken in several laboratories to maximize the accelerator time if there is a time limit. Nevertheless, only one laboratory remains responsible for the results.
- It is possible to select an existing unique irradiation device to analyze a limited number of special targets and keep a standard quality level of the output data and even the standard form.
- Intercalibration and intercomparison can become independent of software, and the mutual control may become easier and lead to more reliable results.

As for the application of PIXE in an aerosol research project, the methodology must be well fitted to the aim. It is very important to take maximum care of the sampling conditions and of the manipulation of the samples to avoid a misinterpretation. If there are not enough data, it seems to be more correct only to present the methodology than to try to derive some conclusions from the insufficient data set.

The problem of the costs and also some technical delays in eastern Europe can be reduced with a suitable interlaboratory cooperation. It could become possible to receive important data from the territory, by employing the potential of local PIXE systems more effectively and taking advantage of the possibility to analyze the samples near the sampling points, i.e., with only a short-distance transport. Remote data processing can become a significant help.

REFERENCES

1. Johansson, S.A.E. and Campbell, J.L. PIXE —*A Novel Technique for Elemental Analysis,* John Wiley & Sons Ltd., Chichester, 1988.
2. Johansson, T.B., Akselsson, R., and Johansson, S.A.E. X-ray analysis: Elemental trace analysis at the 10^{-12} g level, *Nucl. Instrum. and Methods.* 84, 141-143, 1970.
3. Kral, J. and Voltr, J. Ion beam equipment modification for external beam operation, *Nucl. Instrum. and Methods in Phys. Res. B.* 85, 760-763, 1994.
4. Potocek, V. Mass calibration and intercalibration of Prague PIXE set-up, *Nucl. Instrum. and Methods in Phys. Res. B.* 85, 145-149, 1994.

5. Potocek, V. and Nejedly, Z. PIXE data evaluation in Prague, *Nucl. Instrum. and Methods in Phys. Res. B.* 85, 611-615, 1994.
6. Johansson, G. Modifications of the HEX program for fast automatic resolution of PIXE-spectra, in *Proton Induced X-ray Emission — Mass Calibration, Computer Analysis and Applications to Work Environment Aerosols,* LUTFD2/(TFKF-1001)/1-111, 1981.
7. Maenhaut, W. and Vandenhaute, J. Accurate analytic fitting of PIXE spectra, *Bull. Soc. Chim. Belg.* 95, 407-418, 1986.
8. *AXIL X-ray Analysis Package — User's Guide,* November, 1981.
9. Campbell, J.L., Maenhaut, W., Bombelka, E., Clayton, E., Malmqvist, K., Maxwell, J.A., Pallon, J., and Vanderhaute, J. An intercomparison of spectral data processing techniques in PIXE, *Nucl. Instrum. and Methods in Phys. Res. B.* 14, 204-220, 1986.
10. Campbell, J.L., Higuchi, D., Maxwell, J.A., and Teesdale, W.J. Quantitative PIXE microanalysis of thick specimens, *Nucl. Instrum. and Methods in Phys. Res. B.* 77, 95-109, 1993.
11. Potocek, V., Brenner, R., Hodik, F., and Voltr, J. Interlaboratory simultaneous multielement analysis of aerosol samples from Sumava Mountains, *J. Radioanal. Nucl. Chem., Articles,* 149, 205-216, 1991.

6 Characterization of Atmospheric Aerosols and Aerosol Studies Applying PIXE Analysis

Mikio Kasahara

CONTENTS

Atmospheric aerosols have various characteristics depending on the differences in their generation and transformation processes. The properties of aerosol particles are described by a number of physical and chemical factors. Concentration, particle size, and chemical composition are most essential to understanding the behavior of aerosol particles in the atmosphere. Other characteristics, such as shape, density, optical and electrical properties, and reactivity, become important in some specific situations. For example, the optical property of aerosol particles is essential in the problem of global warming.

The micro analytical technique has made remarkable progress during the last two or three decades. Particle-induced X-ray emission (PIXE) is one of the most powerful analytical methods. The analysis of atmospheric aerosols is one of the most suitable fields for PIXE analysis.

We have investigated the characteristics of atmospheric aerosols under the various environmental conditions using PIXE analysis. We have also tried to apply the PIXE technique to atmospheric environmental studies, such as the gas-to-particle conversion and wet scavenging of the particulate pollutants by rain drops. In this chapter, the outline of the PIXE method and the sampling method of atmospheric aerosols for PIXE analysis are described, and some experimental results, obtained mainly in our research group, using PIXE analysis are presented.

145

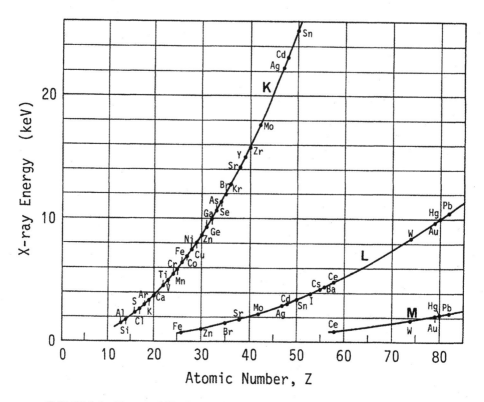

FIGURE 6.1   Energy of Kα, Lα, and Mα X-rays as a function of atomic number.

6.1   OUTLINE OF PIXE ANALYSIS

When energetic protons, or α-particles, collide with atoms, electrons in the inner atomic shells are preferentially removed and vacancies are created in the shell. An electron in the outer shell falls down into the vacancy, and the difference of its binding-energy is transformed to the characteristic X-ray or Auger electrons. Therefore, it is possible to identify a kind of element and determine its quantity in a sample by measuring the energy and the yield of characteristic X-rays. Figure 6.1 shows the energy of Kα, Lα, and Mα X-rays as a function of the atomic number. This analytical method suggested by Johansson et al.[1] is called particle induced X-ray emission (PIXE) analysis.

The greatest advantage of PIXE analysis is in its excellent sensitivity as compared with other analytical methods. Its sensitivity is as far as 1 ng with μg to mg samples. The other advantages of PIXE are that is a nondestructive technique and useful for a wide range of elements with atomic number larger than 13 (A1). On the other hand, the most serious disadvantage of the PIXE analysis is that it is not applicable for the lighter elements, such as C, N, H, and O, which generally occupy more than 80% of the atmospheric aerosols. Simultaneous detection of K- and L-series characteristic X-rays is another disadvantage.

The PIXE technique has been applied to the study of atmospheric aerosols as one of the most important fields.[2,3]

6.2   EXPERIMENTAL ARRANGEMENT AND CALIBRATION

6.2.1   Experimental Set-Up

The experimental system of PIXE analysis consists of an accelerator, an ion beam irradiation chamber, and a measurement system. A beam of proton (or α-particles) from a small accelerator,

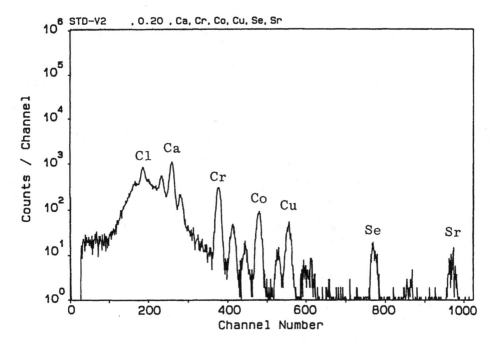

FIGURE 6.2   A typical PIXE spectrum: a mixed sample containing Ca, Cr, Co, Cu, Se, and Sr.

such as Van de Graaff and tandem accelerators, is introduced into a vacuum chamber after passing through an analyzing magnet, an electrostatic Q-lens, and a slit system. In the vacuum chamber, where a large number of samples are usually mounted, the beam is bombarded to the target. The optimum energy range of the ion beam is between 1 and 4 MeV per nucleon for most purposes.[4] Extraction of the ion beam into an air or helium atmosphere has been tried for easy treatment of the sample.

The emitted X-rays are detected by an energy dispersive Si(Li) detector. An absorber film (foil) is situated before the detector to prevent entering of the scattered protons and to adjust the counting ratio of the X-ray. After amplification, signals from the detector are fed into a multichannel pulse analyzer, which is usually combined with a microcomputer to store the data and/or analyze the PIXE spectrum.

6.2.2   Calibration of PIXE Analysis and Detection Limit

A typical PIXE spectrum is shown in Figure 6.2, which was obtained from a mixed sample containing six elements: Ca, Cr, Co, Cu, Se, and Sr.[5] The mixed samples were prepared using a standard solution for atomic absorption spectroscopy (AAS). The x-axis, expressed as the channel number, corresponds to the energy of X-ray, namely the element. The spectrum consists of six large and three (Cl, K, and Br) small characteristic X-ray peaks superimposed on the background. The small peaks of Cl, K, and Br are caused by reagent or solvent in the AAS standard solutions as an impurity in backing material. The background of PIXE spectrum is mainly due to various atomic bremsstrahlung processes.

The calibration of the PIXE, that is, determination of the relationship of X-ray yield and mass thickness of each element, can be performed using a standard sample of known mass thickness. Figure 6.3 shows the X-ray yield plotted as a function of mass thickness.[6] There exists an obvious linear relationship having a slope of unity between them. The sensitivity defined by (PIXE yield per unit dose)/(mass thickness) can be determined theoretically for all objective elements on the basis of at least one sensitivity obtained experimentally. Figure 6.4 illustrates an example of

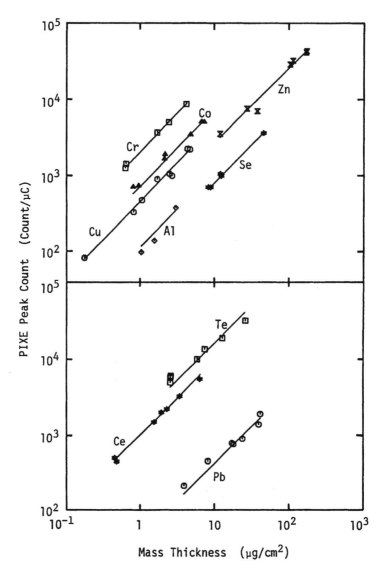

FIGURE 6.3 Relationship between X-ray yield and mass thickness obtained using single-element standard samples.

sensitivity curves of Kα and Lα X-rays.[6] The sensitivity depends upon the beam bombardment and measurement conditions, such as type of ion beam, total dose of beam current, measurement system and geometry, and type of absorber. PIXE spectra of the mixed sample containing S, K, V, Mn, Ni, Zn, and Pb, which were obtained using a different ion beam, absorber, and total dose, are compared in Figure 6.5.[7] From the quantitative evaluation of S/N ratio defined by (net counts of X-ray)/(background counts), it was concluded that the PIXE analysis by the lower beam energy and using thinner absorber was competent for the lighter elements such as Al, Si, and S, and the analysis by the higher beam energy and using thicker absorber was suitable for heavier elements, such as Fe, Zn, and Pb.

In the PIXE analysis of atmospheric aerosols, the detection limit, accuracy, and overall analytical efficiency of the analysis depend upon not only the PIXE analytical conditions but also the sampling method of aerosol particles and the numerical analysis of the PIXE spectrum as shown in detail below:

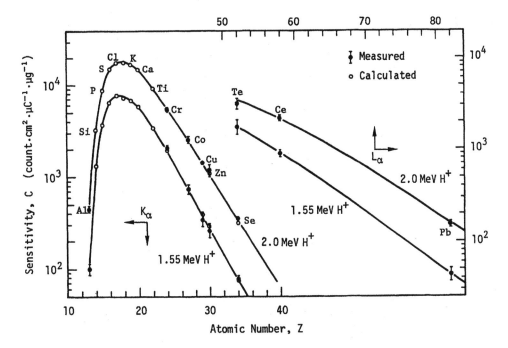

FIGURE 6.4   Example of sensitivity curves of Kα and Lα X-rays.

1. Sampling method of aerosol particles
     Backing material (impurity, background, strength, stability)
     Sample amounts (collection method, sampling time, flow rate)
     Uniformity of sample
2. PIXE analytical conditions
     Ion beam (type, energy, current, size, shape)
     Amount of X-ray yield (accumulated charge, sample amounts)
     Measurement system and geometry (detector, arrangement)
     Absorber (type, thickness)
3. Numerical analysis of PIXE spectrum

The detection limit of PIXE analysis depends on a number of factors, such as the sort of backing material, the kind of ion beam, the beam energy, the total dose of irradiation, and the geometry of the measurement system. The detection limit (D_t), expressed as the mass thickness, can be evaluated by $3\sqrt{N_b}/C$, where C is the sensitivity and N_b is the total background count.[8] The detection limit (D_c), expressed as the concentration in atmospheric air, depends on the sampling conditions and is given by $D_t \cdot (A/Qt)$, where A is the sampling area, Q is the flow rate, and t is the sampling time. An example of D_c is shown in Figure 6.6, which was estimated assuming the accumulated charge of 20 μC by the 2.0 MeV proton beam, the sampling area of 12.5 cm², and the total sampling air volume of 10.8 m³.[9] In the estimation, however, N_b was replaced by 10 in case of N_b value smaller than 10. The minimum detectable limit of Ca(Z = 20) – Zn(Z = 30) was 0.8 – 2.6 ng/m³. The detection limit can be improved by reforming the irradiation system, the measuring geometry and the sampling conditions.

6.3   SAMPLING METHOD OF ATMOSPHERIC AEROSOLS FOR PIXE ANALYSIS

The mass size distribution of atmospheric aerosols is generally described as a bimodal distribution. There is a striking difference in the elemental composition between the fine and coarse particle

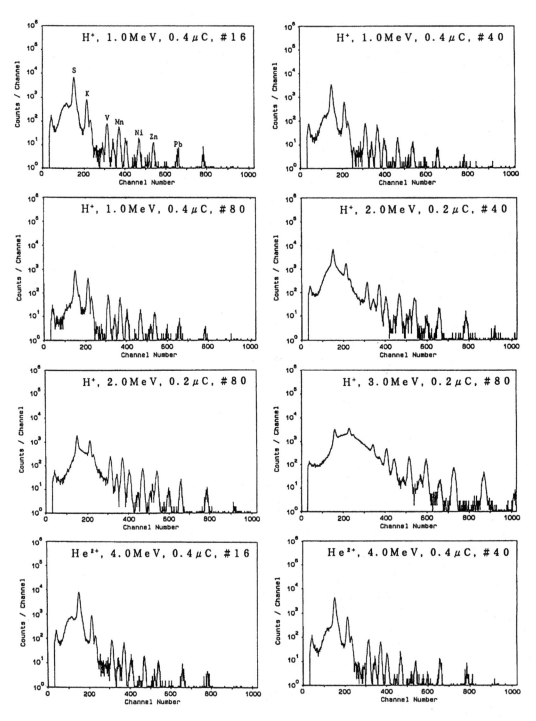

FIGURE 6.5 PIXE spectra of multielement standard sample measured under various conditions.

fractions according to their origin. They are usually divided at about 1 to 2 μm. Therefore, it is recommended to collect the aerosols, classifying the fine and coarse fractions at about 1–2 μm. In the PIXE analysis, the sample is also required to be uniform.

In selecting the sampling method, we should examine thoroughly several points, such as the number of stages classified in size range, uniformity, concentration of particles in the atmosphere,

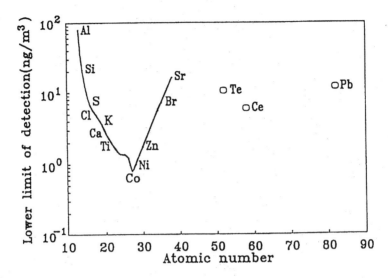

FIGURE 6.6 Example of detection limit of PIXE analysis.

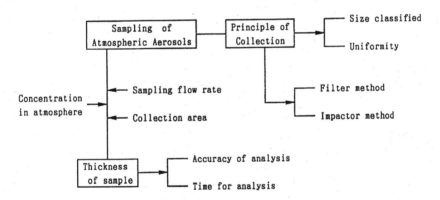

FIGURE 6.7 Sampling of atmospheric aerosols for PIXE analysis.

required amount (mass thickness) and collection area of sample, and sampling flow rate, required time-resolution and accuracy of analysis as shown in Figure 6.7. The atmospheric aerosols are usually separated using an impaction type sampler or filtration method. With the impaction type sampler, it is generally difficult to get the uniform sample without an ingenious contrivance. On the other hand, the filter method is inferior in the classification of particles. Some kinds of impaction-type aerosol samplers have been developed for the PIXE analysis.[10-12] Most of the newly developed samplers can classify particles into 4 to 10 size fractions, and some are devised to make uniform samples.

Furthermore, PIXE samples must have a proper mass thickness. In the case of a thick sample, the intensity of the ion beam is reduced as it passes through the sample and a part of the X-rays emitted from the sample are absorbed. The reduction of ion beam intensity and the absorption of X-rays lead to the degradation of X-ray yield. The effects of mass thickness of aerosol sample on the degradation of X-ray yields were calculated and are illustrated as a function of thickness in Figure 6.8.[7] In the calculation, the elemental compositions of aerosol samples were assumed to be the values listed in the figure. The ion beam intensity is reduced more severely by the elements with larger atomic numbers (Z). On other hand, the absorption of X-rays by the sample is more severe in the lighter elements. In total, X-ray yield is reduced more strongly in the lighter elements. A correction for the reduction of X-ray yields in a thick sample is required when the mass thickness exceeds about $100 \, \mu g/cm^2$ for the lighter elements and about $1 \, mg/cm^2$ for the heavier elements.

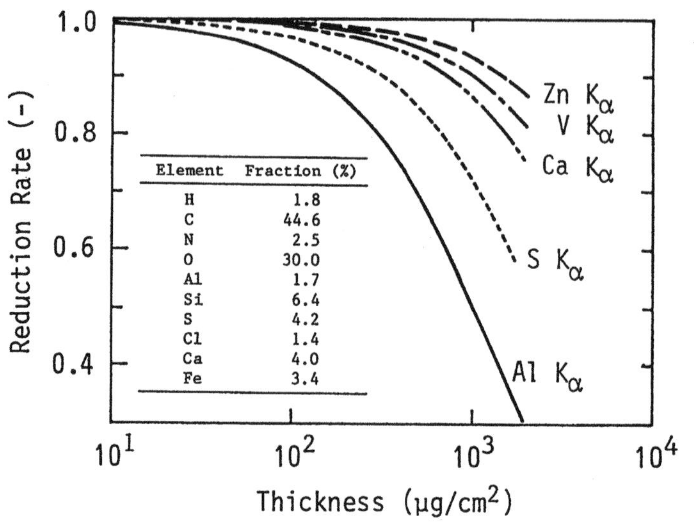

FIGURE 6.8 Effects of mass thickness of atmospheric aerosol sample on the degradation of X-ray yield.

It is desirable that the mass thickness of the collected sample ranges between 10 and 100 μg/cm^2 in order to maintain accuracy and prevent the degradation of X-ray yields caused by the change of ion beam intensity and the absorption of X-rays within the samples.

The two stacked filters method is commonly used when the PIXE sample of atmospheric aerosols is collected. It consists of two filters with different pore sizes, 8.0 and 0.4 μm in most cases. They are connected in series to classify the aerosol particles into fine and coarse fractions. In the selection of filters, we must test their aptitude from various points of view. We use the Nuclepore filter because they generate fewer bremsstrahlung X-rays, less impurity, and show high endurance against the ion beam bombardment.

The flow rate is concerned not only with the time resolution of the sample but also with the cutoff size of the filter. The filtration efficiency of the Nuclepore filter can be estimated for the various sampling conditions using an efficiency model.[13] When the sample air is passed through the Nuclepore filter of 47 mm diameter with 8 μm pore size at a flow rate of 25 L/min (surface velocity — 33 cm/s), the 50% cutoff size is estimated to be about 1 to 1.2 μm, depending on particle density. According to the experimental study on the collection efficiency of Nuclepore filters by Kemp et al.,[14] aerodynamic diameter for the 50% collection efficiency was 1.65 and 1.50 μm at the surface velocity of 25 and 50 cm/s, respectively, in the case of 8.0 μm pore size filter.

At the flow rate of 25 L/min, the sampling time required to collect the desirable mass thickness of 10 to 100 μg/cm^2 can be ensured, in most cases, within several hours, as shown in Figure 6.9, which illustrates the approved sampling time as a function of mass concentration of atmospheric aerosols.[15] In fact, the sampling time was adjusted according to the particle mass concentration between 1 to 8 hours, with a standard of 2 to 4 hours.

6.4 CHARACTERISTICS OF ATMOSPHERIC AEROSOLS WITH PIXE ANALYSIS

Chemical composition is a key to understanding the behavior of aerosol particles in the atmosphere. Therefore, the information about the chemical composition of aerosols, especially with high time resolution, is essential in atmospheric aerosol studies.

Chemical composition data have been used to set the present state of the particulate air pollution, such as the pollution level, the time variation, and the spatial distribution, and to elucidate the behavior of atmospheric aerosols. The source contribution analysis applying a receptor model is one of the most successful breakthroughs in the application of the elemental concentration data.

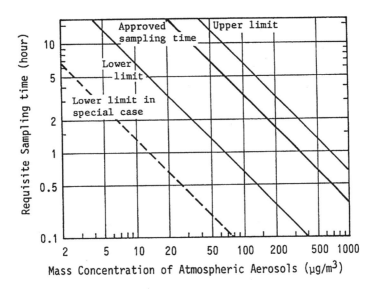

FIGURE 6.9   The approved sampling time as a function of mass concentration of atmospheric aerosols.

6.4.1 STANDARD SAMPLING METHOD AND STANDARD PIXE ANALYTICAL CONDITIONS

1. Standard sampling conditions of atmospheric aerosols. Since 1986, atmospheric aerosols for PIXE analysis have been sampled under various meteorological and environmental conditions, including episodic air pollution events, such as heavily polluted air, Kosa dust, photochemical smog, and extremely clean air, etc. The majority of samples were collected using the two stacked filters method.

2. Standard conditions of PIXE analysis. The PIXE analysis of aerosol samples was usually performed with a proton beam of 6-mm diameter and 2.0-MeV energy from a tandem Cockcroft accelerator at the Department of Nuclear Engineering, Kyoto University. Beam intensities from 5 to 100 nA were employed, and the total dose was usually between 5 and 50 μC.

X-rays with energies of 1.4 to 14.8 keV were detected by a Si(Li) detector, which had a resolution of 152 eV at 5.9 keV. The important elements in atmospheric aerosols, such as Al, Si, S, Cl, K, Ca, Fe, Zn, and Cu by Kα-line and Pb by Lα-line, are included in this energy region. The target and detector were set at 90° and 135°, respectively, with respect to the direction of the ion beam. An absorber of 39.3 μm thick Mylar film was set between the target and detector to control the counting efficiency of lighter elements. In most cases, the concentrations of 15 elements (Si, S, Cl, K, Ca, Ti, V, Cr, Mn, Fe, Ni, Cu, Zn, Br by Kα, and Pb by Lα) were determined. More sensitive measurement for lighter elements, such as Al, was performed with 1.0 MeV proton and 16 μm Mylar absorber as occasion demanded.

The calibration was carried out with each experimental set-up.[6] The relationship between X-ray yield and mass thickness was measured first, using the 18 single element standard samples that we prepared with a vacuum deposition method. The sensitivity curves, as shown before in Figure 6.4, were obtained experimentally and theoretically for all objective elements.

6.4.2 CHARACTERISTICS OF URBAN AEROSOLS

1. Elemental concentrations and sources of roadside particles. An example of PIXE spectra of coarse and fine particles is shown in Figure 6.10.[16] The aerosol particles were collected with the two stacked filters sampler by the side of a main road in the Tokyo metropolitan area. This sampling site is known as the worse air pollution area in Tokyo because of automobile exhaust. The traffic density is about 250,000 cars/day at the sampling point, and 70% of them are large automobiles.

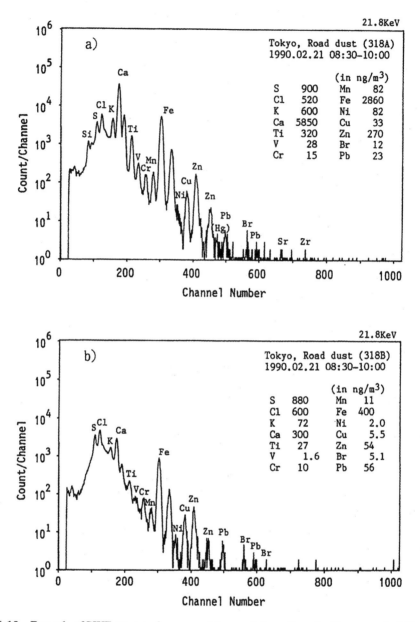

FIGURE 6.10   Example of PIXE spectra of coarse and fine particles collected with two stacked filter sampler by side of main road in Tokyo.

In Figure 6.10, the range of measured X-ray energy was expanded to 21.8 keV from 14.8 keV as the standard condition in order to check the concentration level of the heavier elements, and the signals were stored in the 1024 channels according to the X-ray energy.

The concentrations of Ca, Ti, and Fe in the coarse fraction were extremely high in comparison with the concentrations far from the road. They originated mainly from soil dust. Ca is expected to come from asphalt. Most sulfur compounds in the atmosphere are generally recognized to be formed by gas-to-particle conversion of SO_2 gas and to exist in the fine fraction. In the case of roadside samples in Tokyo, sulfur also exists in the coarse fraction at nearly the same level. Sulfur in the coarse fraction is presumably caused by sulfuric acid particles exhausted from diesel automobiles.

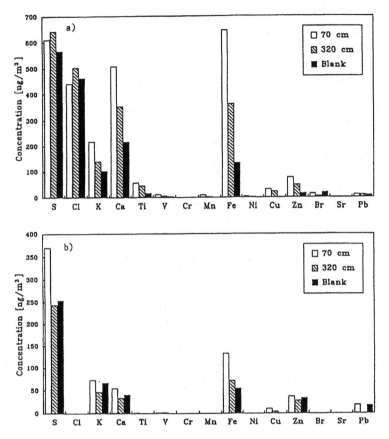

FIGURE 6.11 Comparison of elemental concentrations of aerosol particles collected at three different points on roadside in Kyoto.

At a crossing near the Kyoto University main campus, roadside aerosols were simultaneously sampled at three different points in order to examine turbulence of soil dust caused by cars. The first and second sampling points were located at the boundary of the roadway and sidewalk, with different sampling heights of 70 and 320 cm, respectively. The third point (blank) was behind a fence about 6 m from the road. The elemental concentrations at the three points are compared in Figure 6.11.[16] The concentrations of K, Ca, Ti and Zn in the coarse fraction, and Fe in both fractions were clearly dependent on the sampling point and high in order of the point No. 1, No. 2 and No. 3. Ca, Ti and Fe are main components of soil dust. Ca also originates from asphalt. Zn particles in the coarse fraction are produced by the abrasion of tire. Some parts of Fe are included in the brake pad shaved particles. This figure demonstrates that the flying up of soil dust by car have an important effect upon the particulate pollution at the road side.

2. Seasonal change in Japan. Atmospheric aerosols classified into fine (<1.2 μm) and coarse (1.2 to 10 μm) fractions were continuously sampled every 12 hours for 13 months from October 1, 1992, to October 31, 1993, at the Uji campus of Kyoto University. The sampling site is in a residential area in the middle of town, 10 km southeast of Kyoto city. Figure 6.12 shows the change of monthly average concentrations of the gravimetric mass (Mass), Si, C, Cl, Fe, and Pb.[17]

The seasonal change of the mass concentration of atmospheric aerosols commonly has two peaks, in spring and early winter in Japan.[18] The peak in spring is caused mainly by Kosa dust which is transported from deserts in central Asia. Therefore, the concentrations of Si, K, Ca, Ti, Mn, and Fe, especially in the coarse fraction, which are main components of soil dust, increased considerably in April and showed values 5 to 10 times greater than their minimum concentrations

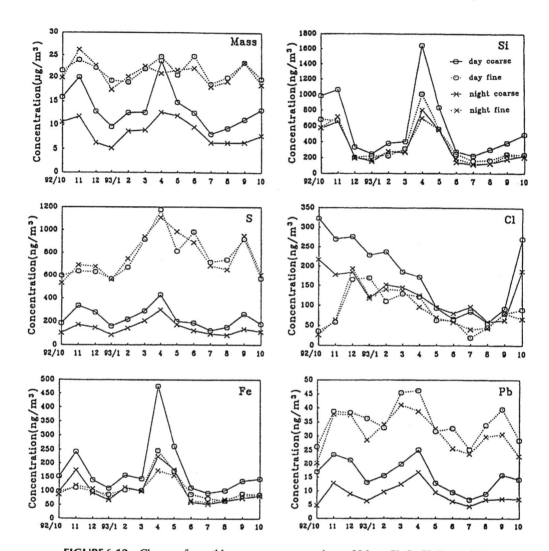

FIGURE 6.12   Change of monthly average concentrations of Mass, Si, S, Cl, Fe, and Pb.

in January or July. On the other hand, the peak in early winter is caused by stagnation of the air mass under the calm and stable meteorological conditions. The elemental concentrations in the coarse particles were higher in daytime than at night. There is, however, no clear difference in the fine particles between day and night measurements. This suggests that the coarse particles are influenced by human activity.

The annual average concentrations of Mass and 15 elements in the coarse and fine fractions are shown in Figure 6.13.[17] The most dominant element among the 15 elements analyzed was Si in the coarse particles and S in the fine fractions, and they exceeded 3.5% of the total mass concentrations of coarse and fine particles, respectively. The value of 3.5% as the element S corresponds to 10.5% as the sulfate.

The elements originating mainly in natural sources, such as Si, Ca, Ti, and Cl, are dominant in the coarse fraction. On the other hand, the elements released mainly from anthropogenic sources or formed from gaseous pollutants in the atmosphere, such as S, Pb, Zn, and Br, are dominant in the fine fraction. These tendencies correspond to the chemical characteristics of atmospheric aerosols reported by a large number of investigators.[19]

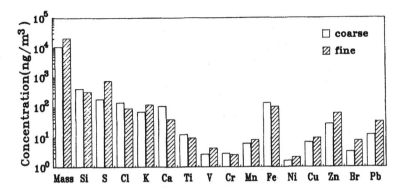

FIGURE 6.13   One-year average concentrations of Mass and 15 elements in coarse and fine particles between November 1992 and October 1993.

The sum of the concentrations of 15 elements accounted for only about 15% of the gravimetric mass concentration (Mass) on the average. It is concluded that the atmospheric aerosols consist mainly of the lighter elements, such as H, C, O, N, and Al, and they usually occupy more than 80% of the total aerosols.

6.4.3   CHARACTERISTICS UNDER EPISODIC ENVIRONMENTAL CONDITIONS

Typical episodes of winter heavy air pollution and the Kosa event have been observed between December 27 and 29, 1987, and in April 1988, respectively.[20] The mean elemental concentrations of atmospheric aerosols collected during two events are tabulated in Table 6.1 together with those under the usual meteorological conditions. In Figure 6.14, the time variation of some elemental concentrations during the Kosa event is illustrated by the contribution of soil dust, estimated using a CMB (chemical mass balance) receptor model. Furthermore, in Figure 6.15, the concentration levels under the three episodic air pollution events are compared with each other.[21] The first is the two Kosa events in 1989 and 1990. The second event is photochemical smog in summer that led to a remarkable reduction of visibility. The third event is in a very clean atmosphere, in which visibility farther than 50 km was observed.

The concentrations of Si, K, Ca, Ti, and Fe in both coarse and fine fractions which originated mainly from soil dust increased tremendously at the Kosa, and there was a very high correlation between these elemental concentrations. The maximum concentration of total Si (fine + coarse) exceeded 50 μg/m³ in the case of the most typical Kosa event. During the Kosa event, S concentration in the coarse fraction also increased and changed in a way similar to the element originating in Kosa dust. This may suggest that S in the coarse fraction is derived from a sulfur compound, such as $CaSO_4$, contained originally in Kosa dust or from the secondary particles condensed onto the surface of soil dust while in long-range transport. Anthropogenic components, such as Zn, Pb, and V, and also S in the fine fraction, changed in different ways than the soil dust components, as shown in Figure 6.14.

During a heavy air pollution event in early winter, the concentrations of S, Zn, Pb, and Cu in the fine fraction and Si in the coarse fraction were higher than those in the usual atmospheric air. Heavy air pollution in early winter in Japan is due to calm and stable meteorological conditions, which allow air pollutants to accumulate in the stagnant air mass.

In the case of photochemical smog, the concentration of S in the fine fraction increased remarkably. Photochemical smog is also likely to be formed under calm and stable meteorological conditions. Therefore, Pb and Zn increased similarly during the photochemical smog event. In the very clean atmosphere, the concentration of each element was lower in comparison with the polluted atmosphere.

TABLE 6.1
Comparison of Elemental Concentrations in Atmospheric Aerosols Collected During Kosa Event, Heavy-Smog Event, and Under Normal Conditions

	Coarse particle (8.0 μm Nuclepore filter)			Fine particle (0.4 μm Nuclepore filter)		
	Usual (n = 20)	Heavy-smog 1987/12 (n = 30)	Kosa 1988/4 (n = 17)	Usual (n = 20)	Heavy-Smog 1987/12 (n = 30)	Kosa 1988/4 (n = 17)
Si	2880	5520	16700	920	<20	5170
S	320	470	600	770	1670	1040
Cl	260	610	970	<20	160	32
K	230	280	940	130	460	360
Ca	560	730	2130	64	<5	330
Ti	57	67	190	10	12	52
V	5.5	3.7	6.1	3.9	8.6	5.1
Cr	12	5.9	14	2.9	1.9	1.3
Mn	17	30	56	8.4	26	20
Fe	460	870	1980	100	290	540
Ni	5.9	3.7	0.9	3.3	5.0	2.9
Cu	8.0	16	15	7.2	26	7.9
Zn	35	98	56	53	160	59
Br	3.0	4.7	<1	5.7	15	6.8
Pb	11	35	9.1	33	140	37

(in ng/m^3)

6.4.4 PARTICLE SIZE FRACTIONATED AEROSOLS

The mass size distribution of atmospheric aerosols is generally described as a bimodal distribution spreading from smaller than 0.1 μm to larger than 10 μm. The filtration method adopted in our aerosol sampling is inferior in the classification of particles. On the other hand, although the impaction type sampler can usually separate the particles into 4 to 10 size fractions, it is generally difficult to get a uniform sample.

A rotating cascade impactor developed by the University of Vienna is designed such that the aerosol particles deposit uniformly onto a rotating substrate and can be classified into 10 size fractions within the size ranges between 0.015 to 16 μm.[22] Using the rotating cascade impactor, atmospheric aerosols were collected simultaneously in central Vienna and suburban Vienna in May 1994 by an aerosol research group from the University of Vienna. The elemental concentrations of these samples were analyzed by the PIXE method in Kyoto University. The size distribution of gravimetric mass (Mass), S, Ca, Fe, Zn, and Pb are illustrated in Figure 6.16.[23]

On the other hand, atmospheric aerosols divided into 13 size ranges between 0.016 and 50 μm were sampled using a low-pressure Andersen air sampler beginning in November 1994 in Kyoto. The deposited aerosol particles were extracted into ultrapure water using an ultrasonic cleaner after measuring the gravimetric mass. The extract was passed through a 0.2 μm pore Nuclepore filter to divide it into soluble and insoluble components. The PIXE sample of the soluble component was prepared by the drying method. The elemental concentrations of both components were determined by PIXE analysis. The size distribution of Mass, S, and Fe are shown in Figure 6.17.[24]

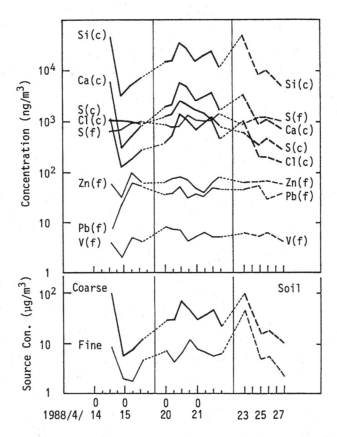

FIGURE 6.14   Time variation of elemental concentrations in atmospheric aerosols and soil source contribution during Kosa event.

Most of the size distributions of Mass and 15 elements had a similar tendency in Vienna and Kyoto. In general, S, Zn, and Pb skewed to the smaller size range and had only one peak in the fine particle region. Most S is thought to exist in the form of sulfate. Since the sulfate is produced by gas-to-particle conversion, S is dominant in the fine fraction. Most Zn and Pb aerosols also must exist in the fine state, because they are mainly formed by the condensation of their vapors which are easily produced in the combustion process because of their low melting points. On the other hand, Si, Ca, Fe, and Ti skewed to the larger size range and had nearly one peak in the coarse particle region. These elements are contained mainly in soil dusts generated by mechanical force. The mass size distribution of K, V, Cr, Mn, Ni, Cu, and Br was represented as the bimodal distribution similar to that of Mass.

Water solubilities of Mass and each element are shown in Figure 6.17 as a function of particle size. Figure 6.18 shows the overall insoluble fractions of 15 elements.[25] The insoluble fraction of Fe was very high, with an average of 0.85. The insoluble fractions of Si and Ti were also high and exceeded 0.5 in most cases. In contrast, the insoluble fractions of S, Cl, K, Ca, and Zn were very small and were less than 0.15. S and Cl existed almost totally in the soluble state. The soluble component was dominant in the fine fraction, and the insoluble component was dominant in the coarse fraction. This might be due to the larger specific surface area of the smaller particles.

6.4.5   Source Contributions Using Chemical Composition Data

In the application of chemical component data, the source contribution analysis using a receptor model is most successful. The so-called receptor model[26] is based on the elemental concentration

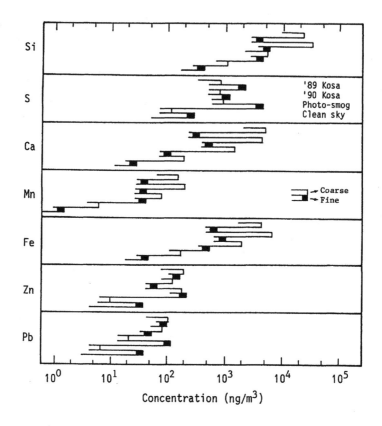

FIGURE 6.15 Comparison of elemental concentrations during some episodic air pollution events.

data of aerosol particles measured at the receptor and/or sources and can be used to identify the source type or estimate the source contribution of particulate pollutants. The chemical mass balance (CMB) method is the most popular receptor model and can quantitatively evaluate the contribution of the assumed source type.

For example, the contribution of soil dust during a Kosa event is shown at the bottom of Figure 6.15. The maximum source contribution of soil dust amounted to 140 μg/m³ as the six-hour average. Since the elemental concentration measured on an hourly basis varied greatly every hour and the total mass of atmospheric aerosols sometimes exceeded 300 to 400 μg/m³ on an hourly basis, the maximum contribution of soil dust may exceed 300 μg/m³ as a one-hour average during the typical Kosa event.

Figure 6.19 shows an example of source contribution analysis to the urban aerosols obtained by the CMB method.[21] It was estimated using the size-classified (<1.1, 1.1 to 11, and >11 μm) data set, including the concentrations of inorganic ions, elemental and organic carbons, as well as elemental compounds. Most particles smaller than 1.1 μm consist of anthropogenic aerosols. Secondary particles, such as sulfate, nitrate, ammonium, and organic aerosols, contribute about half in the <1.1 μm size range. On the other hand, the contribution of soil dust amounted to more than 75% for the coarse particles large than 1.1 μm. These results suggest that the lighter elements (or ions), including at least C and N, which are not measurable by the PIXE method, are indispensable to a source contribution analysis of particulate pollutants, and that the atmospheric aerosols should be classified into at least fine and coarse fractions in the sampling.

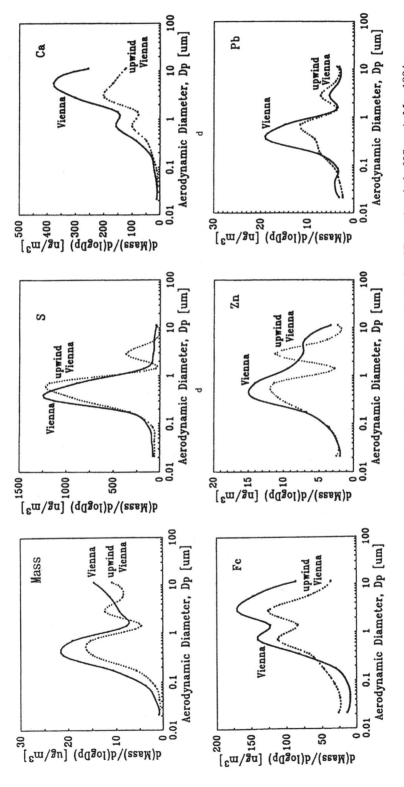

FIGURE 6.16   Particle size distributions of Mass, S, Ca, Fe, Zn, and Pb: Central Vienna and suburban Vienna (upwind of Vienna): May 1994.

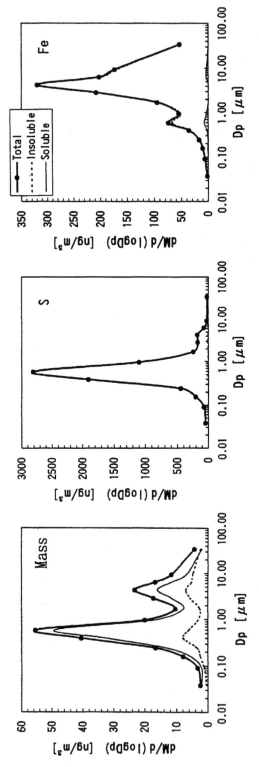

FIGURE 6.17 Particle size distributions of Mass, S and Fe: Kyoto: July 1995. Samples are separated into soluble and insoluble components.

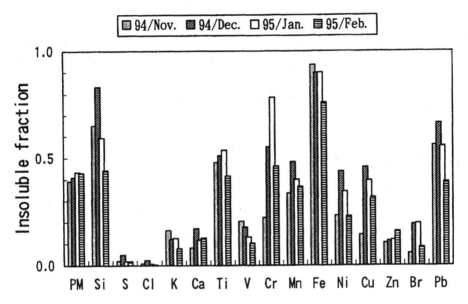

FIGURE 6.18 Overall insoluble fraction of 15 elements in atmospheric aerosols.

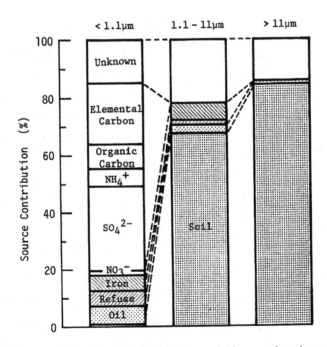

FIGURE 6.19 Source contributions for size-classified atmospheric aerosols estimated by CMB method.

6.5 APPLICATION OF PIXE ANALYSIS TO AEROSOL STUDIES

6.5.1 APPLICATION TO CONTAMINATION CONTROL IN A CLEAN ROOM

Contamination by chemical substances in a clean room is regarded as one of the greatest problems in semiconductor manufacturing. It is generally recognized that chemical substances are detrimental because they contaminate the silicon wafer. In order to determine the source of contaminants and devise a control method, it is essential to know the chemical composition of contaminants in the

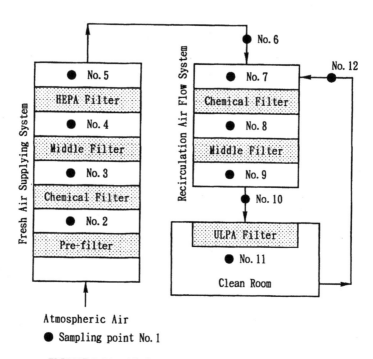

FIGURE 6.20 Air flow system of semiconductor factory.

clean room. We used the PIXE analysis to examine the air cleanliness levels and the source of contaminants in the clean room and also to obtain information on the improvement of an air flow system in the clean room.

The air flow system of the factor studied consists of an inner recirculation air flow system and a fresh air supply system from outdoors, as shown in Figure 6.20.[27] The air is cleaned by passing through the prefilter, chemical filter, middle filter, and HEPA filter in the air supply system and chemical filter, middle filter, and ULPA filter in the recirculation system.

Aerosol particles were collected in a 5-mm-diameter spot at the center of a 25-mm-diameter Nuclepore filter with 0.1 μm pore size by attaching a thin Teflon® film such that all of the sampled particles were bombarded by the ion beam with 6 mm diameter in the PIXE analysis. Sampling points are shown in Figure 6.20. Sampling time was adjusted according to the particle mass concentration to between 2.0 to 4293 hours. The concentrations of 38 elements in the sample were determined with the PIXE method. X-rays of 38 elements with 1.4 to 27.8 keV energy were detected by a Si(Li) detector, instead of 1.4 to 14.8 keV in the standard analysis.

The elemental concentrations at sampling points No. 5, 11, and 12 are shown Figure 6.21. The solid line in the figure means detection limits for each sampling and analytical condition. At point No. 5, downstream of the HEPA-filter in the fresh air supply system, a few particles which were thought to originate in soil dust, existed. In the clean room after passing through the ULPA filter, quite small amounts of Al, Si and Ti were detected. The filter medium may be one of the possible origins of such particles. Return air from the clean room contained all kinds of elements at fairly high concentrations rather than in fresh air at point No. 5.

In another test, aerosol sampling was carried out in the device manufacturing process laboratory that consisted of two testing rooms, an etching room, two sputtering rooms, a photo room, and a wet cleaning room. The etching room was equipped with two kinds of reactive ion etching systems of Mo (RIE-1) and Nb (RIE-2). The sputtering systems of Nb/Al, Mo (Sup-1), and SiO_2 were installed in the two sputtering rooms. The concentration of 17 elements in the aerosol particles sampled in each room and near the RIE-1, –2, and Sup-1 equipment was measured by PIXE analysis.[28] The concentrations of aerosol particles sampled in the RIE-1 exhaust duct were considerably higher than those

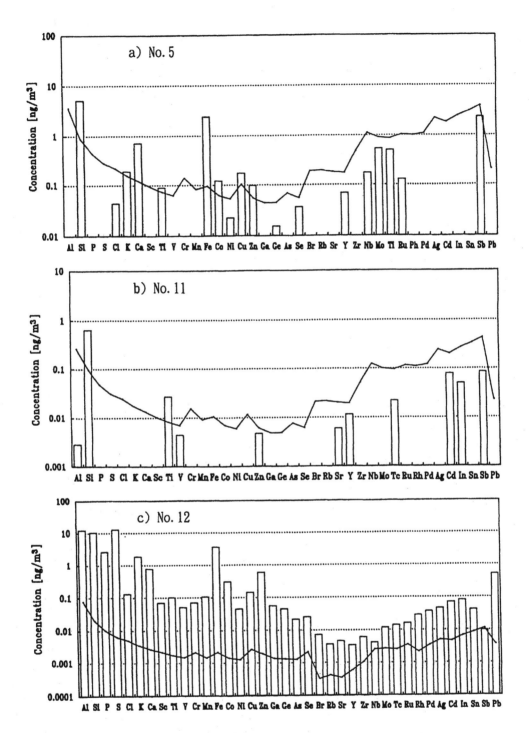

FIGURE 6.21   Elemental concentrations at sampling points, No. 5, No. 11, and No. 12 in air flow system of semiconductor factory. (a) No. 5: Downstream of prefilter in fresh air supply system. (b) No. 11: Downstream of ULPA filter in clean room. (c) No. 12: Return air flow from clean room.

sampled downstream of the HEPA filter. Mo and Nb, used as an etching material, were detected around the RIE-1 and RIE-2 equipment. The concentrations of Mo and Nb rose to 0.022 ng/m^3 near the RIE-1 and 0.26 ng/m^3 near the RIE-2, respectively. Most elements showed higher concentration near the

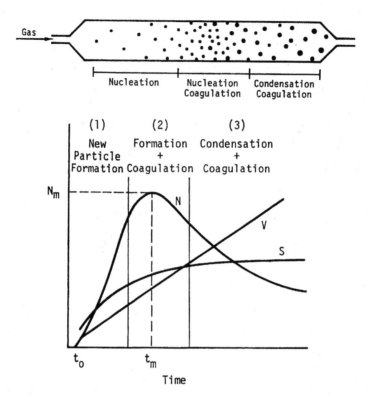

FIGURE 6.22   Illustration of aerosol formation and growth processes in flow-type smog chamber.

RIE-2 than near the RIE-1. This is thought to have been caused by the large differences in the opening and shutting frequency of the chamber. The concentration of Mo at the sampling point Sup-1 where Mo was produced during sputtering amounted to 0.51 ng/m³. The concentration of S and Fe were very high in the exhaust air from the sputtering room. This was presumably caused by the lubrication oil of the pump. It was also suggested that human perspiration and breath constituted the main source of Cl generation in the clean room.

It is proven that PIXE analysis was applied successfully to investigate the potential source of contamination within the clean room. Advanced application of PIXE analysis is expected to elucidate in more detail the behavior and origins of the small particles, especially how the particles are generated in the clean room.

6.5.2   Determination of Conversion Rate of SO_2 Gas to H_2SO_4 Particles

The secondary particles formed by gas-to-particle conversion in the atmosphere play an important role in atmospheric aerosols, especially in the fine particles. The gas-to-particle conversion process is divided into two main categories: the homogeneous nucleation (self-condensation) and the heterogeneous nucleation (condensation onto the small particles). Figure 6.22 illustrates the gas-to-particle conversion process in a flow-type smog chamber in which only gaseous reactants exist initially.[29] At first as a result of the homogeneous nucleation, the particle number concentration increases rapidly. Following the increase of preformed particles, the heterogeneous nucleation and coagulation become important and lead to the reduction of particle number concentration.

It is well known that SO_2 can form sulfuric acid particles in the atmosphere through sunlight irradiation. The conversion of SO_2 to H_2SO_4 particles is the most dominant path in the secondary particle formation in the atmosphere, and sulfuric acid gives rise to the deposition of acid particles and the reduction of visibility. The conversion rate of gas to particle is essential to evaluate the

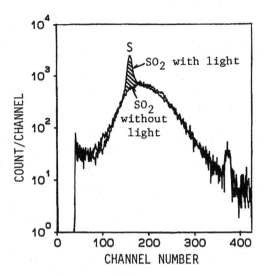

FIGURE 6.23 PIXE spectra of aerosol particles formed by reaction of SO_2 with and without UV light.

effects of the secondary particles and to develop a model simulating the transport and transformation of gaseous and particulate pollutants.

We tried to apply PIXE analysis to determine the conversion rate of SO_2 to H_2SO_4.[30] Sample gas adjusted SO_2 concentration and relative humidity was introduced into the flow-type smog chamber to be irradiated by UV light. The particle number concentration was monitored with a condensation nuclei counter. After the particle number concentration reached equilibrium, the particle size distribution was measured with an electrical aerosol size analyzer, and then sulfuric acid particles formed photochemically in the smog chamber were collected onto the 0.2 μm pore size Nuclepore filter. SO_2 concentration, relative humidity, and reaction time were changed as the experimental parameters.

PIXE analysis of the filter samples was performed with a 2.0 MeV proton beam from a tandem accelerator. Figure 6.23 shows the two PIXE spectra of the samples collected during the gas-to-particle conversion study. One of them was obtained with the irradiation of UV light, and the rest without the irradiation. Therefore, the difference between the two spectra gives the amount of sulfuric acid particles formed photochemically by UV irradiation. The first order conversion rates of SO_2 to H_2SO_4, which were estimated from the amount of H_2SO_4, the initial concentrations of SO_2, and the reaction time ranged from 0.007 to 0.035 h^{-1} under the conditions of dry (<2% r.h.) to 80% r.h., 5 to 20 ppm initial SO_2 concentration, and 75 to 300s reaction time. Since it is possible to heighten the sensitivity of sulfur more than tenfold in PIXE analysis by improving the measurement conditions and sampling method, the PIXE method may be applied to determine the conversion rate of SO_2 to H_2SO_4 in the actual atmosphere.

6.5.3 WET SCAVENGING OF AEROSOL PARTICLES BY RAINDROPS

In the last several years, the importance of global environmental problems has been recognized worldwide. Acid rain is one of the most important global environmental problems. In order to study the scavenging mechanism as a final goal, the characteristics of chemical components in rainwater were examined as a function of the amount of rainfall.[31]

Rainwater was sampled with each 0.1 mm rainfall from the beginning of the rain by using a newly developed rain collector. The pH value and electrical conductivity (EC) of rainwater were measured immediately after sampling. Preparation of the samples for PIXE analysis was carried out after storing the rainwater in a refrigerator until just before the analysis. The rainwater was

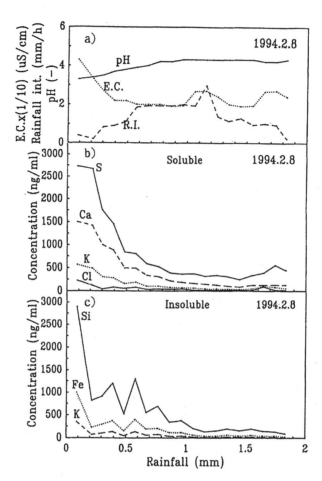

FIGURE 6.24 Changes as a function of amount of rainfall. Sampling: during Kosa event on February 8, 1994. (a) pH, Electrical conductivity, and rain intensity. (b) Elemental concentrations of S, Ca, K, and Cl in soluble component. (c) Elemental concentrations of Si, Fe, and K in insoluble component.

filtered by passing it through a 25-mm-diameter Nuclepore filter with 0.2 μm pore size. Filtrate and residue were considered to be the soluble and insoluble components, respectively. The PIXE sample of the soluble component was prepared from filtrate as follows: 20 μl of filtrate was dropped on the nonhole Nuclepore film using a micropipet and dried by an infrared lamp. In total, 1 to 10 drops were repeatedly dropped and dried according to the concentration level estimated from the color of the residue. The sample size was limited to 4 mm diameter so that the whole sample could be bombarded by the 2.0 MeV proton beam with a 6-mm diameter.

The rain samples were gathered at the 13 precipitation events between September 1993 and November 1994. The changes of pH, EC, rainfall intensity, and the concentration of some elements in the soluble and insoluble components shown in Figure 6.24 were obtained from the rain sample collected during the Kosa event. The figure is plotted as a function of the accumulated amount of rainfall from the beginning of rain.

The elemental concentrations and EC changed with a similar tendency. They decreased quickly from the beginning of rain to about 0.3 to 0.5 mm rainfall and then decreased gradually. This suggests that the scavenging of aerosol particles by rain drops advances very effectively at the beginning of rainfall. The concentrations of all the 15 elements analyzed in both soluble and insoluble components decreased in a similar way with the amount of rainfall. The concentrations of S in the soluble component and Si in the insoluble component reached about 3 μg/ml in the first

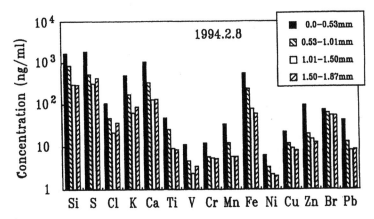

FIGURE 6.25   Total (soluble + insoluble components) concentrations of 15 elements in rainwater averaged with each 0.5 mm rainfall from the beginning of rain.

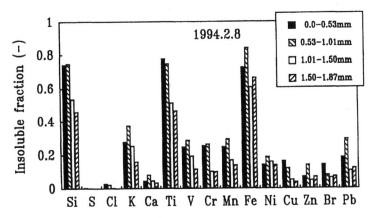

FIGURE 6.26   Insoluble fractions to total (soluble + insoluble) averaged with each 0.5 mm rainfall from the beginning of rain.

0.1 mm of rainfall. After 1.0 mm rainfall, the concentrations of most of the elements in both the soluble and insoluble components decreased to less than one tenth of their initial concentrations.

The total concentrations, that is, the sum of the soluble and insoluble components of 15 elements are illustrated in Figure 6.25. The values are averaged with each 0.5 mm rainfall from the beginning of the rain. The concentration of S, Si, Ca, K, and Fe are relatively high. It is reported that the contents of Si, Ca, and Fe in the Kosa dust were high and ranged between 20 and 30%, between 4 and 7%, and between 2 and 4%, respectively.[32] The average concentration ratio of Si, Ca, and Fe obtained in the present work was 4:2:1 with similar ratios in the literature. On the other hand, as a cause of the high concentration of sulfur, it is possible that S is also contained as $CaSO_4$ in the Kosa dust and that SO_2 gas is deposited on the surface of Kosa dust during the transportation to make the secondary particles.

Figure 6.26 shows an insoluble fraction defined by the ratio of the concentration of the insoluble component to the total concentration. The insoluble fractions of Fe, Ti, and Si, which are originated mainly in the soil dust, were large and exceeded 0.75 in the early stage of rain. On the other hand, the insoluble fractions of S, Cl, Ca, and Zn were smaller than 0.1 throughout rain. In the case of S and Cl, almost all existed in the soluble state in rain water. The insoluble fraction had generally a tendency to decrease with the amount of rainfall. This means that the insoluble components are more easily caught by rain drops. The qualitative nature of the insoluble fraction of each element was roughly similar to the data in the literature.[33]

6.6 SUMMARY

PIXE analysis has such advantages as excellent sensitivity, nondestructive technique, and simultaneous measurement for multielements. It has been successfully applied to aerosol studies for the last two decades. In this chapter, an outline of the PIXE method was described, and then examples of its application to aerosol studies, such as the characterization of atmospheric aerosols and the explication of the behavior of aerosol particles in the atmosphere, were introduced, using experimental data obtained by our research group.

It is beyond doubt that PIXE analysis is useful to aerosol studies, but unfortunately, it can only analyze metal components having an atomic number larger than about 13 (Al). That is, it cannot analyze the lighter elements, such as H, C, O, N, and Al, which usually occupy more than 80% of atmospheric aerosols. New techniques have been developed for analysis of the lighter elements, such as PIGE (particle-induced gamma-ray emission). PESA (proton elastic scattering analysis), and FAST (forward alpha scattering technique). It is required to establish a system that can analyze the lighter elements simultaneously.

REFERENCES

1. Johansson, T.B., Akelsson, R., and Johansson, J.A.E. *Nucl. Instr. Methods,* 84, 3141-3143, 1970.
2. Cahill, T.A. *Nucl. Instr. Methods,* B49, 345-350, 1990.
3. Koltay, E. *Int. J. PIXE,* 1, 93-112, 1990.
4. Cahill, T.A. *Ann. Rev. Nucl. Part. Sci.,* 30, 211-252, 1980.
5. Kasahara, M., Ogiwara, H., Yamamoto, K., Park, J.H., and Takahashi, K. *Int. J. PIXE,* 4, 155-164, 1994.
6. Kasahara, M., Takahashi, K., Sakisaka, M., and Tomita, M. *Nucl. Instr. Methods,* B75, 136-139, 1993.
7. Kasahara, M., Yoshida, K., and Takahashi, K. *Nucl. Instr. Methods,* B75, 240-244, 1993.
8. Swietlicki, E. and Bohgard, M. *Nucl. Instr. Methods,* B3, 441-445, 1984.
9. Unpublished data.
10. Bonani, G., Satish, J., Wanner, H.U., and Wolfli, W. *Nucl. Instr. Methods,* B3, 493-497, 1984.
11. Akelsson, K.R. *Nucl. Instr. Methods,* B3, 425-430, 1984.
12. Cahill, T.A., Feeney, P.J., and Eldred, R.A. *Nucl. Instr. Methods,* B22, 344-348, 1987.
13. Spurny, K.R., Lodge, J.P., Jr., Frank, E.R., and Sheesley, D.C. *Environ. Sci. Tech.* 3, 453-463, 1969.
14. Kemp, K. and Kownacka, L. *Nucl. Instr. Methods,* B22, 340-343, 1987.
15. Kasahara, M. Estimation of source contribution of atmospheric aerosols, Ministry of Education, Culture and Science, Japan, Grant No. 05452399, pp. 74, 1990. [in Japanese]
16. Kasahara, M., Shinoda, K., Takahashi, K., and Yoshida, K. *Int. J. PIXE,* 3, 313-318, 1993.
17. Gotoh, N. MS Thesis, Kyoto University, 1994. [in Japanese]
18. Kasahara, M., Takahashi, K., and Choi, K.-C. *Atmos. Environ.,* 24A, 457-466, 1990.
19. e.g., Hidy, G.M., et al. *J. Air Poll. Contr. Assoc.* 25, 1106-1114, 1975.
20. Kasahara, M., Takahashi, K., Choi, K.-C., and Yoshida, K. *Aerosols,* Pergamon Press, 938-941, 1990.
21. Kasahara, M., Choi, K.-C., and Takahashi, K. *Int. J. PIXE,* 2, 665-678, 1992.
22. Klaus, N. and Berner, A. *Staub–Reinhalt. Luft,* 45, 168-170, 1985.
23. Horvath, H., Kasahara, M., and Pesava, P. The size distribution and composition of the atmospheric aerosol at a rural and nearby urban location, *J. Aerosol Sci.,* in press, 1995.
24. Unpublished data.
25. Kasahara, M., Park, J.H., and Yamamoto, K. Characterization of atmospheric aerosols separated by particle size and water solubility using PIXE analysis, *Nucl. Instr. Methods,* in press, 1995.
26. Miller, M.S., Friedlander, S.K., Hidy, G.M., *J. Colloid Interface Sci.* 39, 165-176, 1972.
27. Kasahara, M., Tada, A., Moriya, M., Masuda, A., and Kubo, K. *Proc. of 4th Int. Aerosol Conf.,* 526-527, 1994.
28. Ro, T., Ishigro, T., and Kasahara, M. The future practice of contamination control, *Mechanical Eng. Pub. Lim.,* 329-332, 1992.
29. Kasahara, M. and Takahashi, K. *Aerosols — Formation and Reactivity,* Pergamon Press, 1172-1175, 1986.

30. Kasahara, M., Tada, A., and Takahashi, K. *J. Aerosol Research,* Japan, 9, 141-151, 1994. [in Japanese]
31. Kasahara, M., Ogiwara, H., and Yamamoto, K. Soluble and insoluble components of air pollutants scavenged by rain water, *Nucl. Instr. Methods,* in press, 1995.
32. Institute of Hydrospheric — Atmospheric Science, Nagoya University, Kosa, Kokon-shoin, pp. 327, 1991.
33. e.g., Giusti, L., Yang, Y.-L., Hewitt, C.N., Hamilton-Taylor, J., and Davison, W.*Atmos. Environ.,* 27A, 1567-1578, 1993.

7 Direct and Near Real-Time Determination of Metals in Aerosols by Impaction-Graphite Furnace Atomic Spectrometry

Joseph Sneddon, Suneetha Indurthy, Mark V. Smith, and Yong-Ill Lee

CONTENTS

7.1   INTRODUCTION

Increasing awareness and concern about the deleterious effects of atmospheric and industrial pollution caused by metals or metallic compounds has led to a need and desire for a rapid, direct, *in situ,* and real-time detection system. The detection and quantitative determination of metals or metallic compounds in aerosols (a solid or liquid particle in a gaseous medium, e.g., air) suffers from a lack of promptness. Typically, a collection stage on a filter system (cellulose or glass fiber filters are the most common) of 0.5 hours to several days, followed by sample preparation (digestion or dissolution), and then the subsequent determination by a variety of analytical techniques, including atomic spectroscopy, is commonly used. This long sampling period is needed to obtain a measurable amount of metal when concentrations are low. Typically, they are low to sub $\mu g/m^3$. While this method is routinely used and provides valuable information, it is tedious, time consuming, and can lead to losses (and reduced accuracy) due to storage, transportation, and contamination. A review of these standard sampling and sample preparation methods combined with various atomic spectroscopic techniques for metal determination is available (Sneddon, 1983), as well as the use of graphite furnace atomic absorption spectrometry (GFAAS) in determining metals in air (Noller et al., 1982). Furthermore, the standard methods are in retrospective and do not include the possibility that a short duration, high concentration, incident exposure, which, when averaged over the sample collection time period, may not appear to be dangerous.

Clearly, there is a need for a system that is capable of providing a quantitative determination of low concentrations of metals in air on a real-time or near real-time basis. Atomic spectroscopic techniques (atomic emission, atomic absorption, and atomic fluorescence spectrometry) are excellent methods for trace metal determination due to several factors, including low sensitivity and detection limits, no, minimal, or easily corrected interferences, high specificity, good (low) precision, acceptable accuracy, widespread availability, and low cost per sample after the initial cost of instrumentation. However, they are primarily regarded (and give best performance) as solution or solid sampling techniques. They have been applied to determine metals in air after collection and digestion or dissolution but have not been widely applied to direct air analysis for metal determination. Several atomic spectroscopic techniques have shown potential for directly determining metals in air including electrostatic precipitation-graphite furnace atomic absorption spectroscopy (AAS) (Torsi et al., 1982; Sneddon, 1989b, 1991; Torsi et al., 1996) and laser-induced breakdown (emission) spectrometry (LIBS) (Cremers et al., 1983; Essien et al., 1988a, 1988b; Radziemski, 1994).

Another promising technique is that of combining a single-stage impactor to a graphite furnace for the direct collection of particles in air for subsequent determination by atomic spectroscopic techniques, primarily atomic absorption spectrometry in a near real-time (a few minutes) manner. The technique has been referred to as impaction-graphite furnace atomic absorption spectrometry (I-GFAAS). This chapter will present and discuss this technique as it is applied to the direct and near real-time determination of metals in air. This area has been discussed by Sneddon, 1986, 1988, 1990, and Sneddon et al., 1995.

7.2 IMPACTION

Impaction techniques have been widely and extensively used in industrial hygiene for the collection and sizing of aerosols (Marples, 1970; Fuchs, 1978). The size of particles in an aerosol is considered important from a health point of view. In general, particles in the one to fifteen micron size are carried more efficiently into the lungs. Particles larger than fifteen microns are stuck or cannot pass through the nose, and particles smaller than one micron pass through the body and are excreted. However, accumulation in the body may play a role in characterizing the toxicity of a metal and a subsequent health hazard.

7.2.1 BASIC PRINCIPLES

An impactor is an instrument in which an aerosol issuing from a narrow jet impinges on an impaction plate or surface. Aerosol particles are deposited on this surface because of their inertia. One of the first instruments was described by May in 1945. It consisted of four jets and four sampling plates. The jets became progressively smaller, so that as the speed of the aerosol increased, the finer or smaller particles were collected or impacted on the impaction surface and removed from the aerosol. The aerosol then continued to the next smaller jet and impaction plate. The impaction plates were usually the same material and size. The final product would be a size grading of the aerosol. In the case of the instrument developed by May in 1945, this would be four sizes (actually ranges) of aerosol based on the four series of jets and impaction plates. This instrument was the forerunner to commercial systems, including the Anderson Impactor (Anderson Impactors, Inc., Atlanta, Georgia). These systems were inexpensive and relatively straightforward to use. Several new and

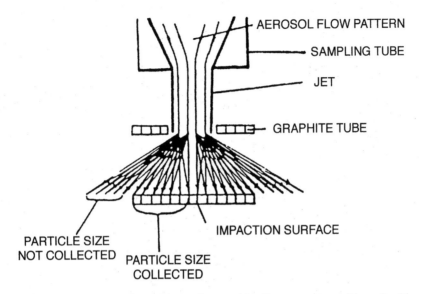

FIGURE 7.1 Schematic cross section of impaction-graphite furnace system. (From Sneddon, J., 1984, *Analytical Chemistry,* 56, 1982–1986. With permission.)

more sophisticated cascade-type systems are available, including 10-stage piezoelectric systems which feature *in situ* electronic weighing and give complete mass (concentration and size distribution) information in a few minutes (Californian Measurements, Sierra Madre, California). In these cases, the information on particle sizes was considered the major focus, as outlined earlier. The use of impaction combined with a graphite furnace uses the principle outlined above. However, in this case, a single stage impactor consisting of a single jet and graphite furnace (impaction plate) are used. The basic principle is shown in Figure 7.1.

The aerosol or air sample is drawn by vacuum at a known flow rate or speed through the single-stage impactor jet. Particles in the aerosol are deposited on the impaction (graphite furnace) surface. This single stage impaction-graphite furnace system will separate particles in an aerosol into two sizes: particles larger than a certain aerodynamic size are removed from the aerosol and deposited (impacted) on the graphite furnace and smaller particles will pass through the impactor-graphite furnace system. The particles removed from the aerosol and collected on the graphite tube can then be added to a graphite furnace atomization unit for subsequent quantification by (primarily) graphite furnace atomic absorption spectrometry.

7.2.2 THEORETICAL CONSIDERATIONS

The movement of an aerosol in a single stage I-GF system is extremely complex: the motion relies on many factors including flow rate, jet diameter, and the distance from the jet exit to the graphite furnace. These factors will be critical in determining the particle size collected on the I-GF. An ideal system would collect and remove all particle sizes from an aerosol. This is termed *collection efficiency* and is defined as the mass of metal collected on the graphite furnace after entering the system divided by the mass of the metal initially entering the system, expressed as a percentage. Ideally, this would be 100%. However, a single-stage I-GF would not have a 100% collection efficiency and will have a cut-off particle size, d_{50}, which is defined as the particle size at which at least 50% of a certain particle size is collected.

If particle motion in an aerosol is governed by Stokes' Law (Hinds, 1982), then

$$Stk = \frac{\rho(dp)^2 U\, Cc}{9\,\mu\, D} \tag{1}$$

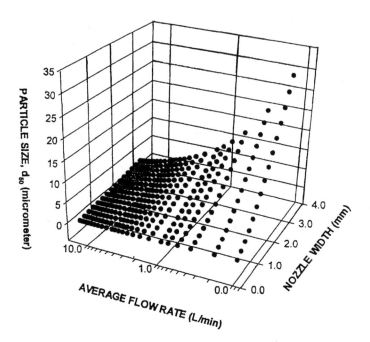

FIGURE 7.2 Theoretical study of particle size collected, d_{50}, versus flow rate and jet or nozzle diameter. (From Lee, Y.I., Smith, M.V., Indurthy, S., Deval, A., and Sneddon, J., 1996a, *Spectrochimica Acta*, 51B(1): 109-116. With permission.)

where U is the average jet exit flow velocity (in cm/s), ρ is the density of the particle (= 1 g/cm³), D is the jet diameter (in mm), and μ is the viscosity of air = 1.81×10^{-4} g/cm at 20°C, dp is the particle diameter, Cc is the Cunningham correction factor (~1), and Stk is the Stokes Number. Assuming $(Stk)^{1/2} = 0.475$ (Hinds, 1982) and rearranging Equation (1) and substituting d_{50} for dp:

$$d_{50} = \left(\frac{9\, Stk\, \mu\, D}{\rho\, U\, Cc} \right)^{1/2}$$

(2)

Using this equation and the constants, a plot of particle size collected, d_{50}, versus average flow rate and jet or nozzle diameter is shown in Figure 7.2. This figure reveals information on the theoretical particle size collected using the geometry of the I-GF system. At low flow rates of around 0.1 L/min and a jet diameter of 1.0 mm, particles of less than 6 µm would not be collected. At a higher flow rate, e.g., 5 L/min, particles of less than 2 µm would not be collected, and at 15 L/min, the size of particles not collected is less than 0.5 µm. In general, increasing the flow rate will decrease the particle size collected. Decreasing the jet diameter to 0.5 mm would decrease the particle size collected, e.g., at a low flow rate of 0.1 L/min, the particle size collected would drop to 2 µm (compared to 6 µm at 1.0 mm jet diameter). Increasing the jet diameter to 1.5 mm would increase the particle size collected. At 0.1 L/min, the particle size collected would be approximately 10 µm. Increasing the jet diameter beyond the 1.5 mm diameter would greatly increase the particle size diameter collected, particularly at low flow rates. However, a jet diameter of much greater than 1.5 mm is impractical because it would be difficult to locate the aerosol or air sample in the graphite furnace. Furthermore, as described earlier, particle sizes in the range of 1 to 15 µm are considered to be the most damaging to health because of their efficiency in entering and remaining in the body.

Finally, it is worth noting that these theoretical calculations provide information on the particle size collected but do not take into account the possibility of "bounce-off" errors (Hinds, 1982).

The "bounce-off" is where a particle will literally bounce off the surface, although it should theoretically be collected on the surface. However, they do provide information which allows the construction of a system with a high probability of collecting the potentially dangerous particles in the size range of 1 to 15 μm.

7.3 INSTRUMENTATION

7.3.1 IMPACTION SYSTEMS

Using the theoretical studies previously outlined for design purposes, several systems have been constructed, characterized, and evaluated (Sneddon, 1984, 1985). This system was designed and constructed as a complete unit in which the impactor was connected directly to the graphite furnace atomizer. Later systems separated the collection stage from the analysis stage.

The jet was made of tantalum to prevent melting at the relatively high temperature (~3000 K) obtained by the graphite furnace at the analysis stage. The inside diameter of the jet was 1.00 mm and its outside diameter was 2.00 mm. The jet was pressed into the impactor tube. The impactor tube was made of easily machinable aluminum of 30.0 mm diameter. The inside diameter was initially 20.0 mm and gradually tapered to an inside diameter of 1.00 mm over a length of 110.0 mm. The gradual decrease in diameter prevented any build-up of particles on sharp edges. A 40.0-mm length of the impactor was threaded and was matched with the specially constructed faceplate, which replaced the faceplate of the commercial graphite furnace atomizer. This was to allow the jet-to-impactor distance to be varied. The faceplate was made of 10.0-mm-thick aluminum and had a quartz viewing window for visual alignment of the jet to the graphite tube. Two cylinder outlets of inside diameter 5.00 mm and height of 15.00 mm were pressed into the faceplate. These two outlets were connected via a tee piece to a flowmeter and then to a vacuum pump, with vacuum tubing used in all the connections. This system allowed a maximum flow rate of 15 L/min. All contact surfaces were sealed with a rubber seal and checked periodically to ensure no leaks. A connecting faceplate (D) and sampling faceplate (E) were connected to the impactor. In position (a), the air was sampled, in position (b), the determination was performed, and in position (c), a standard aerosol could be introduced. A rubber stopper with a glass jet (F) was connected to position (c) and to the standardization unit. This standardization unit was an aerosol deposition system as described by Tapia et al. (1984). The analysis (sampling and determination) could be performed by manually moving the faceplate, although, clearly, there was potential for automation.

A second system was constructed by Liang et al. in 1990. In this device the sampling unit was separate from the graphite furnace system. It consisted of the same impactor tube previously described (Sneddon, 1985), and a plastic chamber with the graphite furnace was inserted into this chamber. Air was drawn by vacuum through the ends of the chamber. After collection at a known flow rate and sampling time, the system was disconnected and the graphite tube removed and placed on the graphite furnace atomization unit for determination. This system was used for graphite furnace AAS and laser-excited atomic fluorescence spectrometry.

A third and most current system is shown schematically in Figure 7.3a and Figure 7.3b (Lee et al., 1996a). It was constructed from a nylon block. The impactor tube was also constructed from nylon and was 10 cm in length by 22 mm diameter. The impactor tube decreased from an inside diameter of 20 mm to the jet exit diameter over the 10 cm length. Several jet diameters were available (0.5, 1.0, and 1.5 mm). The outside of the impactor tube was threaded to allow a complete fit to the rest of the system. This had two advantages: the first was an airtight fit, and the second was that the distance between the jet exit diameter and graphite (impaction) surface could be carefully controlled and varied if required. Three nearly identical impactor tubes were constructed, the only difference being that the jets were of 0.5, 1.0, and 1.5 mm inside diameter. When the 1.0 and 1.5 mm jets were used, the graphite tube entry port had to be slightly enlarged. This new system

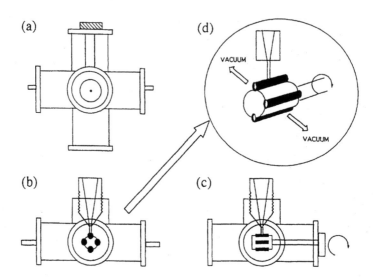

FIGURE 7.3 Schematic diagram of impaction system. (From Lee, Y.I., Smith, M.V., Indurthy, S., Deval, A., and Sneddon, J., 1996a, *Spectrochimica Acta,* 51B(1): 109-116. With permission.)

had a barrel-type fitting which allowed up to four separate graphite tubes to be installed (see Figure 7.3b, part d). An experiment could be performed using one graphite tube by rotating the system to a new graphite tube with the same impactor. This could be repeated a third and fourth time. This was considered to be particularly useful when studying precision from aqueous solutions introduced through the impactor.

7.3.2 DETECTION SYSTEMS

The graphite furnaces used as the collection or impaction surface are, for the most part, from commercial graphite furnace atomization systems. The exception was a laboratory-modified atomization graphite furnace unit used for laser-excited atomic fluorescence spectrometry (LEAFS) (Liang et al., 1990). Initial work was performed using AAS systems which were capable of only single metal determination (Sneddon, 1984, 1985). Recently, this has been extended to multimetal atomic absorption spectrometric systems, (Sneddon et al., 1995; Lee et al., 1996a).

7.4 EXPERIMENTAL RESULTS

7.4.1 EFFECT OF GEOMETRY

Initial work and studies on the I-GF system found that the geometry of the system played an important part in collection efficiency. An ideal collection efficiency would be 100%, but this is not possible with a single-stage impactor. A collection efficiency of 100% would require a system capable of collecting all particle sizes, i.e., a multistage impactor system. Collection efficiency depends on a number of factors related to the system, including geometry (jet exit-to-impaction surface distance, jet diameter, and impaction surface) as well as flow rate and particle size. A rigorous study was undertaken by Lee et al. (1996b) to study the effect of these five parameters on collection efficiency.

Based on this work, Lee et al. (1996b) proposed that the optimum conditions for maximization of collection efficiency for particles of less than six microns is a combination of medium flow rates (around 5 L/min) and jet diameter of 1.0 mm, and a jet-to-impaction distance of 3 mm. The higher the flow rate, the more dense the particles are on the graphite furnace. The results obtained in this recent study by Lee et al. (1996b) confirmed an earlier study by Sneddon (1989b). In this case the work was performed using the early system which had the impactor connected directly to the graphite furnace atomization unit. This earlier study did pinpoint the variables, such as geometry, flow rate, and particle size, which did effect collection efficiency.

7.4.2 STANDARDIZATION

To calculate the concentration of metal in the aerosol or air, a calibration curve is established using aqueous solution standards as follows (Sneddon, 1983, 1985; Lee et al., 1996a):

$$Cm = \frac{Mm}{Va} = \frac{Cstd \times Vstd}{Fr \times St} \tag{3}$$

where Cm is the equivalent concentration of metal in ng/m^3, Mm is the mass of metal in aqueous standard in ng, Va is the volume of air sampled in m^3, Cstd is the concentration of standard in ng/mL (parts per billion (ppb)), Vstd is the volume of standard in mL, Fr is the flow rate in m^3/mL, and St is the sampling time in minutes. An example would be as follows: Cstd = 50 ppb (= 50 ng/mL), Vstd = 10 μL (= 10×10^{-3} mL), Fr = 5 L/min (= 5×10^{-3} m^3/min), and St = 5 min

$$Cm = \frac{50 \text{ ng/mL} \times 10 \times 10^{-3} \text{ mL}}{5 \times 10^{-3} \text{ m}^3/\text{min} \times 5 \text{ min}} \tag{4}$$

$$= 20 \text{ ng/m}^3$$

Thus a calibration curve of absorbance (usually peak area absorbance, but peak height absorbance could be used) versus concentration in ng/m^3 can be established using various concentrations, of aqueous standards. If the conditions change, i.e., flow rate, sampling time, concentration or volume of standard, then a new calibration curve has to be established. It should be noted that standardization is achieved by introduction through the impactor system in the same manner as the air samples.

7.4.3 ACCURACY AND PRECISION

Accuracy can be defined as how close a measurement is to the "correct" answer. It is usually presented statistically and can be established using a number of complementary techniques, including standard additions, internal standards, comparison of the results obtained using this method to a different method, and a standard sample containing a known concentration of the particular metal of interest, typically a National Institutes of Science and Technology (NIST), Standard Reference Material (SRM) (Gaithersburg, Maryland).

Clearly, none of the above methods is available to assess the accuracy of the I-GF system. The NIST urban particulate Standard Reference Material (SRM), 1648, could be added as a powder or slurry, although, at present, this has not been undertaken. An attempt to assess the accuracy of the I-GF system was undertaken by Sneddon (1990) by comparison to standard methods of sampling air by collection on a filter, followed by digestion of the filter, and analysis by flame atomic absorption spectrometry. The experiments involved the use of a filter (0.8 μm pore size) and pump, which sampled the air at a known flow rate, was positioned as close to the I-GF system as possible, and simultaneously sampled air. In both systems, air was sampled at 5 L/min for 5 min. A cap was placed on the end of the filter system during the approximately 2 min it took to analyze the air collected by the I-GF system. After approximately 210 min, 150 L of air had been collected using both systems. The total metal mass collected by the I-GF system was determined by adding the

30 separate analyses, and the total mass content collected by the conventional filter method was determined by digesting the filter paper and analyzing it. Using three separate experiments, the mass collected on the I-GF system was found to be 59 to 69%, compared to the conventional method. These results were quite reproducible. As stated previously, the design and geometry of this single-stage I-GF system would prevent the collection of large particle sizes, and therefore, it is not surprising that a direct 100% comparison was not obtained. The 30 to 40% difference was attributed to the fact that the conventional system will collect all particle sizes, including large particles, whereas the I-GF system will collect particles less than 15 μm in size.

A further factor that could affect accuracy is the fact the system is standardized using aqueous standards. In most cases, it is not acceptable (accurate) practice to use aqueous standards to calculate the concentration of an air sample which will contain a complex matrix. However, for this type of work a factor of two would not be significant in industrial hygiene, i.e., the difference in (say) 5 ng/m^3 and 10 ng/m^3 would not be considered significant.

Precision is defined as the consistency of results in the repetitive analysis of the same sample. Several experiments (three for each method) were performed using the I-GF system using sample introduction by electrothermal vaporization (ETV), a pneumatic nebulization (PN) system, and air (Sneddon, 1990). A precision of 2 to 3% was obtained using the ETV and PN, which is comparable to that obtained by direct sample introduction into a conventional GF-AAS. The air sample precision ranged from 7.6 to 9.9%, a reduction in precision by a factor of four. This highlighted a potential problem with the air sampled, namely that air is not a homogeneous sample. Therefore, the precision obtained, or long-term stability of the concentration of metal in air, will not be constant or (necessarily) expected. Liang et al. (1990) discussed this subject when they analyzed their results of a study of six metals in laboratory air, and a clean room. Further discussion of this is presented in the applications part of this chapter.

7.4.4 Detection Limits, Characteristic Concentration, and Useful Working Range

The detection limits, characteristic concentration, and useful working range obtained using I-GF-AAS are identical to that obtained from conventional GF-AAS. A selection of metals in which the detection limit, characteristic concentration, and useful working range are presented in units of ng/m^3 for selected metals using various wavelengths are shown in Table 7.1, (Sneddon, 1986).

7.5 PRACTICE OF IMPACTION-GF SYSTEM

The advantage of the impaction system is the ability to directly collect particles in air for subsequent quantitation by atomic spectroscopy, most commonly atomic absorption spectrometry. The collection time will vary depending on the flow rate and concentration of metal in air to be determined. The maximum flow rate in the impaction system appears to be around 15 L/min. A collection stage of 5 min at the maximum flow rate will collect 75 L of air. In many instances, this is sufficient to obtain a measurable amount. In some instances, a much longer sampling period is required (see mercury in air, Sneddon, 1989). Early impaction surfaces were connected directly to the graphite furnace atomization unit of an AAS system, but later systems were separate. Using the later systems involved dismantling the impaction system and adding to the graphite furnace and (with practice) took around 30 s, and the actual determination (assuming a predetermined set of experimental conditions of drying, ashing or pyrolysis, atomization, and cleaning) of around 2 min. This gave the results in about 5 min (assuming that a collection stage of a few minutes was adequate and that the standardization had been achieved). This is why the impaction-GF AAS is referred to as near real-time. In practice, the system would be standardized prior to sampling and experimental conditions determined prior to analyses.

TABLE 7.1
Detection Limits, Characteristic Concentration, and Useful Working Range for Selected Metals by I-GF AAS[a]

Metal	Wavelength nm	Detection Limit[b] ng/m^3	Characteristic Concentration[c] ng/m^3	Useful range ng/m^3
Arsenic	193.7*	24.0	3.0	40–400
Barium	350.1	600	—	—
	553.6*	10.0	1.0	20–150
Beryllium	234.9*	2.0	0.3	4–40
Bismuth	223.1*	20.0	3.0	5–40
	227.7	300	100	500–800
Chromium	357.9*	12.0	6.0	20–80
	425.4	60.0	20.0	100–300
	520.8	1000	—	—
Cobalt	240.7*	20.0	8.0	30–100
	391.0	1000	—	—
Iron	248.3*	1.0	0.2	5–90
	372.0	10	20	40–140
	392.0	200	30	300–600
Mercury	253.7*	100	20	200–400
Lithium	323.3	800	—	—
	670.8*	2.0	0.7	3–30
Nickel	232.0*	10.0	1.3	20–200
	341.5	50	10	70–200
	362.5	1000	—	—
Osmium	290.9*	70	18	100–300
	426.1	1000	—	—
Lead	217.0*	20.0	5	40–200
	283.3	30.0	5	40–200
Selenium	196.0*	25.0	5	50–300
	204.0	300	—	
Silicon	251.6*	20.0	5	30–200
	288.2	300	—	—
Silver	328.1*	20.0	5	30–300
	338.3	100	20	150–500
Tin	286.3*	11.0	3.6	20–200
Zinc	213.9*	0.5	0.1	1–15
	376	300	—	—

* Resonance (most sensitive) wavelength.

[a] Obtained using a volume of equivalent to 20 μL, flow rate of 10 L/min, and sampling time of 5 min.

[b] Concentration giving a signal-to-noise ratio of 3.

[c] Concentration which gives 1% absorption (0.0044 absorbance units).

[d] Range of concentration for which the relative standard deviation (precision) is less than 5%.

Source: From Sneddon, J., 1986, *American Laboratory,* 18(3), 43–50. With permission.

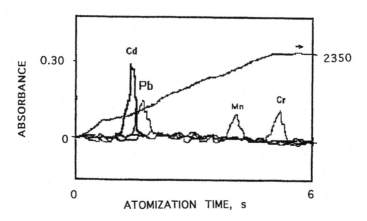

FIGURE 7.4 Absorption profiles of cadmium, chromium, lead, and manganese obtained from cigarette smoke collected by the I-GF system and subsequently determined by multimetal atomic absorption spectrometry. (From Lee, Y.I., Smith, M.V., Indurthy, S., Deval, A., and Sneddon, J., 1996a, *Spectrochimica Acta*, 51B(1): 109-116. With permission.)

7.6   APPLICATIONS

To date, the impaction-GF system has not been widely applied to the determination of metals in air or aerosols. However, sufficient applications and investigations have been performed to show the potential of the system, particularly for determining low concentrations of metals in laboratory air (Sneddon, 1983, 1986, 1987; Liang et al., 1990; Lee et al., 1997).

Recently, the determination of selected metals, specifically cadmium, chromium, lead, and manganese, in cigarette smoke has been achieved using an impaction system and subsequent determination with simultaneous multimetal AAS. Multimetal AAS is not widely used, although there has been increased use and interest in this technique since the early 1990s. A description of multimetal AAS is available from Farah et al. (1993), Deval and Sneddon (1995), and Farah and Sneddon (1995). Tobacco in cigarettes has been extensively studied, primarily to assess the effect of nicotine on the human body. Some work has been performed on the determination of selected metals in various parts (filter, paper, and tobacco) of cigarettes by several analytical techniques, including neutron activation analysis and electrothermal (graphite furnace) atomic absorption spectrometry. In these cases, the cigarette part was first digested, and then presented to a system for quantitation. However, it has been recognized that cigarette smoke may be more dangerous to health because it can be directly and efficiently ingested into the human body. Of particular concern is second-hand cigarette smoke. This has resulted in many restaurants, cinemas, and all domestic line flights completely banning smoking. A typical AAS profile is shown in Figure 7.4 for the simultaneous measurement of these metals in cigarette smoke. Quantitation of these results would be achieved by standardization as described previously in this chapter.

Experiments were performed in the laboratory air prior to smoking, during smoking, and four hours after smoking. The results of this study are shown in Table 7.2.

Results showed variation by as much as a factor of 3, e.g., lead levels in background air are reported as 20 ng/m^3 from Table 7.2. However, results from other experiments showed lead concentrations from 8 to 25 ng/m^3. This variation is somewhat expected because the air sampled is not homogeneous. The results from Table 7.2 show that there is a significant increase in metal concentration in the cigarette smoke from a factor of almost 5 for lead to in excess of 15 for cadmium.

TABLE 7.2
Concentrations of Cadmium, Chromium, Lead, and Manganese Determined in Air[a] Using I-GF-AAS

Metal	Before Introduction of Cigarette Smoke ng/m³	Cigarette Smoke ng/m³	4 hours After Cigarette Smoke ng/m³
Cadmium	8	128	16
Chromium	4	23	4
Lead	20	88	17
Manganese	10	68	14

[a] Air sampled at 10 L/min. for 10 seconds.

Source: From Lee, Y.I., Smith, M.V., Indurthy, S., Deval, A., and Sneddon, J., 1996a, *Spectrochimica Acta,* 51B(1): 109-116. With permission.

7.7   CONCLUSION

Sufficient applications have shown the potential of a single-stage impactor connected to a graphite furnace for the determination of low concentrations of metals in aerosols or air. Recently, the work has concentrated on new designs and applications to cigarette smoke and burned oil (Noble et al., 1997; Hammond et al., 1998).

ACKNOWLEDGMENTS

The authors gratefully acknowledge the generous support of the Thermo Jarrell Ash-Baird Corporation, in particular, Gerald R. Dulude, Zach Moseley, and John J. Sotera. This work was supported by the Louisiana Education Quality Support Fund (LEQSF) Research Program for 1994-96-RD-A-21.

REFERENCES

Browner, R.F. and Boorn, A.W., 1984, *Analytical Chemistry,* 56: 786A-798A.
Cremers, D.A., Radziemski, L.J., Loree, T.R., and Hoffman, N.M., 1983, *Analytical Chemistry,* 55: 1246-1251.
Deval, A. and Sneddon, J., 1995, *Microchemical Journal,* 52(1): 96-100.
Documentation of the Threshold Limit Values for Substances in the Workroom Air, American Conference of Government Hygienists, Cincinnati, 3rd, edition, 1975.
Essien, M., Radziemski, L.J., and Sneddon, J., 1998a, *Proceedings of International Conference on Lasers 87,* STS Press, McLean, Virginia, pp. 908-912.
Essien, M., Radziemski, L.J., and Sneddon, J., 1988b, *Journal of Analytical Atomic Spectrometry,* 3: 985-988.
Farah, K.S., Farah, B.D., and Sneddon, J., 1993, *Microchemical Journal,* 48(3): 318-325.
Farah, K.S. and Sneddon, J., 1995, *Applied Spectroscopy Reviews,* 30(4): 351-371.
Fuchs, N.A., 1978, Chapter 1, *Fundamentals of Aerosols Science,* Shaw, D.T., (Ed.), John Wiley & Sons, New York.
Hammond, J.L., Noble, C.O., Beck, J.N., Proffitt, C.E., and Sneddon, J., 1998, *Talanta,* 47(2):261-266.
Health and Safety at Work Act, 1974, H.M.S.0., London, United Kingdom.
Hinds, W.C., 1982, *Aerosol Technology,* John Wiley & Sons, New York.
Lee, Y.I., Smith, M.V., Indurthy, S., Deval, A., and Sneddon, J., 1996a, *Spectrochimica Acta,* 51B(1): 109-116.
Lee, Y.I., Indurthy, S., Smith, M.V., and Sneddon, J., 1996b, *Analytical Letters,* 29(14): 2515-2524.

Liang, Z.L., Wei, G.T., Irvin, R.L., Walton, A.P., Michel, R.G., and Sneddon, J., 1990, *Analytical Chemistry,* 62(13): 1452-1457.

Marples, V.A., 1970, *Fundamental Study of Inertial Impactors,* Ph.D. Thesis, University of Minnesota.

May, K.R., 1945, *Journal of Scientific Instrumentation,* 22: 187-190.

Noble, C.O., Hammond, J.L., Beck, J.N., Proffitt, C.E., and Sneddon, J., 1997, *Microchemical Journal,* 57:361-369.

Noller, B.N., Bloom, H., and Arnold, A.P., 1982, *Progress In Atomic Spectroscopy,* 3: 81-189.

Radziemski, L.J., 1994, *Microchemical Journal,* 50: 218-234.

Sneddon, J., 1983, *Talanta,* 30: 631-648.

Sneddon, J., 1984, *Analytical Chemistry,* 56: 1982-1986.

Sneddon, J., 1985, *Analytical Letters,* 18(A10): 1261-1280.

Sneddon, J., 1986, *American Laboratory,* 18(3): 43-50.

Sneddon, J., 1987, *Spectroscopy Letters,* 20(6 & 7): 527-535.

Sneddon, J., 1988, *Trends in Analytical Chemistry,* 7(6): 222-226.

Sneddon, J., 1989a, *Analytical Letters,* 22(13 &14): 2887-2893.

Sneddon, J., 1989b, *Applied Spectroscopy,* 43(6): 1100-1102.

Sneddon, J., 1990a, Chapter 12 in *Sample Introduction in Atomic Spectroscopy,* Sneddon, J. (Ed.), Elsevier Publications, Amsterdam, 329-352.

Sneddon, J., 1990b, *Analytical Letters,* 23(6): 1107-1112.

Sneddon, J., 1990c, *Applied Spectroscopy,* 44(9): 1562-1565.

Sneddon, J., 1991, *Analytica Chimica Acta,* 245(2): 203-206.

Sneddon, J., Smith, M.V., Indurtha, S., and Lee, Y.I., 1995, *Spectroscopy,* 10(1): 26-30.

Tapia, T.A., Combs, P.A., and Sneddon, J., 1984, *Analytical Letters,* 17(A8): 2333-2347.

Torsi, G., Desimoni, E., Palmisano F., and Sabbatani, L., 1982, *Analyst,* 107: 96-101.

Torsi, G., Reschiglian, P., Lippolis, M.T., and Toschi, A., 1996, *Microchemical Journal,* 1996, 54(4): 437-445.

8 Mössbauer Study of the Structure of Iron-Containing Atmospheric Aerosols

B. Kopcewicz and M. Kopcewicz

CONTENTS

ABSTRACT

Mössbauer spectroscopy was applied to study the structure of iron-containing particles of atmospheric aerosols. From the temperature dependence of the Mössbauer spectra it was concluded that iron appears in atmospheric aerosol mostly in the form of ultrafine, superparamagnetic particles of Fe_2O_3. The Mössbauer technique permits not only the identification of the chemical compound in which iron appears but also the evaluation of the size of particles in which it is contained. The analysis of the quadrupole splitting distributions provide information on the differences of particle sizes in aerosol collected in different geographical locations. The concentration of iron in air has been calculated from the experimental spectra. The seasonal variation of iron concentration at selected sampling sites is discussed.

Keywords: Mössbauer spectroscopy, atmospheric aerosols, iron-in-air, small particles, superparamagnetism, air pollution, environment.

8.1 INTRODUCTION

Aerosols are a minor, compared to gases, component of the atmosphere, but their role is enhanced by their omnipresence, their interactions with atmospheric radiation, and their participation in chemical reactions and cloud formation processes. Iron is one of the most important elements appearing in the form of atmospheric aerosols. The main sources of iron-containing particulate matter include soil dust, emission from industrial operations, oil- and coal-fired power plants, exhaust from car engines, and natural sources such as volcanic fumes, extraterrestrial particles and meteor showers. In recent years, iron, like other transition metals (copper, nickel and manganese), although long considered to be of marginal importance, started to be the subject of many investigations (Weschler et al., 1986; Dasgupta et al., 1979) because of its significant role in the catalytic oxidation of SO_2 in the atmosphere. It was shown (Graedel et al., 1985), that iron oxide can chemisorb sulfur dioxide, converting it to sulfate at the gas-solid interface. Due to surface coating of insoluble aerosol particles with water soluble materials, their cloud nucleating capacity is enhanced (Parungo et al., 1978). It was also shown (Cho and Carmichael, 1986) that solubility of iron compounds depends on the iron valence and aerosol structure, and that, while Fe_2O_3 dissolves only very slowly, the iron present in pulverized fuel ash promotes its solubility. Therefore, the studies of the size, mass distribution, and concentration of iron-containing particles are of major interest for better understanding of condensation processes and atmospheric precipitation, which, in turn, are essential for cloud physics, health sciences and environment.

In our earlier studies (Kopcewicz and Dzienis, 1971; Kopcewicz and Kopcewicz; 1991), the Mössbauer effect was successfully used to study the properties and concentration of iron in the atmospheric air. It has been shown that Mössbauer spectroscopy permits, under favorable conditions not only the determination of the content of the Mössbauer isotope in the sample, but also identification of the chemical compound in which the Mössbauer isotope appears, and even evaluation of the size of the particles in which it is contained. Mössbauer spectroscopy is a nondestructive technique. The samples remain unaffected by the Mössbauer measurements, so they can be used in other investigations.

In this review, we will concentrate on the method itself rather than on the specific, detailed investigations performed in the last 20 years, the results of which were reviewed recently (Kopcewicz and Kopcewicz, 1991, 1992). However, examples of the application of Mössbauer spectroscopy in aerosol studies will be presented.

8.2   EXPERIMENT

The samples for our studies were prepared from membrane filters which are usually used in the measurement of air radioactivity. The atmospheric aerosol was collected by pumping the air through each filter for 24 h. The volume of air passed through each filter, measured by means of a gas meter, varied from about 600 to 1200 m^3. The filters were dissolved in acetone, evaporated, and the residue was pressed between two thin aluminum foils which were carefully checked for contamination with iron. The aerosol was collected in several sites in Poland: in large cities (Warsaw, Wrocław, Poznań, Łódź), in remote places in the Mazury Lake District (Mikołajki, in northeast Poland), and in the Tatra mountains (Kasprowy Wierch and Zakopane). We analyzed about 1000 samples. In this chapter, we will particularly discuss the results of the aerosol collected in the mountain region. The filters were collected at Kasprowy Wierch, a 1985-m-high summit in the Tatra mountains, and in Zakopane, a town situated in the foothills at 825 m above sea level in a valley bounded by the rocky wall of the Tatras in the south and protected by hills of the Gubalowka range in the north.

The samples were used in the Mössbauer measurements as absorbers. Each consisted of 15 filters obtained from the same site for 15 successive days. The Mössbauer measurements were performed in a transmission geometry. The source was ^{57}Co of about 25 to 50 mCi activity in Pd, Ph, or Cr matrix. Most of the measurements were performed at room temperature. However, the measurements were also carried out in the temperature range from 300K to 75K. The parameters of the Mössbauer spectra, such as line widths, isomer shifts, quadrupole splittings, magnetic hyperfine fields, peak absorptions, and the nonresonant background level, were usually computed by fitting the Lorentzian lineshapes to the experimental data by the least-squares method. In addition to this conventional method of evaluation of the Mössbauer spectra, a more sophisticated approach, which allows the determination of the distributions of hyperfine parameters, was also applied. The distributions of the quadrupole splitting were calculated using a constrained Hesse–Rübartsch method (Hesse–Rübartsch, 1974; LeCaer and Dubois, 1979). Isomer shift data are given relative to α-Fe.

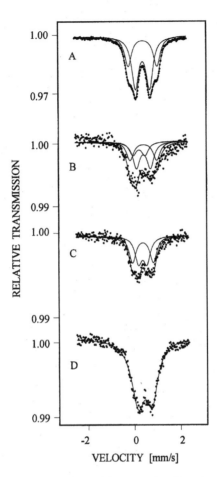

FIGURE 8.1   Typical spectra obtained for aerosol collected at Kasprowy Wierch.

8.3   IDENTIFICATION OF IRON-CONTAINING PHASES

As an example of the analytical potential of the Mössbauer technique for the identification of iron-containing phases, we present in Figure 8.1 the set of Mössbauer spectra recorded for samples collected at Kasprowy Wierch. Figure 8.1A shows two quadrupole doublets with the isomer shift $\delta = 0.37$ mm/s, and the quadrupole splitting $\Delta = 0.58$ mm/s, and $\delta = 0.36$ mm/s and $\Delta = 1.1$ mm/s. These parameters allowed us to identify the iron-containing compounds as γ-FeOOH and $Fe_{1-x}O$ (Wustit), respectively (Greenwood and Gibb, 1971). The Fe^{2+} state has a high ability for oxidation. However, when incorporated into particularly stable crystal configurations, the ferrous compounds may be very stable. They are natural constituents of rocks and minerals. The observation (in this particular case) of a high amount (30% to 40%) of Fe^{2+} in the collected atmospheric aerosol leads to the conclusion that, in this rather unusual case, soil is the major source of the dust collected on filters used for the preparation of the Mössbauer samples.

The spectrum in Figure 8.1B can be fitted with three spectral components: (1) a quadrupole doublet with $\delta = 0.19$ mm/s and $\Delta = 0.50$ mm/s, which corresponds to paramagnetic or superparamagnetic Fe^{3+} species, (2) a quadrupole doublet with $\delta = 0.37$ mm/s and $\Delta = 0.58$ mm/s, which corresponds to γ-FeOOH, and (3) a quadrupole doublet with $\delta = 0.36$ mm/s and $\Delta = 1.1$ mm/s, corresponding to $Fe_{1-x}O$ (Wustit). Figure 8.1C shows two quadrupole doublets which correspond to α-Fe_2O_3 in the superparamagnetic state ($\delta = 0.34$ mm/s, $\Delta = 0.58$ mm/s) and $Fe_{1-x}O$.

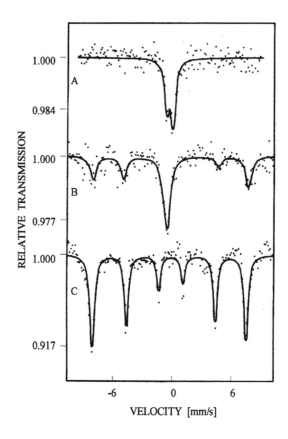

FIGURE 8.2 Mössbauer spectra for aerosol sample: (A) at 300K, (B) at 78K, (C) for bulk α-Fe$_2$O$_3$ at 300K.

For most of the samples collected at Kasprowy Wierch from 1972 to 1975, we observed spectra consisting of a broadened quadrupole doublet with $\triangle$ = 0.63 mm/s (Figure 8.1D).

Such a spectrum measured only at room temperature makes the identification of a chemical compound which contains iron difficult and unreliable. In order to verify the origin of this doublet, the measurements were performed at 78K. The spectrum measured at low temperature (Figure 8.2B) shows a central unresolved doublet, due to quadrupole splitting, of a much smaller intensity than that obtained at 300K (Figure 8.2A) and six lines due to magnetic hyperfine splitting. Two middle lines of this sextet are weak and overlap with the doublet, but four outer lines are very pronounced. The line positions in the magnetic part of the spectrum and the resulting magnet hyperfine field correspond to those observed for the bulk α-Fe$_2$O$_3$ (Figure 8.2C). Such temperature behavior provides evidence that iron appearing in the aerosol collected in such samples is contained in ultrafine particles of α-Fe$_2$O$_3$, which are superparamagnetic. The magnetic properties of α-Fe$_2$O$_3$ have been studied as a function of particle size and temperature (Nakamura et al., 1964; Kündig et al., 1966). It was shown (Kündig et al., 1966) that α-Fe$_2$O$_3$ particles of about 5 to 20 nm (ultrafine particles) at room temperature reveal a Mössbauer spectrum that consists of a doublet similar to that observed in our study (Figures 8.1D and 8.2A). At low temperature, the α-Fe$_2$O$_3$ particles show a weak ferromagnetism, and, therefore, the spectrum measured at 78K consists of a Zeeman sextet, as observed in Figure 8.2B. A Mössbauer study performed as a function of temperature allows the determination of the size of iron-containing particles (see below).

8.4   DETERMINATION OF SIZE DISTRIBUTION OF IRON-CONTAINING AEROSOL PARTICLES

Mössbauer spectroscopy not only permits identification of the compound in which the Mössbauer isotope appears and the determination of its content in the sample, but also the evaluation of the size of the particles in which this isotope is contained. As an example, we show how the size of the Fe_2O_3 particles appearing in the aerosol collected for Kasprowy Wierch could be estimated (Kopcewicz and Kopcewicz, 1978). Figure 8.3 shows the Mössbauer spectra recorded as a function of temperature. As can be seen, the contribution of the magnetically split component of the spectrum gradually increases as temperature decreases, at the expense of the intensity of the quadrupole doublet. Such changes in the Mössbauer spectra are due to magnetic spin relaxation effects. When the ferromagnetic or antiferromagnetic particles are very small ($\cong 10$ nm), the magnetization vector can at a certain temperature become unstable due to thermal agitation. Then these particles behave paramagnetically, but the atomic magnetic moments are still coupled, and it is the large magnetic moment of the particle which fluctuates rapidly between directions. Such particles are called superparamagnetic. The changes as a function of temperature, shown in the spectra in Figure 8.3, are typical of small superparamagnetic particles of α-Fe_2O_3 (Kündig et al., 1966). The disappearance of the magnetic hyperfine structure when the temperature is increased is due to fast relaxation of the magnetic moments of ultrafine particles. In the simple model of superparamagnetism (Kneller, 1969), the relaxation time is given by the formula:

$$\tau_R = \tau_o \exp(KV/kT) \tag{1}$$

where τ_o is the frequency factor, equal to 1.58×10^{-9} s, $K = 4.1 \times 10^4$ erg/cm^3 (the anisotropy constant for α-Fe_2O_3) (Kündig et al., 1966), V is the volume of the ultrafine particle, k is the Boltzmann constant, and T is the temperature. Fast relaxation of the magnetic moment of the particle leads to the collapse of the magnetic hyperfine splitting of the paramagnetic doublet. When the relaxation time, τ_R, is shorter than the Mössbauer observation time (which is the period of Larmor precession of the nuclear magnetic moment of iron in the magnetic hyperfine field acting on the iron nuclei, τ_L, equal to 2.5×10^{-8} s), then the magnetic hyperfine field at the Mössbauer nuclei is averaged to zero, and the spectrum consists of the quadrupole doublet (QS) only. Then the particles are superparamagnetic (Figure 8.3A). For larger particles, the relaxation times exceed the observation time. Such particles show ferromagnetic behavior resulting in the appearance of the magnetic hyperfine structure in the spectra (Figures 8.3B–8.3F).

The size of α-Fe_2O_3 particles can be determined from the temperature measurements as shown in Figure 8.3. The radius of the particles was estimated from the spectrum obtained at 75K, in which 50% of the Mössbauer absorption corresponds to the magnetically split spectral component and 50% to the QS doublet. In this case, it can be assumed that the relaxation time is comparable to the observation time, so the volume of the particle can be evaluated from Equation (1). The radius of the dominating particles of α-Fe_2O_3 estimated in this way is about 5.5 nm. In addition to the dominating size, the cutoff of the size distribution can also be estimated. As can be seen from Equation (1), the relaxation time is comparable at 300K to the observation time for the particle radius of 8.5 nm. Thus, for particles smaller than 8.5 nm, the superparamagnetic behavior dominates at 300K. Since the magnetic lines observed in the spectrum at 300K are negligible (Figure 8.3A), one can infer that the maximum radius of iron-containing particles is larger than 5.5 nm but smaller than 8.5 nm. This means that in this particular sample the size distribution of iron-containing particles collected for Kasprowy Wierch is very narrow. The particle size distribution differs for different sampling sites and reveals characteristic seasonal variations (Kopcewicz and Kopcewicz, 1978).

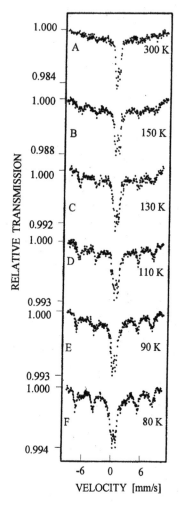

FIGURE 8.3 Mössbauer spectra obtained as a function of temperature for the sample from Kasprowy Wierch.

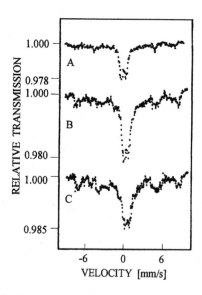

FIGURE 8.4 Typical Mössbauer spectra obtained at 300K for the sample from Wrocław.

In the samples from industrial sites (Wrocław, city in southwest Poland), the size distribution of iron-containing particles is different. In the spectra in Figure 8.4, the magnetic lines corresponding to Fe_2O_3 appear in addition to the QS doublet at 300K. Such a complex spectrum can be due to the coexistence of particles of two dominating sizes: small ones corresponding to the QS doublet (superparamagnetic fraction), and large ones contributing to the magnetic part of the spectrum. The radius of the large particles, estimated in a way similar to that described above, was found to be at least 13.5 nm.

The size of iron-containing ultrafine particles of Fe_2O_3 can also be determined from the dependence of the quadrupole splitting on the particle size (Kündig et al., 1966). The shape of the spectra recorded for most of our samples — a broadened, asymmetric quadrupole split doublet — suggests that the spectra do not consist of one QS doublet, but rather of several doublets with different splittings and isomer shifts. In order to extract reliable information regarding the quadrupole splittings necessary for the estimation of the size of Fe_2O_3 particles, we applied the method which allows the determination of the distributions of the hyperfine parameters, in this case of quadrupole splittings. The histogram method, which permits the calculation of the probabilities of given QS values in the assumed range of the quadrupole splittings, introduced by Hesse and Rübartsch (1974) and improved by LeCaer and Dubois (1979), was used for evaluation of the Mössbauer spectra obtained for samples from Kasprowy Wierch and Zakopane collected in the same period from 1991 to 1992 (Kopcewicz and Kopcewicz, 1995). Figures 8.5A and 8.5B present typical spectra for both sampling sites, which consist of a broadened, fairly well-resolved quadrupole splitting (QS) doublet. The quadrupole splitting distributions, P(QS), extracted from the spectra using the "constrained Hesse–Rübartsch method" (Hesse and Rübartsch, 1974; LeCaer and Dubois, 1979), are shown in Figures 8.5A' and 8.5B'. In order to reproduce well the asymmetry in the line intensities in the quadrupole doublets, clearly seen in the spectra, the linear correlation between the quadrupole splitting and the isomer shift was assumed in the fitting procedure. The typical P(QS) distributions obtained for the samples collected in Zakopane consist of two dominating peaks. However, for the samples from Kasprowy Wierch, the P(QS) distributions also contain a third, much weaker peak at high QS values. The center shift of the spectra (which corresponds to the average isomer shifts calculated when fitting P(QS)), is about 0.25 to 0.32 mm/s for all samples, in good agreement with our earlier results (Kopcewicz and Kopcewicz, 1991, 1994), which confirms that iron appears in aerosol mostly in the form of ultrafine superparamagnetic Fe_2O_3 particles.

Since the P(QS) distributions reveal in most cases a bimodal shape, we analyzed the variation of the QS values corresponding to the main peaks in P(QS) for both sampling sites. The results are shown in Figure 8.6, in which the pairs of QS with the largest probabilities are shown. Each point in the diagrams represents a characteristic QS value for a given sample. Since the pairs of QS values for various samples frequently repeat, many points in Figure 8.6 have a high statistical weight. As can be seen from Figure 8.6, the QS pairs for samples from Zakopane concentrate at the average values $QS_1 = 0.40$ mm/s and $QS_2 = 0.85$ mm/s, while for the samples from Kasprowy Wierch, the QS values are shifted to lower values and are: $QS_1 = 0.20$ mm/s and $QS_2 = 0.65$ mm/s.

The normalized probabilities of appearance of the QS values observed in the P(QS) distributions as a function of the calendar month are shown in Figure 8.7. The quadrupole splittings were divided into groups with characteristic values: for the Zakopane samples, we had two groups with the QS_1 ranging from 0.30 to 0.55 mm/s and QS_2 from 0.60 to 1.00 mm/s; for Kasprowy Wierch, we had three groups with $QS_1 = 0.10$ to 0.25 mm/s, $QS_2 = 0.55$ to 0.70 mm/s, and $QS_3 = 0.80$ to 1.1 mm/s (QS_3 corresponds to the third weak peak in the P(QS) mentioned above). Figure 8.7 reveals that the probability of the QS_1 and QS_2 values is almost equal for the samples collected in Zakopane (Figures 8.7A and 8.7C), while in the case of Kasprowy Wierch, clearly different probabilities for each QS group were observed (Figures 8.7B and 8.7D). Since a QS value is related to a given size of Fe_2O_3 particles (Kündig et al., 1966), the results shown in Figure 8.7 strongly suggest that the aerosol collected in Zakopane contains two distinct sizes of particles with diameters of about 16 nm and less than 10 nm, corresponding to QS_1 and QS_2 values, respectively. The abundance of these

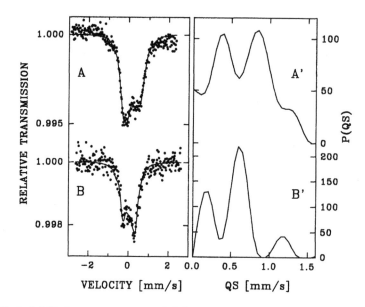

FIGURE 8.5 Typical Mössbauer spectra recorded for aerosol samples collected in Zakopane (A) and Kasprowy Wierch (B) and the corresponding quadrupole splitting distributions, P(QS), (A′ and B′) extracted from the fits of the spectra.

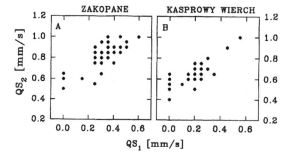

FIGURE 8.6 Diagram of QS1 and QS2 values (discussed in the text) characteristic for each sample collected in Zakopane (a) and Kasprowy Wierch (B).

particles is similar. In the case of Kasprowy Wierch, the QS_1, QS_2, and QS_3 appear with markedly different probabilities, of which the QS_2 has the greatest (Figures 8.7B and 8.7D). Hence, this aerosol consists of particles with three dominating sizes, with clearly different relative abundance, and with the average diameters systematically larger than those observed for the samples from Zakopane. Comparing the results obtained for Zakopane with those for Kasprowy Wierch, we infer that the aerosol collected at Kasprowy Wierch contains some very small particles which were not observed in the samples from Zakopane, but the most abundant particles were larger than those present in the aerosol collected in Zakopane.

It is known that atmospheric particles originating from different sources become mixed by Brownian diffusion and coagulation on a microscale and by atmospheric mixing processes on a larger scale. At Kasprowy Wierch, because of specific localization, we have no local source of iron-containing aerosol except mineral dust, which was not detected in this experiment. The analysis of the hyperfine parameters for the samples collected in Zakopane and Kasprowy Wierch suggests that the larger aerosol particles observed at Kasprowy Wierch may be formed on their way through the atmosphere, both the ones from the closest sources in Zakopane and also those originating from distant sources.

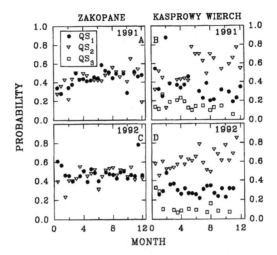

FIGURE 8.7 Probability of appearance of various quadrupole splittings in the spectra recorded for the samples collected in Zakopane (A and C) and Kasprowy Wierch (B and D).

8.5 DETERMINATION OF THE CONTENT OF A MÖSSBAUER ISOTOPE IN THE SAMPLE

The absolute content of the Mössbauer isotope in the sample can be evaluated from the area of the Mössbauer lines in the spectrum if the relevant parameters, such as the recoilless fractions for the source and absorber (f factor), the natural linewidth (Γ), and the cross section for the resonant absorption of γ-rays (σ), are known. In the present investigation, we adapted the method for calculating the f factor from the area under the Mössbauer line (Hafemeister and Brooks Shera, 1966). Since in this method the density of the resonant isotope in the sample is used when calculating the f factor, the content of the resonant isotope in the sample can be inferred from the knowledge of the f factor for the absorbing material. The formula for the absorption area is:

$$A = (\pi/2)\, f\, \Gamma\, L(t) \tag{2}$$

where

$$t = n\, \sigma\, f' \tag{3}$$

is the effective thickness of the absorber containing n atoms/cm^2 of the Mössbauer isotope, f and f' are the recoilless fractions for the source and absorber, respectively, and the $L(t)$ is a function of the absorber thickness defined as:

$$L(t) = \sum_{p=1}^{\infty} \frac{(-1)^{p+1}(2p-3)!!\, t^p}{p!\,(2p-2)!!} \tag{4}$$

The $L(t)$ function has been computed numerically (Hafemeister and Brooks Shera, 1966). When the experimental spectrum consists of i absorption lines of Lorentzian shape with Γ_{jexp} linewidths and the peak absorption (corrected for the nonresonant background) a_{jexp}, the area under the spectrum is:

$$A = (\pi/2)\sum_{j=1}^{i}\left(\Gamma_{jexp} a_{jexp}\right) \tag{5}$$

Comparing Equations (2) and (5) we obtain:

$$L(t) = \left\{ \sum_{j=1} \left(\Gamma_{jexp} a_{jexp} \right) \right\} \Big/ f\Gamma \qquad (6)$$

Equation (6) allows us to evaluate $L(t)$, and hence the corresponding value of the effective thickness t. By using Equation (3), we finally obtain:

$$n = t/\sigma f' \qquad (7)$$

The method described above can be used only when a chemical composition of the iron-containing compound is established and when the f' factor for this compound is known or determined in another experiment.

This method was used to determine the iron concentration in aerosol collected in various sites and periods (Kopcewicz and Kopcewicz, 1978, 1986, 1991, 1994). Here we give an example of the determination of iron concentration and the possible origin of the iron-containing aerosol collected in the Tatra mountain region (Kasprowy Wierch and Zakopane). Topographic differences in the sampling sites that cause turbulent mixing and convective vertical transport could be used as the main controlling atmospheric parameters in the interpretation of the variations of iron concentration in air.

In Figure 8.8, the seasonal variations of ^{57}Fe concentration in air obtained for Kasprowy Wierch and Zakopane are presented. Iron concentration estimated from the Mössbauer spectra is influenced by the systematic error related mainly to the uncertainty of f-factor data for the source and absorber. However, this error affects all the results in the same way and does not distort the relative values of iron concentration at both sampling sites. As can be seen from Figure 8.8, even such short-term measurements offer the possibility to observe trends in the variations of iron concentration at both sampling sites. The average ratio of iron concentration in Zakopane to iron concentration at Kasprowy Wierch is 1.9. The highest and the lowest ^{57}Fe concentrations in air observed in Zakopane were 33.3×10^{-9} g/m^3 and 3.6×10^{-9} g/m^3, respectively. At Kasprowy Wierch, the corresponding values were 11.8×10^{-9} g/m^3 and 2.3×10^{-9} g/m^3. Assuming the natural abundance of ^{57}Fe (2.19% of ^{57}Fe in natural iron), it is possible to convert the ^{57}Fe concentrations into natural iron/m^3 for easy comparison with the data available from other methods, e.g., chemical analysis. The resulting concentrations of natural iron in air observed in the samples collected in Zakopane varied from 15.2×10^{-7} g/m^3 (the highest iron concentration) to 1.6×10^{-7} g/m^3 (the lowest value), and at Kasprowy Wierch — from 5.4×10^{-7} g/m^3 to 1.1×10^{-7} g/m^3. The highest iron concentration for both sites was observed in winter, which suggests that coal combustion is the main source of iron in the air in this area. Zakopane has no local industry, so almost all the coal combustion can be related to house heating in the winter season. The iron concentration curves in Figure 8.8 also show the dynamic state of the atmosphere. The lowest concentration of iron in both sampling sites was observed in spring, when the atmospheric turbulence was higher (foehn season), so the processes of pollution removal were more effective. The variations of iron concentration in summer for both sampling sites were very well correlated, with no dramatic changes, because good mixing of the atmospheric air is related to much greater convective vertical transport than in winter. However, the determination of distant sources of iron, which influence the total iron concentration, require backward air mass trajectory analyses.

8.6   CONCLUSION

Mössbauer spectroscopy offers a unique possibility to determine, in one experiment, the chemical compound in which the Mössbauer isotope is contained, its absolute content, and the size distribution of the aerosol particles. Despite the fact that the Mössbauer effect was observed for 109

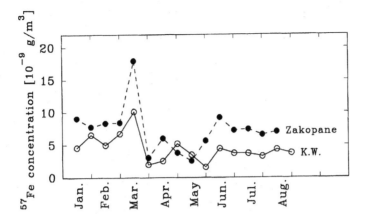

FIGURE 8.8   Seasonal variations of ^{57}Fe concentration in air obtained for Zakopane and Kasprowy Wierch from January to August 1991.

gamma transitions in 90 isotopes of 47 elements, very low concentration of the particles in air, and hence a low content of the Mössbauer isotope collected on filters, limits the applications of this technique in atmospheric research to the ^{57}Fe isotope.

Mössbauer investigations can be rewarding in the search for the sources of iron in air and may be useful in studies of the properties of iron-containing particles in aerosols. Investigation of samples from weakly polluted nonindustrial regions should be especially convenient for studying the various processes occurring in the atmosphere. This, however, would require analysis of numerous samples collected from different sites for a long period of time.

REFERENCES

Cho, S.Y. and Carmichael, G.R., 1986, *Atmospheric Environ.* 20: 1959-1968.

Dasgupta, P.K., Mitchell, P.A., and West, P.W., 1979, *Atmospheric Environ.* 13:775-782.

Graedel, T.E., Weschler, C.J., and Mandich, M.L., 1985, *Nature* 317:240-242.

Greenwood, N.N. and Gibb, T.C., 1971, *Mössbauer Spectroscopy*, Chapman and Hall, London, Chapter 10.

Hafemeister, D.W. and Brooks Shera, E., 1966, *Nucl. Instr. Meth.* 41:133-134.

Hesse, J. and Rübartsch, A., 1974, *J. Phys.* E 7:526-532.

Kneller, E., 1969, in *Magnetism and Metallurgy*, Berkowitz, A.E. and Kneller, E., (Eds.), Academic Press, New York, Vol. I, 366-471.

Kopcewicz, M. and Dzienis, B., 1971, *Tellus* 23:176-181.

Kopcewicz, B. and Kopcewicz, M., 1978, *Tellus* 30:562-568.

Kopcewicz, B. and Kopcewicz, M., 1986, *Hyperfine Inter.* 29:1141-1145.

Kopcewicz, B. and Kopcewicz, M., 1991, *Structural Chemistry* 2:303-312.

Kopcewicz, B. and Kopcewicz, M., 1992, *Hyperfine Inter.* 71:1457-1460.

Kopcewicz, M. and Kopcewicz, B., 1992, *Proc. of the Sixth International Conf. on Ferrites (ICF-6)*, Tokyo, The Japan Society of Powder and Powder Metallurgy, 205-210.

Kopcewicz, B. and Kopcewicz, M., 1994, *Hyperfine Inter.* 91:777-781.

Kopcewicz, B. and Kopcewicz, M., 1995, in print.

Kündig, W., Bömmel, H., Constabaris, G., and Linquist, H.R., 1966, *Phys. Rev.* 142:327-333.

LeCaer, G. and Dubois, J.M., 1979, *J. Phys.* E 12:1083-1090.

Nakamura, T., Shinjo, T., Endoh, Y., Yamamoto, Y., Shinga, M., and Nakamura, Y., 1964, *Phys. Lett.* 12:178-179.

Parungo, F., Ackerman, E., Proulx, H., and Pueschel, R., 1978, *Atmos. Environ.* 12:929-941.

Weschler, C.J., Mandich, M.L., and Graedel, T.E., 1986, *J. Geophys. Res.* 91:5189-5204.

9 Introduction to the Theory of Electron Paramagnetic Resonance and Its Application to the Study of Aerosols

Nicola D. Yordanov

CONTENTS

9.1   INTRODUCTION

Electron paramagnetic resonance (EPR)* spectroscopy is a physical method that detects molecules which contain unpaired electrons. It is a relatively new spectroscopic method developed experimentally in 1944 (Zavoiski, 1945). After extensive theoretical and experimental work, it became a standard analytical method with applications in physics, chemistry, biology, geology, archeology, etc.

The subjects of study of EPR spectroscopy are molecules containing unpaired electrons. In general, molecules contain paired electrons. However, there is a significant number of molecules containing one or more unpaired electrons. In the order of their increasing number of unpaired electrons, they are:

1. Free radicals. These are molecules containing one unpaired electron.
2. Biradicals. Biradicals are molecules containing two unpaired electrons situated sufficiently far from each other so that there is no interaction between them.

---

* Historically, the same method is designated in the literature as electron spin resonance (ESR) and electron magnetic resonance (EMR).

3. Molecules in triplet-state. These molecules contain two strongly coupled unpaired electrons. The triplet state may be a ground or excited (thermally or optically) state.
4. Molecules with more than two unpaired electrons.
5. Point defects in the solid state. These defects may be due, for example, to high-energy irradiation or localized crystal imperfections. In this case, one, two, or more electrons (or electron holes) may be trapped at or near the defect.
6. Transition metal and rare-earth ions and their complexes.

The first and last systems in this list are the most common.

Compared with other physical methods of analysis, EPR spectroscopy is characterized by an exceptional ability to record only paramagnetic substances in diamagnetic materials and with a very high sensitivity in addition, which is now approximately 10^{-12} to 10^{-13} M. Moreover, EPR spectroscopy is a nondestructive method of analysis. Thus, measurements may be repeated at any time, and the samples may be stored for further investigations.

This chapter briefly describes the basic physical principle of EPR and its applications to the study of paramagnetic aerosols. As will be shown, the advantages of EPR in aerosol study did not become well known following the first investigations more that 40 years ago. The reason for this was the high cost of EPR equipment, its stationarity, and the lack of experienced operators. Now the situation has changed, as have the prospects for the use of EPR in different areas of science. For further details and a deeper introduction into the theory and practice of EPR, readers are advised to consult some of the books available (Abragam, 1970; Alger, 1968; Al'tshuler, 1964; Assenheim, 1967; Ayscough, 1967; Ingram, 1958, 1969; Ikeya, 1993; Mabbs, 1992; Orton, 1968; Pilbrow, 1990; Poole, 1983; Weil, 1994; Wertz, 1972), as well as Specialists periodical reports (SPR–), which have appeared regularly in the last 20 years.

9.2   THE BASIC THEORY OF EPR

9.2.1   ZEEMAN INTERACTION

Molecules containing unpaired electrons exhibit paramagnetic properties due mainly to the electron spin magnetic moment μ_S, which is given by Equation (1):

$$\mu_S = -g\beta M_S \tag{1}$$

where β is called the Bohr magneton ($\beta = eh/4\pi mc$), g is called g-factor (with a value of 2.0023 if the electron is free, otherwise it can vary in a wide region depending of the sample), and $M_S = \pm 1/2$. The magnetic moments of the assembly of all paramagnetic molecules are oriented randomly in space. If placed in a magnetic field (H), however, they tend to be oriented parallel or antiparallel with respect to H. The amount of energy, E, required to disalign μ_S from H is given by Equation (2):

$$E = \mu_S \cdot H = g\beta M_S H \tag{2}$$

which can be rewritten as Equation (3):

$$E = (\pm 1/2)g\beta H \tag{3}$$

Equation (3) represents two energy levels:

$$E_1 = g\beta H/2 \tag{4a}$$

$$E_2 = -g\beta H/2 \tag{4b}$$

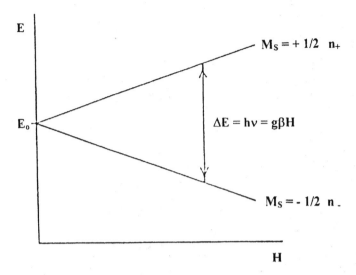

FIGURE 9.1   Diagram of unpaired electron energy levels as a function of the magnetic field strength. (From Yordanov, N.D., Veleva, B., and Christov, R., 1996, *Appl. Magnetic Resonance*, 10: 439. With permission.)

which are displayed as a function of magnetic field in Figure 9.1 with a gap between them

$$\Delta E = h\nu = g\beta H \tag{5}$$

This equation is called electron Zeeman energy and represents the energy of the interaction of the unpaired electron with the external magnetic field. We can rewrite Equation (4) in the form:

$$(\nu/H_o) = \beta(g/h) \tag{6}$$

and we will find that only when it is satisfied, will energy absorption be realized — hence the term "resonance."

It is worth noting that the distribution of the population of these two levels follows Boltzmann distribution (Equation 7):

$$n_+/n_- = - \exp(\Delta E/kT) \tag{7}$$

where n_+ and n_- are populations of the levels with $M_S = +1/2$ and $-1/2$, (ΔE is the energy gap between the two levels, k is Boltzmann's constant, and T is the absolute temperature. Substitution of Equation (5) in Equation (7) gives

$$n_+/n_- = - \exp(g\beta H/kT) \tag{8}$$

Like all other spectroscopic methods, EPR records the net energy absorption from a radiation field when molecules change their energy state. Therefore, if a given sample which is placed in an external magnetic field, is irradiated with electromagnetic energy which satisfies the requirement $\Delta E = h\nu = g\beta H_o$, we will observe spin transitions up and down between the two levels. However, the transition probability is proportional to the population of the levels and since lower energy level is more populated than the upper one, the net effect will be energy absorption. The energy absorption is very fine — for the most commonly used EPR spectrometers with $\nu = 10^{10}$ Hz and at T = 300 K $n_+/n_- = 0.9984$, and only a small fraction of the unpaired electrons actually contribute to the absorption. Thus, very sensitive and sophisticated electronics is required to detect the energy absorption.

When Equation (5) is satisfied and energy absorption is recorded, the position of the absorption maxima along the H-axis will depend only on the g-factor of the sample. Therefore, the g-factor is a value which can characterize a given sample. For example, species with strongly delocalized unpaired electrons are characterized with g which is equal to or near 2.0023, whereas in the case of transition metal complexes, it can vary from 0.5 to 6. In its simple form, the g-factor of transition metal complexes can be represented by the relation (Equation 8)

$$g = g_e (1 - k\alpha^2(\lambda/\Delta) \tag{9}$$

where g_e is the free electron g factor ($g_e = 2.0023$), λ is the spin-orbit coupling constant of the free ion, Δ is the crystal field splitting, α is a constant representing the covalency of the metal-ligand bonds, and k is a constant depending on the orientation of the metal ion in respect to the direction of the external magnetic field. For metal ions in which the d-shell is less than half full $\lambda > 0$ and hence $g < g_e$; if it is half full, (n = 5) $\lambda = 0$ and $g = g_e$; and finally if the d-shell is more than half full, $\lambda < 0$ and $g > g_e$. For example, the g-factor for V^{4+} (n = 1) is 1.98, for Mn^{2+} and Fe^{3+} (n = 5) g = 2.00, and for Cu^{2+} (n = 9) g = 2.2.

In EPR practice, spectra are usually represented by the first derivative of the absorption curve (Figure 9.2). This is made in order to increase the spectral resolution. (Sometimes even second or higher derivatives are used.)

9.2.2 HYPERFINE INTERACTIONS

Up to now, we have considered the interaction between the magnetic moment of the unpaired electron and the external magnetic field only. In fact, there are other local magnetic fields which also interact with the magnetic moment of the unpaired electron. Depending on the source of these local fields, three main kinds of interactions may be observed. *Fine* interaction refers to the interaction between two unpaired electrons; *hyperfine* interaction is that between the magnetic moments of the unpaired electron and nucleus (or nuclei) with nuclear spin other than 0; and *superhyperfine* interactions take place between unpaired electrons and nuclei from the ligand in transition metal complexes. The appearance of these interactions significantly enhances the value of EPR spectroscopy. Most common of them is hyperfine interaction, and this will be considered further.

Let's consider the simplest case of a hypothetical paramagnetic system containing one unpaired electron and one nucleus with nuclear spin I = 1/2, which is placed in the magnetic field. In this case, we can expect three kinds of interactions: *electron Zeeman* interaction, representing interaction of the unpaired electron magnetic moment with the external magnetic field ($E' = g_e\beta_e HM_S$); *nuclear Zeeman* interaction, representing interaction of the nuclear spin magnetic moment with the external magnetic field ($E'' = g_n\beta_n HM_I$); and *hyperfine* interaction, arising between the magnetic moment of the unpaired electron and the nuclear spin magnetic moment ($E\% = AM_SM_I$). The total energy of the interactions of this system, therefore, will be a sum of all of them, which in decreasing order ($E' > E\% \gg E''$) is represented by Equation (10):

$$E = g_e\beta_e HM_S + AM_SM_I + g_n\beta_n HM_I \tag{10}$$

Because $M_S = \pm 1/2$ and also $M_I = \pm 1/2$, Equation (10) must be represented as four Equations (11a through d)

M_S	M_I		
+1/2	+1/2	$E1 = \quad g_e\beta_e H/2 + A/4 + g_n \beta_n H/2$	(11a)
+1/2	−1/2	$E2 = \quad g_e\beta_e H/2 − A/4 − g_n \beta_n H/2$	(11b)
−1/2	−1/2	$E3 = -g_e\beta_e H/2 + A/4 − g_n \beta_n H/2$	(11c)
−1/2	+1/2	$E4 = -g_e\beta_e H/2 − A/4 + g_n \beta_n H/2$	(11d)

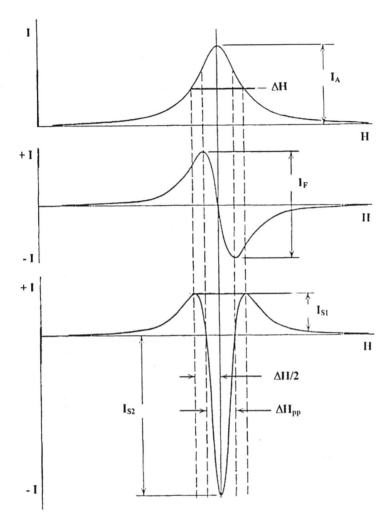

FIGURE 9.2 Lineshapes of: (a) absorption spectrum; (b) first-derivative spectrum; (c) second-derivative spectrum. (From Yordanov, N.D., Veleva, B., and Christov, R., 1996, *Appl. Magnetic Resonance*, 10: 439. With permission.)

In this case, we will have four energy levels (Figure 9.3), and the allowed transitions will be between the levels corresponding to the selection rules $\Delta M_S = \pm 1$ and $\Delta M_I = 0$. These selection rules show that when unpaired electron magnetic moment changes its direction with respect to the external magnetic field ($\Delta M_S = \pm 1$), nuclear spin orientation remains unchanged ($\Delta M_I = 0$). Therefore, two transitions will occur (since E′, E% ≫ E″, the last term will be neglected in further considerations):

$$\Delta E_1 = E_1 - E_4 = \ g_e\beta_eH + A/2 \tag{12a}$$

$$\Delta E_2 = E_2 - E_3 = -g_e\beta_eH - A/2 \tag{12b}$$

In EPR spectroscopy, the microwave frequency is usually kept constant (i.e., $h\nu_o$ is constant) and H is varied. Therefore, the resonance will appear at two fields:

$$H_1 = (h\nu_o/g_e\ \beta_e) + (hA/2g_e\ \beta_e) = H' + a/2 \tag{13a}$$

$$H_2 = (h\nu_o/g_e\ \beta_e) - (hA/2g_e\ \beta_e) = H' - a/2 \tag{13b}$$

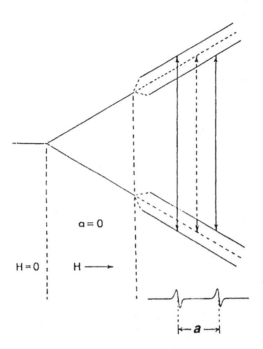

FIGURE 9.3 Energy levels of the hypothetical paramagnetic atom containing a nucleus with spin $I = 1/2$ and the corresponding first-derivative spectrum. (From Yordanov, N.D., Veleva, B., and Christov, R., 1996, *Appl. Magnetic Resonance*, 10: 439. With permission.)

where $a = hA/2g_e \, \beta_e$ is called the hyperfine splitting constant. The EPR spectrum in this case is represented in Figure 9.3. The distance between the two lines in it is just the hyperfine splitting constant, and the point equidistant between them determines the g-factor of the sample.

In cases when an unpaired electron interacts with a nucleus having nuclear spin (I) other than zero, the number of the hyperfine lines (N) is given by the relation $N = 2I + 1$. For example, the nuclear spin of ^1H is 1/2 and $N = 2$, for ^{14}N ($I = 1$) $N = 3$, for ^{55}Mn ($I = 5/2$) $N = 6$, for ^{51}V ($I = 7/2$) $N = 8$, etc. The area under the hyperfine lines in an absorption EPR spectrum are equal. In real cases, the orbit of the unpaired electron is often delocalized over many atoms, and thus it may be coupled to a number of magnetic nuclei. Obviously, the more nuclei that interact with the unpaired electron, the more complex the EPR spectrum will become. The analysis of such spectra may be easy if two common cases are considered, in which the unpaired electron interacts with:

a. A set of equivalent nuclei. Equivalent nuclei are those having one and the same nuclear spin (I) and hyperfine splitting constant (a). The maximal number of the hyperfine lines (N) is given as $N = 2nI + 1$ (n is the number of equivalent nuclei) with a binomial distribution of their intensity.

b. A set of nonequivalent nuclei. Nonequivalent nuclei are those for which interaction is more tightly with one nucleus (or sets of equivalent nuclei) and less tightly with another. In this more common case, the maximal number of the hyperfine lines (N) is given as $N = (2n_1I_1 + 1)(2n_2I_2 + 1) \ldots (2n_kI_k + 1)$ where n_k is the number of equivalent nuclei from the k-th group with a corresponding nuclear spin. The intensities of these lines is complicated, but it is binomial in each set $(2n_mI_m + 1)$.

Figure 9.4 illustrates both possibilities. Since the analysis of EPR spectra gives information about the structure of paramagnetic species, the problem is important and is solved by using computers and special software for spectra simulation. Using iterative procedures, one can obtain the "best fit" between the experimental and theoretical spectra.

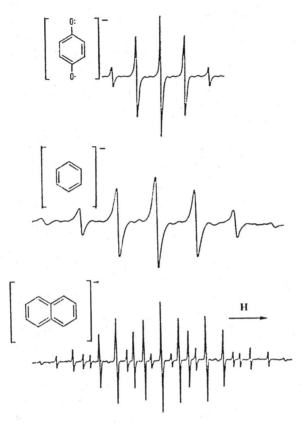

FIGURE 9.4   EPR spectra of: (a) benzosemiquinone; (b) benzene anion-radical; (c) naphthalene anion-radical.

9.3   EPR INSTRUMENTATION

EPR spectrometers are complicated electronic instruments which are typically commercially available. The simplest EPR spectrometer involves three basic parts — a radiation source, a sample
placed between the poles of a magnet and absorbing radiation when Equation (5) (or (10)) is
satisfied, and a detector unit measuring the intensity of the absorbed portion of radiation. Keeping
Equation (6) in mind, there are two ways to reach the resonance condition — to keep the frequency
constant and to vary the intensity of the magnetic field, or to keep the magnetic field constant and
to vary the frequency of the radiation source. The first option is commonly accepted because of
difficulties in varying the frequency in a wide range. In view of this, electromagnets are used in
combination with a klystron or, more recently, a Gunn diode as a microwave source. Spectrometers
operate in several frequency regions, as shown in Table 9.1. Most of them are on the market, but
the most commonly used EPR spectrometers are those operating at 10 GHz (X-band). Until recently,
all of these spectrometers, including the most common X-band instruments, were of the stationary,
laboratory type, with an electromagnet weighing from 500 to 5000 kg. Prices varied broadly,
depending on the spectrometer configuration — starting from $50,000 and going up to $1,000,000.
In addition, the operation of these spectrometers required practice. This was a big disadvantage,
independent of the fact that computers became standard equipment on the spectrometers. In the
last decade, however, the situation has changed, and now it is possible to obtain compact, fully
computer-controlled portable spectrometers whose total weight, including computer, is less than
50 to 60 kg (Ikeya et al., 1989; Kojima et al., 1989, 1993; Lyniov, 1991; Maier et al., 1993;
Nakanishi, 1993; Yamanaka et al., 1991), some of which are available on the market (Lynev, 1991;

TABLE 9.1
Microwave Frequency Bands and Magnetic Fields
for EPR Spectrum with g = 2

Band Designation	Frequence Band Range (GHz)	Typical EPR Frequency (GHz)	Typical EPR Magnetic Field (mT)
L	0.39–1.55	1.5	54
S	1.55–3.90	3.0	110
C	3.90–6.20	6.0	220
X	6.20–10.90	9.5	340
K	10.90–36.00	23	820
Q	36.00–46.00	36	1300
V	46.00–56.00	50	1800
W	56.00–100.00	95	3400
D	100.00–200.00	150	5400
G	200.00–300.00	250	9000

Maier, 1994; Sumitomo, 1992). Moreover, some spectrometers are equipped with automatic sample changers (Kojima et al., patent; Maier, 1994; Sollier et al., 1989), thus increasing the number of samples that can be analyzed in a nonstop protocol.

9.4   APPLICATION OF EPR SPECTROSCOPY TO THE STUDY OF AEROSOLS

The application of EPR to the study of aerosols is related to at least four milestone achievements. The first was the discovery (Ingram et al., 1954a, b; Uebersfeld et al., 1954; Winslow, 1955) that upon carbonization of organic materials at relatively low temperatures (500 to 900 K), systems with high concentrations of unpaired electrons are formed. These papers opened a broad area of EPR study. Also, it is worth noting that most of the carbonaceous products exhibit carcinogenic activity, and studies on their qualitative and quantitative aspects are important for monitoring and improving the environment.

9.4.1   STUDY OF AEROSOLS IN CIGARETTE SMOKE

The second milestone, and in fact the first application of EPR in the field of aerosols, was concerned with the EPR study of paramagnetic products in cigarette smoke published in the period 1958 to 1961 (Austen et al., 1958; Ingram et al., 1961). The goal of these studies, performed on a model system, was to look for a dependence between the number of the unpaired electrons and the carcinogenic activity of tobacco smoke. For these studies, a specially constructed apparatus (Figure 9.5) was used in which air was pulled through burning cigarettes with a flow rate close to that of real smoking. The smoke aerosols (containing mainly cigarette tar) were trapped in an EPR sample tube immersed in liquid nitrogen (T = 77 K), thus playing the role of a finger-shaped trap for the aerosols. After enough material was collected in the trap, it was transferred to a cooled EPR cavity and the first measurement was performed. In this case, it is assumed that the EPR response is the sum of all carbonaceous materials collected, including all free radicals formed in the process of smoking and not recombining with other ones before trapping. In the next experiment, the EPR sample tube was refrozen, kept for some time at room temperature, and again frozen at liquid N_2 temperature. It is assumed that the EPR response in this second experiment is due only to carbonaceous products characterized by long-lasting paramagnetic properties. The difference between the EPR spectra is attributed to short-lasting, active, free radicals which may interact with matter when in contact with it.

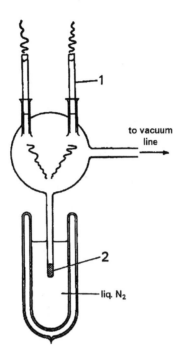

FIGURE 9.5 Sketch of the apparatus for collection of cigarette smoke aerosols for study by EPR. (1) cigarettes, (2) EPR sample tube immersed in liquid N_2.

Another EPR experiment on the same topic 10 years later (*JEOL News,* 1968) was concerned with the effectiveness of a cigarette filter in stopping the cigarette smoke aerosols. It was shown that if the length of the filter is more than 20 mm it can stop the paramagnetic products almost fully if the cigarette is only half-smoked. If it is nine-tenths smoked, the EPR signal due to paramagnetic aerosols (and cigarette tar) is equal for all lengths of the filter. The same paper reported that using the EPR technique, the signal could be divided into at least two components.

For the next 15 years, no progress was made on the topic discussed. After that, the problem of cigarette aerosols content was studied in more detail (Pryor et al., 1983a, b). At least four different free radicals (labeled A, B, C, and D) were separated, using chemical methods as well as differences in their EPR response, and studied in detail. It is claimed that the free radical marked "D," which is approximately 80% of the content, has a structure similar to the charge transfer quinone–hydroquinone complex (Prior et al., 1983 b).

9.4.2 STUDY OF AEROSOLS IN URBAN AIR

Dzuba et al. (1988) made the third important step in the application of EPR spectroscopy to the study of aerosols. These authors were first to study the EPR spectrum of material collected on a filter through which atmospheric air was passed. Three different signals are found, depending on temperature. The EPR spectrum at 300 K is a superposition of one broad singlet signal with a narrow line (Figure 9.6a). At 77 K, the broad signal becomes wider, whereas the linewidth of the narrow line is decreased. An EPR sextet is recorded as well (Figure 9.6b). The authors have attributed the broad line to species containing Fe^{3+}, the sextet spectrum to Mn^{2+}, and the narrow signal to carbonaceous products.

The fourth milestone which may be considered among the first practical applications of EPR spectroscopy was the systematic study of aerosols in the urban air of Sofia, Bulgaria, for a five-month period (Yordanov et al., 1996). Due to different sources (mainly exhaust gases of diesel engines), the quantity of carbonaceous products sometimes may be significant in urban air (Hitzenberger et al.,

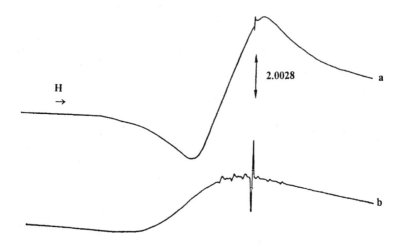

FIGURE 9.6   EPR spectra of aerosols recorded at: (a) T = 300 K, (b) 77 K.

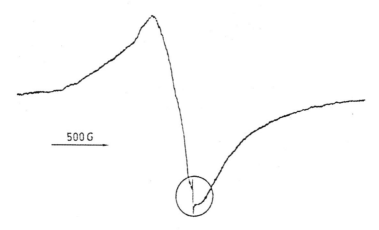

FIGURE 9.7   Typical room-temperature EPR spectrum for aerosols in urban air. The signal corresponding to the carbonaceous products is circled.

1993; Mueller et al., 1982; Scheff et al., 1990; Williams et al., 1989a, b). The subject of the EPR studies was the content of the filters collected during a five-month period in winter. The winter season was specially chosen because the meteorological conditions during the cold period of the year usually increase the pollution of the city's surface air layer due to inversions and calm conditions of low wind (Bluskova et al., 1983). In addition, buildings in the downtown area are heated by steam produced by heating power plants situated outside the town and operating on natural gas and/or black oil. Small private houses in other areas of the town burn oil, coal, or wood in their own heating facilities. The materials under investigation were collected at two stations, numbered 1 and 2 (these stations are part of the air pollution network of the National Institute of Meteorology and Hydrology at the Bulgarian Academy of Sciences), situated downtown (No. 1) and about 4 km away (No. 2).

Figure 9.7 shows typical EPR response of the material collected on the filter. The broad line centered at $g = 2.0$ is obviously due to Fe^{3+} and partly Mn^{2+} ions present in the soil. Other components containing Si, Ca, Al, K, etc., are EPR silent. The narrow singlet line at $g = 2.0028$ and marked with a circle is the EPR signal due to the carbonaceous products, and its intensity was monitored in all reported studies.

The intensity of the EPR signal was found to vary broadly from day to day and between stations. This fact may be attributed to variation in the number of sources and the intensity of their emissions, as well as to different atmospheric conditions for pollutant dispersion. In view of this, an attempt was made to find the relationship between elemental carbon particles found by EPR and other air pollutants, such as SO_2, NO_2, and dust, as well as some other meteorological parameters — temperature at 10 m above the ground, cloud cover, wind speed, and humidity, which were measured simultaneously at both stations. It was accepted in these considerations that: (1) the main source of NO_2 in the urban atmosphere is transport, whereas SO_2 is a pollutant coming mainly from thermal power plants and coal burning in home heating facilities; (2) dust may be used as an average measure for aerosol concentration on the surface layers; (3) sources of carbonaceous products in urban air are engines of diesel vehicles, coal, wood, and other organic burning. The average month concentrations of pollutants and EPR detectable substances (in relative units) for both stations are presented in Figure 9.8. The statistical analysis shows that correlation coefficients are small — the highest value found for dust concentration is less than 0.5. In view of this, the average month values were chosen as more representative (Figure 9.9). As seen from this figure, the EPR recorded values are higher for station No. 2 both in the morning at 8 LST and in the afternoon at 14 LST for the entire measurement period. The average EPR values for days when precipitation occurred were usually 10 to 50% lower compared to those for dry days. This can be expected because of the removal of pollutants by precipitation. In January, the month with the highest precipitation, the EPR values were the lowest for station No. 1 but not for station No. 2. An explanation could be a possible difference in the pollution source intensities between the stations becoming accentuated by low temperatures during this month.

Up to now only arbitrary units were used for EPR estimations. Theoretically, EPR spectroscopy can be used to estimate the absolute number of unpaired electrons in the sample under investigation (for a review, see Yordanov, 1994). However, the procedure carries very high uncertainties in the estimations (Yordanov et al., 1994). In order to get quantitative magnitudes of the studied CP, the data were compared with standards prepared from sucrose pyrolyzed at 550°C. Details of the procedure are described elsewhere (Yordanov et al., 1996). The average quantity of the carbonaceous products in the period of study was found to be 10 µg/m³ (station No. 2) and 5.1 µg/m³ (station No. 1), with a variation within ±0.6 µg for the period of the five-month study. In the same time, the maximal and minimal values were 52 µg/m³ (station No. 2) and 0.4 µg/m³ (station No. 1). The estimated values are very close to those found by other conventional analytical methods. For example, in Athens, Greece, 4.2 µg/m³ was reported for the average quantity of carbonaceous products (Scheff et al., 1990), total carbon 21.1 µg/m³. Approximately the same average values ranging from 1.2 up to 23 µg/m³ of EC were found in the air in some regions of the U.S. (Mueller et al., 1982).

9.5 NATURE OF THE PRODUCTS FORMED UPON PYROLYSIS

The problem of products formed in the thermal processes of hydrocarbons is significant because it is found that some of them exhibit carcinogenic properties (see Rademacher, 1976). It is assumed that the first step of the process includes dehydrogenetive polymerization and condensation reactions (Figure 9.10), and at relatively low temperature (300 to 500 K) pitches are formed, which at increased temperature (500 to 1000 K) form coke. The final product which may be obtained at 2500 to 3000 K is graphite. These products are separated into two groups — polycyclic aromatic hydrocarbons (PAH) containing 5 to 6 condensed aromatic rings as well as most of the carbonaceous products (CP), referred to also as organic carbon (OC) or elemental carbon (EC). All of them, and mainly PAH, exhibit carcinogenic activity (Rademacher, 1976). In urban air, due to different sources, but mainly exhaust gases of diesel engines, the quantity of PAH and CP sometimes reaches significant levels (Williams et al., 1989; Mueller et al., 1982; Scheff et al., 1990; Hitzenberger et al., 1993), and their control is essential. The material collected on the filters (usually of large volume)

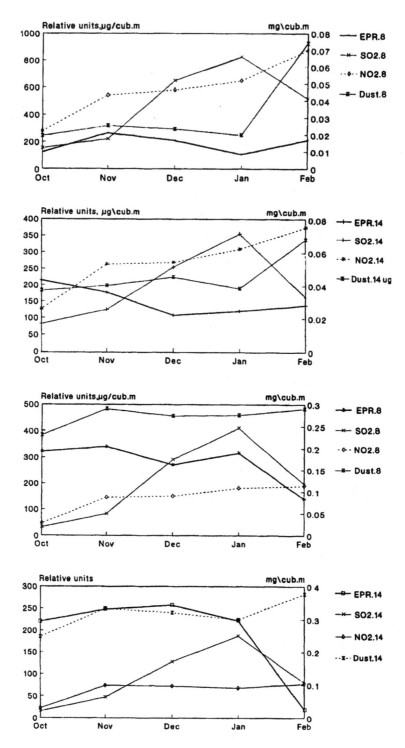

FIGURE 9.8 Average monthly values of the EPR response, SO_2, NO_2, and dust for Station No. 1 (a, b) and No. 2 (c, d) at 8 and 14 LST sampling times, respectively.

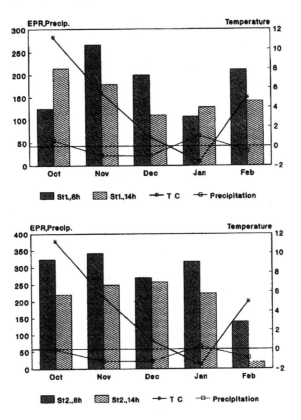

FIGURE 9.9 Average monthly values of the EPR signal intensity, temperature, and precipitation sum during the period of the experiments.

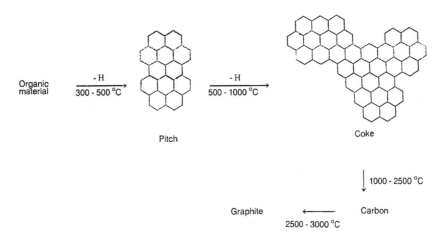

FIGURE 9.10 Scheme of the thermal conversion of organic materials into carbonaceous products.

is principally distributed between PAH and CP. These may be separately determined using thermal separation of carbon in OC and PAH followed by catalytic oxidation to CO_2, which after reduction to CH_4 is further determined by different gas analyzing methods. The procedure starts with increasing the temperature of the collected material to 600°C at which PAH becomes volatile. PAH vapors are removed in a stream of helium and are further oxidized. The next step includes subsequent removal of CP remaining on the filter which is also converted to CO_2 in an O_2 – He stream. The results of carbon analysis thus obtained may be reported as a total carbon or divided into CP and organic carbon (OC).

The structure of the paramagnetic species which are recorded by EPR is a topic for discussion. Some authors associate them with the free radicals of the odd–alternate aromatic hydrocarbons such as perinaphthyl or benzantranyl (Lewis et al., 1981). In order to check this possibility, several neutral free radicals of the odd–electron hydrocarbons were prepared and examined (see Gerson, 1970; Reddoch et al., 1969; Lewis et al., 1969, 1985; Singer et al., 1984). The solution EPR spectra of all these free radicals exhibit relatively well-resolved hyperfine splitting and a spectral width of about 50 G. The EPR spectrum of charcoal, however, is a structureless singlet line with a linewidth of about 5 G (the linewidth of charcoal strongly depends on the absorbed oxygen and varies from approximately 1 G in vacuum up to 15 G in air or oxygen). The singlet EPR spectra of some aromatic cation radicals like anthracene, perylene, naphthacene, etc., were found upon their formation on the surface of a strongly calcinated silica/alumina adsorbent of the type used as a catalyst in the cracking of hydrocarbons (Flockhart et al., 1962). It was also found that the EPR signals were relatively stable during the time the cation radical was in an adsorbed state and that the conversion was quantitative. In further studies by the same authors (Burns et al., 1986), they reported that the lower limit of determination is down to 1 ng of PAH. The only problem remaining is that, in all cases, only a singlet EPR signal with equal parameters was recorded. Therefore, this method cannot be used for separation of different PAHs. It is also essential to note that the procedure for sample preparation of PAH investigation includes its dissolution in some solvents, and the sample (if it is unique) is destroyed. Another procedure for converting PAHs into EPR-active cation radicals is a dissolution in concentrated sulfuric acid (Fraenkel, 1964). One example of such a procedure is given in Figure 9.11, where an EPR spectrum of a cation radical of perilene obtained by dissolution of the diamagnetic material in concentrated sulfuric acid is formed. This method was used (Dzuba et al., 1991) for studying the content of the collected aerosols, and some increase in the EPR response of the treated material compared to the untreated (dry one) was found.

Also assuming a constant ratio between PAH and EC, these authors tried to find a correlation between them. These studies are very interesting in their insight into the structure of carbonaceous materials and especially the possibilities to use EPR spectroscopy for the study of their content in aerosols. To obtain more definite evidence about their structure, perhaps more promising is the use of ENDOR (electron nuclear double resonance) spectroscopy, which is characterized by about 1000 times higher resolution than EPR (see Kevan et al., 1976; Dorio et al., 1979; Kurreck et al., 1988; Yordanov, 1989). Several well-resolved ENDOR lines are obtained in the free radicals spectra of odd–electron hydrocarbons when in solution (Lewis et al., 1985; Singer et al., 1984), and only one or two such transitions are found in the solid samples of charcoal or coals (see Retcofsky et al., 1982; Ueberfeld, 1984; Yordanov et al., 1989). It is also claimed that EPR-active species are those with a well-developed surface of condensed aromatic rings on which unpaired electron defects may be strongly delocalized. This assumption is also supported by the fact that the number of the unpaired electrons increases simultaneously with the increase of carbon in the sample (Ingram, 1969). The ENDOR studies of coal (Yordanov et al., 1989) seems to support this idea, showing that the value of the weak hyperfine splitting due to protons decreases upon the swelling of the material. If this is the case, the lifetime of such strongly delocalized unpaired electrons will be extremely long. This assumption is supported by some reports showing that there is no difference in the number of unpaired electrons in charcoal produced today compared to that found in the pyramids in Egypt (Ingram, 1969).

FIGURE 9.11   EPR spectrum of the cation radical of perilene obtained by dissolution of perilene in concentrated sulfuric acid.

9.6   CONCLUSION

The advantages of the high selectivity of EPR spectroscopy to estimate the quantity of only paramagnetic CP has been shown. All other products, such as PAH and mineral dust, present in the sample are EPR-silent and could very easily be neglected. The total time for a single analysis is about 5 min, compared with several hours for separation and analysis using other conventional procedures. The low cost of the analysis is also remarkable. The procedure is a little bit more time consuming for an extended study in which, in addition to the normal paramagnetic CP, other PAHs are also converted in a paramagnetically active form. In this case, total carbon content, as well as a separation between the EPR-active and silent species, may be done. The evidence described in this chapter suggests the EPR method as a very powerful and useful tool with high potential for estimations of carbonaceous pollutants present as aerosols in urban air. The initial steps have been taken, and EPR seems promising for increasing the quality of human environment monitoring.

ACKNOWLEDGMENTS

The financial support of the National Foundation for Scientific Research (project X–419) is gratefully acknowledged.

REFERENCES

Abragam, A. and Bleany, B., 1970, *Electron Paramagnetic Resonance of Transition Ions*, Oxford University Press, U.K.

Alger, R. S., 1968, *Electron Paramagnetic Resonance*, Wiley, New York.

Al'tshuler, S. A. and Kozyrev, B. M., 1964, *Electron Paramagnetic Resonance*, English translation, Academic Press, New York.

Assenheim, H. M., 1967, *Introduction to Electron Spin Resonance*, Plenum, New York.

Austen, D. E. G. and Ingram, D. J. E., 1958, *Trans. Faraday Soc.*, 54: 400.

Ayscough, P. B., 1967, *Electron Spin Resonance in Chemistry*, Methuen, London.

Bluskova, D., Zlatkova, L., Lingova, S., Modeva, Z., Subev, L., and Teneva, M., 1983, in *Climate and Microclimate of Sofia,* Bulgarian Academy of Science Publishing House, Sofia, p. 119.

Burns, D. T., Salem, M. A., Baxter, R. I., and Flockhart, B., 1986, *Analytica Chimica Acta,* 183: 281.

Centro Spektr, Minsk, Belorussia, Model PS-100X.

Dorio, M. M. and Freed, J. H., 1979, *Multiple Electron Resonance Spectroscopy,* Plenum, New York.

Dzuba, S. A., Puskin, S. G., Tsvetkov, Yu. N., 1988, *Dokl. AN USSR,* 299: 1150.

Dzuba, S. A., Tirishkin, A. M., Puskin, S. G., Senkevitch, S. I., Tsvetkov, Yu. N., 1991, *Dokl. AN USSR,* 321: 127.

Flockhart, B. D. and Pink, R. C., 1962, *Talanta,* 9: 931.

Fraenkel, G. K., 1964, *J. Chem. Phys.,* 40: 3307.

Gerson, F., 1970, *High Resolution ESR Spectroscopy,* Wiley, New York.

Hitzenberger, R. and Puxbaum, H., 1993, *Aerosol Sci. Technol.,* 18: 323.

Ikeya, M. and Furusawa, M., 1989, *Appl. Radiat. Isot.,* 40: 845.

Ikeya, M., 1993, *New Applications of Electron Spin Resonance,* World Scientific, Singapore.

Ingram, D. J. E. and Bennett, J. E., 1954, *Phil. Mag.,* 45: 545.

Ingram, D. J. E., Tapley, J. G., Jackson, R., Bond, R. L., and Murnaghan, A. R., 1954, *Nature,* 174: 797.

Ingram, D. K. E., 1958, *Free Radicals as Studied by Electron Spin Resonance,* Butterworths, London.

Ingram, D. J. E., 1961, *Acta Med. Scand. Suppl.,* 369: 43.

Ingram, D. J. E., 1969, *Biological and Biochemical Applications of Electron Spin Resonance,* Adam Hilger, London.

JEOL News, 1968, Vol. 6B, No. 1, 4.

Kevan, L. and Kispert, L. D., 1976, Electron Spin Double Resonance Spectroscopy, Wiley, New York.

Kojima, T. and Tanaka, R., 1989, *Appl. Radiat. Isot.,* 40: 851.

Kojima, T., Haruyama, Y., Tachibana, H., Tanaka, R., Okamoto, J., Hara, H., and Yamamoto, Y., 1993, *Appl. Radiat. Isot.,* 44: 361.

Kojima, T., Uematsu, T., Tanaka, R., and Aramaki, S., Jap. Patent No.1-138484, 533-536.

Kurreck, H., Kirste, B., and Lubitz, W., 1988, *Electron Nuclear Double Resonance Spectroscopy of Radicals in Solution,* VCH Publishers, New York.

Lewis, I.C. and Singer, L. S., 1969, *J. Phys. Chem.,* 73: 215.

Lewis, I. C. and Singer, L. S., 1981, in *Chemistry and Physics of Carbon,* Walker, P. L. and Thower, P. A. (Eds.), Marcel Dekker, New York, Vol. 17, p. 17.

Lewis, I.C. and Singer, L. S., 1985, *Magnetic Res. Chem.,* 23:698.

Lyniov, V. N., 1991, in *Electron Magnetic Resonance of Disordered Systems,* N. D. Yordanov (Ed.) World Scientific Publ., Singapore, p. 53.

Lyons, M. J., Gibson, J. F., and Ingram, D. J. E., 1958, *Nature (London),* 181: 1003.

Mabbs, F. E. and Collison, D., 1992, *Electron Paramagnetic Resonance of d-Transition Metal Complexes,* Elsevier, Amsterdam.

Maier, D. and Schmalbein, D., 1993, *Appl. Radiat. Isot.,* 44: 345.

Maier, D. C., 1994, *BRUKER Report,* 140: 23.

Mueller, P. K., 1982, in *Atmospheric Life Cycles,* Wolff, G.T., Klimisch, R. A. (Eds.) Plenum, New York.

Nakanishi, A., Sagawara, N., and Furuse, A., 1993, *Appl. Radiat. Isot.,* 44: 357.

Orton, J. W., 1968, *Electron Paramagnetic Resonance. An Introduction to Transition Group Ions On Crystals,* Iliffe Books, London.

Pilbrow, J. R., 1990, *Transition Ion Electron Paramagnetic Resonance,* Clarendon Press, Oxford.

Poole, Ch. P., Jr., 1983, *Electron Spin Resonance. A Comprehensive Treatise on Experimental Techniques,* 2nd Ed., Wiley, New York.

Pryor, W. A., Hales, B. J., Premovic, P. I., and Chirch, D. F., 1983, *Science,* 220: 425.

Pryor, W. A., Prier, D. G., and Chirch, D. F., 1983, *Environ. Health Perspect.,* 47: 345.

Rademacher, P. and Gilde, H.-G., 1976, *J. Chem. Edu.,* 53: 757.

Reddoch, A.H. and Paskovich, D. H., 1969, *Chem. Phys. Lett.,* 3: 351.

Retcofsky, H. L., Hough, M. R., Maguire, M.M., and Clarkson, R. B., 1982, *Applied Spectroscopy,* 36: 187.

Scheff, P. A. and Valizis, C., 1990, *Atmospheric Environment,* 24A: 203.

Singer, L. S. and Lewis, I.C., 1984, *Carbon,* 22: 487.

Sollier, T. J. L., Mosse, D. C., Chartier, M. M. T., and Joli, J. E., 1989, *Appl. Radiat. Isot.,* 40: 961.

SPR — Specialists periodical reports, *Electron Spin Resonance,* series of volumes, Royal Society of Chemistry.

213

Sumitomo Special Metals Co. Ltd., Japan, 1992, Models Spin-X and Spin-XX, *EPR Newsletter,* 4 p. 10.

Uebersfeld, J., Etienne, A., and Combrisson, J., 1954, *Nature,* 174: 614.

Uebersfeld, J., 1984, in *Magnetic Resonance. Introduction, Advanced Topics and Applications to Fossil Energy,* Pertakis, L. and Fraissard, J. P. (Eds.), Riedel, Dordrecht, The Netherlands, p. 165.

Weil, J. A., Bolton, J. R., and Wertz, J. E., 1994, *Electron Paramagnetic Resonance. Elementary Theory and Practical Applications,* Second edition, Wiley, New York.

Wertz, J. E. and Bolton, J. R., 1972, *Electron Paramagnetic Resonance. Elementary Theory and Practical Applications,* McGraw-Hill, New York.

Williams, D.J., Milne, J. W., Roberts, D. B., and Kimberlee, M.C., 1989, *Atmospheric Environment,* 23: 2639.

Williams, D.J., Milne, J. W., Quigley, S. M., Roberts, D. B., and Kimberlee, M.C., 1989, *Atmospheric Environment,* 23: 2647.

Winslow, F. H., Baker, W. O., and Yager, W. A., 1955, *J. Am. Chem. Soc.,* 77: 4751.

Yamanaka, C., Ikeya, M., Meguro, K., and Nakanishi, A., 1991, *Nucl. Tracks.,* 18: 279.

Yordanov, N.D., 1988, in *Coordination Chemistry and Catalysis,* Ziolkowski, J. J. (Ed.), World Scientific Publishing, Singapore, p. 313.

Yordanov, N. D., Duber, S., Zdravkova, M., and Budinova, T., 1989, *FUEL,* 69: 818.

Yordanov, N. D., 1994, *Appl. Magnetic Res.,* 6: 241.

Yordanov, N.D., Veleva, B., and Christov, R., 1996, *Appl. Magnetic Resonance,* 10: 439.

Zavoiski, E., 1945, *J. Phys. U.S.S.R.,* 9: 245, 1945; ibid. 10: 170.

10 Analysis of Environmental Aerosols by Multiphoton Ionization

Vladimir V. Gridin and Israel Schechter

CONTENTS

INTRODUCTION

The substantial influence of aerosols on the environment is recognized and appreciated. The atmosphere is a sink for chemical species emitted from or produced by numerous sources, such as natural cycling by plants, soils, water resources, and various human activities.

Airborne particulate materials are subjected to both photochemical transformations and long-range global transport, before they undergo wet or dry deposition processes onto the earth's surface. Individual particle analysis of airborne pollutants provides important insights into the source, formation, and environmental impact (both positive and negative), etc.

Either the loss or gain of significant amounts of semivolatile compounds may occur during the sampling of aerosols. This, in turn, can alter a determination of their chemical composition. Quantitative analysis of particulate matter, however, is hindered by an obvious uncertainty of the interaction volume, when such microparticles are probed by quite sophisticated analytical tools.

In fact, this volume represents the active interaction space from which the desired analytical readouts are obtained. Each analytical technique, however, may largely either under- or over-estimate the original chemical composition and fractional constituents' concentrations in an attempt to quantify specific physical/chemical parameters of the aerosols studied.

We do not aim to review all the material that ought to be relevant to our presentation. One can appreciate the difficulty of that task by simply acknowledging the topics involved: sampling of aerosols, polycyclic aromatic hydrocarbon contaminated aerosols, analytical methods relevant to environmental studies, multiphoton ionization processes, and the associated research tools.

Instead we choose to present an overview of most of the subjects, while narrowing our presentation to a multiphoton ionization-based fast-conductance analytical scheme with regard to screening of such environmentally important aerosol pollutants as motor vehicle exhausts and tobacco smoke. Hence, material presentation in this chapter is organized according to the following breakdown of the relevant issues involved:

1. Application of laser-induced processes to analysis of aerosols.
2. Multiphoton ionization-based analytical methods.
3. Coupling and application of conductance measuring instrumentation to MPI.
4. Recent renewable water droplet analysis.

10.1 APPLICATION OF LASER-INDUCED PROCESSES TO ANALYSIS OF AEROSOLS

The most frequently used analytical techniques for studying airborne particulate matter have been reviewed by Grasserbauer,[1] Van Grieken et al.,[2] Van Grieken and Xhoffer,[3] Jambers et al.,[4] and Pui.[5]

In particular, an ever-growing number of advanced methods are based on various laser-induced processes, some of which were successfully utilized in studying health hazards arising from coal mine dusts, fossil fuel combustion, steel industries, and chemical plants. Among these are:

On-line laser microprobe mass spectrometry[6-10]
Time of flight mass spectrometry[11-15]
Fourier transform laser microprobe mass spectrometry[16-18]
Micro-raman spectroscopy[3,19,20]
Time-resolved fluorescence spectroscopy[21-23]
Laser-induced fluorescence in graphite furnace technique[24-27]
Laser-induced breakdown spectroscopy[28-30]

Review coverage of subjects related to aerosol sampling, particle size distribution, and composition, together with environmentally relevant specification of various physical and chemical properties of the airborne organic pollutants could be found, for example, in the book edited by Hansen and Eatough.[31]

10.2 MULTIPHOTON IONIZATION-BASED ANALYTICAL METHODS

Briefly speaking, MPI processes involve interaction of matter with a very dense photon field: $\sim 10^{29}$ photons $cm^{-2}s^{-1}$. Such intense photon fluxes became achievable due to advanced developments in laser technology.

There are several interaction schemes for the laser-induced material excitation. A particular choice is governed by the ability to channel MPI for both sensitive and selective probing of the material state studied. Some of the extensively used schemes are: single-color off-resonance and resonance enhanced excitations, multicolor absorption, CW (or pulsed) laser desorption coupled to a frequency tunable ionizing irradiation, laser polarization dependent MPI, etc.

It is beyond the scope of this chapter to review voluminous scientific literature on multiphoton ionization processes and their analytical applications. We trust that the forthcoming choice of cited material (by no means far from a comprehensive one) should, nevertheless, provide sufficient guidance to those aspects of MPI-related phenomena that are in the mainstream of our presentation.

General features of MPI as well as its capacity as a very sensitive analytical tool have been addressed in References 32 through 38. In particular, the material state selectivity offered by MPI-based methods was recently discussed by Boesl[37] and Anderson.[38]

Several recent applications of multiphoton ionization (MPI) or resonant enhanced MPI[39] in mass spectrometry (MS) of airborne particulate matter have been reported by Dale et al.,[40] Fei et al.,[41] Alimpiev et al.,[42] Gittins et al.,[43] and Weickhardt et al.[44]

Unfortunately, despite relatively high material selectivity and spectral resolution offered by various MS techniques, they are limited, at times, by the quite severe vacuum conditions needed for most of the accurate time-of-flight measurement schemes. This limitation is of a much lesser importance, however, for such aerosol analysis tools as micro-Raman spectroscopy,[3,19,20] on-line laser microprobe mass spectrometry,[6-10] Fourier transform infrared spectroscopy,[45] and tandem mass spectrometry of individual airborne microparticles.[46]

10.3 COUPLING AND APPLICATIONS OF CONDUCTANCE MEASURING INSTRUMENTATION WITH MPI

Extensive research efforts have been devoted to merge MPI with appreciably lower-cost conductance measuring techniques operating from a moderate vacuum[47,48] to ambient pressure conditions.[49-63]

Charged particle mobility measuring analytical approaches could be roughly classified according to their environmentally relevant applications to pollution screening and monitoring. These are the gas, liquid and solid phases probing techniques. Quite generally, the ionization threshold energy, $E_{ION}(m,M)$, of a particular molecule, m, is expected to depend on a material matrix, M. Here M specifies the nearest environment of m and: E_{ION} (m, M = gas phase) > $E_{ION}(m,M$ = liquid phase) > $E_{ION}(m,M$ = solid phase). Hence, for instance, in a single color off-resonance ionization scheme, the MP process is just a coherent, simultaneous absorption (within the lifetime of the virtual states involved; ~10^{-15}s) of N monochromatic photons, so that Nhv $\geq E_{ION}(m,M)$. Such multiple absorption of photons produces electrons and positive ions inside the interaction volume, V_{INT}.

In a simple case of no material fragmentation and/or electron trapping processes involved, these are the only products of MPI events. Subsequent analysis of charged material species by time-of-flight (vacuum matrix) or mobility measuring (dense matrix) instrumentation provides a principal experimental readout of any MPI-based technique. It should be noted that material fragmentation is expected to occur due to excitation of molecular vibronic states (life-time of ~10^{-12}s) when a laser frequency tuning is set to probe the real intermediate molecular states (lifetime of ~10^{-8}s) in a stepwise, resonant, MP absorption scheme.

The first MPI experiments on atoms and polyatomic (iodine) molecules were, respectively, conducted by Delone et al. in 1965[64] and Petty et al. in 1975.[65] In 1980–1981, Siomos and Christophorou[52] and Siomos et al.[50] reported a novel MPI arrangement suitable for liquid phase studies of polyatomic molecules in dilute solutions. Relevant to environmental research goals, considerable investigative effort has been devoted to studying various aspects of MPI-based techniques and their applications to detecting polycyclic aromatic hydrocarbon (PAH) molecules in polar and nonpolar liquids.[57,62,66-70]

Trace detection of organic contamination on solid surfaces is also of a significant environmental interest, though, especially when the analysis is carried out at ambient atmospheric conditions. Several recent reports address organic depositions on metals,[60,71] single crystals,[72] and aerosol particles.[73]

A variant of MPI conductance technique operational at ambient atmospheric conditions has been examined in our group.[74-77] It was shown[74,75] that a condensation of water vapors (at the level of ambient moisture) onto soil substrate is sufficient for conducting MPI-based conductivity measurements for trace analysis of organic contamination.

Clearly, this contribution opens up the possibility of applying similar experimental schemes to any porous substrate of interest, e.g., glass/paper filters, plant leaves (open field vegetation), certain food products, etc. Preliminary field tests were also conducted and reported. In these applications, the data acquisition facilities are relatively inexpensive as well as user friendly.[76,77]

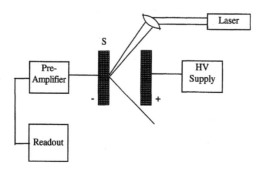

FIGURE 10.1   A block diagram of an MPI–FC facility. Laser irradiation is used for MPI of organic traces positioned at the conducting substrate, S. The high-voltage electrode enables spatial charge separation and transport. Readouts are then recorded and analyzed.

We conclude this section by briefly discussing a single-color MPI-based conductance scheme. For reasons of clarity, the forthcoming presentation is kept within the mainstream of the subject described. A block diagram of an MPI-based fast-conductance facility is shown in Figure 10.1. The principal operation of such an experimental set-up is relatively straightforward[49-63] and is specified in more detail later (see Section 10.4).

Assume PAH aerosols inside an interaction volume, V_{INT}, where they are subjected to UV irradiation by a monochromatic laser energy flux, J. Suppose J is such that a single-color, off-resonance, ionization scheme takes place. Let α stand for the aerosol's source of origin, e.g., chemical plant chimney exhausts, combustion by-product aerosols, traffic, or agricultural-related pollutants, etc. Then n_α represents a variable number density of a particular aerosol α. The latter might be of a very complex composition and contain several organic traces, β, with a fractal number density, f_β, such that $\Sigma f_\beta = 1$; here, for a single-color, off-resonance case, the sum runs over UV active composites only.

Let the total detected photoionization charge, Q, result from the background matrix contribution, Q_B and from n_α-dependent term, $Q_P = Q_P(n_\alpha)$. The former is due to the aerosol sampling matrix within V_{INT}. Note that the experimental scheme of Section 10.4 corresponds to an aqueous sampling matrix for the airborne particulate materials studied. Clearly, random fluctuations in Q_B are expected to occur. Its time-averaged value could be found by gathering appropriate statistics of the blank ($n_\alpha = O$) system readouts. In any case, S/N should be kept as high as possible in order to obtain significant limits of detection for airborne pollutants.

This is especially true for most of the open-field experimental conditions, where aerosol intake is subjected, for example, to fluctuating weather conditions. Later (refer to Section 10.4) we show that even in a single-color MPI scheme there exists a straightforward background elimination procedure which allows establishing a kind of "universal," largely sample matrix independent, calibration curve for detection of PAH airborne pollutants.

Since Q_B is not related to the monitored aerosol intake, n_α, we define: $\Psi \equiv Q - Q_B$. Then an effective order of MPI process, k, is given by $k = \partial Log\Psi/\partial LogJ$. Clearly, k is expected to be β-dependent too. For the present case we write: $Q_P = n_\alpha \Sigma f_\beta J^k \Gamma_\beta^{(k)}$. Here $\Gamma_\beta^{(k)}$ stands for the k-th order cross section of a single-color, off-resonance MPI of a particular pollution constituent, β. Quite obviously, for a one-component aerosol pollutant, $f_\beta = 1$. For a more general case, however, we obtain:

$$\Psi \equiv Q - Q_B = Q_P = n_\alpha \Sigma f_\beta J^k \Gamma_\beta^{(k)} \qquad (1)$$

It is noteworthy that a single-color ionization scheme lacks the trace selectivity provided by commercially available frequency tunable laser sources. Hence, different β constituents of a multiple-component organic aerosol are indistinguishable in such a scheme. Nevertheless, their combined readout is governed by Equation (1), and the *very presence* of airborne pollutants still might be quite conveniently (on-line) detected and monitored.

In order to do any better, however, a routine, resonance-enhanced, MPI scheme is needed. Needless to say, that in such a case, the S/N ratio is expected to increase sharply due to a standardized elimination and minimization of, mostly off-resonant, background signals, Q_B.

10.4 RECENT RENEWABLE WATER DROPLET ANALYSIS

In this section, we summarize some of our recent investigations of MPI-based detection of airborne particulate organic matter.[78,79] Development of this approach was envisioned by several methodological considerations:

a. Mansoori et al.[10] suggested glycerol droplets in their on-line TOF–MS analysis of particulate matter. Their objective was to create a reproducible local environment around the analyte. Such an aerosol-containing droplet provides a well-defined morphology of interaction volume. Hence, a certain degree of standardization in on-line applications of this laser-induced desorption/ionization method was shown to emerge.

b. Recent reports[80-83] suggest efficient applications of water microdroplets for renewable *in situ* sampling, e.g., of such gaseous chemicals as nitrogen dioxide and chlorine. It is also stated that single molecule limits of detection are feasible in analytical studies conducted at ambient pressure conditions.

c. In a recent study by Ogawa and co-workers,[70] a high-sensitivity MPI–FC method was performed on minute liquid samples. There, in the case of pyrene contamination, a detection limit of 0.3 pg was reached.

d. Some of the MPI–FC experiments conducted on moist/wet porous substrates[74,75] might be viewed as aqueous film sampling of environmental PAH depositions. In fact, a tiny water droplet, geometrically confined by a small metallic loop, should play a triple role: (1) to renewably sample PAH pollutants; (2) to standardize and define the interaction volume; (3) to provide sufficient electrical conductivity for minimizing the so-called space charge effects, appearance of which causes ill-functioning of FC instrumentation.[75]

10.4.1 EXPERIMENTAL SET-UP

The experimental arrangement for aerosol analysis by MPI method is shown in Figure 10.2. It was chosen to comply with the above noted (a) through (d) guiding points. Its principal operation is as follows: PAH contaminated water droplet is attached to the -ve electrode. All laser-induced MPI processes occur at the droplet surface and/or in its bulk. Photoelectrons are released there. A portion of them arrives directly at the +ve electrode and is responsible for the fast initial rise of the detected photocurrent.

Additional channel for transferring such electrons is due to the substantial electronic affinity of ambient oxygen molecules present in the chamber. These, therefore, mediate detection of photoelectrons and modify their arrival at the +ve electrode during a much longer time span than the aforementioned, by far faster, unassisted ones.

A bit more specifically, (refer to Reference 78 for a detailed description), the off-resonance MPI processes were induced here by a pulsed nitrogen laser emitting 1.5 mJ pulses at 337.1 nm in 0.6 ns. Measured laser energy, with the focused beam, was 120 µJ/pulse.

A free-hanging water droplet, defined by a small (0.7 mm diameter) copper wire loop, was renewably produced via a stainless steel needle. Under stationary conditions, a mean sampling volume, $V_s \cong 14$ µl, was obtained. The copper loop and the needle were in electrical contact with

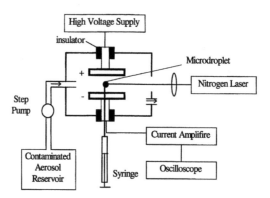

FIGURE 10.2 Schematics of MPI–FC experimental set-up used for renewable water droplet sampling of combustion by-product aerosols. The principal components of the aerosol contamination circle are also indicated. (From Gridin, V.V., Litani-Barzilai, I., Kadosh, M., and Schechter, I., *Anal. Chem.*, 1997, 69, 2098. With permission.)

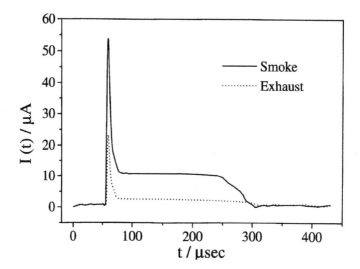

FIGURE 10.3 Photoionization current, I = I(t), obtained by MPI–FC from renewable water droplets for (a) exhaust gas of a gasoline powered motor, and (b) cigarette smoke. (From Gridin, V.V., Litani-Barzilai, I., Kadosh, M., and Schechter, I., *Anal. Chem.*, 1997, 69, 2098. With permission.)

the negative electrode that, in turn, was attached to the current preamplifier, as shown in Figure 10.2. A typical amplification gain was 10^7 to 10^8 V/A.

Positive bias voltage (2500 V) was supplied to the upper electrode of Figure 10.2. The data were collected by a digital oscilloscope triggered by a photodiode (rise-time of 20 ns). Photoionization current, I(t), was routinely obtained by averaging more than 50 laser pulses. I = I(t) is, generally, proportional to the amount of organic neutrals present within the microdroplet sampler. Typical[49-63] MPI–FC data, from on-line contaminated renewable water droplets, are shown in Figure 10.3. The area specified by the I = I(t) curve is just the total charge, Q (see Section 10.3 above).

10.4.2 SINGLE-COMPONENT CONTAMINATION

First, pyrene was chosen to test the experimental set-up. A special method of pyrene deposition was applied in order to classify this case as a monotonously increasing droplet contamination by a single-PAH-constituent aerosol.

External depositions of pyrene were made using 2 µl portions of pyrene/n-hexane solutions in the concentration range, X, from 0.1 ng/ml to 1 mg/ml. Such microshuts were pulled in by surface tension forces. Hexane quickly evaporates, leaving behind the pyrene contaminated water droplet. Since, under normal conditions, the aqueous solubility of pyrene is of the order 0.1 to 0.2 ppm,[84] pyrene microparticle contaminated microdroplets could also result.

In fact, we used the pyrene aqueous solubility threshold to establish a possible influence of the undissolved portion of sampled aerosols on the calibration plots obtained by this MPI–FC technique. In this regard, one needs a standard data processing routine, which should be capable of providing matrix independent calibration curves. This issue is dealt with in the following subsection.

10.4.3 GEOMETRICAL CONSIDERATIONS

In order to comply with the above stated requests, a desirable data processing approach was introduced for analysis of particulate material samples contaminated with pyrene.[74-77] A number of sporadic effects associated with a hardly reproducible sample preparation routine were dealt with and reasonably well accounted for.

It should be noted, however, that moderate variations of laser beam focusing onto solid particulate samples did not result in any significant alterations of the aforementioned calibration curves. This, most definitely, was not so for the microdroplet sampling geometry. Let us elaborate on this issue.

Within an off-resonance MPI scheme at fixed J, there exists (for single,[74,75] as well as multiple[76,77] constituent particulate material contamination) a matrix independent calibration procedure, which interrelates the MPI–FC readouts, Q, and the number density of UV active organic neutrals, n_α. Proceeding in a similar way here too, recall that by Equation (1), with $f_\beta = 1$, one obtains:

$$\Psi = n_\alpha J^k \Gamma^{(k)} \tag{2}$$

According to Equation (2), at fixed J (i.e., fixed laser focusing conditions inclusive), $\Psi = \Psi(n_\alpha) \sim n_\alpha$. Let $\Psi(n_\alpha^*)$ be the value of $\Psi(n_\alpha)$ obtained at, e.g., the maximal dose, n_α^*, of a particular contaminant α; clearly, then, $\Psi(n_\alpha)\Psi(n_\alpha^*) = n_\alpha/n_\alpha^*$. According to this, one might expect a sort of "substrate independent" calibration curve to emerge.

In fact, the very failure to achieve this goal has proved to be quite crucial and helped establish a novel, MPI–FC based, scheme of measuring minute aqueous solubility of poorly dissolving PAH compounds.

The ratio $\Psi(n_\alpha)/\Psi(n_\alpha^*)$ as a function of n_α/n_α^* is shown in Figure 10.4 using logarithmic coordinates. The data are shown for two laser irradiation spots that were either deliberately focused on the droplet's outer surface or, alternatively, within its middle region. In what follows, we refer to the former and latter focus adjustments as *on-surface* and *off-surface,* respectively.

A quite dramatic difference in the slopes of these calibration curves is self-evident. Note also that for both laser beam arrangements shown, there occurs a substantial variation in the slopes near the aqueous solubility threshold of pyrene, X_E. Such a sudden change in the calibration slope for the microdroplet sampled pyrene was noted[78] and inspired further investigation of the apparent capability of MPI–FC to be useful for studying aqueous solubility, surface excess, and surface adsorption of PAH materials.[79]

The slope variation found in the on-surface focusing geometry was modeled using Langmuir adsorption isotherms and Gibbs formulation of the surface excess approaches.[85] The X_E figures[79] obtained for pyrene, perylene, phenanthrene, and anthracene check well with those reported by more traditional experimental techniques.[84] In this regard, it should be noted that even for the same trace level, a variation in the irradiation angle of the incoming laser beam might result in an alteration of the calibration curve. This has been recently reported by Ogawa et al.[70] for pyrene in small liquid droplets.

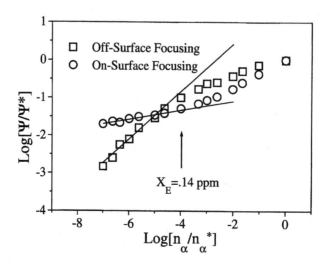

FIGURE 10.4   Plot of $\Psi(n_\alpha)/\Psi(n_\alpha^*)$ versus n_α/n_α^* for pyrene contaminated water droplets in *on-surface* (dotted circles) and *off-surface* (dotted squares) laser beam focusing arrangements.

A qualitative reasoning for such laser beam geometry dependent calibration slopes is as follows. The interaction volume is clearly defined by the laser beam spot. Any reduction in J results, according to Equations (1) and (2), in a corresponding decrease of Ψ, such that $\Psi \sim J^k$.

For $X > X_E$, the bulk concentration of pyrene (within the microdroplet sampler) remains constant. Therefore, any further depositions of pyrene should cause formation of its microparticles therein. Given a positive surface excess of pyrene in water, the microparticles' volume distribution is expected to favor the droplet's outer shell to its interior design region. Naturally, any concentration-dependent increase in $\Psi(n_\alpha)$ for this regime should be associated with the organic neutrals positioned near the droplet's surface region. Hence, for the off-surface beam alignment, such additional pyrene molecules would be subjected to a lesser flux density than before. The apparent decrease in the calibration slope is then imminent.

The opposite is true for on-surface focusing. There, the interaction volume becomes enriched by UV active neutrals. Hence, relative to the $X < X_E$ concentration regime, the calibration curve becomes steeper. Moreover, the findings of Figure 10.4 suggest that the microparticles of pyrene tend to "pile-up" at the droplets' air/liquid interface.

With respect to the off-surface beam alignment, it was found that, as long as the respective $X < X_E$ regime was explored for all the single-constituent-PAH "aerosols" tested, the calibration slopes were not significantly different from the one drawn in Figure 10.4 through the lower concentration part of the appropriate pyrene data. This, however, was not the case for the on-surface laser irradiation. The ultimate PAH/free water surface interactions involved seem to influence the MPI–FC readouts in such geometry.[79]

In accordance with the above, it was accepted[62,68,70] that MPI–FC should be regarded as the air/liquid interface analytical tool.[78,79] Once a particular MPI–FC set-up had been calibrated with a single constituent aerosol at a desirable laser beam spot position, the dissociation status of unknown aerosols could be readily deduced. In other words, fine geometrical adjustments of incoming flux density with respect to the interaction volume of interest provide important physical characteristics of the *as sampled* aerosol material.

As seen in Figure 10.4, the greatest sensitivity of the instrumentation was achieved when an off-surface (on-surface) geometry was applied for the $X < X_E$ ($X > X_E$) concentration regime. It would be interesting to explore such concentration regimes, where the monolayer or even multilayer PAH adsorption should prevail.

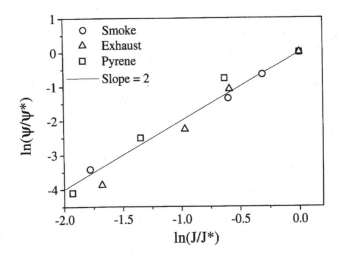

FIGURE 10.5 Plot of normalized photocharges, $\Psi(J)/\Psi(J^*)$, as a function of the normalized laser beam energy, J/J^*; here $J^* = 0.12$ mJ. The Slope = 2 line indicates a pure two-photon process. (From Gridin, V.V., Litani-Barzilai, I., Kadosh, M., and Schechter, I., *Anal. Chem.*, 1997, 69, 2098. With permission.)

Qualitatively speaking, even for the on-surface beam alignment, an effective leveling-off of the calibration slope might result. The reason being that the number of organic neutrals inside the interaction volume should then eventually saturate.

A further speculation would be that due to the inherent insulating properties of a full PAH multilayer droplet coverage, a reduction of MPI–FC readouts might take place there too. A study of these features is incomplete as yet and requires more thorough substantiation of the provided preliminary results.

Now, having calibrated such an MPI–FC set-up by pyrene aerosols and keeping closer to the main theme, we proceed with experimental results obtained on environmentally important airborne pollutants.

10.4.4 AEROSOL SAMPLING

The environmental aerosols used were those found in typical motor vehicle exhaust[86,87] and cigarette smoke.[88] Here we only briefly outline the reported aerosol sampling procedure; for further details refer to Reference 78.

Organic traces present in (a) the exhaust gases of an idling gasoline-powered motor vehicle engine, as well as those found in (b) free-burning cigarette smoke were routinely collected. They, on-line, entered the aerosol contaminated reservoir. Then a step pump was used to introduce the contaminated air into the measurement chamber in a continuous manner. Both of these complex pollutants contained many PAH traces and their overall combined effective response to a fixed line laser excitation was more likely to exhibit the off-resonance MPI features.

Inspecting Equation (1), one observes that for a given aerosol α, when n_α is fixed, a similar ionization order for all its UV active constituents would imply: $\Psi = \Psi(J) \sim J^k$. Hence, at fixed n_α, Log $\Psi \sim k$LogJ. The slope of 2 in Figure 10.5 compares well with the data, indicating an effective 2-photon ionization process, as would be for the most UV-active PAH contaminants at 337.1 nm.

For such combustion by-product aerosols, it was found that the off-surface beam geometry was the most suitable. In Figure 10.6, a plot of Log$[\Psi(n_\alpha)/\Psi(n_\alpha^*)]$ as a function of Log$[n_\alpha/n_\alpha^*]$ is shown for this case for both aerosols studied. The single-constituent calibration slope (referred to as the off-surface case of the $X < X_E$ pyrene data of Figure 10.4) is also drawn for comparison. A close similarity between these results is self-evident. Hence, it was concluded that, for the material

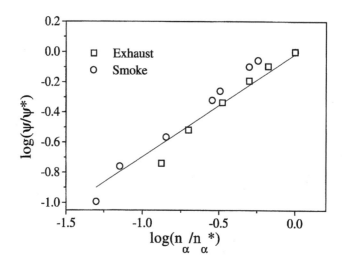

FIGURE 10.6 The data presentation here is similar to that of Figure 10.4. The slope of the *off-surface* low concentration data, X < X_E, of a single-constituent (pyrene) aerosol is shown and checks well with the results obtained for the combustion by-product aerosols studied. (From Gridin, V.V., Litani-Barzilai, I., Kadosh, M., and Schechter, I., *Anal. Chem.*, 1997, 69, 2098. With permission.)

quantities of the combustion by-product aerosols studied, an unsaturated dissociation regime (i.e., X < X_E) should prevail. Note also that, for the off-surface focusing scheme, a nearly sample independent calibration curve resulted.

Observe, however, that for fixed J, according to Equation 1, such plots should produce calibration curves of slope 1. Despite one's anticipation, the slope of the best linear fit through the data is *smaller* than 1. A likely explanation of such an observation seems to be related to the ambient sampling conditions of the set-up. The forthcoming reasoning was experimentally substantiated by tuning over to the on-surface focusing geometry. Then, in accordance with one's anticipation for the X < X_E droplet contamination regime (refer to Figure 10.4), the calibration slope was found to decrease to 0.12 (see Figure 10.4) thus suggesting only a minute (if any) amount of MPI-active microparticles to be present at the air/liquid interface.

For the ambient condition operational MPI–FC facility, an ever-increasing amount of ambient air aerosols (nonorganic "dust" particles) should be accommodated by the microdroplets anyway. These would act upon the incoming photo flux as the light-scattering (absorbing) matrix. Therefore, the ambient condition detection mode ought to be subjected to a Beer–Lambert type decrease of the energy flux available for MPI.

Another explanation seems to be quite plausible also: the probability of the MPI-released electrons escaping the liquid is much smaller for those produced within the droplet's bulk than for those on its surface. Hence, the proportionality relation $\Psi = \Psi(n_\alpha) \sim n_\alpha$ might be drastically altered. Research efforts to discriminate between possible causes for such slope reduction are of considerable importance. The goal would be to introduce appropriate compensation algorithms for the aforementioned sample independent calibration curves.

10.5 SUMMARY AND CONCLUSION

Some of the laser-based analytical techniques used in analyzing airborne hazards were briefly reviewed. Various research steps leading to the application of MPI to on-line screening of airborne particulate organic materials by means of a reasonably low-cost experimental technique were explained. Several environmentally relevant issues associated with the inherent capacity of MPI–FC to be applicable to on-line study of the aerosol aqueous dissociation status were specifically addressed.

Generally speaking, the main conclusion is that MPI-based analytical tools are appropriate for analysis of PAH-polluted aerosols. However, a more detailed and profound investigation of the physical phenomena involved is still necessary. The first steps already carried out point out the feasibility of the method. The liquid droplet sampling is especially promising for on-line aerosol analysis, although other sampling techniques (e.g., liquid jet) are also relevant to environmental applications. The main drawback of this method, namely the lack of selectivity, can only partially be corrected by using the more expensive tunable lasers. Thus, a possible alternative could be the combination of several laser methods, such as MPI and LIF. It is expected that such hybrid methods can provide better aerosol analysis, especially when complex pollution is involved.

ACKNOWLEDGMENTS

This work was supported, in part, by the Israeli Ministry of the Environment, by The James–Franck Program for Laser Matter Interaction, and by the Technion V.P.R. fund.

REFERENCES

1. Grasserbauer, M., *Microchim. Acta*, 1983, *Part III*, 415.
2. Van Grieken, R., Artaxo, P., Bernard, P., Leysen, L., Otten, P., Storms, H., Van Put, A., Wouters, L., and Xhoffer, C. *Chem. Anal. (Warsaw)*, 1990, 35, 75.
3. Van Grieken, R. and Xhoffer, C., *J. Anal. At. Spectrom.*, 1992, 7, 81.
4. Jambers, W., De Bock, L., and Van Grieken, R., *Analyst*, 1995, 120, 681.
5. Pui, D.Y.H., *Analyst*, 1996, 121, 1215.
6. Sinha, M.P., *Rev. Sci. Instrum.*, 1984, 55, 886.
7. Marijnissen, J., Scarlett, B., and Verheijen, J., *J. Aerosol Sci.*, 1988, 19, 1307.
8. McKeown, P.J., Johnston, M.V., and Murphy, D.M., *Anal. Chem.*, 1991, 63, 2069.
9. Murray, K.K. and Russel, D.H., *J. Am. Soc. Mass. Spectrom.*, 1994, 5, 1.
10. Mansoori, B.A., Johnston, M.V., and Wexler, A.S., *Anal. Chem.*, 1994, 66, 3681.
11. Cotter, R.J., *Biomed. Environ. Mass Spectrum*, 1989, 18, 512.
12. Van Vaeck, L., Van Roy, W., Gijbels, R., and Adams, F., in *Laser Ionization Mass Analysis,* Vertes, A., Gijbels, R., and Adams, F. (Eds.), Wiley, Chichester, 1993, pp. 7-126.
13. Harrington, P., Street, T., Voorhees, K., Radicati di Brozolo, F., and Odom, R.W., *Anal. Chem.,* 1989, 61, 715.
14. Dotter, R.N., Smith, C.H., Young, M.K., Kelly, P.B., Jones, A.D., McCauley, E.M., and Chang, D.P.Y., *Anal. Chem.,* 1996, 68, 2319.
15. Hankin, S.M., John, P., Simpson, A.W., and Smith, G.P., *Anal. Chem.,* 1996, 68, 3238.
16. Brenna, J.T., Creasy, W.R., McBain, W., and Soria, C., *Rev. Sci. Instrum.,* 1988, 59, 873.
17. Van Vaeck, L., Van Roy, W., Struyf, H., Adams, F., and Caravati, P., *Rapid Commun. Mass Spectrum.,* 1993, 7, 323.
18. Marshall, A.G. and Verdun, F.R., *Fourier Transform in NMR, Optical and Mass Spectrometry,* Elsevier, Amsterdam, 1990, pp. 225-278.
19. Xhoffer, C., Wouters, L., Artaxo, P., Van Put, A., and Van Grieken, R., in *Environmental Particles,* Buffle, J. and Van Leeuven, H.P. (Eds.), Lewis, Chelsea, MI, 1992, Vol. 1, 107-143.
20. Hachimi, A., Poitevin, E., Krier, G., Muller, J.F., Pironon, J., and Klein, F., *Analysis,* 1993, 21, 77.
21. Allegrini, I. and Omenetto, N., *Environ. Sci. Technol.,* 1979, 13, 349.
22. Niessner, R., Robers, W., and Krupp, A., *Fresenius J. Anal. Chem.,* 1991, 341, 207.
23. Lewitzka, F. and Niessner, R., *Aerosol Sci. Techn.,* 1995, 23, 454.
24. Enger, J., Malmsten, Y., Ljungberg, P., and Axner, O., *Analyst,* 1995, 120, 635.
25. Sjostrom, S., *Spectrochim. Acta Rev.,* 1990, 13, 407.

26. Sjostrom, S. and Mauchien, P., *Spectrochim. Acta Rev.*, 1993, 15, 153.
27. Smith, B.W., Blick, M., Spears, K., and Winefordner, J., *Applied Spectrosc.*, 1989, 43, 376.
28. Ottesen, D.K., Wang, J.C.F., and Radziemski, L.J., *Applied Spectrosc.*, 1989, 43, 1967.
29. Zhang, H., Singh, J.P., Yueh, F.-U., and Cook, R.L., *Applied Spectrosc.*, 1995, 49, 1617.
30. Xu, L., Bulatov, V., Gridin, V.V. and Schechter, I., *Anal. Chem.*, 1997, 69, 2103.
31. Hansen, L.D. and Eatough, D.J. (Eds.), *Organic Chemistry of the Atmosphere*, CRC Press, Boca Raton, Florida, 1991.
32. Letokhov, V.S., *Laser Analytical Spectrochemistry*, Adam Hilger, Bristol, PA, 1986.
33. Lambropoulos, P. and Smith, S.J. (Eds.), *Multiphoton Ionization*, Proceedings of the 3rd International Conference, Iraklion, Crete, Greece, Sept. 1984; Springer-Verlag: Berlin, 1984.
34. *Laser Applications to Chemical Analysis*, Technical Digest Series, Optical Society of America, Washington, D.C., 1990; Vol. 2.
35. Brophy, J.H. and Rettner, C.T., *Optics Lett.*, 1979, 4, 337-339.
36. Rettner, C.T., Brophy, J.H., *Chem. Phys.*, 1981, 56, 53-61.
37. Boesl, U., *J. Phys. Chem.*, 1991, 95, 2949.
38. Anderson, S.L., in *State-Selected and State-to-State Ion-Molecule Reaction Dynamics, Part I: Experiment*, Ng, C-Y. and Baer, M. (Eds.), Advances in Chemical Physics Series, Vol. LXXXII, John Wiley & Sons, 1992, pp. 177-212.
39. Ashfold, M.N.R., *Analytical Proc.*, 1991, 28, 416.
40. Dale, M.J., Jones, A.C., Pollard, S.J.T., and Langridge-Smith, P.R.R., *Analyst*, 1994, 119, 571.
41. Fei, X., Wei, G., and Murray, K.K., *Anal. Chem.*, 1996, 68, 1143.
42. Alimpiev, S.S., Belov, M.E., Mlinsky, V.V., and Nikiforov, S.S., *Analyst*, 1994, 119, 579.
43. Gittins, C.M., Castaldi, M.J., Senkan, S.M., and Rohlfing, E.A., *Anal. Chem.*, 1997, 69, 286.
44. Weickhardt, C., Boesl, U., and Schlag, E.W., *Anal. Chem.*, 1994, 66, 1062.
45. Griffiths, P.R. and de Haseth, J.A., *Fourier Transform Infrared Spectrometry*, ed. Elving, P.J., Winefordner, J.D., and Kolthoff, I.M. (Eds.), Wiley, New York, 1986.
46. Reily, P.T.A., Gieray, R.A., Yang, M., Whitten, W.B., and Ramsey, J.M., *Anal. Chem.*, 1997, 69, 36.
47. Schechter, I., Schroder, H., and Kompa, K.L., *Anal. Chem.*, 1992, 64, 2787.
48. Schechter I., Schroder, H., and Kompa, K.L., *Anal. Chem.*, 1993, 65, 1928.
49. Hall, G.E. and Kenney-Wallage, G.A., *Chem. Phys.*, 1978, 28, 205-214.
50. Siomos, K., Kourouklis, G., and Christophorou, L.G., *Chem. Phys. Lett.*, 1981, 80, 504-511.
51. Voigtman, E., Jurgensen, A., and Winefordner, J.D., *Anal. Chem.*, 1981, 53, 1921-1923.
52. Siomos, K. and Christophorou, L.G., *Chem. Phys. Lett.*, 1980, 72, 43-48.
53. Frueholz, R., Wessel, J., and Wheatley, E., *Anal. Chem.*, 1980, 52, 281.
54. Voigtman, E. and Winefordner, J.D., *Anal. Chem.*, 1982, 54, 1834-1839.
55. Vauthey, E., Haselbach, E., and Suppan, P., *Helvetica Chim. Acta*, 1987, 70, 347-353.
56. Holroyd, R.A., Preses, J.M., Bottcher, E.H., and Schmidt, W.F., *J. Phys. Chem.*, 1984, 88, 744-749.
57. Yamada, S., *Anal. Chem.*, 1991, 63, 1894.
58. Sander, M.U., Luter, K., and Troe, J., *Ber. Bunsenges, Phys. Chem.*, 1993, 97, 953-961.
59. Johnson, M.E. and Voigtman, E., *Anal. Chem.*, 1992, 64, 551-557.
60. Ogawa, T., Yasuda, T., and Kawazumi, H., *Anal. Chem.*, 1992, 64, 2615.
61. Yamada, S. *Anal. Chim.* Acta, 1992, 264, 1.
62. Inoue, T., Masuda, K., Nakashima, K., and Ogawa, T., *Anal. Chem.*, 1994, 66, 1012.
63. Marshall, A., Ledingham, K.W.D., and Singhal, R.P., *Analyst*, 1995, 120, 2069.
64. Delone, N.B., *Usp. Fiz. Nauk*, 1965, 115, 361; *Soviet Phys. Usp.*, 1965, 18, 169.
65. Petty, G., Tai, C., and Dalby, F.W., *Phys. Rev. Lett.*, 1975, 34, 1207.
66. Ogawa, T., Kise, K., Yasuda, T., Kawazumi, H., and Yamada, S., *Anal. Chem.*, 1992, 64, 1217.
67. Li, Y.Q., Inoue, T., and Ogawa, T., *Anal. Sci.*, 1996, 12, 691.
68. Ogawa, T., Sumi, S., and Inoue, T., *Instr. Sci, & Techn.*, 1995, 23, 311.
69. Chen, H., Inoue, T. and Ogawa, T., *Anal. Chem.*, 1994, 66, 4150.
70. Ogawa, T., Sumi, S. and Inoue, T., *Anal. Sci.*, 1996, 12, 455.
71. Kawazumi, H., Yasuda, T. and Ogawa, T., *Anal. Chim. Acta*, 1993, 283, 111.
72. Katoh, R. and Kotani, M., *Chem. Phys. Lett.*, 1990, 166, 258.
73. Zhan, Q., Voumard, P., and Zenobi, R., *Rapid Commun. Mass Spectrom.*, 1995, 9, 119.
74. Gridin, V.V., Korol, A., Bulatov, V., and Schechter, I., *Anal. Chem.*, 1996, 68, 3359.

75. Gridin, V.V., Bulatov, V., Korol, A., and Schechter, I., *Anal. Chem.,* 1997, 69, 478.
76. Bulatov, V., Gridin, V.V. Polyak, F., and Schechter, I., *Anal. Chim. Acta,* 1997, 343, 93.
77. Gridin, V.V., Bulatov, V., Korol, A., and Schechter, I., *Instrum. Sci. Technol.,* 1997, 25, 321.
78. Gridin, V.V., Litani-Barzilai, I., Kadosh, M., and Schechter, I., *Anal. Chem.,* 1997, 69, 2098.
79. Gridin, V.V., Litani-Barzilai, I., Kadosh, M., and Schechter, I., *Anal. Chem.,* 1997, 70, 2685.
80. Liu, H. and Dasgupta, P.K., *Anal. Chem.,* 1995, 67, 4221.
81. Cardoso, A.A. and Dasgupta, P.K., *Anal. Chem.,* 1995, 67, 2562.
82. Barnes, M.D., Whitten, W.B., and Ramsey, J.M., *Anal. Chem.,* 1995, 67, 418A.
83. Liu, H. and Dasgupta, P.K., *Anal. Chem.,* 1996, 68, 1817.
84. Futoma, D.J., Smith, S.R., Smith, T.E., and Tanaka, J., *Polycyclic Aromatic Hydrocarbons in Water Systems,* page 16, Table 2 and references therein; CRC Press, Boca Raton, Florida, 1981.
85. Adamson, A.W., *Physical Chemistry of Surfaces,* Wiley-Interscience, New York, 1976 and references cited therein.
86. Watson, A.Y., Bates, R.R., and Kennedy, D. (Eds.), *Air Pollution, The Automobile and Public Health,* National Academy Press, Washington, D.C., 1988, *Part II.*
87. Heywood, J.B., *Internal Combustion Engine Fundamentals,* McGraw-Hill, New York, 1988, *Ch. 2.*
88. Guerin, M.R., Environmental tobacco smoke, in *Organic Chemistry of the Atmosphere,* Hansen, L.D. and Eatough, D.J. (Eds.), CRC Press, Boca Raton, Florida, 1991, Chapter 3.

Section III

Single Particle Analysis

11 Liesegang Ring Technique Applied to the Chemical Identification of Atmospheric Aerosol Particles

Josef Podzimek and Miroslava Podzimek

CONTENTS

11.1 INTRODUCTION

In 1896, Raphael E. Liesegang observed a phenomenon which can be briefly described in the following way: a crystal of silver nitrate placed on a glass slide covered with a gelatin layer, which dissolved the diluted potassium dichromate, formed a precipitate of silver dichromate in concentric rings around the crystal. Since then, many scientists have proposed their theories and have conducted experimental studies to explain this phenomenon.

This study reviews the history of the Liesegang ring (LR) observation and its application to the chemical identification of aerosol particles. It also describes the laboratory investigations and discusses the main theoretical explanations of the observed periodic precipitation. Essential to the practical application of LR technique for detecting several important constituents of an atmospheric aerosol is the determination of the ring magnification factor which depends on several environmental parameters and on the sampling and evaluation process. Finally, this study describes possible future evolution of the LR technique and its broader application.

11.2 ATTEMPTS TO EXPLAIN LIESEGANG RING FORMATION IN THE PAST

As early as 1899, Wilhelm Ostwald pointed out that supersaturation and the existence of a metastable concentration limit for the formation of nuclei possibly play important roles. Ostwald considered the case when silver nitrate diffused into gelatin containing potassium dichromate and supposed that silver dichromate formed immediately, but that it remained in a supersaturated solution. When

1-56670-040-X/99/$0.00+$.50
© 1999 by CRC Press LLC

silver dichromate precipitated a short distance behind the diffusion front, the silver dichromate in the solution diffused toward the nuclei. This explained the clear space between the rings through which the silver nitrate must travel before repeating the cycle. In spite of several objections such as Hatschek's argument (1912, 1914) that supersaturation cannot exist in the presence of a solid phase, two observations support the idea of supersaturation: (1) rapidity of ring formation and (2) rapidity of crystal growth from the time the rings first appear. However, one also has to prove that the velocity of the diffusion of the outer electrolytes is greater than the rate of crystal growth (natural or seeded).

Despite Hatschek's objections, Wilhelm Ostwald's supersaturation theory had many defenders, such as Notboom (1932), Morse (1930), Van Hook (1938, 1940, 1941), and others. Not surprisingly, several other scientists proposed theories to better explain the observed phenomena.

Bradford (1922) theorized that the clear space between rings was due to the adsorption of the inner electrolyte by the precipitate. The absorbing capacity depended upon the degree of dispersion of the precipitate. Bradford claimed that large precipitated particles cause little adsorption and no ring formation. Also, extremely small particles will not support the ring formation due to the hindered diffusion process. Hatschek (1925) and Morse (1930) showed that rings with few large precipitate crystals can be grown, and Dhar (1929) found the silver dichromate absorbed little potassium dichromate. In conclusion, adsorption undoubtedly occurs; however, it cannot account for the formation of Liesegang circles alone. Dhar and his colleagues published several articles (1922, 1924, 1927a, b) in which they tried to prove that the precipitated substance is produced first as a colloidal dispersion — not as a supersaturated solution. Their coagulation theory showed how the precipitate rings form by the coagulation of the colloid by an excess of diffusing electrolytes. Also, they showed that many precipitates absorb their own solutions and that this phenomenon occurred to a greater extent than adsorption of the electrolyte. Both factors combine and cause precipitation of the peptized solution on the precipitate. However, the repetition of the process requires that the silver nitrate must diffuse through this area at a sufficient concentration. Besides, the gelatin is not an essential component in the process, although it plays an important role.

In 1925, Wolfgang Ostwald's research emphasized the importance of the soluble product formed in the reaction, which the previous theories ignored. If we apply Wolfgang Ostwald's ideas to our case of identifying sodium chloride nuclei in a gelatinous substrate with silver nitrate, we can use the following scheme for a one-dimensional model with the corresponding equation of mass action (see Figure 11.1):

$$AgCl = \frac{K(NaCl)(AgNO_3)}{NaNO_3} \tag{1}$$

From our simplified sketch in Figure 11.1, we see three different diffusion waves interfering with one another. Two are related to the diffusion of the original reactants (NaCl and AgNO₃). A precipitate of AgCl forms on the surface of contact. Continuous precipitation stops when the concentration of silver nitrate is so small and that of sodium nitrate is so great that it prevents the precipitation. The precipitation can start only if the right-hand side of the equation describing the mass action reaches a certain value, depending on the temperature and concentration of the ions. Owing to the high concentration of Na and Cl ions, after a time, the wave of NaCl catches up and eventually overtakes the sodium nitrate wave, and the process repeats. Ostwald cited the experimental evidence that alteration or destruction of the right pattern ensued if the electrolyte formed in the reaction (NaNO₃) was added to the gelatin. The main objections to the diffusion wave hypothesis stem from the presence of ring systems in the case when there was no third electrolyte.

In the literature on LR formation, one finds experimental verifications of the theoretical models (Christiansen and Wulff, 1934) and the more detailed explanation of LR phenomena (Mikhailev, et al., 1934; Hughes, 1935; Wormser, 1946; Veil, 1947, 1948, 1950, 1951; Prager 1956).

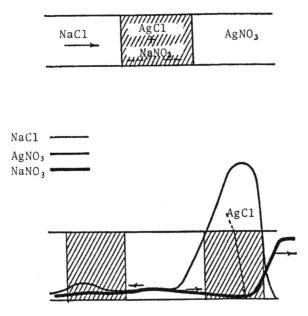

FIGURE 11.1 Ideal Liesegang rings (L.R.) formed around NaCl particle according to the diffusion wave theory (see, for example, Stern, 1954, p. 87). In the lower part of the figure are the concentration profiles of the inner and outer electrolyte and of the reaction products (AgCl and NaNO₃) of two rings formed in the sensitized gelatin layer.

Nikoforov and Kharamonenko (1938) and Raman and Ramaiah (1939) tried the wave mechanical model for Liesegang ring formation. More recently, Mathur (1960, 1961) used an analogy between propagation of a diffusional wave in a Liesegang ring and light. Without a thorough analytical justification, he tried the wave function $c = A \exp(-\alpha\, x) \cos t\, (\beta\, x + \overline{\omega}t + y)$ and obtained the product of the concentrations of the external (c) and internal (c') electrolytes in the n-th ring.

$$c_n c_n' = c_0 c_0' \exp(-\alpha'\beta'\,\pi/2)\, \exp\left(-\alpha' x_n\right) \exp\left(-\alpha'\Delta x_n\right) \sin\left(\beta'\Delta x_n\right) \cos\left(\beta x_n\right), \qquad (2)$$

where c_0 and c_0' are the initial electrolyte concentrations, x_n is the distance between the starting point of the diffusion and the n-th ring; $\Delta x_n = x_n - x_{n-1}$; α, α', β, β' are constants. The formation of each successive ring in the precipitate is explained by the condition $c_n\, c_n' \geq H$, where H is the solubility product. The main feature of the solution was the occurrence of two types of periodicity corresponding to the two trigonometric functions. Lurie (1965) showed that similar ring microstructure can be found from a recording microphotometer study of Liesegang rings. The open question is still the discrepancy between the measured and theoretically calculated values, the neglect of the third electrolyte, and the peptization of the ions.

A more complex explanation of the one-dimensional LR phenomena was presented by Shinohara (1970), and the question of LR chemical instability was discussed by Gold (1964) and Flicker and Ross (1974). Finally, a critical review of the newer investigations was presented by Vaidyan et al. (1981) and Budtov and Yanovskaya (1982). Brun and Gradyshev (1983) focused on irreversible reactions during counter diffusing reactants, and Smith (1984) again on Ostwald's supersaturation theory.

Reviewing all the theories, we observe they all have strong points. Each theory, however, fails because some of the necessary conditions for a certain theory are not fulfilled in experimental results. Common to all theories is the critical solubility product, supersaturation, or a similar quantity. It is highly probable that all the processes described above are involved in the Liesegang ring formation; however, some of those processes might influence the ring formation in a decisive way. We believe, for example, that diffusion can play a decisive role for the simple reason that it acts over much longer distances than adsorption. The spacing of rings is a very important parameter depending on the rate of penetration of the diffusing agent. This process, however, is also influenced by the gelatinous medium and other parameters, such as temperature, humidity, and light. Küster (1913) found no bands over silver chromate in the dark, but Davies (1922) observed well-developed bands. Other authors (Köhn and Mainzhausen, 1937) relate the influence of light only to certain ions, and Isemura (1939) believed that light affected the ring formation only if the inner electrolyte reacts with gel. In general, it was assumed that all main precipitation reactions can occur rhythmically when subjected to proper conditions, and that the gelatinous medium was not essential to ring formation.

11.3 IDENTIFICATION OF SPECIFIC AEROSOL PARTICLES IN THE ATMOSPHERE

The spot test (LR) technique was introduced into environmental studies by Winckelman (1931) and into the investigation of atmospheric aerosol particles important to the explanation of the cloud formation by Seely (1952), Fedele and Vittori (1953), Vittori (1954, 1955), Farlow (1954), Pidgeon (1954), Lodge (1954), and Lodge and Fanzoi (1954). Most of the investigators used the Seely modified spot test technique, which was based on the identification of large chloride particles deposited in the impactor on a glass slide covered by the sensitized gelatin layer. This layer was prepared from 8% gelatin (b.w.) with a small addition of glycerol. Afterward, the slides were dipped for 20 seconds into a 5% solution of mercurous fluosilicate (Hg_2SiF_6) or nitrate ($HgNO_3$). Circular spots developed around chloride-containing particles (droplets), consisting mainly of relatively insoluble mercurous chloride, and their size is directly related to the dry salt particle volume. Instead of mercurous fluosilicate, other investigators used silver nitrate ($AgNO_3$), forming the halos around salt nuclei containing mainly sliver chloride. Farlow (1957) developed a technique making use of an $Ag_2Cr_2O_7$ solution for sensitizing the gelatin layer which usually had a thickness between 17 and 200 µm.

The spot (halo) sizes are affected by the thickness of the gelatin layer, time elapsed between the sampling and the evaluation, and, in the case of micron-size particles, by their fading rates (Pidgeon, 1954). Besides the salt particle identification, other techniques, based in principle on the LR formation, have been suggested for the detection of other constituents of the atmospheric aerosol. Lodge and Fanzoi (1954) described the method for the detection of sulfate, nitrate, and calcium ions in particles of minimum mass of 10^{-16} g. Lodge and Tufts (1956) suggested another technique suitable for proving the presence of ammonium nitrate particles. Vittori and his fellow workers introduced the spot test (LR) technique in cloud physics around 1954 and, for ascertaining chloride nuclei in the atmosphere, they used their specific reaction with silver nitrate. However, in spite of a thorough investigation of the applicability of this technique at different atmospheric conditions by Rau (1955, 1956) and Podzimek (1959), several questions remained unanswered. The main concern was about the effect of environmental humidity on the magnification factor converting the circular spot size into a size or volume of a pure dry nucleus. Later, several authors started to use polyvinyl alcohol substrate (instead of gelatin) with the sensitizing substance for the particle detection (Rinehart, 1970; Ueno and Sano, 1971, 1972).

Other major concerns were the purity of the gelatin (polyvinyl alcohol), the uniform thickness of the gelatin layer on the glass substrate, and the sampling and evaluation procedure. An attempt was made to determine a suitable time for taking the sample and for sample evaluation at specific environmental conditions (e.g., temperature, humidity, illumination). On these points, Pidgeon (1954) recommended the storage of coated slides in the dark and at subzero temperatures; Rau (1955, 1956) and Podzimek (1959) preferred the dipping of the gelatin-covered slides, just before sampling, into the sensitizing reagent solution (e.g., 5% $AgNO_3$ b.w.) and evaluating the micron-sized particles in an optical microscope at a suitable illumination shortly after sampling. For submicron chloride particle detection, special illumination (UV light) and, eventually, coloring of the gelatin layer were recommended.

During the past 45 years, many particle identifications using the spot test technique have been performed on the ground, close to the sea surface, and in the free atmosphere. Lodge (1955) measured the concentration and size distribution of particles over Puerto Rico; Metnieks (1958) performed measurements of sea salt nuclei on the Atlantic coast of Ireland; and Durbin (1959) and Durbin and White (1961) took measurements in an airplane over England. The results of in-flight chloride sulfate particle measurements over northern Bohemia and Slovakia have been published by Podzimek (1959), Podzimek and Černoch (1959, 1962), and Vrkočová and Podzimek (1969). During the following years, many investigations into the nature of tropospheric large (0.1 μm < r < 1.0 μm) and giant (r >1.0 μm) nuclei took advantage of the spot test technique and focused on the nature, size distribution, and transport of sea salt nuclei over the Texas sea shore (1973, 1980), in the Caribbean region (1967), and in Argentina and Brazil (Caimi, 1976; Caimi and Benzaquén, 1975; Caimi and Ramos, 1982; Caimi, 1984). In Japan, similar studies have been performed by Ueno and Sano (1971).

The above-mentioned investigations contributed to our knowledge of the generation and transformation of salt particles over the sea and the transport of these particles over the continent. The results of salt particle measurements around and inside tropospheric clouds (Podzimek and Černoch, 1961) were quite interesting. An important contribution to our knowledge of the sea salt particle distribution over seas in large scale is represented by the investigations by Kikuchi and Yaura (1970), who identified the giant sea salt nuclei by the spot test technique during a voyage from Tokyo to Syowa station in Antarctica. These data completed the extensive and systematic studies of marine aerosol by Toba (1966). The spot test technique was also used by Rinehart (1971) in a nonurban environment of south-central New Mexico for the detection of water-soluble sulfate particles. Most of the giant particles were of a mixed nature and contained, besides a soluble component, an insoluble fraction.

In the 1970s, several improvements in the simple spot test technique, which was not very suitable for particles in the submicron size range, were suggested. For submicron sulfate particles, a coating of a substrate (e.g., impactor plate or electron microscope grid covered by formwar thin film) with the vapors of barium chloride was recommended. The spots, formed at an appropriate humidity, were evaluated in an electron microscope (Bigg and Ono, 1974; Bigg et al., 1974; Bigg, 1975; Mamane, 1977; Mamane and de Pena, 1978). This technique, in spite of some heterogeneity of spots, enabled the evaluation of the diffusion spots of sulfate and nitrate ions. It has been used for the detection of sulfate particles in the lower stratosphere and the upper troposphere (balloon measurements by Bigg and Ono, 1974), around the clouds over central Queensland (King and Maher, 1976), in the urban and desert atmosphere of the Eastern Mediterranean (Mamane et al., 1980), over Hawaii (Parungo et al., 1981), and in the urban atmosphere of Nagoya (Okada, 1985). Similar measurements of particles containing SO_4^{2-} and NO_3^- were performed on a cruise ship, and the results followed the temporal and spatial variation of marine aerosol over the Atlantic Ocean (Parungo et al., 1986).

A technique similar to that described by Bigg et al. (1974) has been used for the detection of submicron sulfate particles collected at Mizuho station and in East Queen Maud Land in Antarctica (Yamato et al., 1987a, b). The most important finding was the uniformity and monomodality of sulfuric acid particle size distribution. These particles make spots due to the reaction in the thin film of vapor deposited calcium. Conversion of the spots into the size of sulfuric acid droplets yielded the average size of about 0.08 μm. The sulfate-containing particles were identified in a thin layer of vapor-deposited $BaCl_2$. These two methods were later applied to the aerosol particle investigation in the vicinity of tropopause folding over the Sea of Japan (Yamato and Ono, 1989).

Sulfuric acid particles were predominant in the lower stratosphere and much larger than those in the troposphere. Their size distribution modus was at about 0.3 μm radius. The acidity of sulfate-containing particles increased with increasing altitude in the middle troposphere. Very few nitrate-containing particles (determined by the reaction with Nitron) were found close to the tropopause. (This technique was used by Mamane and Pueschel, 1980.)

Another interesting application of a quasi-spot technique was described by Pueschel et al. (1980) and used for the identification of nitrate-, sulfate-, and chloride-containing particles in the Antarctic stratosphere (up to 18 km). Particles and droplets were inertially deposited on gold wires coated by Nitron, barium chloride, and silver nitrate, and the spots and reaction products were examined in a scanning electron microscope. In this way, a negative correlation between condensed nitrate and ozone concentration in high altitude was found.

Recently sulfate- and nitrate-containing particles were sampled simultaneously onboard a research vessel and an airplane on the east coast of North America and over the North Atlantic Ocean (Kopcewicz et al., 1991). Using the spot test technique, a solar radiation effect on the aerosol composition was detected. Sulfuric acid particles prevailed near noon, while ammonium sulfate-containing particles with radii between 0.1 μm and 0.35 μm were dominant at night. Concentrations of sulfate-containing particles measured on the ship were in mean about 600 cm^{-3} with a considerable drop in the concentration (down to 25%) above the open sea.

11.4   DETERMINATION OF THE RING MAGNIFICATION FACTOR

There is still some controversy about the determination of the mentioned LR magnification factor even in the simplest case of the identification of sodium chloride particles or solution droplets.

The "isopiestic" method, described by Woodcock and Gifford (1949) and later modified by other investigators (Metnieks, 1958; Toba, 1966) is based on the determination of the NaCl particle size on an untreated glass slide corresponding to the critical relative humidity (around 73% for the accelerated growth of the hygroscopic nucleus). In the 1950s, this technique was compared to the then often-used spot test technique (Fournier d'Albe, 1955). Rau (1956) and Podzimek (1959) determined the spot size of salt solution droplets or particles by their sedimentation rate in a settling tube at a controlled relative humidity (usually between 60% and 75%). In this way, Podzimek (1959) obtained magnification factors (supposedly referred to dry particle size) between 5.7 and 6.3, with larger values for smaller particles. These values were a little lower than the magnification factors determined by Rau (1956), however, in mean, larger than the particle sizes deduced from the comparison of the "isopiestic" technique. At the same time, Durbin (1959), using a different preparation of gelatin layers, obtained a much lower magnification factor (2.7).

A detailed analysis of the magnification factor determination for NaCl solution droplets falling on a sensitized polyvinyl alcohol layer (4% b.w. in water) was published by Ueno and Sano (1971). They used different concentrations of AgNO$_3$ (between 2 ml and 7 ml added to approximately 30 ml solution) for the calculation of the relationship between the spot size, d$_s$, and 5% NaCl solution droplet diameter, D$_d$. If a film of uniform thickness and of homogeneous AgNO$_3$ concentration was assumed and if the circular spots resembled small cylinders, they obtained the relationship

$$D_d^3 = K d_s^2 \tag{3}$$

where K is a constant. Comparison with the size distribution of 5% NaCl solution droplets deposited in oil yielded results justifying the applicability of Equation (3). For different AgNO$_3$ concentrations, the factor K assumed values between 2 and 10.

Another attempt to improve the determination of the magnification factor of the spot test technique was made by Preining et al. (1976) and by Yue and Podzimek (1980). Instead of the direct particle measurement or the particle size determination from their settling speed, they used an aerosol centrifuge and related the spot size to the particle aerodynamic diameter or radius. The

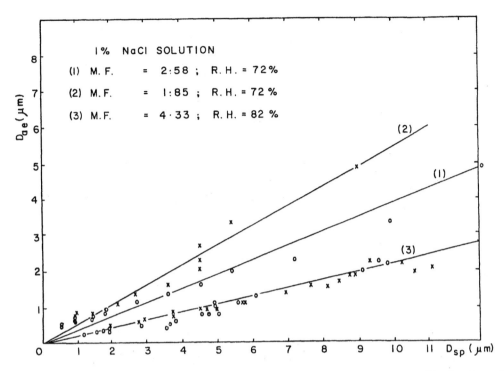

FIGURE 11.2   Effect of the storage relative humidity on the Liesegang ring (L.R.) magnification factor for salt particles sampled by a centrifuge at a relative humidity of about 72% and 82%. Crosses mark samples placed after exposure over distilled water; rings feature samples stored in dry environment.

aerosol centrifuge was calibrated by latex particles of known size and at well-controlled humidity. It was concluded that the environmental humidity and the humidity maintained during the sampling and during the initial period of aerosol storage affect, in a decisive way, the magnification factor. A considerable difference was found between the magnification factor corresponding to lower relative humidity (around 72%) maintained during the aerosol passage from the generator (nebulizer) to the aerosol centrifuge and factors obtained at a higher relative humidity (about 82%). Figure 11.2 shows that the storage of samples taken at lower relative humidity (72%) strongly affects the magnification factor (Yue and Podzimek, 1980) in comparison to high relative humidity (82%). In the latter case, there was no difference between the slopes of magnification factor lines plotted for samples stored over water or in a dry environment.

Comparing the old determination of the magnification factor of spot test technique (based mainly on the sedimentation technique) to those done more recently, one finds a considerably lower value calculated from aerosol centrifuge experiments. One of the possible explanations is the effect of the humid air thermodynamics and changing humidity in the aerosol centrifuge. Figure 11.3 shows that some of the large LR of the deposited sodium chloride particles have incompletely dissolved cores or they have tiny NaCl crystals close to the ring's edge.

11.5   SAMPLING PURE AND MIXED PARTICLES AND SAMPLE EVALUATION

Several techniques have been applied to aerosol particle sampling and LR identification in the atmosphere. The inertial impactor was most frequently used for particle deposition (Gerhard and Johnstone, 1955; Kikuchi and Yaura, 1970; Podzimek, 1973, 1980, 1990; Bigg and Ono, 1974; Yamato et al., 1987; Yamato and Ono, 1989). Rinehart (1971) sampled sulfate-containing particles with an Andersen impactor. For large and giant condensation nuclei, several investigators used

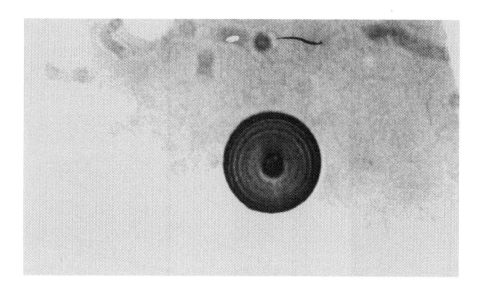

FIGURE 11.3 Liesegang ring (L.R.) with incompletely dissolved salt particle in the center. Outer ring diameter is 18.5 μm.

simple impaction on slides exposed to the high air flow (e.g., airplane sampling by Podzimek and Černoch, 1958 and 1961), and Mamane and Noll (1985) used inertial impaction of particles in a rotating device. Later virtual impactors were used for the same purpose. The particle inertial deposition had been analyzed by Podzimek (1959), by Jaenicke et al., (1971), and by others.

In essence, for coarse aerosol particles and haze elements (with r >1.0 μm), the particle effective radius, r_{50}, for different impactor operation regimes and 50% collection efficiency ($\eta = 0.5$) is

$$r_{50} = \sqrt{\frac{g\mu_a W\,Stk\,\eta}{4\rho_p v_w}} \cong \sqrt{0.564\,\frac{\mu_a W^2 L}{4\rho_p q}} \quad \text{with Stk} = \frac{4r_p^2 \rho_p v_w C}{9\mu_a W} \qquad (4)$$

where W is the impactor rectangular slot width placed at a distance 2W above the sampling glass slide, L is the slot length, and the volumetric flow rate at a mean speed, v_w, is q. The air dynamic viscosity is μ_a and ρ_p is the particle density. The Cunningham slip correction, C, is, in this case, equal to 1.0, and for particles smaller than 1 μm an appropriate Cunningham correction has to be applied (Fuchs, 1964, p. 26).

The sampling of aerosol particles in cascade inertial impactors might be affected considerably by particle bouncing at the deposition plate, which might depend on the nature of the particle and substrate, temperature, and relative humidity during the sampling. Some interesting results have been obtained for the deposition of micron-size salt particles in the UNICO impactor (Podzimek et al., 1991); however, a systematic investigation of this effect ought to be done in the future. These facts, in addition to those related to the sensibilized gelatin layer thickness and slide storage before and after sampling might complicate the conversion of the ring size into salt particle volume or mass (Gerard et al., 1989a, b). In general, high purity of hot filtered gelatin (5 to 8% b.w.) covering the glass slide in a uniform layer of mean thickness between 20 μm and 100 μm is recommended. More reproducible results were obtained with gelatin layers sensitized one-half hour or one hour before sampling by dipping the slides several seconds in an 8% solution of $AgNO_3$ for the identification of Cl⁻-containing particles and storing them at high relative humidity. Salt particle sample processing was performed by UV radiation in the laboratory for 5 to 20 minutes or by exposing them to sun radiation 20 to 30 minutes.

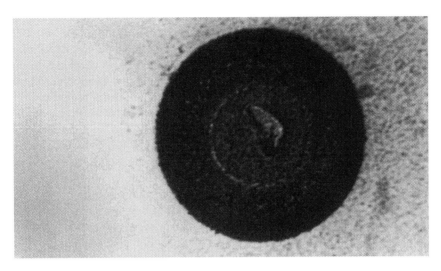

FIGURE 11.4   Haze element Liesegang ring (L.R.) of salt particle with insoluble core. Outer ring diameter is 48.0 μm.

One of the great advantages of the time-consuming LR technique for sample evaluation is the identification of mixed particles. Besides the morphology of the large insoluble carrier particle, the active spots on its surface can be identified, or the location of the micron and submicron particles embedded in a haze element can be determined (Figure 11.4). In this way, the knowledge of the microstructure of haze and fog elements in a polluted environment will help to establish more realistic models of fog (haze) element growth and to study the transfer of radiative energy in the atmosphere. Several studies have been performed in the past which demonstrate the suitability of the LR technique for the mixed particle microstructure investigation.

In the late 1950s, attention was called to coarse insoluble particles several μm in diameter which showed active spots on their surface. These active spots containing chloride or sulfate ions were made visible by the typical LR pattern in a sensitized gelatin layer (Podzimek, 1959). Later, these studies performed in an airplane were completed by sampling on the ground, in pure marine air and in a polluted marine–urban atmosphere (Podzimek, 1973, 1990), in the mountains (Rinehart, 1971), in the upper troposphere (Podzimek et al., 1995), and in the lower stratosphere (Bigg et al., 1974; Yamato and Ono, 1989).

11.6   POSSIBLE EVOLUTION OF SPOT TEST TECHNIQUES IN THE FUTURE

There is still considerable discrepancy between calculated and observed periodic precipitation propagation, especially if a medium of very high viscosity is assumed. This has been pointed out by several investigators (e.g., Gerard et al., 1986b) who were concerned mainly with the application of simple equations for diffusing and reacting ions. The calculation of magnification factors is often based on assumptions such as the uniform thickness of the sensitized layer under the circular spot with a homogeneously distributed reagent, e.g., all silver ions react with chloride ions to yield silver chloride spots and that the pattern of diffusion field has a simple geometry (e.g., cylindrical or spherical). That explains the great difference often found between theoretically calculated and measured magnification factors for submicron, e.g., salt particles and coarse particles (with radii greater than 5 μm) or salt droplets, especially if a very thin gelatin layer (below 20 μm) is used. The nonuniform gelatin layer thickness under the reaction circular spot has been documented in the past (Podzimek, 1959; Yue and Podzimek, 1980).

The wider application of the LR technique in the future assumes that the suggested procedures of sample preparation and storing will be checked and compared again. Several effects have been investigated in the past, such as sensitized layer thickness, effect of reagent concentration, storing procedures for differently sensitized layers (Podzimek, 1959; Ueno and Sano, 1971; Yue and Podzimek, 1980; Gerard et al., 1989a, b); however, recommendations of general validity for specific situations are needed. Sampling of particles at high altitude requires checking the effect of temperatures below $-20°C$ and the effect of environmental humidity. Environmental humidity also plays an important role in the magnification factor determination, as has been clearly demonstrated in the past (see the above-mentioned articles); however, it would be worthwhile to investigate its effect on different calibration procedures (e.g., aerosol centrifuge or impactor). There are still unanswered questions. For example, will the same amount of salt ions contained in the "dry" particle and in a droplet lead to the same size circular spot if the samples are stored in a container with air saturated by water vapor. We also know very little about the effect of salt solution droplet impaction speed (e.g., in an impactor) on the magnitude of a magnification factor.

There are still problems with an effective sample evaluation. For increasing the spot contrast in the case of identification of chloride ions, most of the investigators used $AgNO_3$ as a reagent and kept the sensitized, gelatin-covered slides in the dark. Before evaluating the slides or photographing the spots in a phase-contrast microscope, the slides were exposed to UV light. Usually the recommended exposure time was between 5 and 30 minutes, also depending on the distance of the lamp from the sample and the intensity of radiation (Ueno and Sano, 1971). There are no standard procedures on how to increase the visibility of spots made by other ions.

Several authors described a semiautomatic or automatic sample evaluation. Podzimek (1990) projected the spot photographs on a screen or disk with a scale and evaluated the magnified pictures in a Zeiss TGZ 3 semiautomatic particle evaluator. A wide application is expected by capturing the image of LR spots on video when the camera plugs directly into the video digitizing board of the computer. Common to all these techniques is the problem of how to select the representative sample field if the whole area (e.g., under the slot of an impactor) is not evaluated. Then the inhomogeneous particle deposition and the particle bouncing, which depends on the environmental parameters (e.g., temperature and humidity), will certainly be investigated in the future.

ACKNOWLEDGMENT

The authors are obliged to Mrs. V. Hudgins from the University of Missouri–Rolla, who ably assisted and prepared the text for printing. This study was partly supported by the University of Missouri Research Board Grant, 1993.

REFERENCES

Bigg, E.K., 1975, *J. Atmos. Sci.,* 32:910.
Bigg, E.K. and Ono, A., 1974, *Proc. IUGG Congress,* Melbourne, p. 144.
Bigg, E.K., Ono, A., and Williams, J.A., 1974, *Atmos. Environ.,* 8:1.
Bradford, S.C., 1922, *Kolloid — Z.,* 30:364.
Brun, B.B., and Gradyshev, G.P., 1983, *Zh. Fiz. Khim.,* 57:1337.
Budtov, V.P., and Yanovskaya, N.K., 1982, *Zh. Fiz. Khim.,* 56:2464.
Caimi, E.A., 1975, *Geoacta,* 7:51.
Caimi, E.A., 1984, *J. Res. Atmos.,* 18:243.
Caimi, E.A. and Benzaquén, R., 1975, *Meteorologica,* 6-7:475.
Caimi, E.A. and Ramos, J.A., 1982, *Ciencia e Cultura,* 35:84.
Christiansen, J.A. and Wulff, I., 1934, *Z. Phys. Chem.,* 26B:187.
Davies, H.S., 1922, *J. Am. Chem. Soc.,* 44:2638.
Dhar, N.R., 1929, *Chemistry and Industry,* 878.
Dhar, N.R. and Chatterji, A.C., 1922, *Kolloid — Z.,* 31:15.

Dhar, N.R. and Chatterji, A.C., 1927a, *Kolloid — Z.*, 37:2.

Dhar, N.R. and Chatterji, A.C., 1927b, *Kolloid — Z.*, 37:89.

Durbin, W.G., 1959, *Geofis. Pura e Applic.*, 42:11.

Durbin, W.G. and White, G.D., 1961, *Tellus*, 13:260.

Farlow, N.H., 1956, *J. Colloid Sci.*, 11:184.

Farlow, N.H., 1957, *Analyt. Chem.*, 29:883.

Fedele, D. and Vittori, O.A., 1953, *Riv. Met. Aeronaut.* 13:4.

Flicker, M. and Ross, J., 1974, *J. Chem. Phys.* 60:3458.

Fournier d' Albe, E.M., 1957, in *Artificial Stimulation of Rain*, Pergamon Press, London, p. 73.

Fuchs, N.A., 1964, *The Mechanics of Aerosols*, Pergamon Press, Oxford, p. 408.

Gerard, R., Viallard, A., and Serpolay, R., 1989a, *Atmos. Res.*, 22:335.

Gerard, R., Viallard, A., and Serpolay, R., 1989b, *Atmos. Res.*, 22:351.

Gerhard, E.R. and Johnstone, H.F., 1955, *Analyt. Chem.*, 27:702.

Gold, L., 1964, *Nature*, 202:889.

Hatschek, E., 1912, *Kolloid — Z.*, 10:124.

Hatschek, E., 1914, *Kolloid — Z.*, 14:115.

Hughes, E.B., 1935, *Kolloid — Z.*, 72:212.

Isemura, T., 1939, *Bull. Chem. Soc. Japan*, 14:179.

Jaenicke, R., Junge, C., and Kanter, H.J., 1971, Messungen der Aerosolgrössenverteilung über dem Atlantik, *Met. Forsch.-Erg.*, B7:1-54.

Kikuchi, K. and Yaura, S., 1970, *J. Met. Soc. Japan*, 48:377.

King, W.D. and Maher, C.T., 1976, *Tellus*, 28:11.

Köhn, M. and Mainzhausen, L., 1937, *Kolloid — Z.*, 79:316.

Kopczewicz, B., Nagamoto, C., Parungo, F., Harris, J., Miller, J., Sievering, H., and Rosinski, J., 1991, *Atmos. Res.*, 26:245.

Küster, E., 1913, *Kolloid — Z.*, 13:192.

Liesegang, R.E., 1896a, *Naturw. Wochschr.*, 11:353.

Liesegang, R.E., 1896b, *Phot. Archiv.*, 21:221.

Lodge, J.P., 1955, *J. Meteor.*, 12:493.

Lodge, J.P. and Franzoi, H.M., 1954, *Analyt. Chem.*, 26:1829.

Lodge, J.P. and Tufts, B.J., 1956, *Tellus*, 8:184.

Lurie, A.A., 1966, *Kolloid. Zh.*, USSR, 28:534.

Mamane, Y., 1977, *A Quantitative Method for the Detection of Individual Submicron Sulfate Particles*, Ph.D. Thesis, Dep. Meteorology, Pennsylvania State University, University Park.

Mamane, Y., Ganor, E., and Donagi, A.E., 1980, *Water, Air, Soil Pollut.*, 14:29.

Mamane, Y. and Noll, K.E., 1985, *Atmos. Envir.*, 19:611.

Mamane, Y. and de Pena, R.G., 1978, *Atmos. Envir.*, 12:69.

Mamane, Y. and Pueschel, R.F., 1980, *Atmos. Envir.*, 14:629.

Mathur, P.B., 1960, *Bull. Acad. Polon. Sci.*, Ser. Chim., 8:429.

Mathur, P.B., 1961, *Bull. Chem. Soc. Japan*, 34:437.

Metnieks, A.L., 1958, *Geophys. Bull.*, School of Cosmic Phys., Dublin, No. 15.

Mikhailev, P.F., Nikoforov, V., and Shemyakin, F.M., 1934, *Kolloid — Z.*, 66:197.

Morse, H.W., 1930, *J. Phys. Chem.*, 34:1554.

Nikoforov, V.K. and Kharamonenko, S.S., 1938, *Acta Physicochim.*, USSR, 8:25.

Notboom, K., 1923, *Kolloid — Z.*, 32:247.

Okada, K., 1985, *Atmos. Envir.*, 19:743.

Ostwald, Wilhelm, 1897, *Lehrbuch der allgemeinen Chemie*, Engelmann, Leipzig, pp. 778.

Ostwald, Wolfgang, 1925, *Kolloid — Z.*, 36:380.

Pacter, A., 1956a, *J. Colloid Sci.*, 11:150.

Pacter, A. 1956b, *J. Colloid Sci.*, 11:96.

Parungo, F.P., Nagamoto, C., Schnell, R., and Nolt, I., 1981, *Atmospheric Aerosol and Cloud Microphysics Measurements*, HAMEC Project Report III, NOAA Envir. Research Lab., Boulder, pp. 75.

Parungo, F.P., Nagamoto, C.T., and Harris, J.M., 1986, *Atmos. Res.*, 20:23.

Pidgeon, F.D., 1954, *Anal. Chem.*, 26:1832.

Podzimek, J., 1959, *Studia geophys. et geod.*, 3:256.

Podzimek, J., 1967, *Studia geophys. et geod.,* 11:470.

Podzimek, J., 1969, *Annali Ist. Univ. Navale,* Napoli, 38:3.

Podzimek, J., 1973, *J. Rech. Atmos.,* 7:137.

Podzimek, J., 1980, *J. Rech. Atmos.,* 14:241.

Podzimek, J., 1990, *J. Aerosol Sci.,* 21:299.

Podzimek, J. and Černoch, I., 1959, *Geofys, sbornik,* 1958, Travaux de l' Inst. Geophys. 96:439.

Podzimek, J. and Černoch, I., 1961, *Geofis. Pura e Applic.,* 50:96.

Podzimek, J. and Černoch, I., 1962, *Geofys. sbornik, 1961,* Travaux de l' Inst. Geophys. 161:475.

Podzimek, J., Issac, K.M., and Husen, B., 1991, in *Developments in Mechanics,* Vol. 16, Proc. 22nd Midwestern Mech. Conf., UMR, Rolla, p. 376.

Prager, S., 1956, *J. Chem. Phys.,* 25:279.

Preining, O., Podzimek, J., and Yue, P., 1976, *J. Aerosol Sci.* 7:351.

Pueschel, R.F., Snetsinger, K.G., Goodman, J.K., Toon, O.B., Ferry, G.V., Oberbeck, V.R., Livingston, J.M., Verma, S., Fong, W., Starr, W.L., and Chan, K.R., 1989, *J. Geophys. Res.,* 94:11271.

Raman, C.V. and Ramaiah, K.S., 1939, *Indian Acad. Sci.,* 9A:467.

Rau, W., 1955, *Meteorolog. Rundschau,* 8:169.

Rau, W., 1956, *Archiv. Met. Geophys. Bioklim.,* A 9:224.

Rinehart, G.S., 1971, *J. Rech. Atmos.,* 5:57.

Seely, B., 1952, *Analyt. Chem.,* 24:576.

Shinohara, S., 1970, *J. Phys. Soc. Japan,* 29:1073.

Smith, D.A., 1984, *J. Chem. Phys.,* 81:3102.

Stern, K.H., 1954, *Chem. Revs.,* 54:79.

Toba, Y., 1966, Spec. Contrib. Geophys. Inst. Kyoto University, 6:59.

Ueno, Y. and Sano, I., 1971, *Bull. Chem. Soc. Japan,* 44:637.

Ueno, Y. and Sano, I., 1972, *Bull. Chem. Soc. Japan,* 45:626.

Vaidyan, V.K., Ittachan, M.A., and Pillai, K.M., 1981, *J. Cryst. Growth,* 54:239.

Van Hook, A., 1938a, *J. Phys. Chem.,* 42:1191.

Van Hook, A., 1938b, *J. Phys. Chem.,* 42:1201.

Van Hook. A., 1940, *J. Phys. Chem.,* 44:751.

Van Hook. A., 1941a, *J. Phys. Chem.,* 45:442.

Van Hook. A., 1941b, *J. Phys. Chem.,* 45:879.

Van Hook. A., 1941c, *J. Phys. Chem.,* 45:1194.

Van Hook, A., 1944, Liesegang rings, in *Colloid Chemistry,* Vol. 5, Alexander, J. (Ed.), Reinhold, New York, p. 513.

Veil, S., 1947a, *C. R. Acad. Sci.,* 224:1771.

Veil, S., 1947b, *C. R. Acad. Sci.,* 225:804.

Veil, S., 1948a, *C. R. Acad. Sci.,* 227:1359.

Veil, S., 1948b, *C. R. Acad. Sci.,* 226:1603.

Veil, S., 1950, *C. R. Acad. Sci.,* 230:1769.

Veil, S., 1951, *C. R. Acad. Sci.,* 233:45.

Vittori, O.A., 1954, *Riv. Meteorol. Aeronaut.,* 14:17.

Vittori, O.A., 1955, *Archiv Met. Geophys. Bioklim.,* A 8:204.

Vrkočová, J. and Podzimek, J., 1964, in *Aerosols, Physical Chemistry and Applications,* Proc. 1st Nat. Conf. on Aerosols, Liblice, NCSAV, Praha, p. 545.

Wagner, C., 1950, *J. Colloid Sci.,* 5:85.

Winckelmann, J., 1931, *Microchemie,* 12:437.

Woodcock, A.H. and Gifford, M.M., 1949, *J. Marine Res.,* 8:177.

Wormser, Y., 1946, *J. Chim. Phys.,* 43:88.

Yamato, M. and Ono, A., 1989, *J. Meteorol. Soc. Japan,* 67:147.

Yamato, M. and Iwasaka, Y., Ono, A., and Yoshida, M., 1987, *Proc. NIPR Symp. Polar Meteorol. Glaciol.,* 1:82.

Yamato, M., Iwasaka, Y., Okada, K., Ono, A., Nishio, F., and Fukabori, M., 1987, *Proc. NIPR Symp. Polar Meteorol. Glaciol.,* 1:74.

Yue, P.C. and Podzimek, J., 1980, *Ind. Eng. Chem. Prod. Res. Dev.,* 19:42.

12 Single Particle Analysis Techniques

Lieve A. De Bock and René E. Van Grieken

CONTENTS

12.1 INTRODUCTION

In the beginning, environmental analytical research was focused mainly on the analysis and identification of volatile pollutants. Later, it became clear that in addition to these volatile compounds, individual particles or aerosols also have to be considered, since they play a significant role in different aspects of the environment. Visibility problems related to smog in cities like Athens, Los Angeles, and Santiago de Chile — the increased risk of developing lung cancer due to exposure to toxic products like asbestos — the effects of acid rain on forests and building materials — these are only a few examples of the processes in which aerosol particles are involved.

Microanalysis of individual particles can reveal information on the chemical composition and size of the particle, the elemental lateral and depth distribution, and whether a specific element or compound is uniformly distributed over all particles. Based on these investigations, the particle's origin, formation processes, reactivity, transformation reactions, and environmental impact can be estimated.

Generally, the specific problems of certain microanalytical techniques can often be attributed to analysis problems in the small particle size range. Quantitative information may be limited due to the fact that the variable particle geometry introduces several complications not present in the bulk analysis of polished specimen, and because of the lack of suitable standards. Moreover, to ensure the statistical relevance of the results, large numbers of particles per sample should be analyzed, which makes individual particle analysis very time consuming. The loss of information, due to the transformation or evaporation of the most volatile and unstable compounds under vacuum conditions, remains, however, one of the major problems encountered in various microanalytical techniques.

Although different technological advances in instrumentation have been achieved, as well as a refined knowledge of electron, proton and ion optics, and electronics, the majority of analytical techniques remains limited in one way or another, and this implies that several, often complementary techniques should be applied in combination with bulk analysis for the complete characterization of a sample.

This chapter provides an overview of the latest publications on single-particle analysis of individual environmental aerosol particles obtained by the most frequently used microbeam techniques. The overview covers a computer search of the literature (data banks: Enviroline, Environmental Bibliography, and Current Contents) for the period of January 1990 through May 1996. A map illustrating the sites for which single-particle aerosol analyses have been reported in this period is represented in Figure 12.1. Only a brief outline of the theoretical aspects of the different microbeam techniques is represented. More details can be found in previous reviews on this topic by Grasserbauer (1983), Van Grieken and Xhoffer (1992), Xhoffer et al. (1992) and Jambers et al. (1995).

12.2 SINGLE PARTICLE ANALYSIS

12.2.1 Electron Beam Techniques

12.2.1.1 Electron Probe X-Ray Microanalysis and Scanning Electron Microscopy

In electron probe X-ray microanalysis (EPXMA) as well as scanning electron microscopy (SEM), various signals are generated by the interaction of a nanometer-sized electron probe with the sample of interest. Compositional information on the sample can be acquired by detection of the emitted characteristic X-rays using wavelength- or energy-dispersive spectrometers (WDX and EDX, respectively) and is also provided by the backscattered electron image. Detection of backscattered and secondary electrons allows morphological studies. Due to technological advances, the differences between EPXMA and SEM–EDX have been reduced, so that both techniques now can, to some extent, be used for chemical as well as morphological studies.

In the field of single-particle analysis EPXMA and SEM–EDX are classified as the most applied nondestructive microanalytical techniques. The main reason for this popularity is the fact that both techniques are strongly computerized and automated, offering the possibility to analyze several hundreds of particles in a few hours' time, with a relative accuracy of about 5% and a lateral resolution of 0.1 to 5 µm. The in-depth resolution of EPXMA/SEM–EDX is estimated at 0.5 to 5 µm. Studies revealing the detailed morphological characteristics and the element distribution within individual (giant) aerosol particles are accessible by manual analysis.

The advantages of EPXMA/SEM–EDX are being overshadowed by the unfavorable detection limits (0.1%) provided by these techniques: the necessity to work under vacuum conditions and the problems of quantization. However, a CITZAF correction procedure package, proposed by Armstrong (1995), claims to provide reasonably quantitative results in the analysis of unconventional samples, such as individual unpolished particles.

The specific problems associated with the determination of low-Z elements, such as C, N, and O, as well as with the analysis of different kinds of samples (wet, dry, insulating or conductive samples), are being solved by the new generation of scanning electron microscopes (Danilatos, 1994). Moreover, they offer a higher lateral resolution while working with high-magnification secondary electron images and provide much-enhanced image analysis capabilities (Nockolds, 1994).

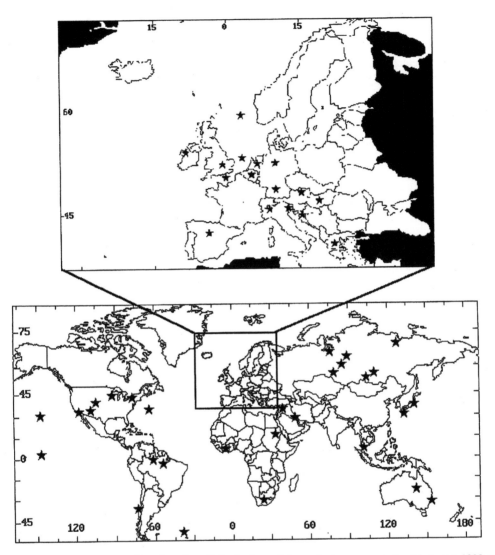

FIGURE 12.1 Aerosol sampling sites for which single particle analysis was carried out between 1990 and 1996 in Europe and in the rest of the world.

12.2.1.2 Scanning Transmission Electron Microscopy

In conventional transmission electron microscopy (TEM), electrons transmitted through the sample are investigated by illuminating the whole imaged specimen area with a fixed defocused electron probe. The magnified image is displayed on a fluorescent screen below. The bright- and dark-field image modes provide morphological information, while the identification of the crystal structure becomes possible by selection of the electron diffraction mode.

In a scanning transmission electron microscope (STEM), the electron probe is focused down to a diameter of 2 to 5 nm and scanned over the specimen. The different signals produced during this probe scanning are recorded by electronic detectors and displayed on a cathode-ray tube. As in TEM, the transmitted electrons can be recorded in the bright- or dark-field image mode. The scattered electrons can provide a BSE or secondary electron (SE) image. The fact that the images are recorded as serial electrical signals implies a flexible on- and off-line image evaluation and processing. X-ray microanalytical information of specific areas of thin and semi-thin specimen can be acquired by supplying the STEM with an EDX spectrometer.

In comparison to EPXMA/SEM–EDX, the imaging and X-ray analysis of smaller specimen areas are possible, due to the production and positioning of smaller electron probes. Moreover, the lateral resolution in STEM for X-ray microanalysis is improved (2 to 5 nm) by using very thin specimens. Detection limits and accuracy are comparable to those found in EPXMA/SEM–EDX.

The application of STEM–EDX in the field of environmental particle research, more specifically in the characterization of individual aerosol particles, has only recently been introduced. On the other hand, studies on the use of electron microscopy in the analysis of aquatic colloids have already been reviewed by Leppard (1992). STEM–EDX was performed for the characterization of mineral and mineral–organic colloids. Simple and powerful aquatic sample preparation techniques for STEM analysis, by which the number of artifacts is reduced, have been developed and described by Leppard (1992) and Perret et al. (1991). Fortin et al. (1993) studied the iron oxyhydroxides present in the sediments of eight lakes using STEM–EDX in combination with electron- and X-ray diffraction, after preparation with Nanoplast resin.

12.2.1.3   Electron Energy Loss Spectrometry

The electron energy loss spectrometry (EELS) instrument was developed by supplying the optical column of a conventional TEM with a magnetic prism spectrometer, positioned between the imaging lenses. The prism energy filter provides two operation modes, a spectrum mode and an image mode. On the one hand, the energy loss spectrum can be sequentially acquired by scanning the spectrum over a small slit and further displayed on a personal computer used as a multichannel analyzer. The energy selected (or filtered) image, on the other hand, can be deflected into the silicon-intensified target (SIT) vidicon camera and from there transferred to an image processing system.

EELS can be performed to identify and also quantify elements present in a sample by means of spectra and element specific image (ESI) collection. This technique provides a lateral resolution of <0.01 μm, an in-depth resolution of 0.01 to 0.1 μm and reveals detection limits similar to EPXMA/SEM–EDX. Compared to X-ray analysis techniques, with elemental coverages of $Z = 11$ to 92, EELS shows a higher sensitivity for the detection of low atomic number elements, down to $Z = 3$. The fact that the technique is based on primary interaction effects results, moreover, in better detectability of low masses. Information on the electronic state and chemical bonding can be deduced from the EEL spectrum. To transmit enough signal and to overcome the production of multiple scattering effects, the samples analyzed by EELS should be very thin.

Recent developments in the data acquisition systems for EELS analysis offer the possibility for parallel spectrum recording (PEELS). Compared with serial EELS the unacceptably long collection times, frequently causing large mass losses in environmental particles, are reduced by a factor of 400 (Maynard and Brown, 1992; Xhoffer, 1993).

12.2.2   Ion Beam Techniques

12.2.2.1   Microparticle-Induced X-Ray Emission

In scanning proton microprobe analysis (SPM) a beam of 1 to 3 MeV protons, focused down to a diameter of 0.5 to 1.0 μm, with a current below 100 pA (to prevent sample damage) is scanned across the sample. With the help of a powerful computer system, the different signals, received from several detectors positioned around the sample, are collected. Deviation of the proton beam due to possible collisions with air molecules is avoided, as in EPXMA/SEM–EDX and STEM–EDX, by high vacuum conditions inside the sample chamber.

Several analytical techniques can be implemented in the SPM unit, but the micro-proton-induced X-ray emission technique (μ-PIXE) is by far the most important one, because of its large reaction cross sections. The X-ray production yield from PIXE is, therefore, very high, and detection of the produced X-rays is obtained by a Si(Li) detector. However, the presence of a Be-window in front of the detector implies again the restriction to detect only elements heavier than sodium. Compared to EPXMA/SEM–EDX, the background in the PIXE spectra in the region from $Z = 15$ onward is 2 to 3 orders of magnitude lower because no continuum radiation is produced by stopping the projectile ions. Together with the high yields, this results in favorable minimal detection limits of tens of ppms in small samples of a few µm in diameter and weighing only a few pg, turning PIXE into a sensitive technique that provides trace element concentrations on a single particle level with a spatial resolution of 0.5 to 10 µm. Generation of real-time X-ray elemental mappings becomes possible by correlation of the X-ray signal intensities with the beam position in the scanned area. In this way, up to 20 preselected elements can be mapped and displayed together during the analysis. Combining µ-PIXE with other nuclear analytical techniques such as RBS (Rutherford backscattering spectrometry), PIGE (particle-induced gamma ray emission), and FAST (forward angle scattering technique), all implemented in the same SPM unit, reveals qualitative analytical information on nearly all elements present in the periodic table. Quantitative data with an accuracy of 10 to 20% can be easily obtained by interception of the proton beam after transmission through the sample. Moreover, density variation maps are available by STIM (scanning transmission ion microscopy), in which the energy loss of ions passing through the thin specimen is measured.

12.2.2.2 Secondary-Ion Mass Spectrometry

The secondary-ion mass spectrometry technique (SIMS) is based on the sputtering and ionization of atoms present in the sample surface by a primary ion beam (O_2^+, O^-, Ar^+, Cs^+, Ga^+, ...) or by molecules generated in a duoplasmatron, with an energy between 1 and 20 keV. Only small fractions of the sputtered atoms are charged and subsequently extracted into a mass spectrometer. Two types of mass spectrometers are commercially available for the separation of secondary ions according to their m/z ratio. One is based on electric/magnetic deflection fields, and the other on the quadrupole/time-of-flight principle (TOF–SIMS). The primary beam intensity in the TOF–SIMS is kept much lower so that, compared to SIMS, only very few secondary ions are detected per cycle. The much higher sensitivity obtained by TOF–SIMS results from the high transmission combined with parallel mass detection as well as from the unlimited mass range. At full transmission, the mass resolution can go beyond 10,000 amu, and the lateral resolution offered by imaging instruments can go down to the submicrometer range.

By sputtering the analyzed surface, information on the distribution of the constituents with depth and the heterogeneity in composition within a single particle becomes available. The detected signals originate from a depth of 1 to 2 nm and with this technique, information from a minimum sample volume of 0.001 µm³ is accessible. The detection limit is about 1 ppm, with an accuracy of 10%. The combination of ion images, available in most SIMS instruments by visualizing the lateral distribution (<1 µm) of the elements, and depth profiles provides a three-dimensional analysis of the sample surface. Additional advantages for single particle analysis are the capability for the detection of all elements ($Z = 1$ to 92), fingerprinting of compounds, and isotope ratio measurements.

12.2.3 Photon Beam Techniques

12.2.3.1 Laser Microprobe Mass Spectrometry

Laser microprobe mass spectrometry (LMMS) uses a high-power, density-pulsed laser beam for the evaporation and ionization of a microvolume of the sample. Usually, a time-of-flight (TOF) mass spectrometer is applied for the separation of the positive and negative elemental and molecular ions, with different mass/charge ratios according to their flight times. The mass resolution achieved by commercially available TOF–LMMS instruments ranges from 850 in the low m/z range to 1 unit for m/z 500. The interpretation and classification of individual particle mass spectra can be obtained by pattern recognition (Harrington et al., 1989; Linder and Seydel, 1989) and library search (Wouters, 1991; Wouters et al., 1993) computer methods. In principal, LMMS is capable of detecting all elements and organic compounds and revealing stoichiometric information. The spatial resolution of the analysis is 0.5 to 3 µm, the in-depth resolution is >1 µm and the detection limits are around a few ppm.

Since no theoretical model can predict the ion yield for a specific specimen as a function of target and laser beam parameters, quantification by LMMS remains a problem. The only solution is to rely on the use of internal reference elements or standard samples which closely resemble the unknown one. Once suitable standards are available, the best precision (15 to 30% for all elements) is obtained for thin samples.

A higher resolving power and mass accuracy was recently achieved by coupling a Fourier transform (FT) ion cyclotron resonance (ICR) mass spectrometer to the laser microprobe technique. In routine FT–LMMS measurements, the mass resolution exceeds 10^6 below m/z = 100 and 10^5 at m/z = 1000 with an m/z accuracy around 1 ppm. Elemental ion detection can be obtained down to 10^8 atoms, but the minimum spatial resolution is around 5 µm, compared to 1 µm in TOF–LMMS. Since this technique has only been available for a few years, no environmental applications have been published yet. However, its possibilities for the speciation of inorganic compounds have already been proven by tests on binary salts, oxides, and different oxysalts, i.e., phosphates, nitrates and nitrites, carbonates, and sulfates (Struyf et al., submitted).

In situ characterization of the size and chemical composition of individual particles becomes possible by performing on-line laser microprobe mass spectrometry (on-line LMMS), or so-called rapid single-particle mass spectrometry (RSMS) (Shina, 1984; Marijnissen et al., 1988; McKeown et al., 1988; Kievit et al., 1992; Mansoori et al., 1994; Murray and Russel, 1994; Prather et al., 1994; Nordmeyer and Prather, 1994; Hinz et al., 1994; Carson et al., 1995; Murphy and Thomson, 1995; Reents et al., 1995; Hinz et al., 1996; Yang et al., 1996). In this technique, aerosol particles are collected immediately into the MS. The scattered radiation obtained from the interaction of each individual aerosol particle with the He–Ne laser beam reveals information on the particle size and triggers a laser which vaporizes the particle and ionizes the fragments. A complete mass spectrum is recorded with a TOF mass spectrometer for each single aerosol particle. This kind of sample analysis reduces contamination or decomposition upon sample surface interaction and decreases volatile component losses owing to the limited time spent by the particles under vacuum conditions. The recent advances in the development and application of the MS for on-line, real-time analysis of single aerosol particles have been reported by Murray and Wexler (1995). To date, the on-line instrumental set-up allows a simultaneous characterization of the size and chemical composition of individual aerosol particles down to 50 to 100 nm in real time (Prather et al., 1994). Reents et al. (1995) reported on the ability to detect particles as small as 20 nm, using a different instrumental configuration. Simultaneous bipolar ion detection from the same airborne particle was accomplished for the first time by Hinz et al. (1996); the scheme of the instrument is illustrated in Figure 12.2. Further research in this field will improve and optimize the different aspects of the on-line particle analysis.

12.2.3.2 Micro-Raman Spectroscopy

The Raman spectroscopic technique was developed in 1928 and has been used mainly for bulk analysis. Much later, microparticle analysis became possible with the introduction of lasers as sources for the excitation of Raman spectra in combination with new developments made in instrumentation optics.

Raman scattering is produced by the inelastic interaction of photons of frequency ν_0 with molecules present in the sample. After collision, photons with different frequencies occur which can be related to the original frequency ν_0 by the expression $\nu_0 \pm \nu'$. The Raman frequency shift ν' can result from molecular rotation or, more commonly, from molecular vibration and reveals the molecular chemical composition, and in some cases this frequency ν' can even provide information on the electronic state of the molecules.

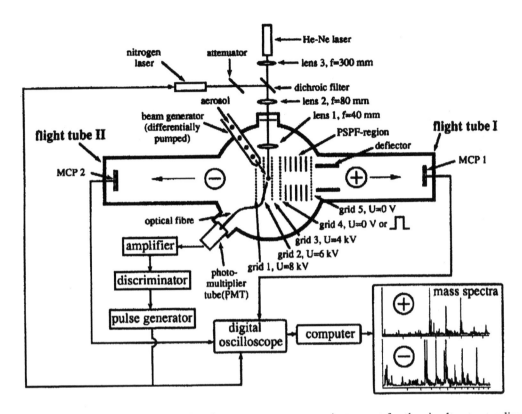

FIGURE 12.2 Scheme of the real-time laser mass spectrometer instrument for the simultaneous on-line detection of + and − ions formed from single aerosol particles. (From Hinz, K.P. et al., *Simultaneous Detection of Positive and Negative Ions from Single Airborne Particles by Real-Time Laser Mass Spectrometry,* 1996, 235. With kind permission of Elsevier Science -NL, Sara Burgerhartstraat 25, 1055 KV Amsterdam, The Netherlands.)

One of the major disadvantages of Raman scattering remains that its effect, compared to elastic scattering (Rayleigh scattering) or the fluorescence effect, is very weak. The difference in scattering cross section is in the order of a factor of 1000 for Rayleigh scattering and even many orders of magnitude larger for the fluorescence effect. Enhancement of the low Raman scattering signal can be acquired by performing resonance Raman spectroscopy. This type of Raman effect is produced by selecting the incident laser beam frequency close to or within an absorption band of the molecules. Referring to the level of incident photon energy with respect to the absorption band, terms like pre- or postresonance are used. Compared to the spontaneous Raman effect, the signal intensity increases but remains linearly proportional to the material concentration, which assures the possibility of quantifying the acquired results. Another way to accomplish the Raman scattering signal enhancement is the use of a high-power, pulsed laser beam. The advantage of the so-called stimulated Raman scattering (SRS) is that the intensity of the scattered photons becomes comparable to those of the exciting laser beam and, therefore, corresponds to a major increase in detection sensitivity. The large signals produced at high concentrations diminish rapidly, with decreasing concentration so that the sensitivity by SRS is at least 20 to 40 times less than for spontaneous Raman scattering (Fung et al., 1994). With SRS, only semiquantitative measurements are possible because the scattering intensity is expressed as an exponential function of the laser intensity and is no longer proportional to the actual concentration.

One of the unique characteristics of micro-Raman spectroscopy in the field of single particle analysis is the possibility to differentiate between irregularly and spherically shaped particles based on the obtained Raman spectrum, since additional peaks are caused by morphological dependent resonances. The basic theory, possibilities, and limitations of Raman scattering in the field of single particle analysis have been discussed by Schrader (1986) and reviewed by Schweiger (1990).

12.2.3.3 Microscopic X-Ray Fluorescence and Microscopic Synchrotron Radiation X-Ray Fluorescence

The development of relatively simple focusing devices for X-rays has made microscopical X-ray fluorescence (μ-XRF) interesting for single particle analysis (Janssens et al., 1994). Conical or tapered capillaries make it possible to focus the X-ray beam to a 10 μm spot with a minimum loss of photon flux. The major advantages of this technique are the good detection limits for heavy elements, simple sample preparation and the possibility to work under atmospheric pressure. The main limitation, at this moment, remains the relatively large beam diameter (<10 μm).

The conventional X-ray source can be replaced by synchrotron radiation to obtain microscopic synchrotron radiation X-ray fluorescence (μ-SRXRF). The high intensity and natural collimation make this kind of radiation ideal for the formation of X-ray microbeams. To date, microbeams with a lateral resolution of 5 to 10 μm are available in second-generation synchrotrons. The low emittance of the synchrotron source offers the possibility to use, next to conical capillaries, Bragg-reflecting or totally reflecting curved mirrors to create X-ray beams of submicrometer dimensions. These sophisticated X-ray optics will be used in the μ-SRXRF dedicated beam-line of the European Synchrotron Radiation Facility (ESRF), Grenoble, France which is now under development and will be operational in 1997. In this beam-line, a high-flux 5 to 10 μm spot will be combined with a lower-flux submicrometer spot. This submicrometer spot will be very useful for the analysis of trace metals in single environmental particles.

12.2.4 Other Microanalytical Techniques

Several other microanalytical beam techniques, such as scanning Auger microscopy (SAM), X-ray photoelectron spectroscopy (XPS), and Fourier transform infrared spectrometry (FTIR), which are frequently used in various research disciplines, have not found their way to environmental particle samples. Their applications in this field are either scarce or nonexistent.

Since SAM is classified as a surface-sensitive technique, it would be very useful to characterize the surface coating and in-depth composition of individual particles. However, major problems are encountered as a consequence of the charging effects caused by nonconductive particles upon electron beam interaction as well as the low detection power of the technique (0.1 to 1%) (Linton et al., 1977). The low sensitivity to high-Z elements is also considered as a significant drawback, because particles with a heavy metal coating are important from an environmental and toxicological point of view. Despite these limitations, SAM does exhibit several advantages over ion microprobe particle characterization, which could reveal useful single particle information, such as: effects of sputtering artifacts are overcome due to the fact that the initial surface spectrum is obtained without sputtering, multielement depth profiles of the particle are more readily acquired, and spectral interferences may be less severe for a few elements.

XPS is, like scanning electron microscopy, a surface-sensitive technique, in which the core-level electrons from the sample are ejected upon X-ray beam interaction. XPS can also be applied for heavy metal coatings, but due to its poor spatial resolution, about 10 μm, most environmental applications are still restricted to bulk analysis. Considerable improvements in the spatial resolution and the sensitivity of XPS are expected in the near future and will certainly lead to a better understanding of the surface composition of individual fly ash particles.

FTIR spectrometry provides a precise and specific method for the identification and quantization of various materials but is limited to compounds with a permanent dipole moment. In the field of single particle analysis, the combination of FTIR with a microscope offers the possibility to analyze individual particles with diameters above 5 μm. Smaller particles can only be identified as particle clusters. Despite the possibility that FTIR microscopy could characterize organic particle content, only few publications in this field were found in the literature (Danler et al., 1987; Kellner and Malissa, 1989).

Atomic emission spectrometry in which particles are vaporized and ionized by an inductively coupled plasma (ICP–AES) can be applied for an on-line elemental analysis of single aerosol particles. Regardless of its poor element dependent detection limits (100 to 1000 ppm), this method is expected to be useful for a variety of purposes (Bochert and Dannecker, 1992; Frame et al., 1996). The laser ablation inductive coupled plasma mass spectrometric technique (LA-ICP-MS) is also promising for the analysis of individual aerosol particles in the future. At the moment, however, the lateral resolution is still limited to 10 μm.

12.3 APPLICATIONS

12.3.1 REMOTE AEROSOLS

Increasing industrialization, together with long-range transport of pollutants, has reduced the possibility to locate and characterize the composition of background aerosols in the absence of any pollution. The Antarctic continent is one of the few locations where these pure aerosols still exist. In the fine and coarse mode fraction of coastal Antarctic aerosols, marine components were identified by EPXMA as the dominant particle type, and only a minor crustal component could be detected (Artaxo et al., 1992; Van Grieken et al., 1992; Artaxo et al., 1990). The small amounts of S present in several NaCl and $MgCl_2$ particles indicated the occurrence of earlier reactions with gaseous S-compounds. In the majority of particles larger than 0.1 μm, S was present as gypsum, and these sulfate particles showed a seasonal variability with maximum number and mass during the summer. Based on manual analysis, some of the aerosols appeared to be internally mixed, containing both marine and silicate components. The trace elemental composition (Z > 11) of the Antarctic aerosol particles was determined by micro-PIXE with detection limits in the 1 to 10 ppm range (Artaxo et al., 1992, 1993). The results confirmed the dominance of sea salt aerosols with NaCl and $CaSO_4$ as major compounds, and Al, Si, P, K, Mn, Fe, Ni, Cu, Zn, Br, Sr, and Pb were detected as minor and trace elements. Factor analysis on the elemental data set revealed four different components: soil dust particles, NaCl particles, $CaSO_4$ with Sr, and finally Br-rich and Mg-rich particles. LMMS investigations revealed, moreover, that typical ions, such as a ammonium, nitrate, and some trace element ions, which are often identified in LMMS spectra of marine aerosol particles from less remote regions, were not present in the coastal Antarctic aerosols (Wouters et al., 1990). To investigate the composition of the early Antarctic atmosphere, ice core particles have been analyzed with μ-SRXRF in the second-generation synchrotron at Brookhaven, New York (Janssens, in press). The spectrum and the metal concentrations of such a particle are represented in Figure 12.3 and Table 12.1, respectively. The concentrations have been calculated on the total mass of this particle, which was 1306 pg, with the iron peak as a reference. Concentrations in the femtogram range illustrate the good detection limits.

Another location which is characterized, by very remote as well as strongly polluted areas is Siberia, until recently almost inaccessible for Western scientists. Since, at the moment, practically no experimental data exist and scientists believe that western and central Siberia could have an important impact on both the Arctic region and the global climate, several studies were set up to determine processes responsible for the production, transport, and fate of both natural and pollution aerosols. One of these studies discusses the chemical characterization by automated EPXMA of individual aerosol particles sampled at two sampling stations in central Siberia in February and August 1992 (Van Malderen et al., 1996). The differences between the two sampling locations, Karasuk and Kljuchi, were small. Comparison between the winter and summer campaigns revealed, however, clear differences in the average size of most of the particle types. The average size of the winter aerosol is, in most cases, lower than that found for the summer aerosol, because a higher

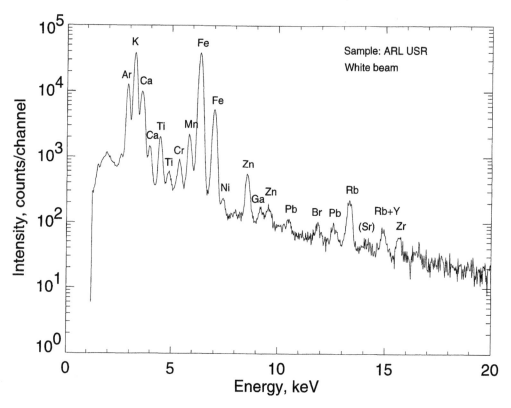

FIGURE 12.3 X-ray spectrum of a 15 μm Antarctic ice core particle analyzed by μ-SRXRF. (From Jambers, W. and Van Grieken, R. *Present and Future Applications of Beam Techniques in Environmental Microanalysis,* 1996, 120. With kind permission of Elsevier Science -NL, Sara Burgerhartstraat 25, 1055 KV Amsterdam, The Netherlands.)

fraction of the particles sampled in the winter, when Siberia is fully frozen and snow-covered, tend to come from anthropogenic sources. The variation in aerosol composition for four different wind sectors appeared to be rather small. This was explained by the fact that the few industrial centers which emit pollutants and act as large point sources, over this enormous area of Siberia, are situated far from the sampling locations. Therefore, the pollutants are well mixed and diluted by the time they are sampled. It was concluded that although the sampling location seemed to be quite remote, many anthropogenic particles were present and that these particles, or at least a significant fraction of them, are transported farther north and eventually deposited in the Arctic region. The heavy metal-containing particles, together with fly ash, could, therefore, be responsible for the major part of heavy metal deposition in the Arctic. Similar investigations above Lake Baikal in the southeast of Siberia were accomplished in June 1992 and September 1993 (Van Malderen et al., 1996). The results pointed to a remarkably clean atmosphere over most areas of Lake Baikal. Nevertheless, in some areas of the lake, the atmospheric pollution caused by industries situated around Irkutsk, Ulan-Ude, and the Baikalsk paper plant could be detected. Although northern parts are unlikely to be affected by atmospheric pollution from the south, care should be taken respecting local pollution in the more southern parts of the lake.

Automated EPXMA was performed to study the composition of individual aerosol particles sampled over the Amazon Basin (Artaxo et al., 1990). Chemical mechanisms occurring in the Amazon Basin's atmosphere and processes involving aerosol and gas emission of the forest, as well as long-range transport effects, were investigated. The main particle types could be classified in three sources of aerosol particles: biogenic, soil dust, and sea salt aerosol particles. Micro-PIXE

TABLE 12.1
Heavy Metal Content Calculated from the X-Ray Spectrum Shown in Figure 12.3

Heavy metal	Content (fg)
Cr	150
Mn	640
Ni	30
Zn	200
Ga	35
Pb	88
Br	48
Rb	340
Sr	35
Y	34
Zr	110

From Jambers, W. and Van Grieken, R., *Present and Future Applications of Beam Techniques in Environmental Microanalysis,* 1996, 120. With kind permission of Elsevier Science -NL, Sara Burgerhartstraat 25, 1055 KV Amsterdam, The Netherlands.

in combination with Rutherford backscattering (RBS) (Artaxo et al., 1992, 1993) showed that the biogenic particles contained high amounts of organic material with K, P, Ca, Mg, Zn, and Si as predominant elements. Zn seemed to be present at 10 to 200 ppm in biogenic particles rich in P and K. The LMMS study (Artaxo et al., 1990) of the biogenic fraction revealed two major particle types: K-rich and P-rich particles, although some of the LMMS spectra were too complex to interpret because of the superposition of mass peaks from many organic ions.

Clear evidence of long-range transport was also found in volcanic loess of the Pahala formation on the island of Hawaii (Begét, 1993). Asian dust grains, mainly composed of quartz and unknown in the basalts of the Hawaiian Islands, were characterized by X-ray diffraction of the bulk samples in combination with SEM–EDX of single mineral grains and X-ray mapping. It appeared that these quartz particles, 1 to 10 μm in size and some even up to 60 μm, were transported over a distance of 10,000 km by late Quaternary windstorms. Their contribution to the total loess deposit was estimated at 1 to 3%. Radiocarbon dates, obtained from many parts of the Pahala formation, gave results ranging from a few thousand to more than 30,000 years BC.

Aerosol particles collected in the upper troposphere (UT) and the lower stratosphere (LS) were studied by single particle analysis on a STEM (Sheridan et al., 1994). These particles seem to play an important role in several atmospheric processes, e.g., by providing surfaces for heterogeneous reactions and participating in the formation of cirrus clouds. Moreover, their composition is indicative for the impact of aircraft emissions and can be used as a tracer for the air exchange between the UT and the LS and the transport within the LS. From the analyzed particles present in the lower stratosphere, 97% contained only O and S and can probably be identified as acidic sulfate. The remaining particles carried soot, C-rich substances, and crustal materials and a sulfur-rich coating on their surface. In the UT, 91 to 94% of the total particle concentration was identified as sulfate particles. The non-sulfate materials included crustal-type materials, hydrated salts, C-rich materials of several types, and metal containing substances of uncertain origin. In the majority of

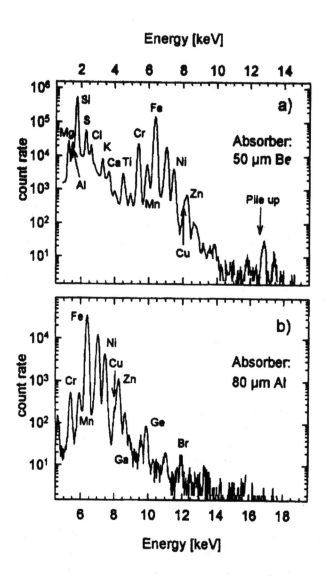

FIGURE 12.4   PIXE spectra of an interplanetary dust particle. The spectra were measured using two different X-ray absorbers to cover the whole Z-range between Mg (Z = 12) and Br (Z = 35). (From Bohsung, J., et al. *High Resolution PIXE Analyses of Interplanetary Dust Particles with the New Heidelberg Proton Microprobe,* 1995, 421. With kind permission of Elsevier Science Ltd., The Boulevard, Langford Lane, Kidlington OX5 1GB, UK.)

UT particles, no sulfur coating could be detected. Changes in morphology were found for the impacted sulfate particles, suggesting an acidity gradient across the tropopause with a higher acidity in the LS than in the UT. Compared to the LS, more sulfate particles with impacted diameters <0.5 μm were detected in the UT.

Interplanetary dust particles (IDP) collected in the stratosphere by the NASA large area collector L2005 were subjected to high-resolution PIXE analysis with the new Heidelberg proton microprobe by Bohsung et al. (1995). Their paper discusses the instrumental set-up and some of the properties which are necessary to perform spatially resolved multielement PIXE analysis of IDPs. A typical PIXE spectrum of an IDP is represented in Figure 12.4.

12.3.2 Marine Aerosols

The characterization of individual North Sea aerosol particles by EPXMA/SEM–EDX in combination with multivariate techniques has been one of the main topics for several years in the University of Antwerp group (Van Malderen et al., 1993). For four years, over 25,000 particles, sampled during successive campaigns aboard the research vessel *R/V Belgica* on the North Sea and English Channel, were identified and classified by hierarchical cluster analysis (HCA) according to their inorganic composition and size (Xhoffer et al., 1991). The North Sea aerosol is composed mainly of sea salt, S-rich particles, silicates, and $CaSO_4$ particles of which the abundances are influenced by meteorological conditions as well as sampling locations. Source apportionment was obtained by principal component analysis (PCA). The marine-derived aerosol particles were found as the first component, which grows with increasing wind speed or for more remote locations. The second and third components appeared to be anthropogenically derived $CaSO_4$ particles and particles with a high content of Si and S, respectively. Further PCA on the latter group differentiated between two sources, namely a mixed marine/continental source and a pure continental one.

The differences in aerosol composition with height above the North Sea were studied by Rojas and Van Grieken (1992). EPXMA on 50,000 individual particles showed that, for continental air masses, the majority of particles could be classified into three types: aluminosilicates, $CaSO_4$, and Fe-rich particles. Due to the turbulent nature of the atmosphere, no clear compositional changes occurred with altitude in the main particle types from western and marine air masses. Nonhierarchical analysis revealed aluminosilicates (coal combustion), Fe-rich particles, and sea salt particles enriched with lead and organic material (oil combustion) as major particle types in these air masses. Combustion or energy-generation processes seemed to be responsible for up to 60% of the analyzed particles. The different particle types distinguished by EPXMA were also found with LMMS (Dierck et al., 1992), but many particles seemed to appear as internal mixtures. Except for the decrease in relative sea salt concentration above the inversion layer, no clear differences were found in particle abundances as a function of height.

Variations in composition of air masses crossing the North Sea due to air–sea exchange processes in the lower troposphere were studied by De Bock et al. (1994). Aerosol and rainwater samples were collected on two research vessels positioned 200 km apart, downwind of each other, in the central area of the North Sea. Based on a combination of automated EPXMA and multivariate techniques, three to eight different particle types could be distinguished, which were apportioned to four major sources. When the air masses crossed the North Sea, a decrease in aluminosilicate particles appeared, associated with a relative increase for NaCl and seawater crystallization products. The particle diameter remained constant. Manual EPXMA was performed on several samples to study the relations between particle composition, origin, and shape. Characterization based on the particle shape was possible for most particle types. Variations in composition and diameter were found in the collected rainwater samples.

Recently the Ca-containing particles, especially $CaSO_4$ particles, which have been encountered in several atmospheric North Sea studies, were investigated to identify the different particle types and to find a correlation between their occurrence and the source regions of the corresponding air masses (Hoornaert et al., 1996). Nonhierarchical cluster analysis (NHCA) on the EPXMA data revealed that $CaSO_4$ in most cases constitutes the largest fraction of the Ca-bearing particles, as illustrated in Table 12.2. The greatest amounts of $CaSO_4$ particles were found for northeastern winds (Figure 12.5), coming from the central part of Germany, indicating the presence of anthropogenic sources in this region. The flue gas desulfurization process, recently studied by Davis et al. (1992) using an electrodynamic balance coupled to a Raman spectrometer, can be considered the major source of $CaSO_4$. Other Ca-containing particles were identified as aluminosilicates, $CaCO_3$, Fe-Ca-rich particles, and $CaSO_4$ or $CaCO_3$ in combination with NaCl. A similar investigation of

TABLE 12.2
Results of the Nonhierarchical Clustering of the Data after Selection of the Ca-Enriched Particles, Identifying CaSO$_4$ Particles to Be the Largest Fraction of the Ca-Bearing Particles. The Highest Amounts of CaSO$_4$ Were Found for Winds Coming from the Central Part of Germany

Cluster no.	Elemental Composition (% wt) (Relative to All Elements Detectable by EPXMA)	Classification	Abund. (%)
	northeast-east		
1	Ca(48) S(45)	CaSO$_4$	49
2	Si(43) Al(19) Ca(17) Fe(12)	aluminosilicates	14
3	Ca(30) Si(20) S(19) Al(12) Fe(8)		11
4	Fe(34) Ca(27) S(23)		9.1
5	Ca(58) S(10) P(5)		7.6
6	Fe(68) Ca(13)		5.6
7	Ca(96)	Ca-rich	4.3
	southeast-south		
1	Ca(48) S(43)	CaSO$_4$	36
2	Si(43) Ca(20) Al(13) Fe(11)	aluminosilicates	17
3	Ca(26) S(26) Si(15) Fe(5)		16
4	Fe(43) Ca(19) S(13) Zn(10)		13
5	Ca(63) P(10) Pb(9)		10
6	Fe(79) Ca(10)		7.8
	southwest-west		
1	Ca(46) S(42)	CaSO$_4$	33
2	Si(42) Al(17) Ca(17) Fe(13)	aluminosilicates	15
3	Cl(57) Na(22) Ca(15)	NaCl-CaCO$_3$	14
4	Ca(83)	Ca-rich	13
5	Ca(31) S(15) Si(14) Fe(13)		12
6	Cl(32) Ca(28) S(17) Na(11)	NaCl-CaSO$_4$	7.4
7	Fe(73) Ca(13)		6.4
	northwest		
1	Cl(42) Na(36) Ca(12) S(8)	NaCl-CaSO$_4$	48
2	Ca(51) S(36)	CaSO$_4$	31
3	Ca(47) Fe(15) Si(14) Mg(8)		15
4	Si(47) Al(27) Ca(17) Fe(9)	aluminosilicates	6.3
	local		
1	Ca(48) S(43)	CaSO$_4$	41
2	Si(42) Al(19) Ca(17) Fe(12)	aluminosilicates	18
3	Ca(21) S(20) Na(18) Cl(6)		14
4	Ca(51) P(11) Pb(6)		11
5	Ca(90)	Ca-rich	9.2
6	Fe(77) Ca(13)		7.7

From S. Hoornaert et al. Gypsum and Other Ca-Rich Aerosol Particles above the North Sea, 1996, 1517. With permission of the American Chemical Society, Washington, D.C.

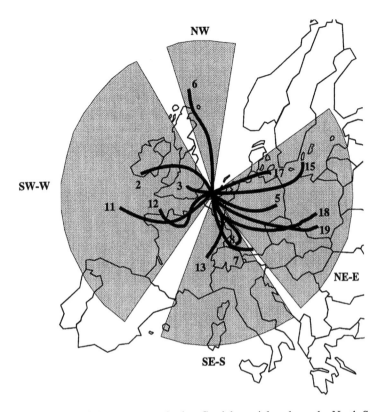

FIGURE 12.5 Identification of the gypsum and other Ca-rich particles above the North Sea: Classification of the different sampling flights into five sectors (the local sector is not shown). The indicated lines are the corresponding 1000 hPa air mass trajectories for air masses arriving at the sampling site. (From Hoornaert, S., et al. Gypsum and other Ca-Rich Aerosol Particles above the North Sea, 1996, 1516. With permission of the American Chemical Society, 1155 16th St. N.W., Washington, D.C.)

45,000 individual North Sea aerosol particles indicated that 11% of these particles contained significant concentrations of one or more of the heavy metals Cr, Pb, and Zn (Van Malderen et al., 1992). Different particle types could be distinguished by performing HCA, NHCA, and Fuzzy clustering techniques, and a clear correlation between the heavy metal aerosol abundance and the origin of the air masses appeared. The highest values for Cr-, Pb-, and Zn-bearing particles were always encountered under southeastern wind directions. The contribution of continental air to the Zn- and Pb-containing particle abundance was 50 times larger than that provided by a marine influence. For Cr, on the other hand, abundances found in the marine sector appeared to be one-third of the values for the continental sectors, suggesting a rather undefined marine source. According to the authors, the recycling of previously deposited material by reinjection into the atmosphere by sea spray could be responsible for these Cr values.

Aerosol particles with diameters above 1 µm are classified as giant aerosols. Compared to the condensation mode particles the amount of these particles in the lower troposphere is small, but their contribution to the atmospheric composition is of great importance. Giant aerosol particles, collected using an impactor rod on top of an aircraft above the Southern Bight of the North Sea, were analyzed by EPXMA and processed by multivariate techniques (Van Malderen et al., 1992). Four different aerosol sources could be distinguished by PFA: aluminosilicate dispersal, combustion processes, industrial activity, and a marine source. Cluster analysis enables classification of the analyzed particles and reveals clear differences between marine and continental air masses. The size distributions of the main particle types could be well fitted by a log-normal distribution with an average diameter of 3 μm. A bimodal size distribution with average size maxima at 4 and 15 μm was found only in the case of aluminosilicate particles. Using the SPM facilities of the Institute for Reference Materials and Measurements (IRMM) in Belgium and of the Lund Institute of Technology in Sweden, the major, minor, and trace elements, with atomic number >15, in individual giant North Sea aerosols were determined by micro-PIXE (Injuk et al., 1993, 1994). Ti, V, and Cr could be detected under the prevailing experimental conditions down to 50 fg. Three different aerosol types could be distinguished: sea salt particles, sea salt combined with high contents of S, K, and Ca, and particles rich in Ti, Cr, Fe, and Ni. The lateral elemental distribution within different giant North Sea aerosol particles was characterized with a constant resolution of 3 μm throughout the depth of the particles. Based on these elemental maps, the individual giant particles could be identified as agglomerates of several large particles.

Xhoffer et al. (1992) studied the composition of individual particles in the surface microlayer and underlying seawater of the North Sea. A comparison was made with the atmospheric and riverine particle data.

Aerosol particles from the Bermuda marine boundary layer were characterized by electron microscopic single particle analysis, in combination with bulk sample analysis using induced neutron activation analysis (Anderson et al., 1996). The results showed the aerosol to be a complex and varying mixture of continental dust, a variety of several S-containing particle types, pure and transformed sea salt, and anthropogenic particle types. Air masses coming from North Africa supplied the highest concentrations of silicate particles and associated types, such as TiO_2 and $CaCO_3$, while the majority of S-bearing particle types were transported from North America. The major associations of the particle types and bulk-sample elements were acquired by PCA. Individual-particle parameters, such as the degree of surface coating or agglomeration of silicates with sulfur; the relative concentrations of silicate types and their corresponding particle sizes within the crustal particle group, appeared, moreover, to depend upon the transport direction.

Individual aerosol particles from the Equatorial Pacific were characterized by Pósfai et al. (1994) using a combination of transmission electron microscopy (TEM) imaging, selected-area electron diffraction (SAED) and energy dispersive X-ray spectrometry (EDS). Sea salt aggregates seemed to be mainly composed of NaCl and Na-Ca sulfate crystals with minor Mg and K. Effects of the relative humidity may be reflected in the variations in composition of the sea salt aggregates. The collected aerosol samples also contained different types of submicrometer, S-bearing particles, which are presumably ammonium sulfates or acidic sulfates. Diatom fragments were often detected in this marine aerosol, and both kaolinite and rutile could be identified as the most abundant crustal particles present. Comparing these results with data obtained from microprobe analysis on individual particles collected simultaneously, a correlation between the change in particle properties and the long-range transport of air masses was found. Giant aerosol particles (>1 μm), which were collected in the same region by Mouri et al. (1995), appeared to be mainly composed of sulfur and may have been formed in the marine boundary layer or transported from the free troposphere. Several modified sea salt particles could also be identified.

Concerning the structure of marine particles, recent laboratory investigations by Cheng (1993) revealed the existence of a shell structure composed of chlorides ($MgCl_2$ and KCl) on the surface of marine aerosol particles. Moreover, the majority of seawater drops, which evaporate and become salt saturated during free fall in the air (RH<60%), seemed to change phase to produce hollow sea salt particles (Cheng et al., 1988). These results were obtained by using a polarizing microscope and an SEM–EDX instrument. The shell structure is valuable for understanding nucleation processes in the atmosphere, since the presence of the chloride film provides a highly hygroscopic surface, which already initiates the condensation of water vapor at an RH of 40%. Thermodynamic analysis by Ge et al. (submitted) showed, moreover, that particles dried from multicomponent aqueous aerosols do not have a homogeneous chemical morphology except at the eutonic point. (The simultaneous precipitation of two salts, at a certain relative humidity, causes the formation of a mixture of solid phases; the corresponding aqueous phase composition is called the eutonic composition.) For the examination of the surface morphology and chemical composition of particles dried from KCl/NaCl, KCl/KI, and $(NH_4)_2SO_4$/NH_4NO_3 mixed solutions at different mole ratios, rapid single-particle-mass spectrometry was used.

Direct evidence of the formation of nitrate on marine particles, which has been frequently stated by different scientists (Bruynseels et al., 1988; Ottley and Harrison, 1992), was provided by Mamane and Gottlieb (1992). They studied the heterogeneous reactions of NO_2 and HNO_3 on individual sea salt and mineral particles under controlled conditions inside a Teflon® reaction chamber by bulk and individual particle analysis. Different particles did react with nitrogen oxides and nitric acid to form nitrates. The formation of these nitrates seemed to depend on the presence of UV radiation. Microscopic investigations associated with a specific microspot technique (Mamane and Pueschel, 1980) proved that in 50% of the studied soil dust particles, mixed nitrate particles were formed upon reaction with NO_2 and HNO_3 and that both gases reacted with 95% of all sea salt particles to produce a surface coating of nitrates. Based on these results, the heterogeneous nitrate formation could be classified as an important removal process for NO_x under solar radiation in the ambient atmosphere. In this context, coarse nitrate-containing aerosol particles collected in a coastal city of Japan were studied by STEM–EDX (Wu and Okada, 1994). The visual identification of nitrate-containing particles was possible by using a vapor-deposited thin film method, upon which they are transformed into needle-shaped crystals. Compositional information was acquired by X-ray analysis. Nitrate was observed in the coarse particle fraction of all samples and chemical modification of sea salt by nitric acid seemed to occur frequently in this coastal urban area. As indicated by the measurements of Mamane and Gottlieb (1992), it was discovered that the dust particles, in this case originating from Asian deserts and loess areas, can be considered important sources of nitrate in urban areas.

In an effort to quantitatively analyze individual salt particles in the smallest size fraction and determine the low Z-elemental content, Xhoffer et al. (1995) performed serial and parallel EELS on standard test aerosols of three inorganic salts, NaCl, $(NH_4)_2SO_4$, and KNO_3. It was found that within less than one minute, radiation damage by the electron beam occurred for all standards. Damage reduction was obtained by liquid nitrogen cooling of the sample holder and a decreasing of the beam current with a factor 10 to 50. Even in the latter case, element losses and structure reorganization could not be avoided. Xhoffer concluded that beam-sensitive samples should be analyzed by PEELS, and only qualitative information can be achieved. Quantitative data on individual aerosol particles, in addition to chemical information, can also be obtained by micro-Raman spectroscopy. In this case, however, the Raman scattering cross sections of all compounds in question should be determined. An experimental approach for this type of measurement was described by Fung and Tang (1991). A very good linearity between the scattering intensity ratio and the molar mixing ratio was found. Sodium nitrate, ammonium nitrate, sodium sulfate and ammonium sulfate aerosols were examined, and their Raman cross sections relative to that of sodium nitrate were identified with uncertainties of less than 10%. Nevertheless, the quantitative interpretation of Raman spectra can become very difficult when morphological and structural resonances, associated with elastic and inelastic scattering, interfere with compositional spectral peaks. Buehler et al. (1991) used Raman spectroscopy in an effort to identify the composition of multicomponent microdroplets by studying the effects of such resonances and microdroplet size on the Raman signal. They found that the Raman signal from spherical microparticles is influenced by the input structural resonances of the incident radiation wavelength. By increasing the signal integration time, the intensity fluctuations, caused by resonances, can be decreased, offering the possibility for the identification of compositional changes. The different resonance Raman scattering effects were also investigated by Fung and Tang (1992), and in another publication, the chemical characterization of both crystalline particles and droplets by resonance Raman spectroscopy was accomplished (Fung and Tang, 1992). Since sulfates and nitrates can be considered major components in atmospheric particles, more specifically in marine aerosols, Fung et al. (1994) determined the detection limits of both components in aerosol particles using the charge coupled device (CCD) detector for Raman photon detection. Inorganic sulfates and nitrates could be traced down to a few picograms in solid particles and to 2.5 millimolar in solution droplets. Sensitivity measurements revealed, moreover, that spontaneous Raman scattering is at least 20 to 40 times more sensitive than SRS. The large signals produced by SRS at high concentrations diminish very rapidly with decreasing concentrations.

12.3.3 Urban Aerosols

In this field, several studies have focused on highly industrialized countries, and only limited data are available from developing countries. Aerosols from the city of Khartoum in the sub-Saharan region in Sudan were investigated using several bulk techniques and automated EPXMA for single particle analysis (Eltayeb et al., 1993). Most of the Khartoum aerosol particles were identified as soil dust and differentiated into aluminosilicates and smaller amounts of quartz and $CaCO_3$. In comparison with literature data on urban aerosol composition, the concentrations of S, Cl, Zn, Br, and Pb in the Khartoum dust particles, relative to Al and crustal rock, are the lowest values ever published. The airborne levels of Br, Pb, and crustal elements seemed to be controlled by traffic emissions. The combination of bulk and single particle analysis was also performed by Cornille et al. (1990) to study both the coarse (>2 μm) and fine (<2 μm) aerosol fraction collected in the Zabadani valley near Damascus, Syria. The results indicated desert soil dispersion to be the major particle source (90% of the total suspended particulate material) in this region, in addition to a minor anthropogenic input. Consequently, the crustal element concentrations were comparable to those in urban and industrial areas and anthropogenic elements were clearly less abundant in this arid region. Based on single particle analysis, the soil dust component could be subdivided into two major components, namely one with a shale-like composition and the other with a limestone composition.

Aerosol sampling 4 km west of Santiago de Chile, followed by automated EPXMA, revealed eight different particle types in both the fine and coarse aerosol mode (Rojas et al., 1990). The coarse mode was mainly dominated by soil dust particles, while anthropogenic particles represented the fine particle mode. Particles rich in S could not be classified into one particle type, but they seemed to be present in six out of eight particle types. LMMS (laser microprobe mass spectrometry) studies indicated that the sea spray particles, which are transported to the city, undergo several transformations and a sulfur enrichment. Using this technique, carbon-rich particles, probably produced by fossil-fuel combustion, could also be identified.

Katrinak et al. (1995) reported on the composition of individual particle types from Phoenix, Arizona. The aerosol in this region consisted mainly of mineral-rich particles in the coarse fraction (crustal source), while the fine fraction was dominated by zero-count particles (elements with Z < 11) and to a lesser extent by mineral-rich particles. The zero-count particles can probably be classified as carbonaceous particles produced by anthropogenic sources, such as the emissions of diesel motor vehicles. Compared with individual particle analysis data obtained 7 to 10 years ago, compositional variations appeared in the Phoenix aerosol. The contribution of S-bearing particles to the fine fraction declined from 30% in 1980 to 17% in 1989 to 1990, while the decrease in Pb concentration probably results from a reduction in the use of leaded gasoline. Due to the recent EEC legislation, which controls the emission of Pb through car exhaust, the total particulate Pb concentration in Antwerp, Belgium, also decreased from 110 ng/m^3 in July 1986 to 74 ng/m^3 in March 1987. Based on an EPXMA particle-induced X-ray emission (PIXE) laser microprobe mass spectrometry study (Van Borm et al., 1990), most of the Pb-containing particles collected near the city of Antwerp were identified in the fine fraction of <1 μm. Pb-containing particles can be subdivided into five main classes. The Pb-sulfates and Pb-halides produced by car exhaust accounted for up to 67% and 28% by mass of the total Pb-containing particles, respectively. The results also suggested that the individual Pb-rich particles are not completely converted to pure Pb-sulfate particles, upon reaction with ammonium sulfate aerosols present in the urban atmosphere. Detection of the ammonium compounds by LMMS revealed that ammonium sulfate coatings were present on nearly all Pb-containing particles. Cl and Br were found to be completely removed from the lead sulfate particles. Wonders et al. (1996) reported on the LMMS analysis of Pb-containing aerosols collected both in the city of Wageningen and at the downwind side of a highway at Ede, in The Netherlands. Most particles were identified as mixed salts of ammonium nitrate and ammonium sulfate and showed a variable content of metals, especially in the case of V and Pb. Aluminosilicate particles, which were collected near the city of Antwerp during another study, showed a vanadium–chromium enrichment in the surface layer, suggesting these particles to be fly ash aerosols (Van Grieken et al., in press). The in-depth analysis of the individual particles sampled on an index grid, which was connected to a freshly sliced indium chip as substrate, was performed by SIMS with a cesium primary-ion beam. This sample preparation method enables a quick location of the particles of interest and prevents loss of surface information due to beam sputtering before the actual analysis.

Depth-resolved analysis was also accomplished on submicrometer aerosol particles (0.3 to 0.8 μm) collected near a busy highway in Germany by a combination of TOF–SIMS and secondary neutral mass spectrometry (SNMS) (Goschnick, 1993). This study revealed that the majority of particles consisted of an inorganic core coated with a 200 nm organic layer. Ammonium sulfate, formed in the atmosphere by reaction of ammonia with sulfuric acid, was identified as the most abundant component inside the aerosol particles. Soot agglomeration, from car exhaust onto the particles' surface, could be responsible for the aromatic organic coating. Preceding this investigation, depth-resolved analyses of salt samples were performed by SNMS and SIMS to evaluate the possibilities of using these techniques in the field of atmospheric microparticles (Fichtner et al., 1991). Chlorides, carbonates, nitrates, and sulfates were selected because of their abundance in environmental materials. Further SNMS investigation on aerosol particles from Karlsruhe (Bentz et al., 1995) made it possible to distinguish between two classes of particles. The first class consisted of submicrometer particles with a 15 nm deep ammonium sulfate coating on a carbon-rich core, while the coarse particles in the second class showed a double-layer system enclosing a geogenic core. A new internal calibration procedure was tested and found to be suitable for the depth-scaling of complex environmental material. TOF–SIMS was used to investigate in detail the carbonaceous compounds of these particles (Bentz et al., 1995).

In a study by Orlic et al. (1994), it was shown that information on the matrix composition and the effective thickness of aerosol particles sampled at the campus of the National University of Singapore can be acquired by RBS. The combination with STIM and PIXE can also reveal the identification and characterization of the size and shape of the particles and their minor and trace elemental composition, respectively. The optimal PIXE sensitivity requires a collection medium for aerosol particles which is both very thin and clean; this could not be found in the different commercially available foils that were tested. Thin pioloform films, made in the university laboratory, appeared to be the most suitable. Minimum detection limits depend strongly on the particle's matrix and were observed to be between 50 and 150 ppm for elements between Ca and Zn. Under the same experimental conditions, the minimum detectable mass was as low as 2 fg.

Fumes, arising from iron and steel industries, are considered the major source of airborne particles containing metal oxides which may be carcinogenic. The toxicity of these metals seemed to be related to their chemical oxidation state. LMMS analysis on particles emitted as dust by the steel industry revealed a correlation between the Cr oxidation state and the particle size. Particles smaller than 1.1 μm were composed of P and Na matrices, whereas large particles (>6 μm) mainly consisted of Ca. Chromium was found to be present in the hexavalent state for both size fractions. Trivalent Cr could only be detected in particles of intermediate size (Poitevin et al., 1992; Hachimi et al., 1993). Chromium speciation by rapid single-particle mass spectrometry is, at the moment, limited to well-defined particles whose spectra can be compared to standard particles having similar size, morphology, and chemical composition (Neubauer et al., 1995).

In combination with nuclear spectroscopy, SEM–EDX analysis can also be used to study radioactive particles. The composition of a Ba hot particle emitted during the Chernobyl accident was determined in this way by Vapirev et al. (1994).

12.3.4 INDOOR AEROSOLS

Indoor air quality research originally focused on determining the effects on public health of the atmospheres inside offices, laboratories, and residences. One of the recent publications in this field discusses the composition of floor dust Pb-bearing particles of 16 residences in the London borough of Richmond, England (Hunt et al., 1993). The assignment of the SEM–EDX analyzed Pb-rich particles to different particle types (road dust, paint) was based on a classification scheme (Hunt et al., 1992) obtained by the analysis of different types of Pb-source particles. In the size range 0 to 64 μm, paint, road dust, and garden soil were identified as the major contributors of Pb-rich particles to floor dust. Pb-containing dust particles with diameters between 64 and 100 μm appeared to be paint. No relation seemed to exist between the age of the houses and the contribution of the major sources. In another study, Gulson et al. (1995) tried to identify the sources of lead in children's blood in an urban environment in Sydney by investigating particulates from three houses that were previously decontaminated by their owners. Based on SEM–EDX and lead isotopic data, it became clear that these houses had been recontaminated by lead paint from adjoining dwellings with deteriorating paint, as well as from unknown sources. Lead paint reached the inside of the houses both by airborne and mechanical transport.

To assess the hygiene hazards in order to improve the health conditions at a shipyard factory in Spain, metal particles coming from welding fumes were investigated by a combination of flame atomic absorption spectroscopy (AAS) and optical and electron microscopy (Bellido et al., 1996). Bulk analysis with AAS showed that the extraction systems are not sufficient, since for an abundant group of samples, the iron and manganese concentrations were above the TLV–TWA values. From the microscopic data, it could be deduced that these particles may reach the gas-exchange region of the lung due to their size (<1 μm) but should not cause microlesions because of their round shape.

The specific harmfulness of coal mine dust to the health of coal workers, on the other hand, has often been related to the quartz content in the respirable fraction of such materials. Because no consistent correlation was found between the experimental toxicity and the total quartz content determined by bulk analysis of coal mine dust, a single particle approach was demanded. LMMS was used successfully by Tourmann and Kaufmann (1993, 1992) to reveal the heterogeneous distribution of potentially toxic constituents such as silica or siderite. Three quartz-rich coal mine dusts, with significant differences in toxicity, but no correlation with conventional data, have been investigated. The acquired analytical data seemed to be, in contrast to bulk analysis, in agreement with toxicity data (Poitevin et al., 1992; Tourmann and Kaufmann, 1993,1992). Further LMMS studies (Tourmann and Kaufmann, 1994) on natural dusts and authentic coal mine dusts showed that only a minute fraction of "pure" quartz (about 1% of the total abundance) may be detected in coal mine dusts. Most of the quartz surface in such highly heterogeneous particulate matter appeared to be covered by aluminosilicates. The LMMS investigation has shown that fibrogenicity of coal dust samples was not correlated with the incidence of pure quartz particles but with the incidence of Fe, Mg(Ca)-containing particles. These particles seem to play a major role in the harm done by coal mine dusts.

In general, and as indicated in the few previously discussed papers, it becomes clear that indoor air pollution can appear as a consequence of both indoor and outdoor factors. However, in addition to serious health effects, aerosol particles can also cause chemical damage or soiling of surfaces by deposition of particulate material or absorption of gases. This aspect grows in importance when we are dealing, for example, with museum atmospheres. In this framework, several campaigns were organized at the Correr Museum in Venice, Italy, to characterize the individual airborne particles inside and outside the museum, because these particles might be responsible for staining famous paintings (De Bock et al., in press). SEM–EDX data processed by multivariate techniques indicated the presence of six to eight different particle types indoors, of which the Ca-rich particles together with the aluminosilicates and organic material appeared to be the most important. Factor analysis produced similar results. A comparison between the in- and outdoor aerosol composition suggested the existence of an indoor Ca-rich particle source, probably the deterioration of the interior wall plaster. The homogeneity study of giant aerosol particles of >8 μm, performed by X-ray mapping, revealed that these particles are heterogeneous and consist mainly of Ca and Si, with smaller particles attached to their surfaces. An illustration of such an X-ray elemental mapping can be found in Figure 12.6a-c. The heterogeneity of these particles was confirmed by micro-PIXE analysis, as illustrated in Figure 12.7, on a limited number of particles. STEM–EDX was used for the analysis of the indoor submicrometer aerosol particles. To prevent particle evaporation on the Formvar Cu grids, the sample holder was cooled with liquid nitrogen. In this way, it became possible to analyze individual aerosol particles of >100 nm. Further research on this topic will presumably provide a way to obtain compositional information on individual particles down to 50 nm.

Until a few years ago, asbestos fibers were frequently used in building materials to improve their strength and make them fire resistant. These days the use of asbestos is prohibited in most countries as a consequence of their threat to human health. EELS in combination with ESI revealed some very encouraging results in a study on the modification of the surface reactivity of chrysotile asbestos fibers (Bergamns et al., 1994). Surface modification, upon reaction with different products could lead to thermally and chemically stable fibers. In this way, their cytotoxic and hemolytic activity could be reduced or could even completely disappear. Carbon maps obtained from the organosilane-coated asbestos fibers showed a heterogeneous distribution of the coating over the fibers, and some of the fibers seemed even to be unaffected by the treatment. Better results were found for $TiCl_3$-modified fibers. Evidence for the chemical reaction between $TiCl_3$ and the chrysotile fibers was discovered in the pre-edge of the oxygen peak in the EEL spectrum, which indicates a bond between oxygen and titanium. Lateral elemental Ti maps of cross-sectioned, treated fibers revealed the presence of titanium inside, as well as on the external surface of the analyzed fibers leading to a complete encapsulation of the material. Studying the distribution of the $TiCl_3$ coating by SIMS after EELS analysis showed clearly both via Ti^+ and TiO^+ distribution images, the uniform reaction of Ti over the selected area of the chrysotile surface. A second indication of a full reaction of $TiCl_3$ with the asbestos fibers was found in the absence of $TiCl^-$ ions. To examine the microstructure of the individual fibers, a higher resolution ion beam would have been necessary. Improvement of the spatial resolution of the ion beam down to less than 0.1 μm can be obtained by the use of a gallium-focused ion beam (Ga–FIB). Moreover, due to the high current density of this FIB, the three-dimensional analysis of microstructures becomes possible. Nihei et al. (1991) used this submicron SIMS technique to analyze both fine particles trapped in a thin film multilayer device, as well as the inorganic microcapsules of TiO_2 powder, encapsulated in a silicon dioxide thin film.

The total composition and depth distribution of the components of combustion particles generated in household fires were investigated by Bentz et al. (1994) using SNMS and TOF–SIMS. This type of particle appeared to have a shell structure with organic halogen compounds and a PAH-content of about 1 ppm in the surface layer. Cl and the organic hydrogen content (H_C) were enriched in the surface region (100 nm), which was found to have an aliphatic character. The concentration of H_C decreased toward the core. Obviously, the condensation of low-volatile H_C with Cl-compounds or a surface reaction with gaseous Cl or HCl is responsible for the organic surface layer enriched with Cl.

In addition to different on-line LMMS studies of artificially generated aerosols (Marijnissen et al., 1988; McKeown et al., 1991; Kievit et al., 1992; Mansoori et al., 1994; Murray and Russel, 1994; Prather et al., 1994; Nordmeyer and Prather, 1994; Carson et al., 1995; Murphy and Thomson, 1995; Reents et al., 1995; Yang et al., 1996), a report was recently published discussing the analysis of aerosol particles from ambient air by a LAMPAS system (laser mass analysis of particles in the airborne state). This set-up, developed at the Institute of Laser Medicine, University of Düsseldorf,

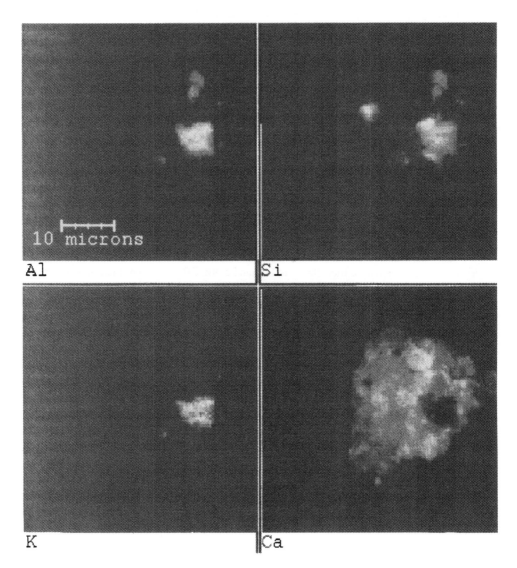

FIGURE 12.6 Heterogeneity study on a giant indoor aerosol particle. (a) SEM–EDX X-ray elemental mappings of Al, Si, K, and Ca in the giant aerosol particle. (b) and (c) Application of an image processing program revealing clearly the association of Ca and Si and the absorption of a few aluminosilicate particles as well as some smaller CaSO$_4$ particles at the particle surface. (From De Bock, L.A., et al. *Microanalysis of museum aerosols to elucidate the soiling of paintings: case of the Correr Museum,* 1996. With permission of the American Chemical Society, Washington, D.C.)

Germany, can detect individual particles with a high repetition rate, a superior mass resolution, which remains constant over the entire mass range, and a very high signal-to-noise ratio. The comparison between the on-line spectrum of an aerosol particle collected from the laboratory room air, showing the typical carbon cluster ion distribution of soot particles, and a candle soot particle analyzed by LMMS is given in Figure 12.8 and shows clearly that the quality of the on-line spectrum is as good as the LMMS off-line spectrum. Although both spectra appeared to be similar, additional information indicating the presence of an aqueous phase surrounding the soot particles can only be identified in the on-line spectrum. The capability of the instrument to distinguish dry and wet particles was previously demonstrated by analysis of aqueous salt droplets (Hinz et al., 1994; Carson et al., 1995; Murphy and Thomson, 1995).

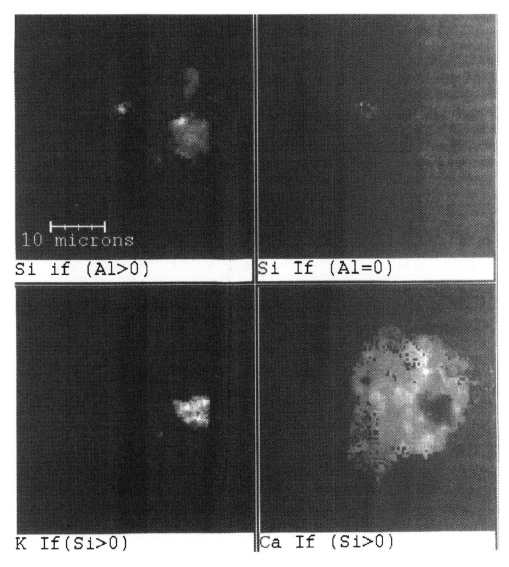

FIGURE 12.6   (continued)

12.3.5 Fly Ash Aerosols

This type of aerosol particle is emitted by high temperature combustion processes in power plants and can often be recognized by its spherical shape. However, the morphology and composition are determined by the burning technology and the coal type. Coal fly ash particles emitted by a Hungarian power station (Sandor et al., 1990) and background aerosols sampled in the middle of the Great Hungarian Plain (Török et al., 1993) were both characterized by single particle analysis using EPXMA. Two unexpected particle groups were detected in the power plant aerosol fraction below 2 μm: Ba-rich particles and As-rich gypsum particles. The first group is probably produced by the cleaning process, employing Ba-containing liquid to reduce the sulfur content of added slag. The As-coating on the second particle group results from the condensation in the cooled stack gas. The composition of the background Hungarian aerosol particles (0.3 to 20 μm), together with recorded air back trajectories, refers to an anthropogenic origin. Another type of fly ash particles, produced by municipal waste combustion, was studied by Sandell et al. (1996) to determine the composition, morphology, and distribution of lead-bearing phases. Knowledge on this matter could

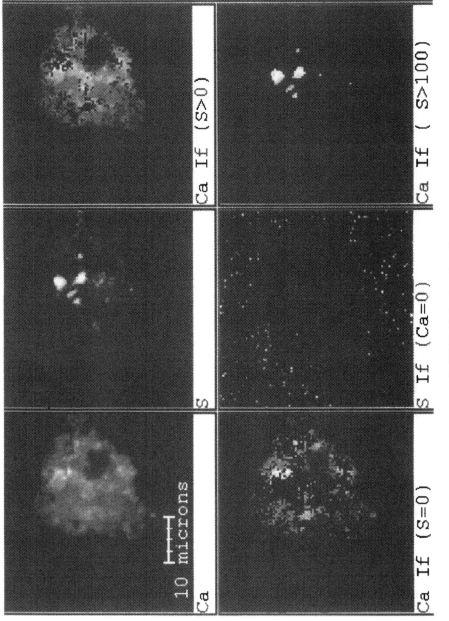

FIGURE 12.6 (continued)

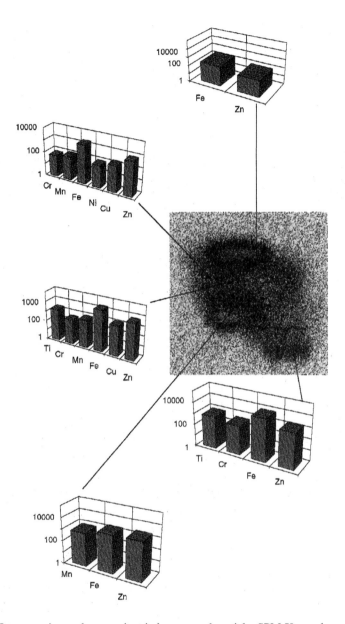

FIGURE 12.7 Heterogeneity study on a giant indoor aerosol particle. SPM X-ray elemental intensity mappings of different elements present in one particle. (From De Bock, L.A., et al., *Microanalysis of Museum Aerosols to Elucidate the Soiling of Paintings: Case of the Correr Museum,* 1996. With permission of the American Chemical Society, Washington, D.C.)

help to improve the process of combustion, including control and transport of heavy metals into the atmosphere. The SEM–EDX/EPXMA data on individual particles indicated the presence of Pb both at the interior and exterior of the particles. At the interior, Pb was found in two forms including Pb-rich inclusions and in lower concentrations in chloride-rich crystals. All Pb-containing phases were marked by other elements such as K, Na, S, Ca, Cl, and Zn. In contrary to previous studies, lead was identified throughout all size fractions not only in the smaller-size ash particles.

Fly ash particles taken from different types of power plants and incinerators have also been analyzed with the μ-SRXRF-module at Brookhaven (Tuniz et al., in press). It was possible to

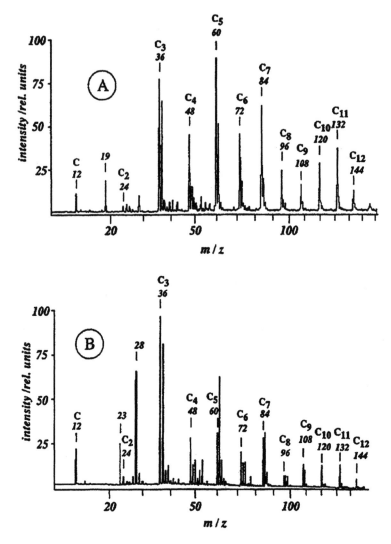

FIGURE 12.8   (a) LAMPAS on-line mass spectrum of a real aerosol particle from ambient laboratory air (diameter ~1 μm). The spectrum shows the characteristic soot fragmentation pattern of small-sized air particles. (b) LAMMA 500 off-line mass spectrum of candle soot sampled on a foil for verification. (From Hinz, K.P. et al. Laser-induced mass analysis of single particles in the airborne state, 1994, 2075, in De Bock, L.A., et al. *Microanalysis of Museum Aerosols to Elucidate the Soiling of Paintings: Case of the Correr Museum,* 1996. With permission of the American Chemical Society, Washington, D.C.)

measure the elemental composition of individual fly ash particles with sizes down to a few micrometers and to obtain two-dimensional maps of the elemental distributions in 10 μm sections of large particles. Micrometer and submicrometer analysis and mappings of individual particles, however, will only be possible in the third-generation synchrotrons.

The spatial trace element distribution in single coal fly ash particles was studied by Jaksic et al. (1991) using the Oxford Scanning Proton Microprobe facilities. Micro-PIXE and RBS measurements provided information on the minor and trace elemental composition, matrix components, and the thickness of the particles, and clear evidence for trace element enrichment at the particle surface. The aluminosilicate fly ash particle analysis was, however, complicated by the inhomogeneous character of the particles. Another micro-PIXE/RBS study on individual fly ash particles, collected at the inlet and outlet ports of the electrostatic precipitator of a coal-fired power station, was performed by Caridi et al. (1993). Their investigation was focused on 1 µm particles, because, for this size particle, the collection efficiency of the electrostatic precipitator is very poor. The importance of the chemical composition effect on the electrostatic precipitation is probably demonstrated by the difference in particle types characterized at the inlet and outlet. Further SPM research on this specific topic, performed by Cereda et al. (1995, 1995), revealed direct evidence of the increasing concentration trend of trace elements with decreasing particle size. Particularly in the case of S, enhanced concentrations between 1600 and 2300 ppm were detected at the electrofilter outlet compared to the S inlet values of 500 to 1300 ppm. This result was accounted for by the condensation of sulfates on the particle surface when particles are passing through the electrofilter. The electrostatic precipitator also acts selectively with the particle composition: particles with higher concentrations of minor and trace elements are less efficiently collected and more likely to escape the electrofilter. A nonuniform distribution of minor (P, Ca, Ti, Fe) and trace elements (S, Mn, Ni, Cu, Zn, Ga) was found for the particles collected at the precipitator inlet and indicates a poor degree of coalescence upon combustion for the first element group and surface segregation and evaporation-condensation mechanisms for the latter one.

The effects of areal density variations on the elemental intensity maps of individual particles obtained by micro-PIXE were evaluated by comparison of theoretically produced X-ray yield maps of spherical particles with experimental micro-PIXE and STIM intensity maps of fly ash particles (Bogdanovic et al., 1994). The fly ash particles were collected from electrostatic precipitators in Croatia and Greece. STIM was demonstrated to be a very useful technique that might prevent misunderstanding of elemental intensity images obtained by PIXE. Recently, two methods for quantitative PIXE analysis of individual particles in the size range of 5 to 10 µm were also represented (Jaksic et al., 1993).

Fly ash aerosols produced during microdrop fuel oil combustion consist almost entirely of cenospheres, hollow, spherical formations with a spongy structure and a size range of 10 to 200 µm (Smith, 1962). The cenospheres are composed of an amorphous component rich in C, S, and Si, in which also Fe and Al are often present. Microcrystalline phases of sulfates, oxides, and pure metallic elements can be discovered inside the cenosphere hollows. By combining the SEM–EDX, TEM–EDX, and EELS techniques, an effort was made to obtain the physicochemical and structural characterization of the crystalline phases in the cenospheres because of their importance from the environmental and toxicological points of view (Paoletti et al., 1994). SEM–EDX studies revealed for the granulometric spectrum of the cenospheres a log-normal diameter distribution with a geometric mean of 47 µm. The diameter of the holes, irregularly distributed over the surface, ranged from less than 1 to 10 µm or more, and the spongy structure seemed to also be present inside the holes. Inside the cenosphere cavity, a complex structure with numerous microparticles was found. Ten percent of the original cenosphere mass appeared to be composed of organic and/or inorganic oxidable components. The structural and compositional analysis of the microcrystalline phases performed by electron diffraction, EDXS, and EELS in TEM showed the presence of sulfates and metal oxides. The oxide phases can be classified into three groups: V/Fe, Fe/Ni, and V/Ni. A crystalline structure was found for the second and third group, while the first one seemed to be amorphous. Fe was sometimes replaced by different trivalent ions, and Ni by different divalent ions.

12.3.6 BLACK CARBON AEROSOLS

Due to the development of X-ray detectors with a removable or ultra thin polymer window, the detection of light elements such as N, O, and C becomes possible by SEM–EDX analysis. The first applications in this field have only been published recently. Fruhstorfer and Niessner (1994) discuss the simultaneous chemical and dimensional characterization of aerosol particles in the size range of 0.05 to 10 µm by windowless SEM–EDX detection, and this with the help of particle standards. They concluded that by working under these conditions, all elements with Z > 4 can be determined, and 80% of the total aerosol particles (soot particles included) can be identified. Soot particles appeared to have a trimodal size distribution and, based on high-resolution SEM micrographs, the associated morphology of each size class could be determined. Airborne carbonaceous particulate matter collected on top of St. Paul's Cathedral in London was investigated under similar conditions by Hamilton et al. (1994). More than 50% of the particles contained some carbon, but the highest carbon contents were found in particles smaller than 1 μm. These results are consistent with studies by Ricardo (1987). In a study focused on the mixing characteristics and water content of aerosol particles from Los Angeles and the Grand Canyon, it became clear that aerosols from both locations were, to some extent, externally mixed and showed different hygroscopic characteristics for a given size (Zhang and McMurry, 1993). Analysis performed using a SEM equipped with a thin window X-ray detector revealed that about 60 to 70% of the particulate carbon is associated with the less hygroscopic particles. The more hygroscopic particles contain sulfates, nitrates, and some carbon. These results were confirmed by a study on aerosol particles collected in Minneapolis and analyzed by STEM equipped with a similar detector (McMurry et al., 1996). The less hygroscopic particles did not grow to any measurable extent when humidified to approximately 90% relative humidity. Carbon was detected as the main element in these particles, among which the shape differed from chain agglomerates to spheres, flakes, and totally irregular particles. The more hygroscopic particles appeared to be liquid spheres consisting of sulfur, oxygen, and sometimes Na and K. Carbon was sometimes found in the more hygroscopic particles.

The ultrafine atmospheric particle fraction of an ambient aerosol sample, studied by Maynard and Brown using PEELS (Maynard and Brown, 1992), seemed to be dominated by carbon-rich particles. Quantitative information became available on atomic concentrations down to 4%, with an accuracy better than 20%. Both Katrinak et al. (1992) and Xhoffer (1993) investigated individual carbonaceous urban aerosol particles from Phoenix, Arizona, USA. The particles incorporated 10 to 100 aggregated spheres and showed diameters below 2 μm. The energy loss near edge fine structure (ELNEFS) of the C–K edge reflected, in both studies (Katrinak et al., 1992; Xhoffer, 1993), the presence of a mixture of graphitic and amorphous carbon within and among the individual aggregate particles. The graphitic structures were assumed to be part of the primary spheres and the amorphous domains of the condensed hydrocarbons. Surface coatings containing sulfur and nitrogen appeared on the carbonaceous aggregates collected during the summer months. The deposition of sulfates and nitrates from the atmosphere, as end products of photochemical reactions, could probably be responsible for these kinds of coatings. Visual evidence of the sulfur coating was obtained by ESI (Xhoffer, 1993; Bergmans et al., 1994). The carbon structural variations were found to be similar for both the coated and uncoated carbonaceous particles, suggesting comparable pre-emission sources.

Conventional EM analysis (transmission and scanning) (Parugo et al., 1992) of aerosol samples collected by an aircraft at different altitudes in the Kuwait oil fire plumes in May 1991 revealed an increase in the small sulfate particle concentration at higher levels and at larger distances from the fire source. This suggested the oxidation of SO_2 gas followed by homogeneous nucleation. In those areas, large soot, salt, and dust particles were converted into cloud condensation nuclei (CCNs) due to an S-coating. These CCNs could play an important role in the formation of clouds, haze, smog, and fog. Air trajectory studies indicated that aerosols, especially black carbon particles which have a tendency in the absence of rain showers to remain airborne for a long time, could have reached China. Whether the severe rainfall in China at the end of May and the beginning of June 1991 was related to the fire plumes in Kuwait requires further research.

Gaudichet et al. (1995) investigated the composition of aerosol particles emitted during savannah fires at Lamto, Ivory Coast. Single particle STEM–EDX analysis pointed out the presence of submicrometer soot particles, salt condensates, vegetation relicts, and soil-derived particles inside the plumes of biomass burning. The soot particles were always associated with K and, together with their morphology, they can be identified as a specific tracer of the burning of vegetation-derived fuels. The presence of S and/or Cl inside these particles could control their hygroscopic characteristics and influence their role as CCN.

12.3.7 BIOLOGICAL AEROSOLS

Biogenic aerosol particles are known for their effects on air hygiene and health (hay fever, asthma, bacterial infections, etc.), but they seem to also play an important role in cloud physics since they can act as ice nuclei after water uptake. A method to sample and identify all airborne particles above 0.2 μm (bio- and non-bioaerosol fractions) has been described by Matthias-Maser and Jaenicke (1993). An optical light microscope was used for the detection of the coarse bioaerosols after staining with a protein dye which turned them bluish. The small particles were characterized by SEM–EDX based on three criteria: morphology, elemental composition, and behavior during EDX. The biological particles appeared to have a special morphology associated with a special elemental composition, e.g., P and S together with K and Ca, and occasionally showed some instability under the electron beam. In this way, the atmospheric aerosol could be classified into six groups, each representing biological or nonbiological particles. Further investigations revealed the size distributions of the primary biological aerosol particles in an urban/rural influenced region in Germany (Matthias-Maser and Jaenicke, 1995). Aerosol particles sampled during high asthma periods in Brisbane, Queensland, Australia, and analyzed by TEM/SEM–EDX and AAS appeared to be mainly composed of soot and incompletely burned hydrocarbons, crustal matter, spores, and soil bacteria (Glikson et al., 1995). Evaluation and quantification of the contribution from pollen and fungal spores shows that fungal spores dominate the bioaerosol fraction in the 2 to 10 μm range, and their number is very high in Brisbane from the end of April through May to mid-June. However, the contribution of biological material to the total particulate mass reaches only 5 to 10% even at peak periods. Electron microscopy indicated that crustal and motor vehicle emission material can act as carriers for dispersed cytoplasmic allergenic material released from pollen and fungal spores. Direct evidence for S-enrichment due to atmospheric reactions in minerals and spores (0.5 to 10 μm), sampled at Research Triangle Park, North Carolina, USA, was found by Mamane et al. (1992).

Concerning the characterization of S-compounds in individual aerosol particles, Qian and Ishizaka (1993) reported on the characterization of methane sulfonic acid using SEM, transmission electron microscopy, and thin film chemical tests in aerosols sampled over Sakushima Island, Japan, and a new method for the determination of the S-mass content in single aerosol particles using SEM was developed and described by Pardess et al. (1992).

12.4 CONCLUSION

This chapter clearly illustrates how the need to characterize individual particles, understand their formation processes, and estimate their environmental impact has led to the application of many micro- and trace analytical techniques. Microanalysis on the single particle level has provided interesting and almost unique information, which was hitherto inaccessible by bulk analysis. Since each microanalytical technique still has its own drawbacks and no technique is capable of revealing the complete characterization of the sample, a combination of several complementary techniques to solve the problem is still required.

Although mostly environmental applications performed by off-line techniques were discussed in this review, there exists a tendency now toward particle analysis in the on-line configuration, in which the particles are collected directly into the instrument. Problems concerning sample preparation, such as contamination or decomposition upon sample–surface interaction as well as loss of components due to vacuum conditions (e.g., SEM–EDX/EPXMA and SPM analysis), are eliminated. On-line analysis will consequently be able to provide time information on each particle since particles from different events are distinguishable. Mass spectrometry and atomic emission spectroscopy are, to date, the only two methods capable of on-line analysis of single particles. Since these techniques have only been developed recently, their applications are still being refined.

ACKNOWLEDGMENTS

This work was partially financed by the Belgian State–Prime Minister's Services for Scientific, Technical, and Cultural Affairs, in the framework of the Impulse Programme on Marine Sciences (under Contract MS/06/050). L.D.B. acknowledges financial support from the Instituut voor de bevordering van het Wetenschappelijk–Technologisch Onderzoek in de Industrie (IWT).

LIST OF ACRONYMS

AES	Atomic Emission Spectroscopy
EELS	Electron Energy Loss Spectrometry
EPXMA	Electron Probe X-ray Microanalysis
FAST	Forward Angle Scattering Technique
FTIR	Fourier Transform Infrared Spectroscopy
FT-ICR-LMMS	Fourier Transform Laser Microprobe Mass Spectrometry
INNA	Induced Neutron Activation Analysis
LAMMS/LMMS	Laser Microprobe Mass Spectrometry
LAMPAS	Laser Mass Analysis of Particles in the Airborne State
PEELS	Parallel Electron Energy Loss Spectrometry
μ-PIXE	Micro-Proton Induced X-ray Emission
PIGE	Particle Induced Gamma Emission
RBS	Rutherford Backscattering
RSMS	Rapid Single Particle Mass Spectrometry
SAM	Scanning Auger Microscopy
SEM/EDX	Scanning Electron Microscopy/Energy Dispersive X-ray Analysis
SIMS	Secondary Ion Mass Spectrometry
SPM	Scanning Proton Microprobe
SRS	Stimulated Raman Scattering
STEM	Scanning Transmission Electron Microscopy
STIM	Scanning Transmission Ion Microscopy
TEM	Transmission Electron Microscopy
WDX	Wavelength Dispersive X-ray Analysis
XPS	X-ray Photoelectron Spectroscopy

REFERENCES

Anderson, J.R., Buseck, P.R., Patterson, T.L., and Arimoto, R., 1996, *Atmos. Environ.* 30: 319-338.

Armstrong, T.J., 1995, *Microbeam Analysis* 4: 177-200.

Artaxo, P., Van Grieken, R.E., Watt, F., and Jakšic M., 1990 *Proceedings of the Second World Congress on Particle Technology,* Society of Powder Technology, Kyoto, 421-426.

Artaxo, P., Rabello, M. L. C., Watt, F., Grime, G., and Swietlicki, E., 1993, *Nucl. Instrum. Meth. B* 75: 521-525.

Artaxo, P., Rabello, M. L. C., Watt, F., Grime, G., Swietlicki, E., Knox, J., and Hansson, H-C., 1992, *J. Aerosol Sci.* 23: S373-376.

Artaxo, P., Rabello, M.L.C., Maenhaut, W., and Van Grieken, R.E., 1992, *Tellus* 44B: 318.

Begét, E.J., Keskinen, M., and Severin, K., 1993, *Sediment. Geol.* 84: 189-197.

Bellido-Milla, D., Hernandez-Artiga, M.P., Hidalgo de Cisneros, J.L., and Muñoz-Leyva, J.A., 1996, *Appl. Occup. Environ. Hyg.* 10: 921-926.

Bentz, J.W.G., Goschnick, J., Schuricht, J., Ache, H.J., Zehnpfennig, J., and Benninghoven, A., 1995, *Fresenius J. Anal. Chem.* 353: 603-608.

Bentz, J.W.G., Goschnick, J., Schuricht, J., and Ache, H.J., 1995, *Fresenius J. Anal. Chem.* 353: 559-564.

Berghmans, P., Injuk, J., Van Grieken, R.E., and Adams, F., 1994, *Anal. Chim. Acta.* 297: 27-42.

Betz, J.W.G., Fitchtner, M., Goschnick, J., and Ache, H.J., 1994, *Abstracts from Fourth International Aerosol Conference*, August 29-September 2, 1994, Los Angeles, California, Ed. R.C. Flagan, American Association for Aerosol Research, Kemper Meadow Drive, Cincinnati, OH, USA.

Bochert, U.K. and Dannecker, W., 1992, *J. Aerosol Sci.* 23: S417-420.

Bogdanovic, I., Fazinic, S., Jakšic, M., Grime, G.W., and Valkovic, V., 1994, *Nucl. Instrum. and Methods B* 85: 732-735.

Bohsung, J., Arndt, P.A., Jessberger, E.K., Maetz, M., Traxel, K., and Wallianos, A., 1995, *Planet. Space Sci.* 43: 411-428.

Bruynseels, F., Storms, H., and Van Grieken, R.E., 1988, *Atmos. Environ.* 11: 2593-2602.

Buehler, M.F., Allen, T.M., and Davis, E.J., 1991, *J. Colloid Interface Sci.* 146: 79-89.

Caridi, A., Cerede, E., Grime, G.W., Jakšic, M., Braga Marcazzan, G.M., Valkovic, V., and Watt, F., 1993, *Nucl. Instrum. and Methods B* 77: 524-529.

Carson, P. G., Neubauer, K. R., Johnston, M. V., and Wexler, A. S., 1995, *J. Aerosol Sci.* 4: 535-545.

Cereda, E., Marcazzan, G.M.B., Pedretti, M., Grime, G.W., and Baldacci, A., 1995, *Nucl. Instrum. and Methods B* 99: 414-418.

Cereda, E., Marcazzan, G.M.B., Pedretti, M., Grime, G.W., and Baldacci, A., 1995, *Atmos. Environ.* 29: 2323-2329.

Cheng, R.J., 1993, *Proc. of the Twelfth Annual Meeting of the American Association for Aerosol Research*, Oak Brook, Illinois, October 11-15.

Cheng, R.J., Blanchard, D.C., and Cipriano, R.J., 1988, *Atmos. Environ.* 22: 15-25.

Cornille, P., Maenhaut, W., and Pacyna J.M., 1990, *Atmos. Environ.* 24A: 1083-1093.

Dangler, M., Burke, S., Hering, S.V., and Allen, D.T., 1987, *Atmos. Environ.* 21: 1001-1004.

Danilatos, D.G., 1994, *Mikrochim. Acta* 114/115: 143-155.

Davis, E.J., Rassat, S.D., and Foss, W., 1992, *J. Aerosol Sci.* 23: S429-432.

De Bock, L.A., Van Malderen, H., and Van Grieken, R.E., 1994, *Environ. Sci. Technol.,* 28: 1513-1520.

De Bock, L.A., Van Grieken, R.E., Camuffo, D., and Grime, G.W., *Environ. Sci. Technol.,* in press.

Dierck, I., Michaud, D., Wouters, L., and Van Grieken, R.E., 1992, *Environ. Sci. Technol.,* 26: 802-808.

Eltayeb, M.A.H., Xhoffer, C.F., Van Espen, P.J., and Van Grieken, R.E., 1993, *Atmos. Environ.* 27B: 67-76.

Fichtner, M., Lipp, M., Goschnick, J., and Ache, H.J., 1991, *Surf. Interf. Anal.* 17: 151-157.

Fortin, D., Leppard, G.G., and Tessier, A., 1993, *Geochim. Cosmochim. Acta* 57: 4391-4404.

Frame, E.M.S., Takamatsu, Y., and Suzuki, T., 1996, *Spectroscopy* 11: 17-22.

Fruhstorfer P. and Niessner, R., 1994, *Mikrochim. Acta* 113: 239-250.

Fung, K.H. and Tang, I.N., 1991, *Appl. Spectrosc.* 45: 734-737.

Fung, K.H. and Tang, I.N., 1992, *Appl. Spectrosc.* 46: 159-162.

Fung, K.H. and Tang, I.N., 1992, *J. Aerosol Sci.* 23: 301-307.

Fung, K.H., Ire, D.G., and Tang, I.N., 1994, *J. Aerosol Sci.* 25: 479-485.

Gaudichet, A., Echalar, F., Chatenet, B., Quisefit, J.P., Malingre, G., Cachier, H., Buat-Menard, P., Artaxo, P., and Maenhaut, W., 1995, *J. Atmos. Chem.* 22: 19-39.

Ge, Z., Wexler, A.S., and Johnston, M.V., 1996, *J. Coll. Int. Sci.*, submitted.

Glikson, M., Rutherford, S., Simpson, R.W., Mitchell, C.A., and Yago, A., 1995, *Atmos. Environ.* 29: 549-562.

Goschnick, J., Fichtner, M., Lipp, M., Schuricht, J., and Ache, H.J., 1993, *Appl. Surf. Sci.* 70/71: 63-67.

Grasserbauer, M., 1983, *Mikrochim. Acta Part III:* 415-448.

Gulson, B.L., Davis, J.J., and Bawden-Smith, J., 1995, *Sci. Total Environ.* 164: 221-235.

Hachimi, A., Poitevin, E., Krier, G., Muller, J.F., Pironon, J., and Klein, F., 1993, *Analysis* 21: 77-82.

Hamilton, R.S., Kershaw, P.R., Segarra, F., Spears, C.J., and Watt, J.M., 1994, *Sci. Total Environ.* 146/147: 303-308.

Harrington, P., Street, T., Voorhees, K., Radicati di Brozolo, F., and Odom, R.W.P., 1989, *Anal. Chem.* 61: 715-719.

Hinz, K-P, Kaufmann, R., and Spengler B., 1996, *Aerosol Sci. Technol.* 24: 233-242.

Hinz, K-P, Kaufmann, R., and Spengler, B., 1994, *Anal. Chem.* 66: 2071-2076.

Hoornaert, S., Van Malderen, H., and Van Grieken, R.E., 1996, *Environ. Sci. Technol.,* 30: 1515-1520.

Hunt, A., Johnson, D.L., Thornton, I., and Watt, J.M., 1993, *Sci. Total Environ.* 138: 183-206.

Hunt, A., Johnson, D.L., Watt, J.M., and Thorton, I., 1992, *Environ. Sci. Technol.,* 26: 1513-1523.

Injuk, J., Breitenbach, L., Van Grieken, R.E., and Wätjen, U., 1994, *Mikrochim. Acta* 114/115: 313-321.

Injuk, J., Van Malderen, H., Van Grieken, R.E., Swietlicki, E., Knox, J.M., and Schofield, R., 1993, *X-ray Spectrometry* 22: 220-228.

Jakšic, M., Bogdanovic, I., Cereda, E., Fazinic, S., and Valkovic, V., 1993, *Nucl. Instrum. and Methods B* 77: 505-508.

Jakšic, M., Watt, F., and Grime, G.W., 1991, *Nucl. Instrum. and Methods B* 56/57: 699-703.

Jambers W., De Bock, L., and Van Grieken, R.E., 1995, *Analyst* 120: 681-692.

Janssens, K., Vincze, L., Rubio, J., Adams, F., and Bernasconi, G., 1994, *Anal. At. Spectrom.* 9: 151.

Janssens, K. in press.

Katrinak, K.A., Anderson, J.R., and Buseck, P.R., 1995, *Environ. Sci. Technol.*, 29: 321-329.

Katrinak, K.A., Rez, P., and Buseck, P.R., 1992, *Environ. Sci. Technol.*, 26: 1967-1976.

Kellner, R. and Malissa, H., 1989, *Aerosol Sci. Technol.* 10: 397-407.

Kievit, O., Marijnissen, J.C., Verheijen, P.J.T., and Scarlett, B., 1992, *J. Aerosol Sci.* 23: S301-S304.

Leppard, G.G., 1992 in *Environmental Particles* Vol. I, Eds. J. Buffle and H.P. Van Leeuwen, Lewis, Chelsea MI, 231-289.

Lindner, B. and Seydel, U., 1989, in *Microbeam Analysis 1989*, Ed. P.E. Russell, San Francisco Press Inc., San Francisco, 286-292.

Linton, R.W., Williams, P., Evans, Jr., C.A., and Natusch, D.F.S., 1979, *Anal. Chem.* 49: 1514-1521.

Mamane, Y, Dzubay, T.G., and Ward, R., 1992, *Atmos. Environ.* 26A: 1113-1120.

Mamane, Y. and Gottlieb, J., 1992, *Atmos. Environ.* 9: 1763-1769.

Mamane, Y. and Pueschel, R.F., 1980, *Atmos. Environ.* 14: 629-639.

Mansoori, B.A., Johnston, M.V., and Wexler, A.S., 1994, *Anal. Chem.* 66: 3681-3687.

Marijnissen J., Scarlett B., and Verheijen P., 1988, *J. Aerosol Sci.* 19: 1307-1310.

Matthias-Maser, S. and Jaenicke, R., 1993, *J. Aerosol Sci.* 24: 1605-1613.

Matthias-Maser, S. and Jaenicke, R., 1995, *Atmos. Research* 39: 279-286.

Maynard, A.D. and Brown, L.M., 1992, *J. Aerosol Sci.* 23: S433-436.

McKeown P.J., Johnston M.V., and Murphy D.M., 1991, *Anal. Chem.* 63: 2069-2073.

McMurry, P.H., Litchy, M., Huang, P., Cai, X., Turpin, B.J., Dick, W.D., and Hanson, A., 1996, *Atmos. Environ.* 30: 101-108.

Mouri, H., Okada, K., and Takahashi, S., 1995, *Geoph. Research Let.* 22: 595-598.

Murphy, D.M. and Thomson, D.S., 1995, *Aerosol Sci. Technol.* 22: 237-249.

Murray, K.K. and Russel, D.H., 1994, *J. Am. Sco. Mass Spectrom.* 5: 1-9.

Murray, V.J. and Wexler, A.S., 1995, *Anal. Chem.* 67: 721-726.

Neubauer, K.R., Johnston, M.V., and Wexler, A.S., 1995, *Int. J. of Mass Spectrometry and Ion Processes* 151: 77-87.

Nihei, Y., Satoh, H., Tomiyasu, B., and Owari, M., 1991, *Anal. Sci.* 7: 527-532.

Nockolds, C.E., 1994, *Microbeam Analysis* 3: 185-189.

Nordmeyer, T. and Prather, K.A., 1994, *Anal. Chem.* 66: 3540-3542.

Orlic, I., Watt, F., Loh, K.K., and Tang, S.M., 1994, *Nucl. Instrum. and Methods B* 85: 840-844.

Ottley, C.J. and Harrison, R.M., 1992, *Atmos. Environ.* 9: 1689-1699.

Paoletti, L., Diociaiuti, M., Gianfagna, A., and Viviano, G., 1994, *Microchim. Acta* 114/115: 397-404.

Pardess, D., Levin, Z., and Ganor, E., 1992, *Atmos. Environ.* 26A: 675-680.

Parungo, F., Kopcewicz, B., Nagamoto, C., Schnell, R., Sheridan, P., Zhu, C., and Harris, J., 1992, *J. Geophysical Research* 97: 15867-15882.

Perret, D., Leppard, G.G., Müller, M., Belzile, N., De Vitre, R., and Buffle, J., 1991, *Water Res.* 25: 1333-1343.

Poitevin, E., Krier, G., Muller, J.F., and Kaufmann, R., 1992, *Analysis* 20: M36-39.

Pósfai, M., Anderson, J.R., and Buseck, P.R., 1994, *Atmos. Environ.* 28: 1747-1756.

Prather, K. A., Nordmeyer T., and Salt, K., 1994, *Anal. Chem.* 66: 1403-1407.

Qian, G-W. and Ishizaka, Y., 1993, *J. Geophysical Research-Oceans* 98: 8459-8470.

Reents, W.D., Downey, S.W., Emerson, A.B., Mujsce, A.M., Muller, A.J., Siconolfi, D.J., Sinclair, J.D., and Swanson, A.G., 1995, *Aerosol Sci. Technol.* 23: 263-270.

Ricardo Consulting Engineers, 1987, *A New Study of the Feasibility and Possible Impact of Reduced Emission Levels from Diesel Engined Vehicles for the Transport and Road Research Laboratory (TRRL)*, Report DP 87/0927.

Rojas, C.M. and Van Grieken, R.E., 1992, *Atmos. Environ.* 26A: 1231-1237.

Rojas, C.M., Artaxo, P., and Van Grieken, R.E., 1990, *Atmos. Environ.* 24B: 227-241.

Sandell, J.F., Dewey, G.R., Sutter, L.L., and Willemin, J.A., 1996, *J. Environ. Engin.* 122: 34-40.

Sándor, S., Török, S., Xhoffer, C., and Van Grieken, R.E., 1990, *Proceedings of the Twelfth International Congress for Electron Microscopy*, Eds. Peachy, L.D., and Williams, P.B., San Francisco Press, San Francisco, 245-255.

Schrader, B., 1986, in *Physical and Chemical Characterisation of Individual Airborne Particles*, Ed. Spurny, K.R., Ellis Horwood Last, Chichester, U.K.

Schweiger, G., 1990, *J. Aerosol Sci.* 21: 483-509.

Sheridan, P.J., Brock, C.A., and Wilson, J.C., 1994, *Geophys. Res. Letts.* 21: 2587-2590.

Shina, M.P., 1984, *Rev. Sci. Instrum.* 55: 886-891.

Smith, W.S., 1962, *Atmospheric Emission from Fuel Oil Combustion, An Inventory Guide*, U.S. Department of Health, Education and Welfare, PHS Publication n. 999-AP-2.

Struyf H., Van Vaeck, L. and Van Grieken, R.E., *Anal. Chem.*, submitted.

Török, Sz., Sándor, Sz., Xhoffer, C., Van Grieken, R.E., Mészáros, E., and Molnar, A., 1993, *Idojaras* 96: 223-233.

Tourmann, J.L. and Kaufmann, R., 1992, *Analysis* 20: 65S.

Tourmann, J.L. and Kaufmann, R., 1993, *Int. J. Environ. Anal. Chem.* 52: 215-227.

Tourmann, J.L. and Kaufmann, R., 1994, *Ann. Occup. Hyg.* 38: 455-467.

Tuniz, C., Jones, K.W., Rivers, M.L., Sutton, S.R., and Török, S., *Environ. Sci. Technol.*, in press.

Van Borm, W., Wouters, L., Van Grieken, R.E., and Adams, F., 1990, *Sci. Total Environ.* 90: 55-66.

Van Grieken, R.E., Injuk, J., Owari, M., and Van Espen, P., *Abstracts of Fourth International Aerosol Conference*, in press.

Van Grieken, R.E., Artaxo, P., and Xhoffer, C., 1992, *Proceedings of the Fiftieth Annual Meeting of the Electron Microscopy Society of America*, Eds. Bailey, G.W., Bentley, J., and Small, J.A., San Francisco Press, San Francisco, 1482-1483.

Van Grieken, R.E. and Xhoffer, C., 1992, *J. Anal. At. Spectrom.* 7: 81-88.

Van Malderen, H., Van Grieken, R.E., Bufetov, N.V., and Koutzenogii, K.P., 1996, *Environ. Sci. Technol.*, 30: 312-321.

Van Malderen, H., Van Grieken, R.E., Khodzher, T., Obolkin, V., and Potemkin, V., 1996, *Atmos. Environ.* 30: 1453-1465.

Van Malderen, H., Hoornaert, S., and Van Grieken, R.E., 1996, *Environ. Sci. Technol.*, 30: 489-498.

Van Malderen, H., De Bock, L., Injuk, I., Xhoffer, C., and Van Grieken, R.E., 1993, in *Progress in Belgian Oceanographic Research*, Royal Academy of Belgium, Brussels, 119-135.

Van Malderen, H., Rojas, C., and Van Grieken, R.E., 1992, *Environ. Sci. Technol.*, 26: 750-756.

Vapirev, E.I., Tsacheva, Ts., Bourin, K.I., Hristova, A.V., Kamenova, Ts., and Gourev, V., 1994, *Radiation Protection Dosimetry* 55: 143-147.

Wonders, J.H.A.M., Houweling, S., De Bont, F.A.J., Van Leeuwen, H.P., Eeckhaoudt, S.E., and Van Grieken, R.E., *Internat. J. Environ. Anal. Chem.*, in press.

Wouters, L., 1991, PhD Thesis, University of Antwerp, Belgium, 51-512.

Wouters, L., Artaxo, P., and Van Grieken, R.E., 1990, *Int. J. Environ. Anal. Chem.* 38: 427-438.

Wouters, L., Michaud, D., and Van Grieken, R.E., 1993, *Mikrochim. Acta* 110: 31-40.

Wu, P.-M. and Okada, K., 1994, *Atmos. Environ.* 28: 2053-2060.

Xhoffer, C., 1993, PhD Thesis, University of Antwerp, 189-204.

Xhoffer, C., 1993, PhD Thesis, University of Antwerp, 205-233.

Xhoffer, C., Bernard, P., and Van Grieken, R.E., 1991, *Environ. Sci. Technol.*, 25: 1470-1478.

Xhoffer, C., Jacob, W., Buseck, P.R., and Van Grieken, R.E., 1995, *Spectrochim. Acta part B* 50: 1281-1292.

Xhoffer, C., Wouters, L., and Van Grieken, R.E., 1992, *Environ. Sci. Technol.*, 26: 2151-2162.

Xhoffer, C., Wouters, L., Artaxo, P., Van Put, A., and Van Grieken, R.E., 1992, in *Environmental Particles Vol. I*, Buffle, J. and Van Leeuwen, H.P., Eds., Lewis, Chelsea, MI, Vol. 1: 107-143.

Yang, M., Reilly, P.T.A., Boraas, K.B., Whitten, W.B., and Ramsey, J.M., 1996, *Rapid Commun. in Mass Spectrometry* 10: 347-351.

Zhang, X. Q. and McMurry, P. H., 1993, *Atmos. Environ.* 27A: 1593-1607.

13 The Analysis of Individual Aerosol Particles Using the Nuclear Microscope

Ivo Orlić

CONTENTS

13.1   INTRODUCTION

Nuclear microscopy (NM) is a novel analytical technique that simultaneously uses several ion beam techniques, all with micron or even submicron resolution. The principal technique is particle induced X-ray emission (PIXE), often referred to as micro-PIXE. From its very beginning in the 1970s, PIXE has been extensively used in various applications. It is a truly multielemental analytical technique with a sensitivity in the ppm range. Absolute detection limits as low as 10^{-17} g can be achieved using a focused proton beam. PIXE is ideally suited for environmental studies where sample masses are typically small and multielemental analysis is required. As such, it has been accepted as one of the key analytical methods for the analysis of fine aerosols in a worldwide network.[1] According to Cahill's survey,[2] analysis of atmospheric aerosols is the major application in about a quarter of the roughly 100 PIXE facilities presently in operation worldwide. That number is still increasing because of the high sensitivity of PIXE, multielemental capabilities, relative simplicity of data interpretation, and easy automation for routine analysis of a large number of samples.

With the development of NM, it soon became obvious that complementary ion beam techniques such as Rutherford backscattering (RBS), scanning transmission ion microscopy (STIM), forward scattering or proton elastic scattering (PESA), nuclear reaction analysis (NRA), and others could contribute additional valuable information. Combined, they extend the multielemental capabilities of NM to essentially all elements of the periodic table. For example, PESA is often used to determine hydrogen concentrations in aerosol samples,[1,2] RBS to find the concentrations of matrix elements such as oxygen and carbon, and PIXE for all elements above fluorine. These techniques expanded the applicability of NM into various fields of microbiology (e.g., analysis of single cells), microelectronics and various novel environmental applications, including the analysis of single aerosol particles.

Features that make NM, and particularly micro-PIXE, an attractive alternative for single particle characterization are:

1. PIXE is a multielemental technique with a relatively simple spectra structure.
2. PIXE has a large dynamic range — meaning that major as well as trace elements can be analyzed simultaneously in a single run.
3. The required total sample mass is minimal and can be lower than 10^{-12} g (one pg) for a well-focused beam.
4. With the addition of complementary ion beam techniques such as RBS, STIM, etc., total quantitative analysis of single particles is possible.
5. RBS allows concentration depth profiling.
6. The method is nondestructive.
7. Minimal sample preparation is required.

However, in spite of these attractive features, the NM analysis of single aerosol particles has not yet gained wide popularity. One of the main reasons is the limited resolution which is currently at about 1 μm for routine measurements. As the most important fractions of anthropogenic aerosols fall in the fine aerosol category (particles smaller than 1 μm), it would be certainly advantageous to have a proton microprobe with around 0.1 μm spatial resolution. Such a high resolution NM is under development in Oxford and Singapore[3] but presently such resolution is not achievable. As a consequence, the NM analysis of single aerosol particles has focused on the analysis of the so called "gigantic" or "super-micron" particles (particles with a diameter larger than 1 μm) and fly ash particles, while sub-micron particles are typically analyzed by electron probe X-ray microanalysis (EPXMA) technique.

A general overview of various microanalytical techniques used for single particle analysis can be found in recent reviews by R. Van Grieken, Wendy Jambers, and co-workers.[4,5,6] More specifically, analysis of single aerosol particles using NM are extensively discussed in reviews by the author[7] and G. Grime.[8]

In spite of the recent developments and large interest in NM, one can say that the NM technique is still in its infancy, with only about fifty installations worldwide. This is due to the complexity of the instrumentation and the relatively high initial price of the equipment (presently amounting to several million dollars), the bulk of which is for the purchase of an accelerator. NM laboratories are, therefore, inevitably linked to big universities and/or research institutes and as such, nurture a wide range of research programs and applications. There are recent reports on the development of a 'desk-top' nuclear microscope[9] but these are not yet in widespread use. In terms of industrial applications, there are presently only a few dedicated and highly automated NM installations, mainly in geological research. (CSIRO Laboratories in Australia,[10,11] University of Guelph, Canada.[12,13]) A growing number of installations dedicated to microelectronics applications is found in Japanese institutes like JAERI[14] and GIRIO.[15]

13.2   METHODS

13.2.1   Particle Induced X-Ray Emission (PIXE)

A range of reactions takes place when a high-energy ion beam impinges on a target. One of the most significant processes is the production of X-rays utilized for PIXE. The basic principles of X-ray spectroscopy were laid down in 1913, when Moseley[16] discovered that materials irradiated with X-rays emit photons of energies that are characteristic for the target element. Soon, it was realized that the technique offers the possibility of a systematic, multielemental analysis of a given sample. The method gained popularity and was developed into various fields of X-ray spectroscopies of which photon excitation, or X-ray fluorescence (XRF) was exclusively used until the late 1960s, when the development of semiconductor detectors and ion accelerators made PIXE feasible.

When a target is bombarded with photons or charged particles (ions or electrons) inner shell electrons may be excited to unfilled orbitals or the continuum. In a very short time, the atom will return to its "ground state," i.e., an electron from a higher shell will transfer to the vacant electron site. The emitted electromagnetic radiation (photon) carries an amount of energy equivalent to the energy difference between the two shells. Hence, the energy of the photon will be well defined and determined by the atomic number of the target atom. As in PIXE, one is primarily interested in the emission spectra that are induced by creation of vacancies in the inner shells (K and L shells), the characteristic X-rays will have several important features:

- The number of electron shells involved, and, therefore, the number of emission lines that contribute to the X-ray spectra is low. Hence, the characteristic X-ray emission spectra are simple, which in turn simplifies their interpretation.
- The energy levels, and, therefore, the transition energies, are essentially insensitive to the chemical state of the atom.
- The relatively high energy of the X-rays affords effective analysis of material in bulk form.

These are the major features that made PIXE and XRF well-established analytical techniques, as evidenced by their widespread use and the fact that a number of different commercial instruments are available on the market. The speed, accuracy, sensitivity, and flexibility, coupled with the multielemental capability, have made them invaluable tools in nearly all branches of science.

The work done by Johansson and co-workers[17] from the Lund Institute of Technology is often considered a cornerstone of the modern PIXE. They applied PIXE to the analysis of airborne dust and demonstrated convincingly that the combination of a MeV proton excitation and X-ray detection with a solid-state Si(Li) detector constitute a very powerful multielemental analytical tool for trace element analysis. Since then, the method has developed to a mature analytical technique that is being used in many research and applied areas.

A comprehensive review of PIXE technique and its application can be found in the excellent textbooks edited by Johansson et al.[18,19] More specifically, numerous reviews dealing with applications of PIXE in environmental sciences are subjects of historic reviews by Cooper[20] and Johansson,[21] and more recently of Winchester,[22] Malmqvist.[23,24] In an excellent review, Maenhaut[25] compares PIXE with five analytical techniques currently used for trace element analysis in aerosols, Neutron activation (NAA), XRF, atomic absorption (AAS), and inductively coupled plasma atomic emission/mass spectrometry (ICP–AES/MS). The applicability of ion beam analysis techniques

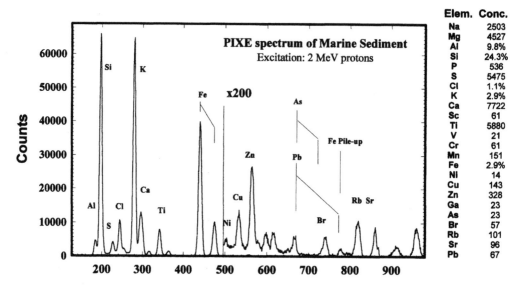

Elem.	Conc.
Na	2503
Mg	4527
Al	9.8%
Si	24.3%
P	536
S	5475
Cl	1.1%
K	2.9%
Ca	7722
Sc	61
Ti	5880
V	21
Cr	61
Mn	151
Fe	2.9%
Ni	14
Cu	143
Zn	328
Ga	23
As	23
Br	57
Rb	101
Sr	96
Pb	67

FIGURE 13.1 PIXE spectrum of marine sediment collected in Singapore.[28] Sample was irradiated with 2 MeV protons for 30 minutes so that the total accumulated charge was approximately 1 μC. Concentrations are given in ppm, unless otherwise stated. (From I. Orlic, J. Makjanic and S.M. Tang, Multielemental analysis of marine sediments from Singapore coastal region by PIXE and XRF analytical techniques, *Proceedings of ASEAN-Canada Cooperative Programme on Marine Science,* Midterm Technical Review Conference, Singapore, 24-28 Oct. 1994, pp. 314-326. With permission.)

such as PIXE, micro-PIXE, γ-spectroscopy, scattering techniques, ion beam tomography, etc., in environmental studies is discussed by Malmqvist.[26,27]

To illustrate the relative simplicity and multielemental capabilities of PIXE, an example of a marine sediment spectrum is given in Figure 13.1. A sediment sample was collected in Singapore coastal waters, dried, and ground in a mechanical mortar.[28] The powder was then made into a pellet (13 mm diameter) and irradiated with 2 MeV protons until the accumulated charge was approximately 1 μC. The X-rays were detected by means of a Si(Li) solid state detector, the pulses from the detector amplified, and registered in a pulse height analyzer.

The spectrum shows a large number of well-separated characteristic K X-ray lines of many elements between Al and Mo as well as the L lines of Pb.[29] The X-ray line intensities are typically extracted by means of spectrum fitting routines, and concentrations quantified by comparing measured intensities with the theoretically calculated yields (for the assumed initial elemental concentrations and sample thickness). Concentrations and thicknesses are then iteratively adjusted until a convergency criteria is reached. Such procedure is possible because the mechanism of X-ray production is well known, and, therefore, X-ray yields can be theoretically calculated for a given sample and experimental conditions. The accuracy of the resulting concentrations varies considerably but can be as good as a few percent.[18] The spectrum shown in Figure 13.1 was analyzed by means of the code GUPIX.[30] The concentrations of matrix elements (carbon, nitrogen, and oxygen) were derived by means of simultaneous RBS analysis (see the following section). The concentrations obtained for 24 elements are expressed in ppm, unless otherwise stated. Errors are not stated, but they are in general between 5 and 10%, depending on the concentration levels and energy of the characteristic X-ray lines.

One of the major obstacles in quantitative PIXE analysis, especially in NM, is sample inhomogeneity and roughness. Almost all theoretical models are based on the assumption that the sample is ideal, i.e., a uniform and flat slab of material with homogeneous distribution of all elements. This is an acceptable approximation for broad-beam PIXE analysis (i.e., when the ion beam is spread over a large area and when the eventual sample's grains are much smaller than the range

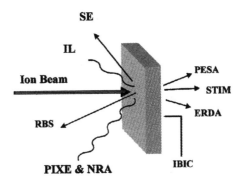

FIGURE 13.2 Major high-energy ion beam interactions with target atoms: PIXE or particle induced X-ray emission, NRA — nuclear reaction analysis, RBS or Rutherford backscattering, IL — Ionoluminescence, SE — secondary electrons, PESA — proton elastic scattering, STIM — scanning transmission ion microscopy, ERDA — elastic recoil detection analysis, and IBIC — ion beam induced current.

of incident protons and/or characteristic X-rays) but often fails in micro-PIXE analysis. The problems encountered in micro-PIXE quantitative analysis, with the special emphasis on single aerosol analysis, will be discussed in more details in Section 13.5.

It should also be mentioned that the background present in PIXE spectra is very low (see Figures 13.1 and 13.14) compared to EPXMA spectra. This is due to the very small interaction between protons and target electrons which results in a low electron bremsstrahlung radiation. As a consequence, PIXE detection limits are two to three orders of magnitude lower than in the analogous EPXMA technique.

13.2.2 COMPLEMENTARY ION BEAM ANALYSIS TECHNIQUES

A number of ion beam interactions occurs during the collision of high-energy ions with a target. The most important processes are schematically depicted in Figure 13.2.

In general, other complementary techniques can extend the elemental coverage PIXE down to hydrogen. In most cases, the additional analyses can be carried out simultaneously with PIXE. The most common techniques are RBS and STIM, which will be briefly discussed in the following sections. Others, such as NRA and proton-tagged NRA (pNRA), ionoluminescence, proton elastic scattering (PESA), elastic recoil detection analysis (ERDA), and secondary electrons (SE), are often used in broad ion beam analysis and only recently have started gaining popularity in NM. As an excellent example of various ion beam analysis (IBA) techniques in NM, a schematic diagram of the Lund set-up is shown in Figure 13.3.[31] In addition to the facilities for traditional PIXE and RBS analysis, this system is equipped with numerous surface barrier detectors (SBD) and scintillators to facilitate NRA, as well as ionoluminescence (IL) and SE. With this addition, light elements (hydrogen, lithium, boron, fluorine) can be analyzed with excellent detection limits (6 to 400 ng/cm²).

Due to complexity of the data acquisition and interpretation, these techniques are not yet common in the NM of single aerosol particles, and they will not be discussed here. More comprehensive reviews can be found in two excellent textbooks on PIXE and NM,[18,32] recent review,[33] as well as in the proceedings of the recent Ion Beam Analysis conferences.[34] The application of techniques such as PESA, forward alpha scattering (FAST), and atomic mass spectrometry (for ¹⁴C) in broad beam aerosol analysis can be found in Cahill's work.[18,35]

13.2.2.1 Rutherford Backscattering, RBS

Rutherford backscattering spectroscopy involves the detection of particles elastically scattered from the target nucleus. For a given scattering angle, the cross section for the process and the energy of the scattered particle are functions of the atomic number and the mass of the target atom. As a

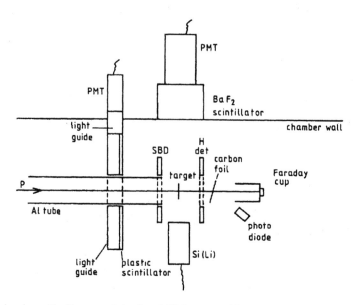

FIGURE 13.3 A schematic diagram of the Lund NM set-up which, in addition to traditional PIXE, RBS, and STIM analysis, uses various nuclear reactions for the simultaneous analysis of hydrogen, lithium, boron, and fluorine, all with one micron spatial resolution. (From K.A. Sjoland, P. Kristiansson, M. Elfman, K.G. Malmqvist, J. Pallon, R.J. Utui, C. Yang, *Nucl. Instr. Meth.*, B130, 1997, pp. 20-24. With permission. Copyright Elsevier Science Publishers.)

consequence, the method can be effectively used to determine concentrations as well as depth profiles of elements of interest. The sensitivity of RBS is often not as high as for PIXE, but it is advantageous for lighter elements and, in that sense, the two methods are complementary. RBS is often used in NM as a tool to determine the concentrations of matrix elements such as C, N, O, etc., as they are needed for proper quantitative PIXE analysis.

RBS also allows the extraction of sample thickness (for thin samples and layers) and major elements concentration depth profiles. If RBS is done simultaneously with PIXE, it is also possible to extract the total accumulated charge from the RBS spectrum, which is needed for the normalization of PIXE spectra and not easily measured by other means. Data reduction is usually done using various simulation codes. In our laboratory in Singapore, the code RUMP[36] and its MS Window version, NUSDAN,[37] are in use for that purpose. The accuracy of the results varies from sample to sample but is in general around 5% if the sample is "ideal," i.e., a uniform and flat series of layers with well-defined and homogeneous distribution of all elements within a layer. Again, this can be a good approximation in broad beam PIXE/RBS analysis but not in NM analysis of single aerosol particles and similar samples with microstructures (see Section 13.5.5 for details). Another problem with RBS analysis is that the scattering of MeV protons on light elements (e.g., C,N,O) is not entirely elastic. At MeV energies, the incident protons are able to penetrate the Coulomb barrier of the target nucleus and cause various nuclear reactions. The cross section then deviates significantly from the Rutherford formula, and accurate databases are needed for quantitative analysis.

An example of an RBS spectrum from a single aerosol Ti rich particle is shown in Figure 13.4 together with the fit obtained by means of the program NUSDAN.[37] The thickness of the particle was computed to be approximately 1.5 µm. It was also found that it is covered by a very thin layer of Pb with a thickness of approximately 0.08 µm. The concentrations of other trace and minor elements are determined from the corresponding PIXE spectra. The spectrum shown in Figure 13.4 is derived from the Ti-rich particle labeled Ti2 in Figure 13.15.

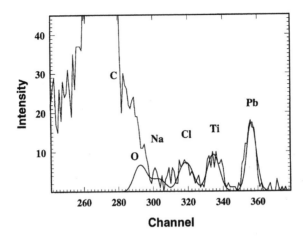

FIGURE 13.4   RBS spectrum of the Ti-rich particle with a distinct Pb surface peak. The full line represents fitted spectrum obtained for 0.08 μm thin Pb layer on top of Ti_2O_3+NaCl particle with a thickness of 1.5 μm. The large C and O lines are the result of the backscattering from the thin pioloform backing material. (From I. Orlic, *Nucl. Instr. Meth.*, B104, 1995, 602-611. With permission. Copyright Elsevier Science Publishers.)

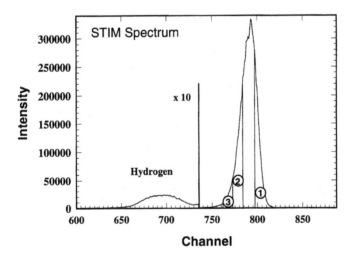

FIGURE 13.5   Off-axis STIM spectrum obtained for the cascade impactor sample with corresponding PIXE, RBS, and STIM maps shown in Figure 13.15. Regions 1, 2, and 3 are formed by ions that suffer small, medium, and high energy loss, respectively. Due to a very low intensity of the hydrogen peak, the vertical scale of the left side of the spectrum is magnified 10 times.

13.2.2.2   Scanning Transmission Ion Microscopy, STIM

As it traverses through the target material, the incident ion loses its energy primarily in collisions with target electrons. Because the energy transfer per collision is very small (due to the large difference between the ion and the electron mass), the incident ion will slowly lose its energy without significantly altering its path. The energy loss will be a measure of the electron density or, to first approximation, to the density of the sample. If the sample thickness is smaller than the range of the incident ions (typically 50 to 100 μm) and the beam is scanned across the sample, the ion energy loss can be mapped by placing a particle detector directly behind the sample. From the obtained energy loss distribution and from the approximately known sample composition (obtained from PIXE and/or RBS results) one can deduce a sample areal density distribution (in g/cm²). The method is very fast and requires very low beam current (fA range). The added bonus is that for such low current, the object and collimator slits are set at very low openings, minimizing lens aberrations and significantly improving resolution. This has enabled a spatial resolution of 100 nm to be achieved.[3] Unfortunately, due to low current requirement, direct STIM is not compatible with the simultaneous operation of PIXE and RBS analysis.

As an example of density mapping of a single fly ash particle, the work of Bogdanovic et al.[38] should be mentioned. The authors used STIM to determine the density distribution, which was then used to facilitate transformation of the PIXE intensity maps to the actual concentration maps. This is of great importance in single particle analysis, because the intensity maps, which are often incorrectly called elemental maps, do not represent the actual concentration distributions. This is because local variations in ion energy loss and X-ray absorption effect the PIXE yields when the beam passes through segments of particles with different densities and topology. The effect can be corrected only if density distribution and particle topology are known. As the authors have demonstrated, it is possible (although time consuming) to determine the density distribution of a single particle. Topology, however, was assumed to be spherical, which is often a good approximation for fly ash particles. Particle topology, as well as internal density could, in principle, also be determined using the STIM tomography technique. In this technique, STIM images are recorded at many different projections (sample orientations) with the aim of reconstructing three-dimensional images of the specimen density. The technique is actually "borrowed" from the medical applications of computer tomography (CT) and modified to accommodate NM needs. Within the past 10 years, STIM tomography has gone through significant development and has been applied to various samples, mainly of a biological nature[39,40,41] but to the best of our knowledge, never to the analysis of single aerosol particles. The method is still in the developmental stage and its major disadvantage is that many projections are required. To overcome this problem, our group applied a novel technique, the maximum likelihood expectation maximization (MLEM) algorithm, to the successful reconstruction of a 3D microstructure, requiring only four projections.[42]

The off-axis STIM (or dark-field STIM) mode is often used for fast mapping of a sample. In this mode, the STIM detector is placed approximately 20° off the beam axis. The recorded proton energy spectrum shows the forward scattered particles which have suffered energy loss by passing through the sample. The technique is effective in identifying particles of various densities. The advantage of the dark-field STIM is that the reduced signal makes the procedure compatible with the simultaneous application of PIXE and RBS. Another advantage is that the hydrogen concentration can be determined from the (p,p) scattering intensity. See Figure 13.5, where an off-axis STIM spectrum of an aerosol sample is shown. The large peak is formed by incidental ions passing through the aerosol sample, referring to a scan shown in Figure 13.15. The high-energy region of the peak (1) is formed by ions passing through a thinner section, and low energy tail (areas 2 and 3) through thicker sections of the sample (higher energy loss). The broad peak to the left, however, is a forward scattered peak from H atoms that can be used efficiently to determine hydrogen concentration in the sample.

13.3 INSTRUMENTATION

13.3.1 ACCELERATOR AND FOCUSING LENSES

The ion beams used in NM are produced in various types of particle accelerators. The energies required, typically several MeV, can be achieved with relatively small accelerators, often single- or double-ended electrostatic accelerators of the Van de Graaff type. For NM analysis, stringent requirements on brightness, energy stability, and beam emittance are the most important parameters. An ultrastable single-ended (Singletron) accelerator has recently been developed by the High

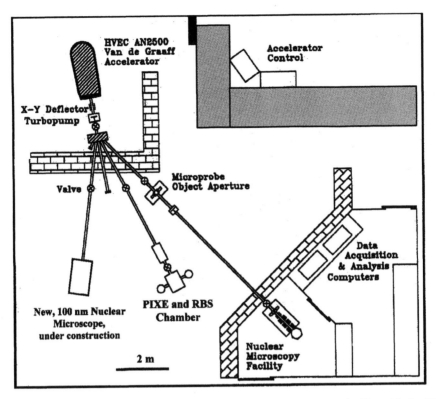

FIGURE 13.6   The layout of the National University of Singapore accelerator facility with the NM set-up, broad beam PIXE, and RBS chamber and the new, 100 nm nuclear microscope, which is presently under construction.

Voltage Engineering, Europe, which seems to satisfy all the requirements of an ideal, dedicated NM accelerator.[43]

Accelerated ions are transported to the irradiation chamber through the beam transport system, which typically consists of vacuum tubes, focusing and bending magnets and deflectors, collimators, and slits. The layout of the National University of Singapore accelerator facility[44] is shown in Figure 13.6.

The 2.5 MeV HVEC Van de Graaff accelerator is equipped with a radio-frequency source capable of delivering up to 50 μA of singly charged protons (or He ions). Accelerated ions are directed down any of the three optional beam lines through a switching magnet. In the NM line, microbeam formation is achieved by projecting a demagnified image of the object slits onto the target. The beam focusing system comprises a coupled triplet of high excitation magnetic quadrupole lenses with prefocus scan coils and is based around an Oxford Microbeams endstation (OM 2000), shown in Figure 13.7, together with the target chamber, zoom microscope, and the Si(Li) detector. The lenses are of integral construction, and tests have shown that these lenses have minimal aberrations. The system routinely provides a microbeam with 1 μm spatial resolution, while the best performance recorded was 400 nm in PIXE and 100 nm in STIM mode, currently the best performance in the world[45,46] for MeV protons.

13.3.2   TARGET CHAMBER

As an example of an NM target chamber, the layout of the National University of Singapore chamber is shown in Figure 13.7. The body is essentially the Oxford Microbeams octagonal stainless-steel target chamber. The sample is positioned in the center by means of the target holder, which is

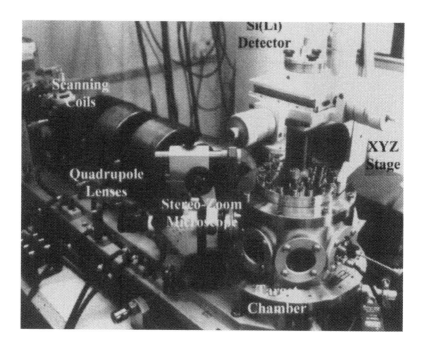

FIGURE 13.7 Photograph of the nuclear microscope endstage showing the scanning coils, quadrupole triplet lenses, stereo-zoom microscope, octagonal target chamber, XYZ sample translator, and Si(Li) detector.

inserted from the top through a quick-release opening and attached to the three axes translator. The sample can be positioned either at 90° or 45° with respect to the beam and viewed through a long-working distance stereo-zoom microscope with magnification of up to 200×. The chamber accommodates two 80 mm² Si(Li) x-ray high-resolution (150 eV at 5.9 keV) detectors. They are both inserted into the vacuum chamber and, if necessary, additional filters are mounted. Further, to reduce counting (or irradiation) time, which is a very important factor when thin samples are analyzed (for example single cells or single aerosol particles), detectors are mounted on a sliding rail. In this way, the distance from the target can be reduced to only several mm, increasing the solid angle subtended by the crystal to as much as 400 msr. This increases the solid angle four times over the usual 100 msr, allowing four times reduced irradiation time.

The chamber also accommodates a number of surface barrier detectors for RBS and STIM analysis. Usually two RBS detectors are used, one passivated implanted planar silicon (PIPS) for high-resolution work with acceptance solid angle of 20 msr (active area of 50 mm²) and the other with large solid angle (200 msr) for thin samples such as single aerosol particles and biological samples.

Mounted behind the target is a carbon Faraday cup to monitor the beam current when thin targets are investigated. The Faraday cup is mounted on a rotatable plinth together with the STIM detector and is adjustable from outside the chamber via a rotary drive. In the normal position (Figure 13.8), the STIM detector is at 20° from the beam axis and is used for fast mapping of thin samples by means of off-axis STIM. As mentioned before, the advantage of the off-axis STIM is that it can be used simultaneously with PIXE and RBS measurements without reducing beam intensity, which is not the case for the direct STIM measurement. More details on the NM experimental set-up can be found in textbooks.[18,19,32]

13.3.3 DATA ACQUISITION SYSTEM

In most NM systems, the focused beam is scanned over the sample by using either magnetic coils or electrostatic deflectors. The scan size can be changed on demand from a very small, few μm

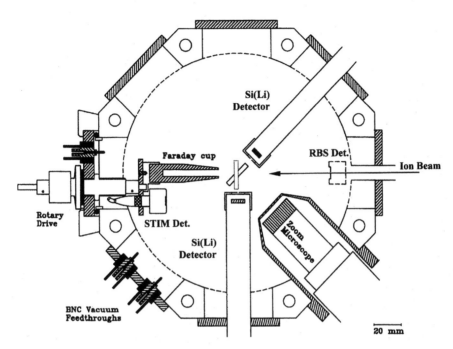

FIGURE 13.8   The schematic layout of the NUS target chamber showing the position of the two Si(Li) detectors with filters, zoom-microscope, target, RBS detector, and the rotary drive arrangement on which the STIM detectors and the Faraday cup are mounted.

size, up to 5 mm. The scan pattern is computer controlled and most data acquisition systems have provisions for the user to define the shape of the scan (either area scan, line scan, scan over predefined masked area, or point analysis). Step size in either x or y direction is digitized (typically between 64 and 1024 steps per scan) and computer controlled so that data streaming from various detectors can be correlated to the actual beam position and the corresponding maps created on-line. For the off-line analysis, data are collected in event-by-event mode (also called list-mode). In that mode, each data pulse is digitized and subsequently stored sequentially in a file in a form of data triples, containing the energy and the actual x and y beam positions for the entire duration of the run. Such files are often very large (20 or more MB). The advantage of list mode files is that they contain a complete record for the entire run. The record can be replayed on demand, and energy spectra and maps of all active detectors for any particular part of the scan can be retrieved at any time.

Because there is a large number of data continuously acquired by various detectors (provisions are typically made for up to 8 detectors), considerable computing power is required to process and to sort all these data into energy spectra and elemental maps. Various data acquisition systems have been developed. Some are based on mainframe processors and workstations, and some are PC-based systems. A comprehensive review by Lowestam on this topic can be found in Reference 47.

13.4   OPTIMIZATION OF EXCITATION AND DETECTION CONDITIONS

Depending on the subject of the study, different groups of elements might be of interest. In that respect, NM offers great flexibility. Excitation and detection parameters, e.g., type of incident ions and their energies, geometry, X-ray filters, etc. All these parameters can be adjusted to optimize the experimental conditions so that the group of elements of interest can be efficiently analyzed. In environmental studies, the elements of interest are usually transition metals, rare earth elements, and heavy metals. Cross sections for X-ray production of these elements are relatively low in comparison to the cross sections for the light elements. Furthermore, the concentrations of light

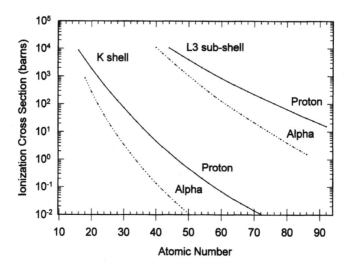

FIGURE 13.9 Ionization cross sections for K and L shells and for 2 MeV proton and α particle excitation.

elements are typically much higher than the metal concentrations. As a result, characteristic X-ray lines of low Z elements, such as Al, Si, Cl, etc., dominate the PIXE spectra. To avoid that, and to enable optimal detection of trace elements of interest, the experimental conditions have to be fine tuned. Some possibilities are discussed briefly in the following section.

13.4.1 INCIDENT IONS AND THEIR ENERGY

The type of ion and its energy will appreciably affect the production of X-rays. These effects are shown in Figure 13.9 where the X-ray production cross sections for protons and alpha particles are shown as a function of atomic number of the target atom. For the same energy (2 MeV), protons have obviously favorable cross sections compared with alpha particles, especially for heavier elements (cross sections are 5 to 30 times higher, depending on the atomic number). This is the reason that, in PIXE analysis, protons are most often used for excitation, while α excitation is preferred for RBS analysis.

Another observation from Figure 13.9 is that the X-ray production cross sections for lighter elements are several orders of magnitude higher than for heavy elements. Fortunately, the cross sections for the L X-ray lines are higher for heavy elements and are used for the detection of heavy metals. However, there is a "blind spot" in PIXE sensitivity for elements with atomic numbers 40 to 60 that can be partially overcome by applying relatively thick filters (see the following section).

$K\alpha$ production cross sections increase rapidly with ion energy, until a saturation level is reached (see Figure 13.10a). Cross sections for low Z elements reach saturation at lower energies, making a difference much smaller at higher energies. Therefore, the advantage of using higher energies becomes obvious. To quantify this observation, cross sections are normalized to the corresponding 2 MeV values and obtained ratios are shown in Figure 13.10b. As expected, the increase is most evident for high Z elements. For example, if the energy of protons is changed from 1 to 4 MeV, the $K\alpha$ production cross sections for Ba will increase more than 100 times.

Unfortunately, there is an upper, practical limit in increasing the energy of incident protons which is at approximately 3.5 to 4 MeV. At higher energies, the intensity of Compton background, which is formed in a detector volume by γ-rays emerging from various nuclear reactions, becomes very high. In such a way, the gain on cross section is diminished by the increase of the background level which ultimately affects the detection limit. For samples that contain carbon (biological and urban aerosol samples), one should keep the proton energy below 4.43 MeV, where carbon has a very strong resonance.

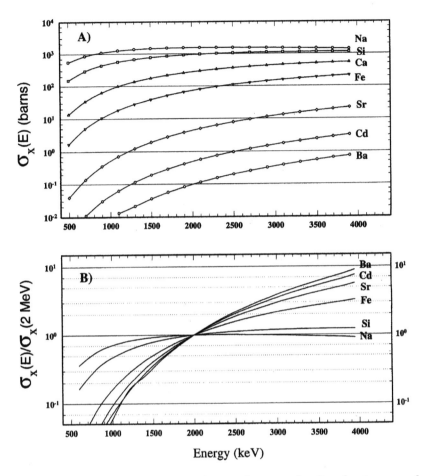

FIGURE 13.10 (a) $K\alpha$ X-ray production cross sections are shown as a function of proton energy for elements between Na and Ba. (b) $K\alpha$ X-ray production cross sections for protons normalized to the corresponding 2 MeV values.

13.4.2 DETECTION EFFICIENCY

The obvious way to increase the counting rate of a detector is to decrease the distance between target and detector. This is usually done by making this distance adjustable (see Section 13.3.2 and Figure 13.8) so that the user can choose a distance that is appropriate for the specimen being irradiated. In NM of single aerosol particles, sample masses are of the order of 10^{-12}g, and it is, therefore, very important to be able to increase the detector acceptance angle as much as possible. As shown in Section 13.3.2, in the Singapore NM facility, the detector can be as close as 5 mm to the sample, increasing the solid angle from a typical 100 msr to approximately 400 msr. Antolak et al.[48] reported that a dedicated NM for high throughput single particle analysis is being installed in the Sandia Laboratories. Their plan is to use an array of four 200 mm² high-purity germanium detectors with the total solid angle larger than 1000 msr.

However, as we have already mentioned in the previous section, X-ray line intensities of lighter elements are "by default" very high and should be suppressed so that X-ray line intensities of transition and heavy elements becomes relatively stronger. This is done by choosing a suitable filter. In Singapore, a "double-magic" filter was found indispensable for almost all environmental studies. The filter is a 150 μm Al foil with the central hole of 0.8 mm diameter, all covered with an additional 38 μm kapton foil. The effect of a central hole in a rather thick filter is to transmit to the detector only a small portion of low energy X-rays which would otherwise be completely

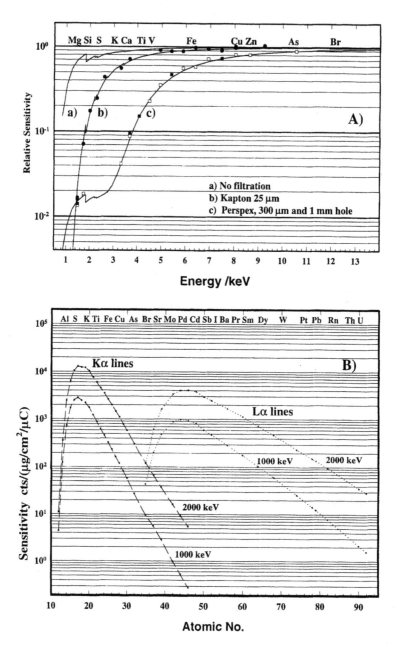

FIGURE 13.11 (**A**) Si(Li) detector efficiency measured for two different absorbers; (a) no additional filtration — bare detector, (b) 25 μm kapton foil, and (c) "magic" filter with the following characteristics, 300 μm perspex with the central hole with diameter of 1.0 mm. Full lines are theoretical efficiencies calculated for detector and filter parameters, and dots represent measured values for thin NIST and Micromatter standards. (**B**) PIXE sensitivity for K and L lines calculated for thin sample and for 1 and 2 MeV incident protons. Sensitivities are calculated for kapton filter of 25 μm thickness and for the 45° geometry.

absorbed. Additional kapton foil is to further reduce still rather intense X-ray lines of Na, Al, and Si. The final efficiency curves for several types of filters are shown in Figure 13.11A. Experimentally measured values are shown as dots, and theoretically calculated values for optimized detector and filter parameters as full lines.[49]

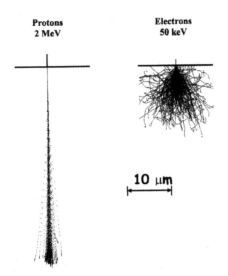

**Protons
2 MeV**

**Electrons
50 keV**

10 μm

FIGURE 13.12   Comparison of 2 MeV proton and 50 keV electron range in Si. While protons have range of approximately 50 μm, electrons have range of 10 μm. Another important feature of proton beam is that lateral spread is almost negligible comparing to a spread of electrons.

Strong absorption effects of the magic filter will significantly reduce otherwise high intensities of light elements so that the final system sensitivity will be well balanced across the entire region of interest (see Figure 13.11B). As a result, an aerosol sample can be irradiated with a higher beam density (nA if a large area is scanned), and the optimum counting rate of 1 to 2 kHz can still be maintained.

Another possibility is to use two detectors, one with a very thick filter, optimized for the high-energy region (5 to 25 keV), and the other with a thin filter (or "magic" filter and small solid angle) dedicated to the low-energy region (1 to 5 keV). The signals from both detectors could be processed and accumulated by the same pulse height analyzer and analyzed almost on-line. Such a dedicated, highly automated system with high throughput for aerosol analysis is installed at the Davis Cyclotron facility,[18,50] one of the leading laboratories in the field of aerosol analysis.

13.4.3 EFFECTIVE RANGE OF PROTONS AND X-RAYS

Two MeV protons have a typical range of 50 to 100 μm in various materials. This makes PIXE using protons a bulk analysis method. This range is much larger than the range of electrons (less than 10 μm), and, in addition, the lateral spread of protons is almost negligible throughout the range (see Figure 13.12). As a result, the superior spatial resolution of EPXMA can only be utilized for particles smaller than 1 μm. For larger particles, NM has the advantage of better resolution and higher penetration depth.

There is, however, another factor which is affecting all X-ray spectroscopy methods: the limited range of the X-ray photons. This range is, in most cases, smaller than the range of protons. This will be reflected in a relatively small volume that contributes to the total X-ray yield. This volume, called the effective volume (or effective range/mass) will strongly depend on the energy of characteristic X-rays and will more significantly affect lighter elements. To illustrate how small the effective volumes (or range of photons) are, the X-ray yields are calculated using the PIXE simulation code TTPIXAN[51] for a typical aluminosilicate matrix, for 2 MeV protons incident at 45° to the sample normal and the X-ray take-off angle of 45°. The results are shown in Figure 13.13 in a form of normalized total X-ray yields for Na, Al, Si, Ca, Fe, and Sr. Yields are shown as a function of the depth at which they are produced. It can be seen that the yield of Na is already reaching saturation at the depth of only 3 μm, Al and Si at 7 μm, and all other elements at 15 to 20 μm.

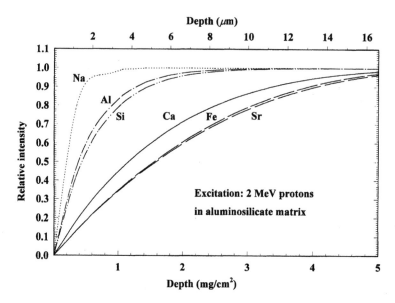

FIGURE 13.13 The normalized total X-ray yields of several elements as a function of sample depth. For convenience, depth is expressed as areal density in g/cm² and in μm, assuming volume density of 2.5 g/cm³. See text for details.

This means that, in spite of the relatively large range of protons, the probing depth for light elements will only be several μm due to strong absorption effects. Such strong absorption effects might pose a serious problem in quantitative analysis, if not taken into account. This topic will be discussed further in Section 13.5.5.

13.5 ENVIRONMENTAL APPLICATIONS OF NM

Humanity, lead by industrial nations and followed by the fast growing economies in Asia and South America, is conducting an unintended global experiment by releasing large amounts of pollutants, such as greenhouse gases and various aerosols, into the atmosphere.[52,53] The outcome of this experiment might be disastrous in terms of health implications and changes in global climate.[54,55] There are predictions that the global temperature will increase at least several degrees C during the next 50 years.[56] So far, research has concentrated mainly on greenhouse gases, but there is strong evidence that aerosols are as important and have to be included in the models of the Earth's "radiation budget."

The way in which the greenhouse gases influence the global mean temperature (14.9°C) is quite well understood. It is known that the mean temperature is the result of a fragile energy balance between the incident solar short-wave radiation and outgoing long-wave thermal radiation, which is partly governed by greenhouse gases.[53] The other important parameter is the global albedo, which accounts for 70% of the total absorbed solar energy. Albedo depends strongly on cloud cover and structure but, if there are no clouds, aerosols will govern the absorption, scattering and reflecting solar radiation. Also, in clouds, aerosols may act as condensation nuclei and influence the drop size spectrum, the lifetime of clouds, and their albedo.[54] Anthropogenic aerosols, especially sulfates and soot particles, were found to be responsible for climate changes.[57] The global increase of cloud reflectivity due to anthropogenic aerosols was estimated to be 2%.[58,59,60] Obviously, there are many ways aerosols may influence climate changes, and that is why analysis and control on a global scale is important.

13.5.1   NUCLEAR MICROSCOPY OF SINGLE AIRBORNE PARTICLES

The ultimate goal of environmental pollution monitoring and control is to identify pollution sources, find their impact on the environment, and, if necessary, reduce their concentration to an acceptable level. In the process of source identification, the analysis of single airborne particles has an indispensable value, and this is where NM and other microanalytical techniques could make a significant contribution. The importance of analysis of single aerosol particles is illustrated in the following example: quartz dust and asbestos are well-known hazardous work environment pollutants. Long exposure to quartz dust could result in silicosis (a damaging lung disease), while asbestos causes lung cancer. Both materials are, therefore, very interesting from the pollution monitoring aspect and yet difficult to analyze by bulk techniques (such as PIXE). The reason is that their major components (Si in quartz and Mg-aluminosilicates in asbestos) would be "obscured" by typically high concentrations of aluminosilicates and Mg originating from soil erosion and/or marine aerosols. Other examples are the antimony, tin, and barium particles found in our study.[61] These elements are usually present in aerosols at very low concentrations, and they must be analyzed by their L-lines, which often overlap the intense $K\alpha$ lines of usually abundant Ca and Ti. That will considerably increase detection limits for these elements by broad beam PIXE analysis. On the other hand, when analyzed in a single particle mode, these particles can easily be identified, and their major and trace elements quantified.

In addition to facilitating source identification, analysis of single airborne particles provides detailed information on particle size distribution. This is a very important parameter which determines aerosol impact on health and climate changes. During the last few decades, an enormous amount of knowledge on atmospheric chemistry has been acquired. It is known that each airborne particle is a kind of miniature floating chemical plant that reacts and transforms in contact with other particles and gases and in which numerous chemical reactions continuously go on. By focusing attention on the individual particles, their "life cycle" and evolution can be better understood.

In terms of the number of analyzed single aerosol particles most of the work has so far been carried out by EPXMA and LAMMA by the group from Antwerp.[62,63,64] Particles are usually analyzed automatically, using the particle recognition and characterization code (PRC),[65] which allows fast analysis of a large number of individual, size-segregated particles (10 to 50 thousands of particles are reported). More comprehensive reviews of EPXMA analysis of single aerosol particles can be found in reviews.[66,67]

EPXMA results for single particles comprise typically 5 to 10 major and minor elements. NM, on the other hand, has a much better sensitivity and can provide concentrations of 20 to 30 major, minor, and trace elements even in a single particle mode.[7] Further, both EPXMA and LAMMA are surface or near-surface methods (1 to 3 µm) and, therefore, could in some cases give a wrong estimation of concentrations since some elements are known to be concentrated on the surface of particles (e.g., sulfur and lead). Protons can, on the other hand, penetrate as deeply as 20 µm in a silicate matrix and will, therefore, provide "bulk" information even for large single particles.

13.5.1.1   Overview

The first attempts to analyze airborne particles dates back to 1973 when McKenzie analyzed large soot particles,[68] but it was almost 10 years later when Vis et al.[69] scanned fly ash particles more systemically to find elemental distributions across their cross sections. The same year, 1981, the Bochum group also reported analysis of aerosol particles using micro-PIXE. Due to a limited beam size of approximately 10 to 15 µm, only intensity profiles of linear scan across cascade impactor samples could be obtained.[70]

Successful micro-PIXE analysis of single airborne particles was reported by Lövenstan[71] in 1988, followed by P. Artaxo[72] and Grime and Watt.[73] Artaxo[74,75,76] analyzed a number of aerosol particles from the Brazilian Antarctic station and at biomass burning sites in the Amazon basin. The results were statistically evaluated using factor analysis which enabled source apportionment. Particles of different origins were discriminated. A sea salt component from the Antarctic region was split into three particle types: one that revealed reactions of sulfur gases with NaCl, the other with $CaSO_4$ particles, and the third with a soil-dust component. In other studies,[77,78,79] Artaxo compared and complemented micro-PIXE with other microanalytical techniques, such as EPXMA and LAMMA.

An active collaboration between Oxford, Zagreb, and Milan resulted in a number of reports on analysis of fly ash particles. Jaksic[80] analyzed more than 20 fly ash particles in an attempt to find if there was significant surface enrichment of trace/major elements. Caridi[81] analyzed more than 60 fly ash particles collected at the inlet/outlet ports of the electrostatic precipitator and divided them into several groups on the basis of major element concentrations. Trace and major elements were analyzed in about one hundred coal fly ash particles in the 0.5 to 5 µm size range by Cereda et al.[82] The particles were grouped into seven categories, according to the concentrations of major and trace components, and some toxic elements were found to be preferentially concentrated on smaller particles.

Traxel and Watjen analyzed a number of urban aerosol particles and explored the abilities of the Heidelberg NM facility to investigate mass ratios of trace elements and detection limits of the method.[83]

Several approaches in quantitative analysis of single fly ash particles were investigated by Jaksic et al.[84] Advantages and disadvantages of point analysis and line scans were investigated by comparing experimentally obtained yields with simulated yields for spherical particles obtained with a modified version of TTPIXAN[85] and a Monte Carlo simulation. The importance of matrix corrections was pointed out. Using the same code, Bogdanovic[86] simulated the X-ray map of a single aluminosilicate particle with a surface layer and with a homogeneous distribution of sulfur in order to prove that sulfur is concentrated on the surface of the single large fly ash particle.

In collaboration between groups from Lund and Antwerp, J. Injuk[87] used three X-ray emission techniques (energy-dispersive X-ray fluorescence, EDXRF, EPXMA, and NM) in the framework of a multipurpose study of air–sea exchange processes in the lower troposphere of the North Sea. In another joint work, Treiger[88] investigated the usefulness of a multivariant statistical evaluation technique, nonlinear mapping, in extracting information for source apportionment from NM results of 58 particles. No matrix corrections were made in either of these studies.

Rousseau et al.[89] from the National Accelerator Centre, South Africa, analyzed a number of medium size fly ash particles (18 to 100 µm) by using a combination of EPXMA and NM measurements. They found relatively good agreement between EPXMA and NM results as well as that trace elements accumulate in fly ash particles by various mechanisms.

To reduce background levels coming from the relatively thick filter materials (Nuclepore or similar), a novel cascade impactor backing material, pioloform, often used as a sample support in electron probe microanalysis, was introduced in Singapore.[90] This film, approximately 0.5 µm thick, was found to be rugged enough to withstand cascade impactor pressure and thin enough to allow virtually background free PIXE, RBS, and STIM spectra acquisition. More than 300 single airborne particles were analyzed using Singapore NM, and more details can be found in References 7 and 90. Some of the results of this study will be outlined briefly in the following sections.

13.5.1.2 Sample Collection and Preparation

Aerosol samples to be analyzed by NM or any other microanalytical technique are typically collected using a single-orifice cascade impactor. To enable analysis of single particles, it is important to avoid heavily loaded substrates. Therefore, the collection time (volume) is reduced. Also, to ensure optimum PIXE detection limits, the sampling substrates should be as thin and as clean as possible. These requirements are important for fast STIM and RBS imaging and successful determination of particle thicknesses and major components like carbon and oxygen. Typically, thin kapton, Mylar, or some polycarbonate foils/filters coated with a thin layer of paraffin (or Apeizon/Vaseline) oil are used as substrate materials in cascade impactors. These materials give satisfactory detection limits for broad-beam analysis. However, our preliminary NM tests with all the above-mentioned materials did not satisfy the double requisite of being sufficiently thin as well as clean and stable under the NM beam. Also, the paraffin/Vaseline coating was found to melt under the well-focused proton beam, making single particles simply disappear from the film. After extensive testing, thin pioloform film, which is often used in electron microscopy as sample support, was found indispensable. The film was prepared by dipping a microscopic glass slide into a 0.5% solution of pioloform powder in chloroform. After drying, the glass slide was immersed in water at a large angle and a pioloform film detached from the glass and set floating on the water's surface from which it was collected on an Al ring with a 15 mm inner diameter. After drying, the films were tested for rigidity and uniformity and placed into a cascade impactor. The thickness of such films was estimated to be 20 to 50 $\mu g/cm^2$. They were found to be rugged enough to withstand the cascade impactor air pressure and also "sticky" enough to keep aerosol particles at their impaction site without any additional paraffin film. Most important, the foil was found to be very stable under the proton bombardment and provided excellent PIXE and RBS spectra with virtually no background (see Figure 13.14).

Some authors are also using stack filter unit samples equipped with a Nucleopore polycarbonate filter for single particle analysis.[91] This, as already mentioned, might not be the best choice as the position of the single particles, which are immersed into Apeizon grease, is not stable because of the local melting of the grease by a well-focused proton beam.

For the analysis of single fly ash particles, which are usually available as bulk powder, some groups use the following procedure: fly ash powder is suspended in a dilute solution of pioloform resin in chloroform. Droplets of the suspension are allowed to flow over a glass slide[82] or are simply dropped on the surface of distilled water[38] to form a thin film, which can then be mounted on a frame suitable for NM. The drawback of this method is that any thin layer of volatile/soluble elements (e.g., S, Br, Cl, etc.) which might be deposited on the surface of individual particles will be removed by this method.

Another variation of this method was recently applied by Rousseau et al.[89] The authors embedded particles into petrographic resin and, after it hardened, polished the surface to approximately half the diameter of the particles. In this way, otherwise strong X-ray yield corrections caused by the geometrical effects (see Section 13.5.5) were avoided.

13.5.1.3 Irradiation and Analysis

There are several common irradiation and analysis techniques used in NM of single aerosol particles. Typically, the sample is first analyzed by using a large area scan engulfing the whole deposit. This is to obtain the bulk composition as well as to get an impression of the particle distribution and density. Once the area with suitable particle density is found, the scan size is reduced to 20 to 50 μm and, depending on the sophistication of the data acquisition system and the aim of the project, the user may apply one of the following common techniques: (a) point analysis, (b) predefined masks, and (c) area scan with the list-mode acquisition. Advantages and disadvantages of each technique are briefly discussed here.

Point analysis — the area of interest is irradiated for a short period of time and fast STIM imaging used to identify particles. Typically, several minutes is enough to obtain a satisfactory image. Major particles are identified and coordinates are recorded. Depending on the degree of automation of the data acquisition system used, the beam can be automatically (or manually) positioned to the center of each particle and irradiated for a preset time or total charge. With a typical current of 100 to 500 pA, satisfactory statistics can be obtained within 5 to 10 minutes. The acquired PIXE and RBS spectra are saved for the off-line analysis and the beam is moved to

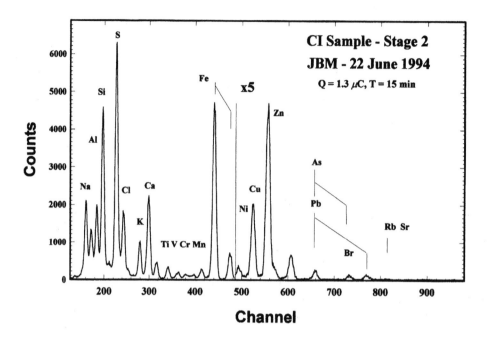

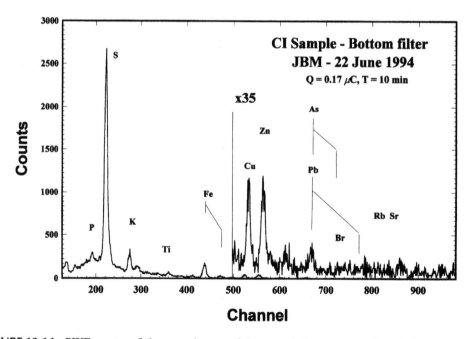

FIGURE 13.14 PIXE spectra of the second stage of the cascade impactor sample and the corresponding bottom filter. Sulfur concentration is clearly much higher on the bottom filter. It can also be noticed that the PIXE background is very low. This is due to a thin pioloform film used for cascade impactor substrate.

the next particle, either manually or automatically. In such a way, hundreds of particles can be irradiated per day. The major drawback of this technique is that the chosen point might not be representative for the whole particle — either in terms of thickness or concentration. Another disadvantage is that the particle size cannot be measured. Moreover, particle heating and subsequent damage and evaporation of volatile particles/elements (NaCl, PbBr, etc.) might be substantial due to a continuous irradiation of the same spot. On other hand, this method offers the fastest analysis.

Predefined masks — this method is similar to the previous one except that the beam is scanned over the whole particle and not positioned at one point only. There are two obvious advantages to this method: First, localized heating of the particle and the beam damage is reduced. Second, as often happens, the acquired elemental maps will show if the particle is actually a cluster formed from several subparticles. If data are acquired in the list mode, the original mask can then be divided into two or more regions to allow for more detailed off-line analysis of all individual particles in the cluster. In this way, the analyst will get better insight into the particle composition and distribution of elements, but of course at the expense of time. The method offers excellent flexibility. One disadvantage is that some of the smaller particles will be missed if only a quick STIM image is used to define all the masks within the field of view. As in the first method, depending on the degree of automation of the data acquisition system, the process could be automated to a high degree.

Area scan — an area of 20×20 to $50 \times 50\ \mu m^2$ is typically irradiated for an hour or more, and all data are acquired in a list-mode. In this way, a large file that contains a detailed history of the whole run, including the energy spectra of all active ADCs and beam coordinates throughout the run, is created. The file can be replayed off-line using digital masks that are engulfing individual particles, and the corresponding PIXE and RBS spectra can be extracted and analyzed. This method offers the highest flexibility. The user can replay the run and change masked areas as many times as needed and, if necessary, analyze even the smallest particles and clusters that are visible within the scanned area. The disadvantage is that the method is time consuming, and if the particle density of the scanned area is low, a considerable charge (beam time) might be wasted on areas that contain no particles.

With the constantly increasing level of automation of NM data acquisition systems, the speed and convenience of single particle analysis will certainly improve in the next few years, allowing for much higher throughput than presently available.

13.5.2 Sample Results

To illustrate the abilities of a state-of-the-art NM analysis of single aerosol particles analysis, some results from the National University of Singapore NM group are presented.

The samples were collected by means of a cascade impactor with four stages and the final, bottom filter. By applying a flow rate of 2 l/min, the equivalent cut-off aerodynamic diameters of particles impacting on the four stages were approximately: >8, 4, 2, and 1 μm. Particles smaller than 1 μm were collected on the bottom, Teflon® filter.

A well-focused beam was scanned over preselected areas of approximately $50 \times 50\ \mu m^2$ and PIXE, RBS, and STIM spectra were collected in list-mode. The total accumulated charge per run was typically 0.5 to 3.5 μC. Individual particles were analyzed off-line by replaying a list-mode file for masks over each particle found within a scan (see the last insert in Figure 13.17).

Such "manual selection" of individual particles (compared to the automated particle recognition and characterization (PRC) method often used in electron probe analysis[92]) is relatively slow but has the advantage of identifying particle agglomerates and analyzing their components. In the course of this study it was realized that particles often form clusters, e.g., particles *Fe2* and *Si2* in Figure 13.17 (see the last insert) are linked together with *Cl1* (marine airborne particle). Also in the same figure, particle *Fe1* seems to be engulfed with another Cl particle, etc. Small particles are often attached to larger ones due to their own adhesion forces or adhesion forces of "intermediate" particles such as water and sea salt droplets. Our study showed that sea salt particles are the dominating group of airborne particles in Singapore air.[95] As these particles are highly hygroscopic, they can act as a "glue" keeping one or several particles together. Such clusters could hardly be isolated with an automated scanning system and as a result, the "average" composition of the cluster would be obtained, together with the size of the cluster instead of individual particles.

The PIXE spectra was analyzed using the program GUPIX[30] running in "matrix" mode with iterated target thickness. Total charge used to irradiate the whole area of the scan was obtained from the corresponding RBS spectra with the use of the code NUSDAN[37] and then multiplied by the relative particle size to obtain the effective charge for that particular particle.

The record for each aerosol particle consists of the following data: particle code, size (area/diameter estimated by assuming circular shape), thickness (actually areal density in $\mu g/cm^2$, mass (pg), and finally the absolute concentrations of up to 29 elements. Absolute concentrations were computed by GUPIX, assuming that each metal (or analyzed element) is present in its most common oxide form (e.g., Fe as Fe_2O_3, Al as Al_2O_3, etc.).

The analyzed particles were categorized into nine groups, according to their major components and weight percentage (abundance). The final results are shown in Table 13.1. An important note has to be made here: the results do not include submicron particles, most of which are sulfates and potassium compounds. According to our bulk analysis studies,[93] sulfates actually contribute up to 60 to 80% of the total aerosol mass in Singapore, but most of it is in the submicron range. See Figure 13.14, in which the spectra of a cascade impactor sample and the corresponding bottom filter are shown.

Marine and soil particles were found to be the most abundant, representing 40% and 46% of the total mass, respectively. They are followed by Fe, Pb, and Ti-rich particles (6.3%, 3.1%, 1.8%, respectively). "Exotic" particles with high concentrations of tin, antimony and barium, most likely coming from nearby incinerators and power plants, were also found (2% in total).

Sulfur, as one of the major anthropogenic pollutants in urban environments, was found to be evenly distributed in all particles (3 to 8% on average) and not in any large, exclusively sulfuric particle. This is expected, because S is released to the atmosphere in the form of SO_2 gas which, through photochemical reactions, undergoes gas-to-particle conversion and forms various sulfates. The initial size of droplets is smaller than 0.1 μm, but they are hygroscopic and thus grow quickly to approximately 0.5 μm in diameter and attach to other airborne particles, forming a thin film over almost all solid particles. As a consequence, all sulfur maps (see Figures 13.15, 13.16, and 13.17) appear diffused and have no contrast or sharp edges. This assumption was proven with the following experiment: one of the cascade impactor samples was gently dipped into a warm (40°C) solution of 5% nitric acid for about 2 minutes, then dipped again into distilled water, and after drying analyzed again by NM. While most of the aerosol particles remained intact on the pioloform film, PIXE spectra showed almost complete absence of the marine component (Na, Mg, and Cl), as well as a high depletion of S, Pb, and Br. This supports the assumption that these elements are concentrated mainly on the surface of larger particles. Such a "shell" around particles will also hinder PIXE quantitative analysis (see Section 13.5.5).

Maps of several elements for three different runs are shown in Figures 13.15, 13.16, and 13.17.

The scan size for the run shown in Figure 13.15 was 100×100 μm^2. A large number of various particles is seen clearly on all elemental maps. Aluminosilicate and NaCl particles are the most abundant. More than 20 Al–Si particles and approximately 40 NaCl particles are well distinguished. Their diameter varies from several μm to 15 μm. Al, as well as Na and Mg maps are not shown because they are almost identical to Si and Cl, respectively. The second largest group are the Ca-rich particles — most likely $CaCo_3$ as a typical soil component (from our previous studies,[28,29] concentration of Ca in sediments in various regions of Singapore varies from 1 to 3%). A surprising result was that some of the Ca particles were found to have high concentrations of Cu, Zn, and S (see particles 1, 2, 3, and 4 on Ca, Cu, and Zn maps) indicating a possible anthropogenic origin, probably an incinerator. Particle #3 is also associated with a large soot particle (see the last insert).

In most cases, the Fe map coincides with the Si map. Typical concentration of Fe in Al–Si particles was found to be 3 to 5%, which is the expected concentration of Fe in soil. Exceptions are particles 1, 2, 3, and 4 on the Fe map. These are not associated with any other element and are most likely iron oxides released into the atmosphere from some of the anthropogenic sources (incinerator, car exhaust, etc.). Similarly, the two largest Ti particles (Ti 1, 2) are stand-alone particles. The same is valid for Ti particles in Figures 13.16 and 13.17. All other, smaller particles

TABLE 13.1
The summary of the analysis of all 314 single aerosol particles. Each record comprises the following: particle code, abundance (weight %), average diameter (μm), estimated average thickness (μg/cm²), average mass (pg), and finally the absolute concentrations of elements. Elements with concentrations below detection limits, such as Sc, V, Cr, Co, As, Se, etc. are not included.

Particle	Abund. (%)	Diam (μm)	Thckns (μg/cm²)	Mass (pg)	Na	Mg	Al	Si	P	S	Cl	K	Ca	Ti	Fe	Cu	Zn	Br	Cd	Sb	Ba	Pb
Al Silicates	46.0	7.6	117.0	67.90	3.9	1.8			0.6	3.4	8.3	2.6	3.8	0.4	3.7	0.11	0.12	0.04	0.07	0.18	0.39	0.01
NaCl	39.8	6.7	80.1	35.81		2.8	2.2	6.2	0.8	7.4		2.4	3.7	0.6	2.5	0.26	0.39	0.07	0.07	0.16	0.34	0.17
Iron	6.3	6.7	79.9	35.50	2.9	1.2	1.9	6.4	0.4	3.5	8.3	1.1	2.2	0.7		0.93	0.29	0.02	0.14	0.22	0.80	0.32
Lead	3.1	6.1	65.2	24.00	4.3	1.6	3.9	6.9	1.4	5.8	12.8	2.5	5.7	3.1	4.0	0.30	0.43	1.71	0.27	1.12	0.82	
Titanium	1.8	5.7	61.0	19.51	5.9	2.1	2.3	5.7	1.1	4.7	16.0	2.3	3.9		4.4	0.14	0.15	0.17	0.33			0.77
Antimony	1.2	6.8	61.0	28.41	4.6	2.2	4.9	12.0	0.8	5.0	9.2	1.4	6.1	4.2	3.4	0.10	2.57	0.25			0.29	0.34
Calcium	0.7	5.3	66.5	18.58	5.1	1.9	1.7	6.1	0.6	7.0	15.1	1.3		0.3	0.7	0.10						
Barium	0.6	5.3	48.4	13.41	5.9	4.4	0.8	6.8	0.3	7.1	15.4	2.0	2.8	0.4	7.5	0.53	0.30		0.24	1.42		
Magnesium	0.6	7.2	42.7	22.39	7.4		0.6	8.5	0.4	5.8	16.3	1.3	1.6	0.1	7.1	0.12	0.08		0.19	0.80	1.72	
Min		5.3	42.7	13.4	2.9	1.2	0.6	5.7	0.3	3.4	8.3	1.1	1.6	0.1	0.7	0.10	0.08	0.02	0.07	0.16	0.29	0.01
Max		7.6	117.0	67.9	7.4	4.4	4.9	12.0	1.4	7.4	16.3	2.6	6.1	4.2	7.5	0.93	2.57	1.71	0.33	1.42	1.72	0.77
Avg		6.4	69.1	29.5	5.0	2.3	2.3	7.3	0.7	5.5	12.7	1.9	3.7	1.2	4.2	0.29	0.54	0.38	0.19	0.65	0.73	0.32
±STD		0.8	20.6	15.3	1.3	0.9	1.4	1.9	0.3	1.4	3.3	0.6	1.5	1.4	2.1	0.26	0.77	0.60	0.09	0.50	0.49	0.25

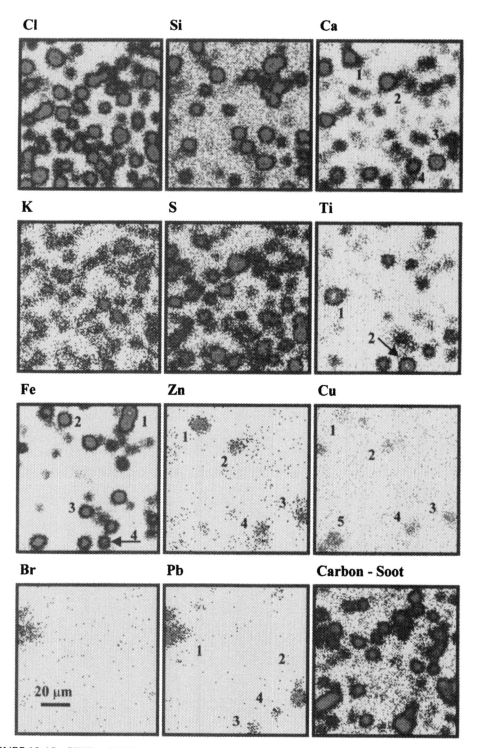

FIGURE 13.15 PIXE and RBS elemental maps of single aerosol particles collected by the cascade impactor. Scan size is $100 \times 100 \ \mu m^2$. PIXE elemental maps of Cl, Si, Ca, K, S, Ti, Fe, Zn, Cu, Br, and Pb, and the RBS map of carbon (soot particles) are shown.

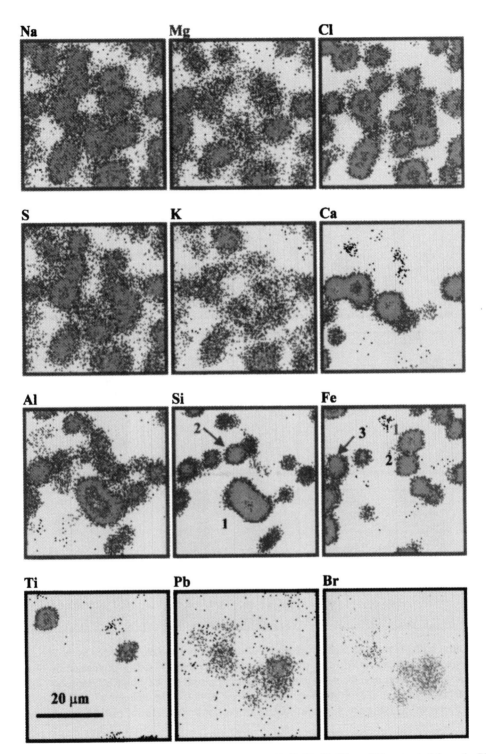

FIGURE 13.16 Elemental maps of Na, Mg, Cl, S, K, Ca, Al, Si, Fe, Ti, Pb, and Br extracted from the PIXE signal. Scan size 50×50 µm². First three images (Na, Mg, and Cl) are well correlated and represent sea-spray particles. Sulfur is also often associated with the sea-spray particles as the submicron sulfuric acid droplets are likely to aggregate with larger NaCl droplets. For more details, see text.

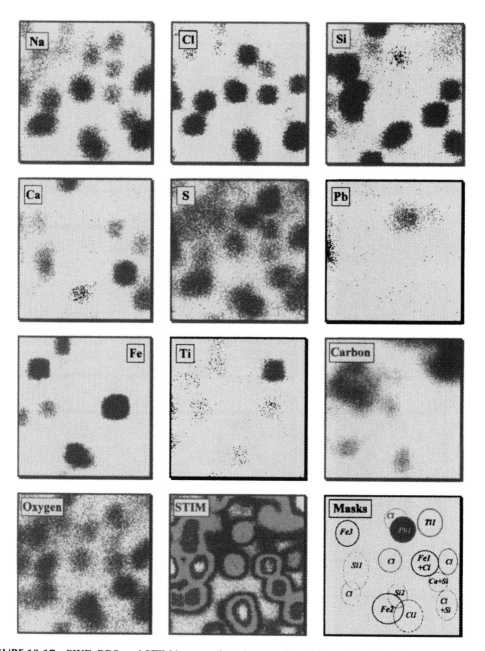

FIGURE 13.17 PIXE, RBS, and STIM images of single aerosol particles collected by the cascade impactor. Scan size $40 \times 40\ \mu m^2$. PIXE maps show: sea spray — Na and Cl, soil — dominated by Si and $CaCO_3$, anthropogenic sources represented by transition and heavy metals (Ti, Fe, Pb). RBS maps show light elements, carbon (soot particles), and oxygen from aluminosilicate particles. The contours of all particles can easily be viewed from the STIM image. In the last insert, contours of several selected particles are shown.

on the Ti map are associated with Al–Si particles (average concentration of Ti in Al–Si particles was found to be 0.4%) (see Table 13.1).

As already discussed in the previous paragraph, sulfur and potassium maps do not show sharp contrast but are rather diffuse due to the fact that these two elements come in very fine particles that form a thin coating over the whole sample.

Lead–bromo–chloride particles are well known car exhaust products (see, for example, D.D. Cohen[94]). They are often associated with large soot particles, as is the case in Figure 13.15. It has been noticed that Pb particles are not always accompanied by Br and Cl particles — see, for example, Pb particle 2, 3, and 4 in Figure 13.15. The depletion of Br, especially during the summer periods, was already noticed by other authors.[5,95,96] The effect is caused by a photochemical reaction in which Pb–Br–Cl car exhaust products react with acidic particles and form a volatile HBr, causing depletion of Br, just as it had already driven off the Cl from filters by forming hydrochloric acid (HCl). Jambers et al.[5] noticed that ammonium sulfate was present as a surface element on the Pb-rich particles.

The last insert in Figure 13.15 shows a number of soot particles extracted from the RBS signal. The carbon signal coming from soot particles is well contrasted from the carbon signal coming from the pioloform substrate, because the thickness of the pioloform film is much thinner than the average soot particle.

13.5.3 QUALITY ASSURANCE

To ensure reliability of the analytical result, it is good laboratory practice to run a set of calibration standards at the beginning of each analysis session. For this purpose, NIST (National Institute of Standards and Technology, USA), IRMM (Institute for Reference Material and Measurements, Belgium), and IAEA (International Atomic Energy Agency, Vienna) produce a number of thin (or nearly thin) aerosol (or aerosol-like) standard reference materials (SRM). A systematic study of 14 of these SRMs by PIXE can be found in a report by Maenhaut et al.[96]

In the NM laboratory in Singapore, the sensitivities of the main objective elements are typically measured using NIST SRM (thin glass films, SRM 1832 and 1833) and standard foils produced by the Micro-Matter Co., U.S.A. The results of such a long-term system calibration are shown in Figure 13.18, in which 23 independently obtained concentrations of Si and Mn during the period January to April 1996 are shown. Concentrations are normalized to the reference values and plotted vs. time (measurement number). Deviations of 3% to 7% can be explained by small variations in the instrumental constant caused by changes in detector positions (solid angles of X-ray and RBS detectors), distance and tilt of the sample, local variations in SRM material concentrations and thicknesses, etc.

The same sets of SRM materials are found very useful to determine system sensitivities. Two examples of determination of system sensitivities are shown in Figure 13.11b.

13.5.4 STATISTICAL EVALUATION OF DATA

For statistical evaluation of a large number of data, it is common to use some of the available receptor models, such as chemical mass balance, principal component analysis (PCA), factor analysis, or cluster analysis.[97,98] These methods, if applied properly, can provide information on source apportionment or enable cluster analysis — i.e., identification of groups of particles of different origin and/or composition. On the other hand, according to some authors, the current factor analysis receptor models are ill-posed[99,100] and have a serious limitation in credibility and applicability in aerosol analysis. According to these authors, it is difficult to expect sensible results from first trials. Data processing usually requires some preprocessing and selection of parameters/data to be analyzed. This was also our experience. The first attempt of principal component analysis performed over all available data/particles failed to clearly isolate clusters. The reason was quite obvious: more than half of the analyzed particles from our samples belong to marine aerosols (major components Na, Cl, and Mg), while the majority of the other particles contain marine components in amounts ranging from a few % up to 40%. This high "background" obscured the other components present in smaller amounts. Therefore, our data set was "filtered" by eliminating the "noise" coming from the marine aerosol particles. The results of PCA over such a reduced data set (156 particles instead of 312) are graphically presented in Figure 13.19b for principal component (PC) #1 vs. PC #5. In this example, PC#1 has high loadings of Al, Si, S, and K (obviously dominated

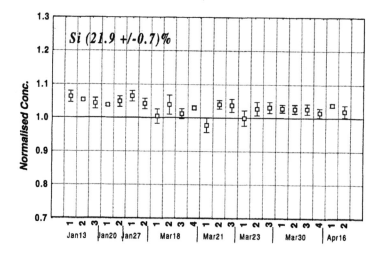

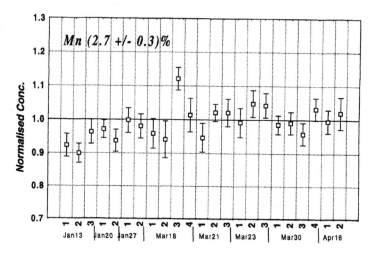

FIGURE 13.18 The results of a quality assurance exercise for Si and Mn. The same standard, NIST 1833 was irradiated every time a batch of aerosol samples was analyzed. As a result, more than 20 spectra were collected and analyzed during the period January to April 1996. Results are normalized to the certified values and plotted against the date. Deviations of 3 to 7% from the certified values are observed.

by aluminosilicate particles), and PC#5 has the highest loading of Ca, Ni, Zn, and Sb (possibly incinerator products). The resultant graphs show very distinct "clusters" of particles with different major elements. In the example, silicon-rich particles form a well-isolated cluster #1; iron-rich particles are grouped in the cluster #2; antimony- and barium-rich particles form clusters #3 and #4, while the majority of lead particles is in the cluster #5 with only two particles in #6. Two lead particles from cluster #6 differ from the rest of the Pb particles in their relatively high concentration of Fe and Mg. Graphs of other PCs show similar clustering, but each emphasizes a different component. When the data set is reduced even further to only iron, silicon and lead-rich particles (all together 86 particles), clusters are even better distinguished (see Figure 13.19a.) Again, the lead particles form two groups; the particles in the cluster #4 contain much higher concentrations of Ti and Fe and no Br at all.

Even though PCA shows distinct clusters, the interpretation of such diagrams is often difficult because the principal components are functions of many parameters and the relationship between PCA clusters and particle families is not obvious.

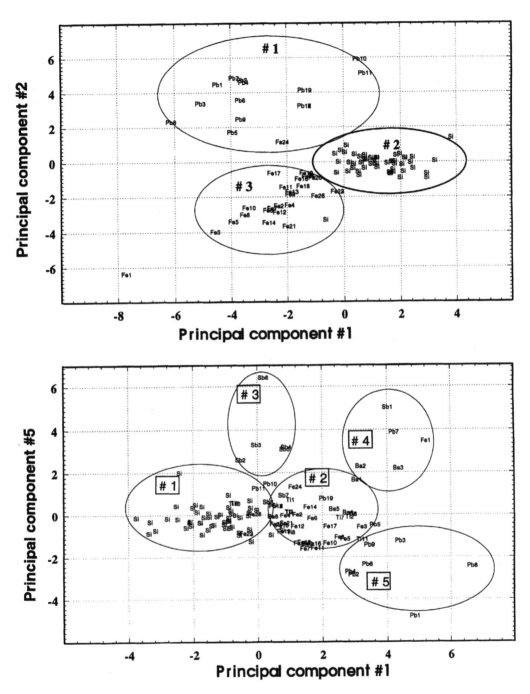

FIGURE 13.19 Results of principal component analysis. For details, see text. (From I. Orlic, T. Osipowicz, F. Watt and S.M. Tang, B104, 1995, 630-637. With permission. Copyright Elsevier Science Publishers.)

Another commonly used statistical method is the hierarchical cluster analysis (HCA). The results of HCA are typically presented in a form of "dendograms" or "tree" diagrams. The particles are grouped in clusters according to their "similarity," a number derived from an analysis of the differences between their elemental concentrations. In order to evaluate the performance of various hierarchical cluster techniques, P.C. Bernard and R. Van Grieken[101] tested seven hierarchical cluster techniques on large data sets from EPXMA analysis of known mineral mixtures. Their conclusion was that the Wardt's method[102] provides the most successful particle classification.

13.5.5 QUANTITATIVE ANALYSIS OF SINGLE AEROSOL PARTICLES

As already discussed, quantitative analysis of single particles is not an easy task. The main obstacles are caused by local variations in X-ray absorption due to geometrical effects. An analytical solution for an ideal, spherical particle was developed in 1991 by Jex.[103] However, the airborne particles are seldom ideal spheres, they come in all shapes and sizes, overlap each other, and form complex clusters. Some elements are often concentrated on the surface (Section 13.5.2). Therefore, a number of approximations are generally used to obtain quantitative data of such complex systems. In principle, it is possible to obtain excellent quantitative results for supermicron particles (particles with diameter larger than 1 μm) from a combination of micro-PIXE, RBS, STIM, and microtomographic 3-D reconstruction techniques. RBS can provide pixel-by-pixel concentrations of the major matrix elements and concentration depth profiles, while STIM tomography can be used to derive shape, internal structure, and areal density. With all these parameters, one can then extract accurate trace and minor element concentrations from the PIXE data. Such a procedure is presently feasible but would certainly be extremely time consuming. Therefore, in practice, simplified models are used, and their accuracy and limitations are briefly discussed in the following paragraphs. We shall consider the accuracy and the limitations in quantitative analysis for the three most often encountered situations in aerosol analysis: uniform intermediate thick sample, simple geometrical shapes (cube and sphere), and geometrical shapes with surface layer.

13.5.5.1 Intermediate Thick Sample

A thin sample in PIXE terminology is assumed to be a very thin slab of matter in which effects of absorption and ion energy loss can be neglected. This is an idealization often used to calculate a so-called thick target correction factor (TTCF). TTCF is defined as the ratio of the X-ray yield emerging from a real intermediate thick sample and the calculated yield of an identical, but "ideal" sample (a sample of the same composition and areal density but without attenuation and energy loss). Such a factor is obviously always smaller than unity. As an illustration, TTCFs are computed for S, Si, Ca, Fe, and Zn in a typical aluminosilicate matrix and for 2 MeV proton excitation. The sample is assumed to be homogenous and flat, with well-defined thickness (area density). Results of computation are shown in Figure 13.20. As expected, the yields of elements with soft X-rays are affected more than those with higher X-ray energies. For example, sulfur K lines emerging from a 4 μm thick sample will have only 50% of the intensity emerging from an equivalent ideal sample. For the same thickness, Fe and Zn K lines will be reduced only less than 5%. This clearly shows the importance of the effect — if corrections are not made, the obtained concentrations will be underestimated by the same factor. The accuracy of PIXE results will, therefore, depend strongly on the choice of the proper sample thickness. Obviously, light elements like Na, Mg, Al, Si, S, etc., will be affected the most. The sample thickness (areal density) is usually estimated either gravimetrically, or iteratively from the PIXE yields, or from the RBS signal. All these methods give good estimates of the areal density for a uniform sample, but not for a nonuniform sample, as for example cascade impactor samples. For such cases, estimated mean areal density will be smaller than the actual particle thickness, implying a wrong estimation of thick target corrections. To estimate the significance of this effect, elemental concentrations of a real aerosol sample collected on a stretched Teflon filter are calculated using Gupix[30] for four different sample thicknesses, i.e., 80, 160, 300, and 600 μg/cm^2 (equivalent to 0.27, 0.53, 1, and 2 μm thickness, respectively, assuming a volume density of 3 g/cm^3). The concentrations obtained (in μg/cm^2) are then normalized to the values obtained for the lowest thickness and shown as percentages in Table 13.2. Deviations larger than 60% are observed for the light elements, and only 5 to 10% for transition metals. Because

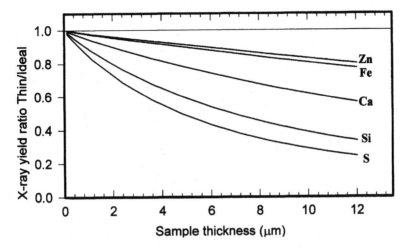

FIGURE 13.20 Thick target correction factor vs. target thickness computed for a typical aluminosilicate matrix and 2 MeV proton excitation for the homogeneous and flat sample. For details, see text.

TABLE 13.2
Relative concentration change calculated by GUPIX for different sample thicknesses. Concentrations are calculated for a real aerosol sample collected on a Teflon filter and normalized to the values obtained for the thinnest sample (80 µg/cm²). Large deviations, especially for the light elements, are observed.

Thickness µg/cm²	Al	Si	S	Cl	K	Ca	Ti	Fe	Cu	Zn	Br
165	9.1	6.0	4.4	2.7	2.4	2.3	1.4	0.9	0.8	0.6	0.5
300	25.9	17.3	12.8	9.4	7.5	7.2	4.9	3.3	3.0	2.7	2.2
600	67.3	43.8	31.9	24.1	18.3	17.6	11.9	8.0	7.2	6.5	5.3

the actual sample thickness is impossible to estimate accurately, these are the ranges of errors one might expect when analyzing aerosol samples by the ion beam.

13.5.5.2 Simple Geometrical Shapes

The transition from a broad beam to NM makes quantitative analysis even more challenging. For illustration, the following experiment is performed: a long, flat aluminosilicate film is cut into ideal cubes and distributed sparsely over a large area, as shown schematically in Figure 13.21a. Using the same experimental conditions as before (see previous section), the total intensity and intensity distributions for Si and Fe are calculated using the computer code TTPIXAN.[42,51,85] Intensities normalized to unit areal density and charge are shown in Figure 13.21. Very asymmetric distributions are found with much stronger intensities on the right (detector) side of the sample. This intensity enhancement is caused by the reduced attenuation of characteristic X-rays emerging through the vertical side of the cube (see the insert in Figure 13.21). Another interesting result is that the Si intensity, which would normally reach saturation at a sample thickness of approximately 8 µm, is still increasing with the sample size. The enhancement effect is calculated for various cube sizes ranging from 0.5 µm to 20 µm, and the results are shown graphically in Figure 13.22 for Al,

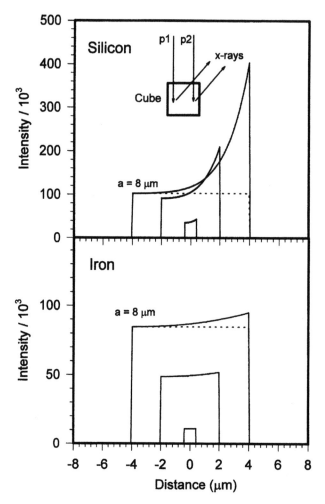

FIGURE 13.21 The simulated Si and Fe intensity distributions computed for the aluminosilicate cubical sample with the size of 0.8, 4, and 8 μm. The enhancement of Fe and especially Si intensity at the right (detector) side is due to the reduced attenuation of characteristic X-ray lines emerging through the vertical side of the cube — see the insert on the top figure and the difference in X-ray exiting path for the incident proton beam p1 and p2.

S, and Fe. X-ray lines of light elements are being enhanced more than 60% compared to the yields emerging from a long, uniform slab of matter of the same area density. This effect will, to some extent, compensate the TTCF caused by the attenuation of the soft X-rays (see Section 13.5.5.1).

A similar simulation for spherical particles shows an almost identical enhancement effect. In this simulation, a hypothetical spherical aluminosilicate particle is irradiated with a proton beam and the characteristic X-ray yields of Al, Si, S, and Fe are computed. The X-ray detector is again positioned at 45° with respect to the incident proton beam, and the beam is scanned over the sphere with approximately 100 steps in each, x and y direction. The obtained X-ray intensity distribution of the Si Kα line is shown in the lower part of Figure 13.23. Obviously, artifacts of this effect will show up when such intensity distributions are directly translated into elemental maps. The maps will have much higher intensities on the right (detector side), giving the impression that the concentrations of light elements are higher on that side. Such effects were discussed by several authors[38] but seldom in a quantitative manner. The group from the Rudjer Boskovic Institute, Zagreb, found good agreement between simulated contour maps for various elements and the corresponding experimental maps obtained for fly ash particles.[80,86] Recently, the group from Max-

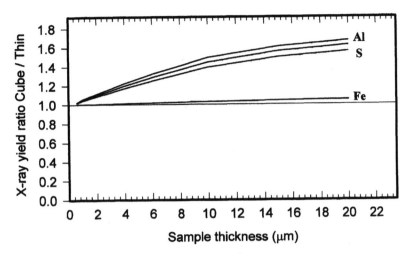

FIGURE 13.22 Intensities of Al, S, and Fe Kα X-ray yields emerging from a cubic aluminosilicate samples normalized to the yields produced in the same, but infinitely long sample of the same thickness. Enhancement effects are observed for characteristic X-ray lines of light elements. All intensities are normalized to the same areal density and the incident charge. See text for details.

Planck Institut für Kernphysic, Heidelberg, also developed computer code for quantitative analysis of spherical particles and discussed the accuracy of the method.[104]

13.5.5.3 Surface Layers on Spherical Particles

Elements of anthropogenic origin like sulfur, lead, bromine, etc., are present in aerosols mainly as submicron fractions. As such, they often form a thin film on larger particles of terrestrial origin (silicates, carbonates, etc.). Bioavailability or toxicity of an element will depend largely on its concentration profile. Clearly, surface elements will be more readily absorbed by the organism than those which are buried in the particles.[105,106] Therefore, to assess the environmental impact of an element or compound, it is important to know its absolute concentration as well as its concentration depth profile. Attempts were made by Cox[107] to study elemental profiles within fly ash particles using SIMS (secondary ion mass spectrometry) as well as by Valkovic et al.[108] using PIXE. Generally, very little NM work is reported on this subject, mainly because of the limitations and difficulties in quantitative analysis of such thin films on micron size particles.

To estimate the effect of such a "shell" on quantitative analysis of individual aerosol particles, a simulation was performed. The same experimental conditions as in the previous example, were used, but this time with a thin (0.5 μm) layer of sulfur on the surface of the aluminosilicate sphere. The computed intensity distributions of elements like Si and Fe remain almost the same, but sulfur intensity acquires an entirely different shape with a dip in the center and a much higher intensity on the detector side. The intensity profiles are shown in Figure 13.24 for particles with 2.5, 5, 10, and 20 μm diameter. For smaller particles, a "crater-like" intensity distribution is obtained, while for the bigger particles the right (detector) side becomes much "shinier" than the other side, which is in the "shade." As observed before, the intensity profiles for Si and Fe are symmetric for the two smaller particles and are more deformed for bigger ones.

Simulation shows that if the sulfur forms a shell around a silicate particle and matrix correction is performed using a "classical" approach (assuming homogeneous distribution of all elements and flat target), the obtained sulfur mass will be overestimated by a factor of two or more, depending on the particle size and composition. Again, we see that there is yet another effect that can hinder the accuracy of quantitative analysis of individual particles.

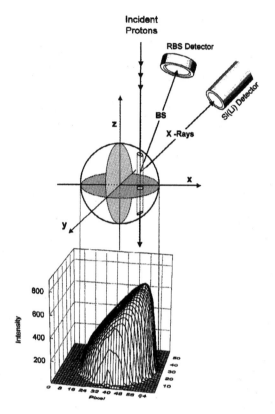

FIGURE 13.23 (Top) Excitation and detection geometry used in calculation of simulated X-ray yields from a spherical particle with the diameter of 10 μm. 2 MeV proton beam was scanned over the x–y plane with the pixel size of approximately 1/100 of the sphere diameter. (Bottom) The simulated Si Kα intensity distribution. The distribution is clearly skewed toward the detector side (that side appears shiner) and will be more so for elements with the lower characteristic X-ray energies.

13.5.5.4 Beam Profile

The elemental maps are convoluted functions of the beam profile and the exact particle projection on the x–y plane. If the beam profile is much smaller than the particle size, the elemental map will, to a good approximation, present a real size of the particle. However, when the beam size is approaching the particle size the image will reflect both, and the particle size will be overestimated. This will significantly affect the quantitative analysis. The problem was already discussed by the author[7] and will be briefly reviewed here using the following example: assume that the beam has a "top-flat" profile with gaussian edges as in Figure 13.25. If such a beam is scanned over the particle of approximately the same size (2 μm in this example), the final convoluted PIXE or RBS images will generally have a line intensity distribution as shown in Figure 13.25b.

For a given example, the particle image will have an area approximately four times larger than the original particle's projection. Therefore, the areal density (thickness) obtained from the corresponding PIXE and RBS spectra will be only one-fourth of the real value. As a consequence, the total mass of the particle obtained from the PIXE and RBS spectra will be correct, but the area densities were underestimated for the size ratio. Effectively, instead of obtaining densities as shown in Figure 13.25a, the observed elemental profiles will be as shown in Figure 13.25c and as such, underestimated particle thicknesses will in turn, result in underestimated matrix corrections.

The beam profile effect can be partially corrected by using appropriate image processing techniques. If the beam profile is known, it is possible to restore the image of a particle. Initial

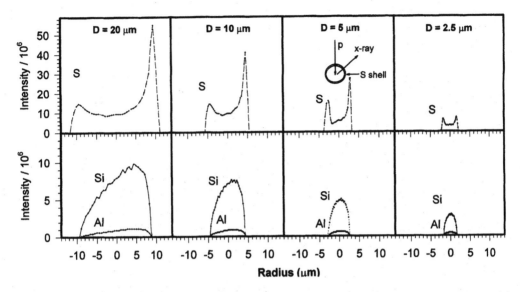

FIGURE 13.24 Sulfur, silicon, and iron intensity profiles computed for the spherical aluminosilicate particles of various diameter covered with a 0.5 μm thin sulfur "shell." If the sulfur is uniformly distributed through the sphere volume, the intensity profiles would be similar to the Si profiles.

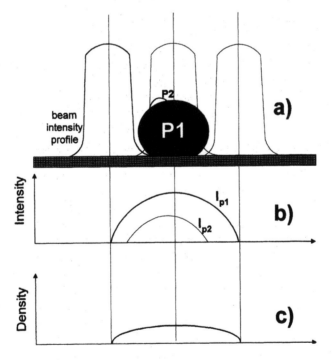

FIGURE 13.25 Proton beam with the profile shown in the inset (a) is scanned over particle P1 and P2. The approximate intensity distributions are shown in (b) for both particle. (c) Broadening of the particle's image will result in much lower thickness obtained from the RBS and PIXE spectra/maps. (From I. Orlic, *Nucl. Instr. Meth.*, B104, 1995, 602-611. With permission. Copyright Elsevier Science Publishers.

work on restoration of NM images was done by P.H.A. Mutsaers et al.[109] and D. Jiang et al.[110] The authors attempted NM image deconvolution by applying Fourier transformation with subsequent filtration to reduce the noise. Recently V. M. Prozesky et al.[111] demonstrated the use of maximum entropy and Bayesian techniques for information recovery in deconvolution of beam profile and spectrum fitting. B. Fisher[112] proposed a new method for measuring beam profile which might be useful in this kind of study.

13.6 DETECTION LIMITS

For point analysis, or even for small area scans, minimum detection limits will generally be similar to those in the broad beam PIXE analysis. This means 1 to 10 ppm for transition metals and somewhat higher for other elements. This is because the effective current densities for small scan sizes (several μm²) are of the same order of magnitude as in a broad beam PIXE. On the other hand, detection limits will be somewhat higher if a larger area is scanned due to a smaller effective charge per particle — if particles are sparsely distributed, most of the beam is "wasted" on the areas not covered by the sample. For a realistic situation, minimum detection limits for a single particle analysis will typically be 10 to 100 ppm for the total accumulated charge of 1 μC and scan size of 25×25 μm².[7,113]

An obvious advantage of micro-PIXE is that such low detection limits can be achieved for small samples which are only several microns in diameter and with a total mass of a pg or less. As a consequence, the absolute detectable mass will be of the order of 10^{-16} to 10^{-17}g. This puts NM in a very unique position of being able to provide major and trace element concentrations for very small samples and at the same time retaining a high spatial resolution.

13.7 SOFTWARE FOR QUANTITATIVE ANALYSIS IN NM

The most commonly used computer programs for PIXE spectra fitting and quantitative analysis are GUPIX,[30] PIXAN,[114] SPEX,[115] etc. RUMP[36] is the most often used program for RBS spectrum fitting. These programs are based on oversimplified assumptions and cannot be directly applied to the complicated situations that appear in environmental, geological, and similar applications of NM. As was demonstrated in several examples in Section 13.5, conversion of PIXE intensity maps to the corresponding concentration maps is often a difficult task. Local concentration/density variations can be very high and cause problems such as severe secondary excitation, completely wrong estimations of X-ray attenuation and energy loss corrections, charge integration problems, etc. All these effects can severely hinder quantitative analysis and lead to misinterpretations. These problems are particularly important for light elements in environmental, geological, metallurgical and similar applications where trace elements are to be analyzed in the matrix dominated by medium and/or high-Z elements with large local variations of concentrations/densities, either in lateral directions or in depth.

New computer programs able to solve such complex analytical problems are needed. Several NM groups are developing suitable software. Ryan et al. developed a "Dynamic Analysis" package which is able to convert intensity maps to more realistic concentration maps.[116,117] The algorithm is very fast, able to do the conversion on-line, and can eliminate some artifacts often encountered in intensity maps caused by pile-up effects and line interference. In Oxford, Grime combined RUMP and PIXAN[115,118] and recently GUPIX[30] in a very useful package developed for off-line RBS and PIXE spectrum analysis. It is also worth mentioning that an advanced version of RUMP which can simulate RBS spectra taken from homogeneous samples, characterized by low energy tails, was developed by Moretto.[119] Other groups are developing methodologies and software for PIXE and STIM tomography; Schofield,[120,121] and Saint and Cholewa.[122] In Singapore the program TTPIXAN, initially developed by the author,[85] has been further extended by Loh and Liew[37,123] to simulate NM elemental maps for a complex 3-D sample composition. The program is now developing further, allowing a 3-D reconstruction by means of an iterative maximum-likelihood procedure.[124] The powerful reconstruction capabilities of the algorithm have been demonstrated[42,124] where good microtomographic reconstruction was achieved using only four projections instead of 100 or more required by classical computer tomography techniques.

13.8   OTHER ENVIRONMENTAL APPLICATIONS OF NM

Hair analysis was one of the first applications of NM in bioenvironmental applications. Initiated in the mid-70s by Cookson et al.[125] and followed by Horowitz,[126] and Valkovic and Vis.[127] In the 1980s, A.J.J. Bos and R.D. Vis from the NM group from the Free University, Amsterdam, did extensive studies on longitudinal and radial distribution of major, minor, and trace elements in different hair samples.[128,129] Other NM groups followed the work, and the results are reviewed by F. Watt.[130] Hyperaccumulation of metals like Co, Cu, Ni, Zn, and Pb by certain plants is a well known but poorly understood phenomenon. Some initial studies of accumulation of Ni by the plant *Senecio coronatus* has recently been reported by Przybylowicz.[131,132] Analysis of growth rings (strontium and fluorine markings) in shells was studied by Coote and Trumpetter,[133,134] elemental analysis of annual rings of oysters and mussels by Swann,[135] Nystrom,[136] Z. Dai,[137] ivory by Prozesky,[138] and tree rings by Gourlay and Grime.[139]

Lake and sediment chemistry was studied by Grime and Davison[140] as well as suspended particles in lake waters[141] and leaching of aluminum from the weathering profile of the African surface.[142]

An excellent review of numerous applications of nuclear microscopy can be found in the book edited by F. Watt and G. Grime,[32] and a more recent book edited by Johansson et al.[18]

13.9   CONCLUSION

Nuclear microscopy is a novel multielemental analytical technique that uses medium-energy ion beams focused down to approximately 0.5 μm spatial resolution. It has a unique ability to simultaneously use several complementary nuclear analytical techniques like PIXE, RBS, STIM, PIGE, etc. When combined, they can provide highly quantitative analytical data for almost all elements of the Periodic Table, while retaining high spatial resolution. This puts NM at the cutting edge of microanalysis.

For analysis of individual aerosol particles, minimum detection limits of approximately 10 ppm can routinely be obtained. In terms of absolute detectable mass, impressive detection of 10^{-17} g can be achieved by the micro-PIXE technique. Further, the detection limits are significantly improved for elements/particles present in very low concentrations (e.g., Sn, Sb, Ba-rich particles). In a broad-beam PIXE analysis, these elements are difficult to detect because of strong interferences of their L lines with K X-ray lines of abundant elements like Ca and Ti.

The success of the application of NM in environmental sciences will ultimately depend on identifying significant problems and our ability to solve these problems. There is still room for improvement in terms of beam stability, focusing, and especially data acquisition and analysis software and hardware.

REFERENCES

1. T.A. Cahill, *Nucl. Instr. Meth.,* B57, 1993, 217.
2. T.A. Cahill, J. Miranda and R. Morales, *Int. J. PIXE,* 1, 1991, 297.
3. F. Watt, *Nucl. Instr. Meth.,* B104, 1995, 647-648.
4. W. Jambers and van Grieken, *Trends in Analytical Chemistry,* 15, 1996, 114-122.
5. W. Jambers, Debock L., and R. van Grieken, *Analyst,* 120, 1995, 681-692.
6. W. Jambers, Debock L., and R. van Grieken, *Fresenius Journal of Analytical Chemistry,* 355, 1996, 521-527.
7. I. Orlic, *Nucl. Instr. Meth.,* B104, 1995, 602-611.

8. G. Grime, The analysis of individual environmental particles using microPIXE and nuclear microscopy. *X-ray Spectrometry,* 27, 221-234, 1998.
9. L. Grodzins, P. Boisseau, H. Glavish, R. Klinkowstein, W. Nett, R. Shefer, *Nucl. Instr. Meth.,* B104, 1995, pp. 1-6.
10. S.H. Sie, *Nucl. Instr. Meth.,* B130, 1997, pp. 592-607.
11. S.H. Sie, W.L. Griffin, C.G. Ryan, G.F. Suter, and D.R . Cousens, *Nucl. Instr. Meth.,* B54, 1991, p 284.
12. J.L. Campbell, G.K. Czamanske, L. MacDonald, and W.J. Teesdale, *Nucl. Instr. Meth.,* B130, 1997, pp. 608-616.
13. G.K. Czamanske, T.W. Sisson, J.L. Campbell and W.J. Teesdale, *Am. Mener.,* 78, 1993, p 893.
14. T. Sakai, T. Hamano, T. Suda, T. Hirao, T. Kamiya, *Nucl. Instr. Meth.,* B130, 1997, pp. 498-502.
15. M. Takai, *Nucl. Instr. Meth.,* B130, 1997, pp. 466-469.
16. H.G.J. Moseley, *Phil. Mag.,* 26, 1913, 1024.
17. T.B. Johansson, K.R. Akelsson and S.A.E. Johansson, *Nucl. Instr. Meth.,* 84, 1970, 141.
18. S.A.E. Johansson, J. Campbell, and C.G. Malmqvist, *Particle Induced X-ray Emission Spectrometry (PIXE),* John Wiley & Sons, Chichester, 1995.
19. S.A.E. Johansson and J. Campbell, *PIXE a Novel Technique for Elemental Analysis,* John Wiley & Sons, Chichester, 1988.
20. J.A. Cooper, *Nucl. Instr. Meth.,* 106, 1973, 525.
21. S.A.E. Johansson and T.B. Johansson, *Nucl. Instr. Meth.,* 137, 1976, 473.
22. J.W. Winchester, *Nucl. Instr. Meth.,* 181, 1981, 367.
23. C.G. Malmqvist, *Nucl. Instr. Meth.,* B3, 1984, 529.
24. K.G. Malvquist, *Nucl. Instr. Meth.,* 104, 1995, 138-151.
25. W. Maenhout, Analytical techniques for atmospheric trace elements, in *Control and Fate of Atmospheric Trace Metals,* 259-301, Eds. J.M. Pacyna and B. Ottar, Publ. Kluwer Academic Publisher, 1989.
26. K.G. Malmqvist, *Nucl. Instr. Meth.,* B85, 1994, 84.
27. K.G. Malmqvist, *Nucl. Instr. Meth.,* 113, 1996, 336.
28. S.M. Tang, I. Orlic, and X.K. Wu, Analysis of Singapore Marine Sediments by PIXE, *Nucl. Instr. Meth.,* B136-138, 1998, pp. 1013-1017.
29. I. Orlic, J. Makjanic and S.M. Tang, Multielemental analysis of marine sediments from Singapore coastal region by PIXE and XRF analytical techniques, *Proceedings of ASEAN-Canada Cooperative Programme on Marine Science,* Midterm Technical Review Conference, Singapore, 24-28 Oct. 1994, pp. 314-326.
30. J.A. Maxwell, J.L. Campbell and W.J. Teesdale, *Nucl. Instr. Meth.* B43, 1989, 218.
31. K.A. Sjoland, P. Kristiansson, M. Elfman, K.G. Malmqvist, J. Pallon, R.J. Utui, C. Yang, *Nucl. Instr. Meth.,* B130, 1997, pp. 20-24.
32. F. Watt and G. Grime, *Principles and Applications of High-Energy Ion Microbeams,* Adam Hilger, Bristol, 1997.
33. K.G. Malmqvist, *Nucl. Instr. Meth.,* B130, 1997, pp. 138-151.
34. Ion Beam Analysis, *Proceedings of the 11th Conference on Ion Beam Analysis (IBA), Nucl. Instr. Meth.,* B85, 1993, as well as *Proceedings of the 12th and 13th IBA Conferences.*
35. T.A. Cahill, J. Zubillaga, The role of ion beam based analysis in global climate research, AIP Conference Proceedings 392, *Applications of Accelerators in Research and Industry,* Part One, 1996, pp. 525-529.
36. L.R. Doolittle, *Nucl. Instr. Meth.,* B9, 1985, 344.
37. K.K. Loh, C.H. Sow, I. Orlic and S.M. Tang, *Nucl. Instr. Meth.,* B77, 1993, 132.
38. I. Bogdanovic, M. Jaksic, S. Fazinic, V. Valkovic, *Nucl. Instr. Meth.,* B85, 1994, 732.
39. P. Formenti, M.B.H. Breese, S.H. Connell, B.P. Doyle, M.L. Drummond, I.Z. Machi, R.D. Maclear, P. Schaaff, J.P.F. Sellschop, G. Bench, E. Sideras-Haddad, A. Antolak, D. Morse, Heavy ion and proton beams in high resolution imaging of a fungi spore specimen using STIM tomography, *Nucl. Instr. Meth.,* B130, 1997, pp. 230-236.
40. A. Sekellariou, M. Cholewa, A. Saint, G.J.F. Legge, *Nucl. Instr. Meth.,* B130, 1997, pp. 235-258.
41. R.M.S. Scofield, H.W. Lefevre, *Nucl. Instr. Meth.,* B77, 1993, p 217.
42. Y.K. Ng., I. Orlic, S.C. Liew, K.K. Loh, S.M. Tang, T. Osipowicz, F. Watt, *Nucl. Instr. Meth.,* B130, 1997, pp. 109-112.

43. D.J.W. Mous, R.G. Haitsma, T. Butz, R.H. Flagmeayer, D. Lehmann, J. Vogt, *Nucl. Instr. Meth.*, B130, 1997, pp. 31-36.

44. F. Watt, I. Orlic, K.K. Loh, C.H. Sow, P. Thong, S.C. Liew, T. Osipowicz, T.F. Choo and S.M. Tang, *Nucl. Instr. Meth.*, B85, 1994, pp. 708-715.

45. G.W. Grime, and F. Watt, *Nucl. Instr. Meth.*, B77, 1993, pp. 495.

46. F. Watt. *Nucl. Instr. Meth.*, B104, 1995, 647-648.

47. N.E.G. Lovestam, *Nucl. Instr. Meth.*, B77, 1993, 71.

48. A.J. Antolak, D.H. Morse, D.W. Heikkinen, M.L. Roberts, G.S. Bench, *Nucl. Instr. Meth.*, B130, 1997, pp. 211-218.

49. I. Orlic, T. Osipowicz and C.H. Sow, L-X-ray production cross sections of medium Z elements by 4He ion impact, *Nucl. Instr. Meth.*, B136-138, 1998, pp. 184-188.

50. T.A. Cahill, R.A. Eldred, D. Shadoan, B.P. Perley, P.J. Feeney, B.H. Usko, and H. Miyake, In Transactions of the Twelfth Intern. Symp. on Appl. Of The Ion Beams In Material Science, Hosie University, Tokyo, Japan, 1978, p 433.

51. I. Orlic, K.K. Loh, S.C. Liew, Y.K. Ng, J.L. Sanchez, and S.M. Tang, *Nucl. Instr. Meth.*, B130, 1997, 133-137.

52. M.K. Owen, D.S. Ensor and L.E. Sparks, *Atmospheric Environment*, 26A, 1992, 2149.

53. T.A. Cahill, *Nucl. Instr. and Meth.*, B75, 1992, 219.

54. S. Bakan and H. Hinzpeter, 1988, *Atmospheric Radiation, Ladolt-Bornstein*, New Series, V/4b. Springer, Berlin.

55. S. Twomey, R. Gall and M. Leuthold, 1987, Pollution and cloud reflectance, in *Interactions between Energy Transformations and Atmospheric Phenomena*, Edited by M. Beniston and R.A. Pielke, Reidel, Dordrecht.

56. M. Hantel, *Climate Modeling*, 1988, Ladot-Bornstein, New Series, V/4c/2. Springer, Berlin.

57. K.Y. Kondratyev and V.I. Binenko, Optical properties of dirty clouds, in *Interactions between Energy Transformations and Atmospheric Phenomena*, Edited by M. Beniston and R. A. Pielke, Reidel, Dordrecht.

58. J.A. Coakley, R.L. Bernstein and P.A. Durkee, 1988, Effect of ship-stacked effluents on the radiative properties of marine stratocumulus: implications for man's impact on climate, in *Aerosol and Climate*, Edited by Hobbs P.V. and McCormick M.P., Deepak, Humpton, U.S.A.

59. J. Heintzenberg, Fine particles in the global troposphere-A review, *Tellus*, 41 B, 149.

60. R. Jaenicke, 1988, *Aerosol Physics and Chemistry*, Ladolt-Bornstein, New Series, V/4b. Springer, Berlin.

61. I. Orlic, T. Osipowicz, F. Watt and S.M. Tang, *Nucl. Instr. Meth.*, B104, 1995, 630-637.

62. C. Xhoffer, P. Bernard, R. Van Grieken, L. Van der Auwera, *Env. Sci.*, 25, 1991, 1470.

63. F. Bruynseels, H. Storms and R. Van Grieken, *Atm. Env.*, 22, 1988, 2593.

64. P.C. Bernard, R. Van Grieken, L. Brügmann, *Marine Chemistry*, 26, 1989, 155.

65. B. Raeyermakers, Ph.D. Dissertation, University of Antwerp, 1986.

66. F. Bruynseels, H. Storms, R. Van Grieken, and L. Van der Auwera, *Atmos. Environ.*, 22, 1988, 2593-2602.

67. J. Injuk, Ph. Otten, R. Lane, W. Maenhaut and R. Van Grieken, *Atmos. Environ.*, 26A, 1992, 2499-2508.

68. C.D. McKenzie, Proc. Aust. Conf. on Nucl. Techniques of Analysis, Lucas Hights, 1976, 95.

69. R.D. Vis, A.J.J. Boss, V. Valkovic and H. Verheul, *IEEE Trans. Nucl. Sci.*, NS-30, 1981, 1236.

70. H.R. Wilde, W. Bischof, B. Raith, C.D. Uhlhorn and B. Gonsior, *Nucl. Instr. Meth.*, 181, 1981, 165.

71. N.E.G. Löwenstam, J. Pallon and H.C. Hansson, *J. Aerosol Sci.*, 19, 1988, 1031.

72. P. Artaxo, W. Menhout, H. Storms and R. Van Grieken, *J. Geophys. Res.*, 95, 1990.

73. G. Grime and F. Watt, *Nucl. Instr. Meth.*, B75, 1993, 495.

74. P. Artaxo, M.L.C. Rabello, R. Watt, E. Swietlicki, J. Knox and H.C. Hannson, *Journal of Aerosol Science*, 23, 1992, S373.

75. P. Artaxo, M.L.C. Rabello, F. Watt, and G. Grime, *Nucl. Instr. Meth.*, B75, 1993, 521.

76. P. Artaxo, Hansson, H.C., Size distribution of Biogenic Aerosol Particles from the Amazon Basin, *Atmospheric Environment*, 29, 1995, 393-402.

77. P. Artaxo and R. Van Grieken, The microanalysis of individual atmospheric aerosol particles by electron, proton and laser microprobe, *Proc. of Second World Congress — Particle Technology*, September 19-22, 1990, Kyoto, Japan.

78. P. Artaxo, M.L.C. Rabello, W. Maenhout and R. Van Grieken, *Tellus, Series B — Chemical and Physical Meteorology,* 44, 1992, 318.

79. P. Artaxo, F. Garab, and M.L.C. Rabello, *Nucl. Instr. Meth.,* B75, 1993, 277.

80. M. Jaksic, F. Watt, G.W. Grime, E. Cereda, G.M.B. Marcazzan and V. Valkovic, *Nucl. Instr. Meth.,* B56/57, 1991, 699-703.

81. A. Caridi, E. Cereda, G.W. Grime, M. Jaksic, G.M. Marcazzan, V. Valkovic and F. Watt, *Nucl. Instr. Meth.,* B77, 1993, 524-529.

82. E. Cereda, G.M.B. Marcazzan, M. Pedretti, G.W. Grime, and A. Baldacci, *Atmospheric Environment,* 29, 1995, 2323-2329.

83. K.R. Spurny, *Physical and Chemical Characterization of Individual Airborne Particles,* Ellis Horwood Limited, Chichester, 1986.

84. M. Jaksic, I. Bogdanovic, E. Cereda, S. Fazinic and V. Valkovic, *Nucl. Instr. Meth.,* B77, 1993, 505.

85. I. Orlic, J. Makjanic, G.H.J. Tros and R. Vis, *Nucl. Instr. Meth.,* B49, 1990, 166.

86. I. Bogdanovic, M. Jaksic, S. Fazinic, V. Valkovic, *Nucl. Instr. Meth.,* B85, 1994, 732.

87. J. Injuk, H. Van Malderen, R. Van Grieken, E. Swietlicki, J.M. Knox and R. Schofield, *X-ray Spectrometry,* 22, 1993, 220.

88. B. Treiger, J. Injuk, I. Bondarenko, P. Van Espen, R. Van Grieken, L. Breitenbach and U. Wätjen, *Spectrochimica Acta,* 49B, 1994, 345-353.

89. P.S.D. Rousseau, W.J. Przybylowicz, R. Scheepers, V.M. Prozesky, C.A. Pineda, C.L. Churms, and C.G. Ryan, *Nucl. Instr. Meth.,* B130, 1997, pp. 582-586.

90. I. Orlic, F. Watt, K.K. Loh and S.M. Tang, *Nucl. Instr. Meth.,* B85, 1994, 840.

91. W. Maenhaut, R. Salomonovic, J. Ptasinski, G.W. Grime, *Nucl. Instr. Meth.,* B130, 1997, pp. 571-576.

92. C. Xhoffer, P. Bernard and R. Van Grieken, *Environ. Sci. Techn.,* 25, 1991, 1470.

93. I. Orlic, B. Wenlan, F. Watt and S.M. Tang, *Environmental Monitoring and Assessment,* 44, 1997, 455-470.

94. D.D. Cohen, J.W. Martin, G.M. Bailey, P.T. Crisp, E. Bryant, R. Rothwell, J. Banks, R. Hyde, *Clean Air,* 28, 1994, pp. 79-88.

95. R.A. Eldred, T.A. Cahill, R.G. Flocchini, *Proceedings of the 71st Annual Meeting of the Air Pollution Control Association,* Publ: APCA, City: Pittsburg, PA, June 1978; paper No 78-69.6, 1978, p. 2.

96. W. Maenhaut, *Scan. Microsc.,* 4, 1990, p. 43.

97. R.C. Henry, C.W. Lewis, P.C. Hopke and H.J. Williamson, *Atmos. Environ.,* 18, 1984, 1507.

98. G.E. Gordon, *Environ. Sci. Techn.,* 22, 1988, 1132.

99. R.C. Henry, *Atmospheric Environment,* 21, 1987, 1815.

100. R.C. Henry, Statistical methods to apportion heavy metals, in *Control and Fate of Atmospheric Trace Metals,* Edited by J.M. Pacyna and B. Ottar, Kluwer Academic Publisher, 1989.

101. P.C. Bernard and R. Van Grieken, *Analytica Chimica Acta,* 267, 1992, pp. 81-93.

102. D.L. Massart and L. Kaufman, *The Interpretation of Analytical Chemical Data by the Use of Cluster Analysis,* John Wiley & Sons, New York, 1983.

103. D.G. Jex, M.W. Hill and N.F. Mangelson, *Nucl. Instr. Meth.,* B49, 1990, 141.

104. P. Arndt, E.K. Jessberger, M. Maetz, D. Reimold, K. Traxel, *Nucl. Instr. Meth.,* B130, 1997, pp. 192-198.

105. R.W. Linton, A. Loh, D.F.S. Natusch, C.A. Evans and P. Williams, *Science,* 191, 1976, 852.

106. R.W. Linton, P. Williams, C.A. Evans, Jr. and D.F.S. Natusch, *Anal. Chem.,* 49, 1977, 1514.

107. X.B. Cox, S.R. Bryan, R.W. Linton, D.P. Groffits, *Anal. Chem.* 59, 1987, 2018-2023.

108. V. Valkovic, J. Makjanic, M. Jaksic, S. Popovic, A.J.J. Boss, R.D. Vis, K. Wiederspahn and H. Verheul, *Fuel,* 63, 1984, 1357.

109. E. Rokita, E.B. May, P.H.A. Mutsaers, and M.J.A. de Voigt, *Nucl. Instr. Meth.,* B130, 1997, 138.

110. D. Jiang, Yu Mao, K. Shen, J. Zhu and F. Yang, The deblurring of SPM images, Unpublished.

111. V.M. Prozesky, J. Padayachee, R. Fischer, W. Von der Linden, V. Dose, and C.G. Ryan, The use of maximum entropy and Bayesian techniques in nuclear microprobe applications, *Nucl. Instr. Meth.,* B130, 1997, pp. 113-117.

112. B. Fischer and S. Metzger, *Nucl. Instr. Meth.,* B104, 1995, 7-12.

113. M. Jaksic, I. Bogdanovic, S. Fazinic, T. Tadic and V. Valkovic, *Nucl. Instr. Meth.,* B104, 1995, 152.

114. E. Clyton, *Nucl. Instr. Meth.,* B22, 1987, 64.

115. G. Grime, and M. Dawson, *Nucl. Instr. Meth.,* B104, 1995, 107-113.

116. C.G. Ryan and D. Jamieson, *Nucl. Instr. Meth.*, B77, 1993, 203.

117. C.G. Ryan and D. Jamieson, C. Churms and J. Pilcher, *Nucl. Instr. Meth.*, B104, 1995, 157.

118. D.N. Jamieson, G.W. Grime and F. Watt, *Nucl. Instr. Meth.*, B40/41, 1989, 669.

119. P. Moretto, L. Razafindrabe, *Nucl. Instr. Meth.*, B104, 1995, 171-175.

120. R.M.S. Scofield and H.W. Lefewre, *Nucl. Instr. Meth.*, B77, 1993, 217.

121. R. Scofield, *Nucl. Instr. Meth.*, B104, 1995, 212.

122. A. Sakellariou, M. Cholewa, A. Saint and G.J.F. Legge, *Nucl. Instr. Meth.*, B104, 1995, 253-258.

123. S.C. Liew, K.K. Loh, S.M. Tang, *Nucl. Instr. Meth.*, B85, 1994, 621.

124. S.C. Liew, I. Orlic, S.M. Tang, *Nucl. Instr. Meth.*, B85, 1994.

125. J.A. Cookson and F.D. Pilling, *Phys. Med. Biol.*, 20, 1975, 1015.

126. P. Horowitz, M. Aronson, L. Grodzins, W. Ladd, J. Ryan, G. Merrian and C. Lechene, *Science*, 194, 1976.

127. V. Valkovic, *Analysis of Biological Materials for Trace Element Analysis Using X-Ray Spectroscopy*, CRC Press, Florida, 1980.

128. A.J.J. Boss, Ph.D. Thesis, Free University, Amsterdam, 1984.

129. A.J.J. Boss, C.C.A.H. Van der Stap, R.D. Vis, and H. Verheul, *Nucl. Instr. Meth.*, B3, 1984, 654.

130. F. Watt, *Nucl. Instr. Meth.*, B77, 1993, 261.

131. W.J. Przybyrowicz, C.A. Pineda, V.M. Prozesky and J. Mesjasz-Przybyrowicz, *Nucl. Instr. Meth.*, B104, 1995, 176-181.

132. J. Mesjasz-Przybyrowicz, K. Balkwill, W.J. Przybyrowicz and H.J. Annegarn, *Nucl. Instr. Meth.*, B89, 1994, 208.

133. G.E. Coote, *Nucl. Instr. Meth.*, B66, 1992, 191-204.

134. G.E. Coote and W. J. Trompetter, *Nucl. Instr. Meth.*, B77, 1993, 501-504.

135. C.P. Swann, K.M. Hansen, K. Price and R. Lutz, *Nucl. Instr. Meth.*, B56, 1991, 683-685.

136. J. Nystrom, and E. Dunca, *Nucl. Instr. Meth.*, B104, 1995, 612-618.

137. Z. Dai, C. Ren, Q. Zhao, P. Wang and F. Yang, *Nucl. Instr. Meth.*, B104, 1995, 619-624.

138. V.M. Prozesky, E.J. Raubenheimer, W.P. Grotepass, W.F.P. Van Heerden, W.J. Przybylowicz, and C.A. Pineda, *Nucl. Instr. Meth.*, B104, 1995, 638-644.

139. I.D. Gourlay and G.W. Grime, *Int. Assoc. of Wood Anatomists J.*, 15, 1994, 137.

140. G. Grime and W. Davison, *Nucl. Instr. Meth.*, B77, 1993, 509.

141. W. Davison, G. Grime and C. Woof, *Limnol. Oceanogr.*, 37, 1992, 1770-1777.

142. M.J. McFarlane, D.J. Bowden, F. Watt and G.W. Grime, Contemporary leaching of the African erosion surface in Malawi, Mineralogical and Geochemical Records of Paleoweathering, edited by Schmitt and Gall, ENSMP Mem. Sc. de la Terre, 18, 1992, 5-14.

14 *In Situ* Chemical Analyses of Aerosol Particles by Raman Spectroscopy

Gustav Schweiger

CONTENTS

14.1 INTRODUCTION

Since the invention of the laser, optical methods have experienced dramatic progress in nearly every field. This holds especially for spectroscopic techniques, such as Raman spectroscopy. If light is scattered on molecules, it is affected by the thermal motion of the molecules. This thermal motion modulates the scattered light in a very characteristic way. We will see that this modulation causes

319

the appearance of frequency shifted lines in the scattering spectrum. These frequency shifted spectral lines are called Raman lines. Their frequency shift is a characteristic property of the scattering molecule and its thermodynamic state. Raman scattering is, therefore, used to identify molecules and their thermodynamic state. However, the Raman effect is a very weak effect, and this prevented a wider application until the arrival of the laser, with its unique spectral power density, which made it an ideal light source for Raman scattering. Not only did the classical linear Raman scattering, which got its name from the linear relation between the intensity of the Raman lines and the power density of the incident radiation, get a big boost, but also a number of new nonlinear effects, such as stimulated and coherent Raman scattering, were detected. This chapter will be limited to linear Raman scattering.

In addition, dramatic progress in light detecting technology laid the groundwork for a rapid increase in the number of applications of Raman scattering techniques. Aerosol analysis is one such field. Raman spectroscopy on aerosols shows one peculiarity which cannot be observed on bulk material. Spherical particles such as microdroplets act as optical resonators of very high quality. If the incident radiation corresponds to a resonance frequency, the amplitude of the electromagnetic field within the particle can increase dramatically. This is important not only for the excitation of nonlinear optical effects but also for linear scattering such as Raman scattering. These resonance properties have gained considerable interest in the last few years. Hill and Chang (1992) recently gave a review of nonlinear optics of droplets. The resonance effects of light scattering on small particles are also treated in detail by Barber and Chang (1990) and Barber and Hill (1990). Raman scattering on small particles was reviewed by Schrader (1986) and on aerosols by Schweiger (1990).

In the following, only linear Raman scattering will be treated. The classical treatment of Raman scattering will be presented in some detail to lay the groundwork for a better understanding of what can be achieved with Raman scattering in principle. The quantum mechanics aspect will be presented only briefly. The theory of Raman scattering on spherical aerosol particles, however, will be presented in detail, with emphasis on the deduction of some relations of immediate practical use for aerosol analysis. The last section will present examples of the application of Raman scattering to aerosol analyses.

14.2 BASIC THEORY OF RAMAN SCATTERING

In the classical treatment of the scattering process, which will be presented first, the source of light scattering is the oscillating molecular dipole moment that is induced by incident radiation. This dipole moment oscillates with the frequency of the incident radiation and, therefore, emits radiation with the same frequency. This process is called elastic light scattering. The molecules, however, are also subject to rotational and vibrational motions. If these motions cause nonsymmetric changes in the polarizability, the dipole radiation is modulated and frequency shifted bands appear in the scattered light. These additional bands are called Raman bands. The scattered light flux is proportional to the fourth power of the frequency. If the induced dipole depends on the orientation of the molecules, which is the case for nonspherical molecules, and the molecules have arbitrary orientation, the scattered radiation is at least partially depolarized. The vector of the electric field of the scattered wave contains components perpendicular to the vector of the electric field of the incident wave and the scattering direction.

The quantum mechanical treatment of the scattering process will show that Raman scattering is connected with changes in the energy state of the molecule. If the molecule is in a higher energy state before scattering than after, the frequency of the Raman line is higher than the frequency of the incident radiation. These lines are called anti-Stokes Raman lines. In the opposite case, the lines are shifted to lower frequencies and called Stokes Raman lines. Such transitions are only possible if a transition moment from the initial energy level to at least one intermediate eigenstate of the molecule and from this intermediate state to the final state is not zero. All eigenstates of the system participate in principle in the scattering process. However, it is important to notice, that in Raman scattering the molecule is not really transferred to the intermediate state as in the fluorescence process. The Raman scattering process is, therefore, much less sensitive to the environment of the molecule than fluorescence. The transition time is approximately 10^{-12}s and, therefore, about four orders of magnitude less than in fluorescence processes. The probability for such a transition is proportional to the square of the transition moment. The transition rate and, therefore, the intensity of the scattered light is proportional to the number of molecules in the initial state. This occupation number is temperature dependent, and the Raman scattering depends, therefore, also on temperature and is different for anti-Stokes and Stokes Raman scattering in contrast to the classical result.

14.2.1 Classical Treatment of Molecular Light Scattering

There are a number of excellent textbooks treating mainly elastic scattering (Kerker (1969), Bohren and Huffman (1983), van de Hulst (1981) as well as Raman scattering (see for example Herzberg, (1945, 1950); Anderson, (1971, 1973); Brandmüller and Moser, (1962), Long, (1977)), so we restrict ourselves to a presentation of the basic relations. The purpose of this chapter is to present the basic physical processes in no more detail than is necessary to understand the most important properties of Raman scattering. Although a complete description of Raman scattering is only possible by its quantum mechanical treatment, most of its properties can be explained at least qualitatively by a classical description. The classical treatment is much easier to comprehend than the rather formal quantum mechanical description. We will present, therefore, the classical electrodynamic analyses of the elastic and inelastic scattering process first.

14.2.2 The Induced Electric Dipole Moment

If an electrical field acts on a distribution of positive and negative charges — represented, for example, by a molecule — the negative charges are pulled by the field in one direction and the positive charges are pulled in the opposite direction and an electrical dipole moment is induced. This induced dipole moment $\bar{p}$ depends on the Coulomb forces holding the charge cloud together. It is, therefore, a material property, and it depends nonlinearly on the applied electrical field $\bar{E}$. The dependence on the applied electrical field can be approximated in most cases to a high degree of accuracy by a power expansion.

$$\bar{p} = 4\pi\varepsilon_0\left(\chi^{(1)}\bar{E} + \chi^{(2)}\bar{E}\bar{E} + \chi^{(3)}\bar{E}\bar{E}\bar{E} + \ldots\right). \tag{1}$$

The quantities $\chi^{(2)}$, $\chi^{(3)}$ are known as the second- and third-order nonlinear optical susceptibilities, respectively.* These susceptibilities are, in general, tensors of increasing order. The first-order optical susceptibility $\chi^{(1)}$ is a second-rank tensor, which is often designated by $\bar{\alpha}$ and called polarizability. We will adapt this designation in this text. The second-order nonlinear optical susceptibility is a third-rank tensor, and so on. The ratio of the susceptibilities is approximately $\chi^{(1)}:\chi^{(2)}:\chi^{(3)} \approx 1:10^{-8}:10^{-15}$. The linear term in Equation (1) is responsible for elastic light scattering and for the linear Raman scattering. Except at very high electric fields, the linear term is sufficient to describe the interaction of an electrical field with matter, and we will restrict our considerations completely to this linear term.

The polarizability tensor $\bar{\alpha}$ represents, to a very good approximation, the response of a specific molecule to the external electrical field $\bar{E}$. The polarizability tensor depends, in general, on the

---

* The factor 4π is used here for convenience (see Equation (8)). The dimension of α is L^3.

thermal motion of the nuclei. These rotational and vibrational thermal motions change the internuclear distances that cause changes in the polarizability. There are other sources affecting the polarizability, electronic motions, intermolecular collisions, photon interaction, etc., which will be excluded here from our consideration, although they can give reason for the appearance of frequency shifted lines in the scattering spectrum.

The effect of internal molecular motions, such as vibration and rotation of the nuclei on the polarizability is weak. We represent, therefore, the dependence of polarizability on the coordinates of the internuclear distance by a power expansion.

$$\alpha_{xx} = \alpha_{xx}^0 + \sum_k \left[\left(\frac{\partial \alpha_{xx}}{\partial x_k} \right)_0 x_k + \left(\frac{\partial \alpha_{xx}}{\partial y_k} \right)_0 y_k + \left(\frac{\partial \alpha_{xx}}{\partial z_k} \right)_0 z_k + \dots \right]. \quad (2)$$

The triple x_k, y_k, z_k represents the displacement of the k-nucleus from its equilibrium position. Quantities at the equilibrium position are marked by the suffix "0." The displacement is more or less a complicated function of time.

It is well known from classical dynamics that the motion of N centers of mass coupled by elastic forces can be represented as the superposition of harmonic oscillations in 3N–6 distinct directions (3N–5 directions for linear molecules). These directions are called the normal coordinates ξ_i of the system. Expressing the polarizability tensor in the normal coordinate one gets:

$$\alpha_{xx} = \alpha_{xx}^0 + \sum_i \left(\frac{\partial \alpha_{xx}}{\partial \xi_i} \right)_0 \xi_i^0 \cos(\omega_i t + \delta_i) + \dots \quad (3)$$

The frequencies of the molecular oscillations (rotations) are designated by ω_i, where the index i refers to a particular eigenmode. The phase factor δ_i is arbitrary and varies from molecule to molecule because the molecular oscillations are not in phase with the incident radiation. In the following, we will assume that the wavelength of the incident light is much larger than the molecular dimensions. This is usually a reasonable assumption for visible light. Therefore, the electric field of the incident wave can be considered to be independent of the coordinates of the electrons and nuclei of the scattering molecule. Furthermore, we assume that the frequency of the incident light is far enough away from any eigenfrequency of the system to neglect resonance effects. Only small displacements from the equilibrium position of the nuclei are caused under this condition (in the following we will consider only the motion of nuclei and neglect electron motions).

14.2.3 Stokes and Anti-Stokes Raman-Lines

The electric field $\bar{E}$ of Equation (1) oscillates with the frequency ω_0 of the incident wave and has in Cartesian coordinates the three components:

$$E_x = E_0 e_x \cos \omega_0 t; \quad E_y = E_0 e_y \cos \omega_0 t; \quad E_z = E_0 e_z \cos \omega_0 t \quad (4)$$

E_0 is the amplitude of the incident wave and $\bar{e}$ is a unit vector in the direction of the electrical field. This unit vector has the components e_x, e_y, e_z.

The polarization induced by an electrical field as described by Equation (1) holds also for time varying fields, such as that of electromagnetic radiation represented by Equation (4). Equations (3) and (4) introduced into Equation (1) give the following expression for the x-component of the dipole:

$$p_x = \underbrace{4\pi\varepsilon_0 E_0 \left[\alpha_{xx}^0 e_x + \alpha_{xy}^0 e_y + \alpha_{xz}^0 e_z \right] \cos\omega_0 t}_{\text{elastic-scattering}}$$

$$+ 4\pi\varepsilon_0 E_0 \sum_i \left[\left(\frac{\partial\alpha_{xx}}{\partial\xi_i}\right)_0 e_x + \left(\frac{\partial\alpha_{xy}}{\partial\xi_i}\right)_0 e_y + \left(\frac{\partial\alpha_{xz}}{\partial\xi_i}\right)_0 e_z \right] \tag{5}$$

$$\cdot \xi_i^0 \frac{1}{2} \left[\underbrace{\cos\left((\omega_0+\omega_i)t+\delta_i\right)}_{\text{anti-Stokes}} + \underbrace{\cos\left((\omega_0-\omega_i)t-\delta_i\right)}_{\text{Stokes}} \right]$$

The two other components of the induced dipole moment can be calculated accordingly. This equation shows that the incident wave induces a dipole moment that oscillates with the same frequency as the incident wave, but also with the sum and difference frequencies $\omega_0 \pm \omega_i$. We already know that the spectral lines that have frequencies $\omega_0 - \omega_i$ are called Stokes-Raman lines, and spectral lines shifted to higher frequencies $\omega = \omega_0 + \omega_i$ are called anti-Stokes-Raman lines. The amplitude of the oscillation that has the same frequency as the incident wave, is proportional to the equilibrium polarizability and the amplitude of the incident wave. The amplitudes of the frequency-shifted components are again proportional to the amplitude of the incident wave, but proportional to the derivative of the polarizability with respect to the direction of the normal coordinates.

14.2.4 PROPERTIES OF DIPOLE RADIATION AND RAMAN SCATTERING

From classical electrodynamics it is known that an oscillating dipole emits electromagnetic radiation at the oscillation frequency of the dipole. The electric field E $(\bar{r},t)$ of this radiation in the far field, where the distance from the dipole is much larger than the wavelength and the dimensions of the dipole, is given by (Stratton, 1941; Jackson, 1975)

$$\bar{E}(\bar{r},t) = \frac{k^2}{4\pi\varepsilon_0|\bar{r}-\bar{r}'|}\left[\bar{r}_0 \times \left[\bar{r}_0 \times \bar{p}(\bar{r},t)\right]\right] = \frac{k^2}{4\pi\varepsilon_0 r}\left((\bar{r}_0 \cdot \bar{p})\bar{r}_0 - \bar{p}\right) \tag{6}$$

where $k = \omega/c$ is the absolute value of the wave vector, c is the speed of light, $\bar{r}_0$ is a unit vector in the direction of propagation of the radiation. The dipole is located at $\bar{r}'$. From this equation one can see that the electrical field of the dipole radiation is perpendicular to the direction of propagation $\bar{r}_0$ and is coplanar to the plane given by $\bar{r}_0$ and $\bar{p}$. The radiation is completely polarized parallel to $\bar{p}$, which means, that the electric field of the scattered radiation has no component perpendicular to the direction of the dipole and the scattering direction. If the polarizability is a scalar, the direction of the induced dipole is the same as the direction of the electric field of the incident radiation. The electric field of the scattered radiation has, therefore, no component perpendicular to the electric field of the incident wave in this case. The scattered radiation is completely polarized parallel to the incident radiation. Equation (5) shows that the frequency shifted Raman lines appear in the scattering spectrum only if the gradient of the polarizability tensor is not zero. In other words, the polarizability must change nonsymmetrically with deviation from the equilibrium position. Changes in the polarizability can be caused by vibrations of the nuclei, the corresponding Raman lines in the spectrum of the scattered light are called *vibrational Raman lines*. If the polarizability changes by the rotation of the molecule, the corresponding lines are designated by *rotational Raman lines*. Frequently Raman lines are caused by simultaneous rotational and vibrational movements, and the corresponding lines are called *vibrational rotational Raman lines*.

The elastically scattered light is in-phase with the incident radiation, the Raman scattered light is not. In the standard situation a large number of scattering molecules participate in the scattering process. Due to the phase relation in elastic scattering, interference effects between the radiation from the individual molecules can play an important role, and the total scattered radiation is not simply the algebraic sum of the individual contributions. This is not so for Raman scattering. The Raman effect is, as shown before, an inelastic scattering process, where the interaction of the incident field with the internal motion of the scattering molecule causes an energy transfer between the incident light field and the molecules. This results in the scattering of frequency shifted photons.

In contrast to infrared radiation, no permanent dipole moment is necessary for elastic or Raman scattering. The light flux $\partial \phi$ radiated into the space angle $d\Omega$ can also be calculated by using classical electrodynamics. The result is (see Jackson, 1975 or Corney, 1977):

$$I = \frac{\partial \phi}{\partial \Omega} = \frac{1}{2Z_0} \left[\frac{k^2}{4\pi\varepsilon_0} \right]^2 \left| \left[\bar{r}_0 \times (\bar{r}_0 \times \bar{p}) \right] \right|^2 \qquad (7)$$

where Z_0 is the free space impedance, $Z_0 = \sqrt{\mu_0/\varepsilon_0}$, μ_0 and ε_0 are the magnetic and electric field constants, respectively. By introducing the radiant flux density

$$D = \frac{1}{2} \sqrt{\frac{\varepsilon_0}{\mu_0}} \left| E_0 \right|^2 = \frac{1}{2} c\varepsilon_0 \left| E_0 \right|^2$$

into Equation (7) we get:

$$\frac{1}{D} \frac{\partial \phi}{\partial \Omega} = k^4 \left| \left[\bar{r}_0 \times (\bar{r}_0 \times \bar{p}_0) \right] \right|^2 = k^4 \left| \bar{p}_0 \right|^2 \sin^2 \theta, \quad \bar{p}_0 = \frac{\bar{p}}{4\pi\varepsilon_0 E_0} = \bar{\alpha}\bar{e} \qquad (8)$$

where θ is the angle between $\bar{r}_0$ and $\bar{p}_0$. The latter represents the induced dipole moment per unit electric field. The scattered intensity is proportional to the fourth power of the frequency and proportional to the radiant flux density of the incident radiation. Its angular variation is given by: $\sin^2 \theta$. The direction of the vector of the electrical field is always parallel to the direction of the induced dipole moment, as already discussed. In other words, the dipole radiation is completely polarized parallel to the direction of the induced dipole moment. These are the well-known characteristics of dipole radiation.

The right-hand side of Equation (8) has the dimension of an area and is called the (molecular) differential scattering cross section. Equation (8) is only correct for elastically scattered light. The total scattered radiation can be found by integrating Equation (8) over the full space angle 4π. For linearly polarized light and a scalar polarizability, the angle θ remains constant and the integration yields the factor $4\pi/3$. The total scattered light per incident power density per molecule, [which is also called total (molecular) scattering cross section σ], is therefore,

$$\frac{\phi}{D} = \sigma = \frac{8\pi}{3} \times \left[k_0^4 \left| \alpha^0 \right|^2 + \frac{1}{4} \sum_i (k_0 \pm k_i)^4 \left| \left(\frac{\partial \alpha}{\partial \xi_i} \right)_0 \right|^2 (\xi_i^0)^2 \right]. \qquad (9)$$

If the polarizability is not a scalar quantity, as for nonspherical molecules, this equation still holds qualitatively. It can be transformed into the correct equation if the scalar quantities are replaced by the appropriate components of the scattering tensor.

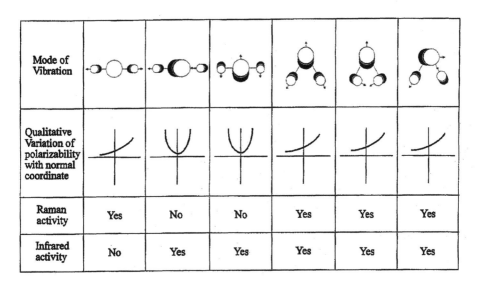

Mode of Vibration						
Qualitative Variation of polarizability with normal coordinate						
Raman activity	Yes	No	No	Yes	Yes	Yes
Infrared activity	No	Yes	Yes	Yes	Yes	Yes

FIGURE 14.1 Oscillation modes, variation of the polarizability and infrared and Raman activities of two and three atomic molecules.

The essence of the classical treatment of light scattering is given by Equation (9). The classical treatment correctly predicts the appearance of frequency shifted lines in the scattering spectrum. The magnitude of frequency shift is also in agreement with experimental findings. Finally, Raman lines will only be generated if the derivatives of the polarizability tensor at the equilibrium position do not disappear. The classical treatment of the Raman process does not predict the intensities of the Raman lines correctly.

14.2.5 SELECTION RULES

From Equation (5) we learn that the induced dipole moment oscillates only at frequencies shifted against the incident radiation, if the polarizability α changes with the coordinate ξ and the polarizability gradient is different from zero at the equilibrium position. The calculation of the dependence of the polarizability on the coordinates of the nuclei is a quantum mechanical problem and far beyond the scope of this chapter. We restrict our discussion, therefore, to an intuitive treatment of this problem (see e.g., Long, 1977). Let us first consider a diatomic molecule. If we increase the internuclear separation, the binding forces decrease until the molecule finally breaks apart. The polarizability is proportional to the charge separation of the molecules caused by an external field. We expect, therefore, that the polarizability increases with increasing molecular separation from the equilibrium position because the binding forces are becoming weaker. The change in polarizability with decreasing internuclear separation is not as obvious. Quantum mechanical calculation shows that in the neighborhood of the equilibrium position, the polarizability decreases with decreasing internuclear distance. During one cycle of the molecular oscillation the polarizability has its minimum at the turning point with minimum intermolecular separation and its maximum at the second turning point, where the separation of the two nuclei has its maximum. The gradient of the polarizability at the equilibrium distance is, therefore, not zero, and the diatomic molecule is Raman active in agreement with observation and quantum mechanical calculations. Figure 14.1 shows the relation between oscillation modes and variation of the polarizability and the consequences for the Raman scattering for linear molecules such as N_2 or CO_2 and nonlinear molecules such as water.

14.2.6 STATE OF POLARIZATION OF SCATTERED LIGHT

Let us consider a nonspherical molecule. The polarizabilty is a tensor in this case, and the direction of the induced dipole will in general be different from the direction of the incident electrical field. Therefore, the scattered radiation is not polarized completely parallel to the plane of the incident wave. The electric field of the scattered radiation has a component parallel to the direction of the electric field of the incident radiation as well as a component perpendicular to it.

In general, the experimentally detectable scattered light is the result of scattering on a very large number of molecules. If we neglect intermolecular interactions, the orientation of these molecules is completely arbitrary. To calculate the state of polarization of the scattered light, we now have to calculate the induced dipole moments for all possible orientations of the molecules and calculate the radiation from these dipoles. In linear Raman scattering, there is no phase relation between the frequency shifted scattered light of the individual molecules. No interference effects have to be considered. Therefore, the total intensity is simply found by calculating the intensity for an arbitrary orientation of the dipole and averaging over all possible orientations. The result of this procedure is that the scattered light always has nonzero components of the electric field in any two directions perpendicular to each other and to the direction of observation. The scattered light is at least partially depolarized.

The anisotropic part of the polarizability and its derivative is usually much smaller than the average isotropic part. The intensity of the scattered light as a function of the scattering geometry is, therefore, often not very different from the dipole radiation. The scattered radiation perpendicular to the oscillation plane of the incident wave is usually much larger than for any other scattering geometry in the case of vibrational Raman scattering. In rotational Raman scattering, the scattered light is highly depolarized. (An example is shown in Figure 14.7.)

14.2.7 QUANTUM MECHANICAL TREATMENT

In the quantum mechanical picture of interaction of radiation with molecules, the reaction of an individual molecule to the radiation field is not predictable. Quantum mechanics deals with probabilities. What can by calculated is the probability of emission or scattering of a photon by a quantum mechanical system. If a molecule, for example, is in a particular energy state E_k, the probability $W_{k\ell}$ per unit time that the molecule changes its energy to a state with energy E_ℓ can be calculated by quantum mechanical methods. Such transitions can be connected with the emission of a photon of energy $\Delta E = E_k - E_\ell = \hbar \cdot \omega_{k\ell}$, where $\hbar = h/2\pi$ and h is Planck's constant, $\omega_{k\ell}$ can be positive (anti-Stokes Raman lines) or negative (Stokes Raman lines) depending on the sign of ΔE and ω_0 is as before the frequency of the incident radiation. The total radiative power $\phi_{k\ell}$ emitted from N molecules is then given by*:

$$\phi_{k\ell} = \hbar\left(\omega_{k\ell} + \omega_0\right) \cdot W_{k\ell} \cdot N \cdot n_k \tag{10}$$

The relative number of molecules in the energy state E_k is n_k, and N is the total number of molecules in the volume under consideration. For an explicit evaluation, we have to calculate the transition probability $W_{k\ell}$.

Equation (10) shows the fundamental difference in the classical analyses. In contrast to Equation (9), in the quantum mechanical treatment the power of the scattered radiation depends on the number of molecules $N \cdot n_k$ in the initial state. We will come back to this point later.

---

* Equation (10) neglects the final spectral range of the light emitted by the system. The energy states in the molecules are not distinct but extend over a certain range due to the uncertainty principle. As a consequence, the photons emitted cover a small but final frequency range. The frequency ω in Equation (10) has to be considered as the most probable frequency emitted by a transition.

The quantum mechanical calculation of the transition probability $W_{k\ell}$ is beyond the scope of this chapter and can be found in standard textbooks on quantum mechanics. The transition probability is given by:

$$W_{k\ell} = \frac{(\omega_{k\ell} + \omega_0)^3}{3\pi\varepsilon_0\hbar c^3}\left|\left\langle\psi_\ell\left|\hat{P}\right|\psi_k\right\rangle\right|^2, \quad \left\langle\psi_\ell\left|\hat{P}\right|\psi_k\right\rangle \equiv \int \psi_\ell^* \cdot \hat{P} \cdot \psi_k d^3 r \tag{11}$$

Ψ_k is the time dependent state function (Ψ_k^* its conjugate complex) of the molecule in the energy eigenstate E_k weakly disturbed by an electromagnetic wave. The state function represents every physical realizable state of the system and can be calculated by standard quantum mechanical methods. $\hat{P}$ is the momentum operator. The momentum operator is the mathematical instruction that extracts the value of the electrical dipole momentum from the state function in this particular case. As a result of quantum mechanical calculations one gets:

$$\left|\left\langle\psi_\ell\left|\hat{P}\right|\psi_k\right\rangle\right|^2 = \left|\frac{1}{\hbar}\sum_j\left\{\frac{\left(\overline{A}\cdot\hat{P}\right)_{kj}}{\omega_{jk} - \omega_0}P_{j\ell} + \frac{\left(\overline{A}\cdot\hat{P}\right)_{j\ell}}{\omega_{j\ell} + \omega_0}P_{kj}\right\}\right|^2,$$

$$\left(\overline{A}\cdot\hat{P}\right)_{kj} = \left\langle\psi_j^{(0)}\left|\overline{A}\cdot\hat{P}\right|\psi_k^{(0)}\right\rangle, \quad P_{j\ell} = \left\langle\psi_\ell^{(0)}\left|\hat{P}\right|\psi_j^{(0)}\right\rangle \tag{12}$$

The angular frequency of the incident radiation is as before ω_0 and $\omega_{jk} = (E_k - E_j)/\|$, where E_k and E_j are energies of eigenstates with eigenfunctions $\Psi_k^{(0)}$ and $\Psi_j^{(0)}$ of the molecule. The corresponding holds for $\omega_{j\ell}$. $\overline{A}$ is the amplitude vector of the incident electromagnetic wave.

14.2.8 THE RULE OF THE INTERMEDIATE STATES

We see from Equation (12) that the matrix elements $\langle\Psi_\ell|\hat{P}|\Psi_j\rangle$ are only different from zero, if at least one of the intermediate states j combines with both the initial state, and the final state, and the transition moments to both states P_{kj} and $P_{j\ell}$ are different from zero. This does not mean that Raman scattering is a two-step process, transition from energy state $E_k \rightarrow E_j$ and from $E_j \rightarrow E_\ell$. In effect, Equation (12) is only valid if the frequency ω_0 of the incident radiation is very different from any eigenfrequency $\omega_{k\ell}$ and no absorption of the incident radiation is possible. The energy of the intermediate states can, therefore, lie above or below the final state. The theory of Raman scattering shows that in a molecular system disturbed by the field of an electromagnetic wave, transitions between energy states are induced without absorption of the incident radiation or emission of a photon. Frequently, one speaks of a virtual absorption and emission process and represents this process symbolically as shown in Figure 14.2. The virtual absorption of a photon of the incident radiation transfers the molecule in the virtual state indicated by the broken line. This, however, is not an eigenstate and the molecule goes back to an eigenstate by an emission of a photon. If the final state is the same as the initial state, the emitted photon has the same energy as the incident photon, one speaks of elastic light scattering. If the final state is different, this is called an inelastic light scattering process. Depending on whether the energy is higher or lower than the initial state, one calls this process Stokes or anti-Stokes Raman scattering. While this describes the Raman scattering process qualitatively quite reasonably, it is not correct. From the result of quantum mechanical analyses, reproduced in Equation (12) we can see that no absorption of the incident radiation takes place and the molecule is in effect never in any kind of virtual state.

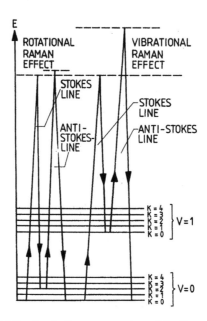

FIGURE 14.2 Symbolic representation of the Raman process.

14.2.9 TEMPERATURE DEPENDENCE

The most important difference in the classical and quantum mechanical treatment of Raman scattering is the dependence of the scattered radiation on the number of molecules in the initial state, as shown by Equation (10). The population of a specific state depends on temperature. According to the Boltzmann distribution, the relative number of molecules n_j in a specific energy state E_j is given by:

$$n_j = \frac{N_j}{N} = \frac{g_j e^{-E_j/kT}}{\sum_j g_j e^{-E_j/kT}} \tag{13}$$

where N is the total number of molecules, g_j is the degree of degeneracy (quantum mechanically distinguishable states with the same energy), and k is the Boltzmann constant.

Usually the total energy E_j of the molecule can be separated into the energy of the electronic state E_e, that of the vibrational state E_v, which depends on the vibrational quantum number v, and E_R, the energy of the rotational state, which depends on the rotational quantum number J.

$$E_j = E_e + E_v(v) + E_R(J) \tag{14}$$

The energy of the electronic state is not of interest here. The vibrational energy is given by:

$$E_v = hc\omega_e(v+1/2) - hc\omega_e x_e(v+1/2)^2 + ... \tag{15}$$

The quantities ω_e and x_e are molecular constants. The dimension of ω_e is 1/length. The index e indicates that the quantities also depend on the electronic state of the molecule. The rotational energy E_R is given by:

$$E_R = hc\left[B_v J(J+1) + D_v J^2 (J+1)^2 + ...\right] \tag{16a}$$

$$B_v = B_e - \alpha_e(v+1/2), \quad D_v = D_e + B_e(v+1/2) \tag{16b}$$

The molecular constants B_e, D_e, and α_e depend slightly on the electronic state.

14.2.10 Vibrational Raman Scattering

The population in the vibration level with energy E_v is, therefore, given by combining Equations (13) and (15) and neglecting the small second term in Equation (15) as follows ($g_j = 1$):

$$n_v = \frac{e^{-hc\omega_e(v+1/2)/kT}}{\sum_{v=1}^{\infty} e^{-hc\omega_e(v+1/2)/kT}} = e^{-hc\omega_e v/kT} \cdot \left(1 - e^{-hc\omega_e/kT}\right) \tag{17}$$

As already mentioned, the weak coupling allows the analyses of vibrational and rotational transitions independently. Quantum mechanical calculations show that the transition moment for transitions with $\Delta v = 1$ is proportional to $\sqrt{v}$ and for transitions with $\Delta v = -1$ it is proportional to $\sqrt{v+1}$, where v is the quantum number of the initial state.

Although at moderate temperatures most of the molecules are in their vibrational ground state, the contribution of molecules in higher vibrational states cannot always be neglected. If the molecule can be considered as a harmonic oscillator, the energy differences between adjacent vibrational levels are all the same and all vibrational Raman lines coincide. The intensity of a Stoke vibrational band for a molecule with different vibrational modes j is, therefore, given by:

$$N_v = N \sum_i \left(v_j^i + 1\right) n_j^i = N \sum_i \frac{\left(v_j^i + 1\right) e^{hc(\omega_e)_j^i (v_j^i + 1/2)/kT}}{\sum_k e^{-hc(\omega_e)_j^i (v_j^i + 1/2)/kT}} \tag{18}$$

The summation has to be carried out over all quantum numbers i for a given vibrational mode j (e.g., all bending modes of a triatomic molecule). The degeneracy is assumed to be $g_i = 1$. The result of the summation is for Stokes and anti-Stokes transitions, respectively:

$$N_{\text{Stokes}} = \sum_i \left(v_j^i + 1\right) n_j^i = \frac{N}{1 - e^{-hc(\omega_e)_j/kT}}, \quad N_{\text{anti-Stokes}} = \sum_i v_j^i n_j^i = \frac{N}{e^{hc(\omega_e)_j/kT_{-1}} - 1} \tag{19}$$

The ratio of the intensities of Stokes to anti-Stokes scattering is finally given by:

$$\frac{\phi_{\text{Stokes}}}{\phi_{\text{anti-Stokes}}} = \left(\frac{\omega - \omega_{jk}}{\omega + \omega_{jk}}\right)^4 e^{hc\omega_e/kT} \tag{20}$$

14.2.11 Rotational Raman Scattering

The coefficients in the transition moment which depend only on the rotational quantum numbers are called Placzek Teller coefficients. For totally symmetric linear or spherical molecules, these coefficients are:

$$b_{J+2,J} = \frac{3(J+1)(J+2)}{2(2J+1)(2J+3)}, \quad b_{JJ} = \frac{J(J+1)}{(2J-1)(2J+3)}, \quad b_{J-2,J} = \frac{3J(J+1)}{2(2J+1)(2J-1)} \quad (21)$$

The coefficient b_{JJ} is only relevant to vibrational rotational transitions (scattering processes with simultaneous change in the rotational and vibrational quantum number). The temperature dependence of the rotational spectrum follows from Equations (13) and (16). Neglecting higher order terms, in Equation (16a) one gets for freely rotating molecules with $2J + 1$ possible orientations:

$$n_j = \frac{Ng_J(2J+1)e^{-BJ(J+1)hc/kT}}{\sum_j g_J(2J+1)e^{-BJ(J+1)hc/kT}} \quad (22)$$

where g_J is the degeneracy of the nuclear spin. The degeneracy is, for all rotational levels, the same for heteronuclear diatomic molecules. For homonuclear molecules, however, it takes different values for odd and even rotational quantum numbers. For O_2 molecules, the statistical weight is $g_J = 1$, and the envelope of the rotational lines is a smooth function. The nuclear degeneracy of nitrogen, however, is $T = 3$, and the statistical weight is $g_J = 3$ for J odd, and $g_J = 6$ for even J. This causes an alternating amplitude variation in the rotational spectrum, as shown in Figure 14.3. The rotational lines are usually only observable in the gas phase. In the liquid state, the intermolecular interaction causes a broadening of the rotational lines which overlap and form a continuum, as for example in the case of alcohol.

14.2.12 Summary of Basic Properties of Raman Spectra

(a) Frequency shift. In the foregoing paragraphs, the relation between the Raman spectrum and the vibrational and rotational energy levels was outlined. One of the main fields of application of the Raman effect is indeed the revelation of molecular structures. Together with infrared spectroscopy, Raman spectroscopy plays a crucial role in the determination of molecular structures. The other classical domain of Raman spectroscopy is the identification of molecules — in other words, the analysis of the chemical composition of an unknown sample. Similar to the infrared spectrum, the Raman spectrum represents a "fingerprint" of the scattering molecule. The Raman spectra of many chemical substances can be found in the literature. Several textbooks are available which contain collections of Raman spectra; examples are Schrader and Meier (1974, 1975, 1976), Brame and Graselli (1976, 1977), Graselli et al. (1981), Freeman (1974), Ross (1972), and Dolish et al. (1974).

Raman spectra comprise relatively few lines if the molecules are composed of only a few atoms (simple molecules); an example is shown in Figure 14.3. The spectra become increasingly complicated, the larger the molecules are. In Figure 14.4, part of the Raman spectrum of dibutylphthalate is shown. The identification of the different chemical components in an unknown sample is only possible as long as the characteristic lines of these components can be identified in the spectrum. This aspect of Raman scattering has been used very successfully in the last 15 years for the determination of the composition of gaseous media, even in such a hostile environment as combustion processes. Ledermann and Sacks (1984) gave a survey of the applications of Raman scattering to flow field and combustion research.

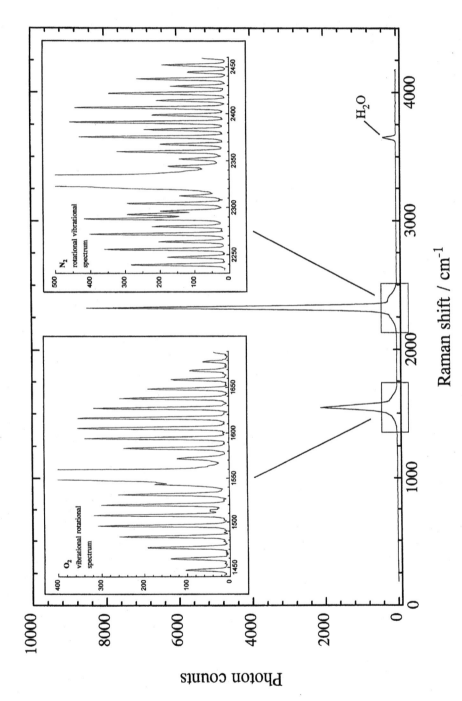

FIGURE 14.3 Raman spectrum of air. The inserts show the vibrational rotational Raman spectra of O_2 receptively N_2.

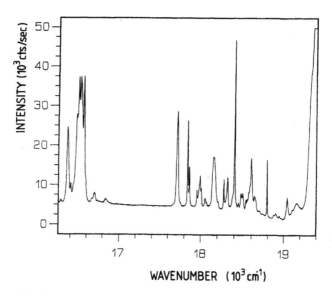

FIGURE 14.4   Part of the Raman spectrum of dibutylphthalate.

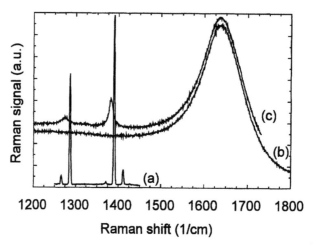

FIGURE 14.5   Part of the Raman spectrum of (a) gaseous CO_2, (b) CO_2 dissolved in water, and (c) pure water (the spectrum is shifted in the y direction for better visibility).

The application of this technique for the identification of molecular composition in gases at elevated densities, in liquids, or solids is not as straightforward as for gases at moderate densities. The reason is the increasing importance of intermolecular forces, which affect the line shift, band shapes, or both. In effect, the Raman spectrum can be used to investigate intermolecular forces as outlined by Srivastrava and Zaidi (1979). Obviously, the frequency shift depends strongly on the chemical substances under investigation. Not only pressure, but also phase changes can affect the Raman spectrum. The frequency shift observed in phase changes depends on the vibrational mode of the particular Raman line and is usually accompanied by intensity changes. An example of phase change is shown in Figure 14.5. The frequency shift makes a discrimination between Raman scattering from the liquid phase and Raman scattering from the gas phase possible. If such a discrimination is possible, the simultaneous determination of liquid phase and gas phase concentrations is possible from Raman scattering.

(b) Intensities. The intensities of Raman lines depend on the number of molecules of a specific chemical component in the scattering volume and on temperature, as shown by Equations (10), (18), and (19). This relationship holds for most gases at moderate pressure and has found widespread application in fluid dynamics and combustion research for the local determination of gas composition and temperature (as mentioned before).

The intensity of Raman scattering is proportional to the local electric field, which may be different from the external incident field. Two reasons for this can be identified, as pointed out by Schrötter and Klöckner (1979). Due to the close proximity of the surrounding molecules, the local electromagnetic field differs from the external incident field. This is an extremely important point for Raman scattering on aerosol particles, where the local field deviates appreciably from the external field. We will discuss this point in the next chapter. For scattering on bulk material, the local field can usually be expressed as a relatively simple function of the refractive index of the medium and the energy flux density of the incident radiation (see, for example, Kaiser, Voßmer-bäumer, and Schweiger, 1992).

If the refractive index depends on concentration (composition), the local field and, therefore, the intensities of the Raman lines also depend on concentration. Another source for deviation from a linear dependence on the incident intensity is due to intermolecular forces. This effect depends on the chemical components in the sample. The effect is usually weak for nonpolar liquids. A quantitative prediction of this effect is difficult because it depends on the distribution and type of molecular neighbors, on the interaction strength between dissimilar neighbors, and on the electronic properties (transition frequencies). Complexes are often formed, especially in aqueous solution, which — in addition to the intensity change — also affect the band shape and line frequencies. We will come back to this point later. The solvent effect is usually different for different Raman lines of the same solute. Several research groups have investigated this effect. Fini et al. (1968) have investigated the solvent effect for some hydrocarbons. Examples are given by Schweiger (1990), where references for further details can also be found. These effects are usually not serious. They are often weak and can usually be evaluated quantitatively by calibration measurements on bulk material under well-defined conditions.

(c) Bandwidth and band shape. Intermolecular forces often affect not only intensities — band intensities can be enhanced or reduced due to the presence of other chemical components — but also the bandwidth. In the gaseous state at moderate pressures, the effect is usually negligible. In the liquid state, however, where the intermolecular distances are short, the bandwidth and shape are usually affected. Examples are shown in Figures 14.5 and 14.6. The Raman band of CO_2 is broader in the liquid state, as we can see from Figure 14.5, whereas the contrary holds for SO_2 (Figure 14.6). Not only the width but also the change in shape caused by a transition from the gaseous to the liquid state can be clearly seen from Figure 14.6. Le Duff and Quillon (1973) and also Tanabe (1854) have shown that the line width can also depend on the solvent. A very interesting case is water, which shows a dramatic increase in the line width by phase change from the gaseous into the liquid state. We will come back to this point later. There are several mechanisms that can affect frequency, bandwidth, and intensities by intermolecular forces, namely resonant transfer of vibrational energy and intermolecular coupling such as Fermi resonances, vibration–rotation interactions, or collisional displacement. The contribution of these effects increases with density, as already mentioned in connection with the effect on line shift. Zerda et al. (1987) have investigated the density effect on the line width of the C–O and C–H mode of liquid methanol.

Intermolecular forces affect not only the intensity, shape, and width of Raman bands, but also temperature. The intensity of the Raman lines depends (as we can see from Equation 10) on the occupation numbers and, therefore, on temperature (Equation 17 or 22). This temperature effect is widely used for gas phase temperature measurements, as already mentioned, and can also be used for the temperature determination in water and aqueous solutions.

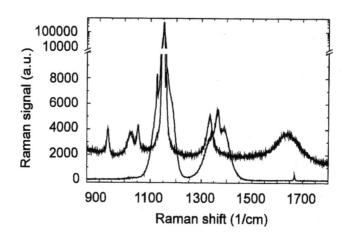

FIGURE 14.6   Part of the Raman spectrum of gaseous SO_2 (lower trace) and of SO_2 dissolved in water. The peak at 943 cm^{-1} is from the ClO_3 ion, which is used as an internal standard.

The quantitative evaluation of Raman spectra is usually quite straightforward for gases but can be complicated for liquids. For application of this technique to aerosol analysis, this is not a disadvantage, because the different effects can be determined by Raman scattering on bulk materials.

(d) Polarization of scattered light: The polarization of the Raman lines depends on the polarizability of the molecules and on the transition. It is different for rotational and vibrational lines. Figure 14.7 shows two records of the rotational vibrational band for gaseous N_2. The spectra were recorded with a polarization filter in front of the entrance slit of the monochromator. In the upper trace, the polarization of the filter was parallel to the polarization of the laser beam, and perpendicular to it in the lower trace. Obviously, the Raman lines related to a change in the rotational quantum number are highly depolarized in contrast to pure vibrational transitions, where the intensity in the upper trace is much higher than in the lower trace. In the liquid state the situation is complicated by intermolecular forces. The interaction processes can affect the symmetry properties of the molecules and change the polarizability, but it can also hinder the rotation of the molecules. Both interaction processes influence the degree of polarization of the scattered light.

14.3   THEORY OF RAMAN SCATTERING ON MICROPARTICLES

14.3.1   QUALITATIVE DESCRIPTION

We know from the theoretical analyses that the amplitude of Raman scattering is proportional to the amplitude of the local electrical field. This is the field in the interior of the particle — the transmitted field. This field can deviate appreciably from the incident field outside the particle. The incident field, often the field of a plane wave, is deformed by the change of the refractive index at the surface of the particle. This change acts on the incident and scattered field as well. Consequently, not only is the transmitted field affected by the boundary of the particle, but also the radiation fields emitted by the induced dipoles. The particle acts as a micro lens on the incident and the emitted radiation. The focusing effect of the incident radiation is displayed in Figure 14.8(a). The ray paths were constructed by geometrical optics. In Figure 14.8(b) the effect of the particle on dipole radiation is shown. The particle causes the refraction of the emitted light into a preferential direction. This projection effect depends on the position of the dipole within the particle.

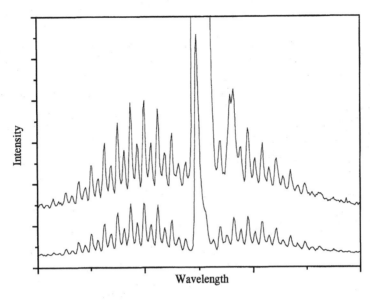

FIGURE 14.7   Rotational vibrational Raman spectrum of N_2 polarized parallel (upper trace) and perpendicular (lower trace) to the polarization plan of the incident beam.

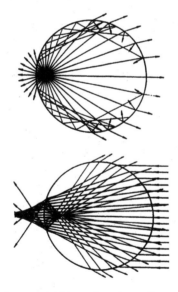

FIGURE 14.8   Effect of the particle on an incident plane wave (lower figure) and on the radiation emitted from a source within the particle.

14.3.2   QUANTITATIVE DESCRIPTION

Geometrical optics can only be applied with sufficient accuracy to describe the interaction between the incident light wave and the particle, if the dimension of the particle is large compared to the wavelength. If this is not so, a rigorous solution of the Maxwell equations has to be used. Such a solution for spherical particles was developed by Gustav Mie (1908) by expanding the electric and magnetic fields into a series of solutions of the wave equation for electromagnetic fields.

14.3.3 MULTIPOLE EXPANSION

In the modern treatment of light scattering, the electromagnetic fields are represented by an infinite series of orthonormal functions called vector spherical harmonics. Each of these functions describes the radiation pattern of an electric or magnetic multipole. The expansion coefficients determine the contribution of the individual multipoles to the total radiation field. Slightly different definitions of the vector spherical harmonics are in use. Bohren and Huffman (1983) and Barber and Hill (1990) include the radial function into the definition, whereas Jackson (1975) and Chew, McNulty, and Kerker (1976) and many other authors designate by vector spherical harmonics only the angular dependent part. We will follow the latter practice here.

Omitting the time dependent factor $\exp(-i\omega t)$ a general solution of the source free Maxwell equations is given by:

$$\overline{E} = \sum_{\ell,m} \left\{ \frac{ic}{n^2\omega} A_E(\ell,m)\overline{\nabla} \times \left[f_\ell(kr)\overline{X}_{\ell m}(\theta,\phi) \right] + A_M(\ell,m)g_\ell(kr)\overline{X}_{\ell m}(\theta,\phi) \right\}$$

$$\overline{B} = \frac{1}{c} \sum_{\ell,m} \left\{ A_E(\ell,m)f_\ell(kr)\overline{X}_{\ell m}(\theta,\phi) - \frac{ic}{\omega} A_M(\ell,m)\overline{\nabla} \times \left[g_\ell(kr)\overline{X}_{\ell m}(\theta,\phi) \right] \right\} \qquad (23)$$

where c is the free space light velocity, ω is the frequency of the electromagnetic wave, n is the index of refraction, $n^2 = \varepsilon \cdot \mu$, where ε is the relative dielectric constant and μ the relative magnetic permeability. $f_\ell(k\,r)$ and $g_\ell(k\,r)$ are spherical Bessel or Hankel functions which depend on the wavenumber k and radial coordinate r. The vector spherical harmonics $\overline{X}_{\ell m}(\theta,\phi)$ are orthonormal vector functions, which depend only on the angular coordinates. The expansion coefficients A_E and A_M have to be calculated from the boundary conditions. They are functions of the dimensionless size parameter, $k \cdot a = (2\pi/\lambda) \cdot a$, where 2a is the particle diameter and λ is the wavelength of the radiation. The size parameter is also called Mie parameter, and the letter x is often used to designate the size parameter.

Chew, McNulty, and Kerker (1976) applied this method to the problem of Raman scattering on microparticles, and their procedure will be sketched in the following. The quantities outside the particle are identified by the index 2, the quantities within the particles by the index 1. From the condition that all fields emitted from the particle must disappear at infinity, it follows that the radial dependence of the fields outside must be represented by spherical Hankel functions $h_\ell^{(1)}(kr)$. The fields within the particle must be represented by spherical Bessel functions $j_\ell(kr)$ to remain finite at the origin, which is chosen to be at the center of the particle. After this specification the procedure is as follows:

Electromagnetic fields, which oscillate at the frequency ω_0 of the incident wave:

- Calculate the expansion coefficients, $\alpha_E(\ell,m)$, $\alpha_M(\ell,m)$ for the incident wave $\overline{E}_{inc}(\overline{r},\omega_0)$, $\overline{B}_{inc}(\overline{r},\omega_0)$
- Expand the transmitted fields (the field within the particle) $\overline{E}_t(\overline{r},\omega_0), \overline{B}_t(\overline{r},\omega_0)$ into a series of multipole radiation as shown by Equation (23). The radial functions are spherical Bessel functions, because the fields have to remain finite at the origin, as mentioned above.
- Expand the scattered fields. $\overline{E}_{sc}(\overline{r},\omega_0), \overline{B}_{sc}(\overline{r},\omega_0)$ The radial functions are spherical Hankel functions.
- Calculate the expansion coefficients $\gamma_E(\ell,m)$, $\gamma_M(\ell,m)$ of the transmitted fields $\overline{E}_t(\overline{r},\omega_0), \overline{B}_t(\overline{r},\omega_0)$ and the expansion coefficients $\beta_E(\ell,m)$, $\beta_M(\ell,m)$ of the scattered fields $\overline{E}_{sc}(\overline{r},\omega_0), \overline{B}_{sc}(\overline{r},\omega_0)$ from the conditions, that the *tangential components* of the electric and magnetic fields are *continuous* on the surface of the particle.

- Calculate the amplitude and orientation of the dipole moments induced by the transmitted field from $\bar{p}(\bar{r}',\omega_R) = 4\pi\varepsilon\varepsilon_0\bar{\alpha}(\omega_0,\omega_R)\bar{E}(\bar{r}',\omega_0)$.

Frequency shifted Raman fields, which oscillate at the frequency ω_R:

- Calculate the expansion coefficients $a_E(\ell,m)$, $a_M(\ell,m)$ for the frequency shifted dipole radiation $\bar{E}_{dip}(\bar{r},\omega_R)$, $\bar{B}_{dip}(\bar{r},\omega_R)$
- Calculate the expansion coefficients $b_E(\ell,m)$, $b_M(\ell,m)$ of the frequency shifted fields $\bar{E}_b(\bar{r},\omega_R)$, $\bar{B}_b(\bar{r},\omega_R)$ within the particles, which has to be added to the dipole field to take into account the boundary, and $c_E(\ell,m)$, $c_M(\ell,m)$ the expansion coefficients of the frequency shifted Raman fields $\bar{E}_R(\bar{r},\omega_R)$, $\bar{B}_R(\bar{r},\omega_R)$ outside of the particle by the same procedure as before.
- Calculate the incoherent sum of the radiation of all induced dipoles.

The results of the procedure are:
Electromagnetic fields oscillating at the frequency ω_0 of the incident wave.
Incident wave:

$$\bar{E}_{inc}(\bar{r},\omega_0) = \sum_{\ell,m}\left\{\frac{ic}{n_2^2\omega_0}\alpha_E(\ell,m)\bar{\nabla}\times\left[j_\ell(k_2r)\bar{X}_{\ell m}(\theta,\phi)\right] + \alpha_M(\ell,m)j_\ell(k_2r)\bar{X}_{\ell m}(\theta,\phi)\right\}$$

$$\bar{B}_{inc}(\bar{r},\omega_0) = \frac{1}{c}\sum_{\ell,m}\alpha_E(\ell,m)j_\ell(k_2r)\bar{X}_{\ell m}(\theta,\phi) - \frac{ic}{\omega_0}\alpha_M(\ell,m)\bar{\nabla}\times\left[j_\ell(k_2r)\bar{X}_{\ell m}(\theta,\phi)\right] \qquad (24)$$

Following common practice one pair of brackets was omitted in the representation of the magnetic field and will be omitted in the following equations.
The expansion coefficients follow from*:

$$\bar{E}_{inc} = E_0\bar{e}_x\cdot e^{ik_2z}$$

and are:

$$\alpha_E(\ell,m) = i^{\ell+1}n_2\sqrt{\pi(2\ell+1)}\left(\delta_{m,-1} - \delta_{m,+1}\right)E_0, \quad \alpha_{,M}(\ell,m) = i^\ell\sqrt{\pi(2\ell+1)}\left(\delta_{m,-1} + \delta_{m,+1}\right)E_0 \quad (25)$$

Transmitted fields:

$$\bar{E}_t(\bar{r},\omega_0) = \sum_{\ell,m}\frac{ic}{n_1^2\omega_0}\gamma_E(\ell,m)\bar{\nabla}\times\left[j_\ell(k_1r)\bar{X}_{\ell m}(\theta,\phi)\right] + \gamma_M(\ell,m)j_\ell(k_1r)\bar{X}_{\ell m}(\theta,\phi)$$

$$\bar{B}_t(\bar{r},\omega_0) = \frac{1}{c}\sum_{\ell,m}\gamma_E(\ell,m)j_\ell(k_1r)\bar{X}_{\ell m}(\theta,\phi) - \frac{ic}{\omega_0}\gamma_M(\ell,m)\bar{\nabla}\times\left[j_\ell(k_1r)\bar{X}_{\ell m}(\theta,\phi)\right] \qquad (26)$$

---

* In contrast to the expansion of a circular polarized wave given by Chew et al. (1976) or Brock et al. (1980), the expansion coefficients are given here for a linearly polarized wave with its amplitude normalized to one, and the MKSA-system is used.

Scattered fields:

$$\overline{E}_{sc}(\overline{r},\omega_0) = \sum_{\ell,m} \frac{ic}{n_2^2\omega_0}\beta_E(\ell,m)\overline{\nabla}\times\left[h_\ell^{(1)}(k_2r)\overline{X}_{\ell m}(\theta,\phi)\right] + \beta_M(\ell,m)h_\ell^{(1)}(k_2r)\overline{X}_{\ell m}(\theta,\phi)$$

$$\overline{B}_{sc}(\overline{r},\omega_0) = \frac{1}{c}\sum_{\ell,m}\beta_E(\ell,m)h_\ell^{(1)}(k_2r)\overline{X}_{\ell m}(\theta,\phi) - \frac{ic}{\omega_0}\beta_M(\ell,m)\overline{\nabla}\times\left[h_\ell^{(1)}(k_2r)\overline{X}_{\ell m}(\theta,\phi)\right] \quad (27)$$

From the boundary condition on the surface of the particle one gets:

$$\overline{n}\times\overline{E}_t = \overline{n}\times\left(\overline{E}_{inc}+\overline{E}_{sc}\right), \quad \overline{n}\times\overline{H}_t = \overline{n}\times\left(\overline{H}_{inc}+\overline{H}_{sc}\right).$$

From these vector equations one gets four scalar equations, which relate the six expansion coefficients. Because two of them $\alpha_E(\ell,m)$, $\alpha_M(\ell,m)$ are already known from the plan wave expansion, the other four can be calculated. This yields:

$$\gamma_E(\ell,m) = \frac{\left(in_1^2/\mu_2k_2a\right)\cdot\alpha_E(\ell,m)}{\left(n_1^2/\mu_1\right)j_\ell(k_1a)\left[k_2a\cdot h_\ell^{(1)}(k_2a)\right]' - \left(n_2^2/\mu_2\right)h_\ell^{(1)}(k_2a)\left[k_1a\cdot j_\ell(k_1a)\right]'}$$

$$\gamma_M(\ell,m) = \frac{\left(i\mu_1/k_2a\right)\alpha_M(\ell,m)}{\mu_1 j_\ell(k_1a)\left[k_2a\cdot h_\ell^{(1)}(k_2a)\right]' - \mu_2 h_\ell^{(1)}(k_2a)\left[k_1a\cdot j_\ell(k_1a)\right]'} \quad (28)$$

where

$$\left[xf(x)\right]' = \frac{d}{dx}\left[xf(x)\right]$$

Only the transmitted fields are needed to calculate the Raman scattering. The explicit calculation of the expansion coefficients for the scattered fields is, therefore, not necessary, but the expansion coefficients are given here for completeness.

$$\beta_E(\ell,m) = \frac{\left\{\left(n_2^2/\mu_2\right)\cdot j_\ell(k_2a)\left[k_1a\cdot j_\ell(k_1a)\right]' - \left(n_1^2/\mu_1\right)\cdot j_\ell(k_1a)\left[k_2a\cdot j_\ell(k_2a)\right]'\right\}\alpha_E(\ell,m)}{\left(n_1^2/\mu_1\right)\cdot j_\ell(k_1a)\left[k_2a\cdot h_\ell^{(1)}(k_2a)\right]' - \left(n_2^2/\mu_2\right)\cdot h_\ell^{(1)}(k_2a)\left[k_1a\cdot j_\ell(k_1a)\right]'}$$

$$\beta_M(\ell,m) = \frac{\left\{\mu_2 j_\ell(k_2a)\left[k_1a\cdot j_\ell(k_1a)\right]' - \mu_1 j_\ell(k_1a)\left[k_2a j_\ell(k_2a)\right]'\right\}\alpha_M(\ell,m)}{\mu_1 j_\ell(k_1a)\left[k_2a\cdot h_\ell^{(1)}(k_2a)\right]' - \mu_2 h_\ell^{(1)}(k_2a)\left[k_1a\cdot j_\ell(k_1a)\right]'} \quad (29)$$

Electromagnetic fields oscillating at the Raman frequency ω_R.
The dipole fields follow from:

$$\overline{B}_{dip}(\overline{r},\omega) = \nabla\times\overline{A}_{dip} \quad \text{with} \quad \overline{A}_{dip} = -i\frac{\omega\mu\mu_0}{4\pi}\overline{p}(\overline{r}')\frac{e^{ik(\overline{r}-\overline{r}')}}{|\overline{r}-\overline{r}'|}$$

If we express $\bar{p}$ by:

$$\bar{p} = 4\pi\varepsilon\varepsilon_0\bar{\alpha}\cdot\bar{E} = 4\pi\varepsilon\varepsilon_0\bar{p}', \quad \bar{p}'\left(\bar{r}',\omega_0,\omega_R\right) = \bar{\alpha}\left(\omega_0,\omega_R\right)\bar{E}\left(\bar{r}',\omega_0\right)$$

we get:

$$\overline{A}_{dip} = -i\frac{k\cdot n}{c}\bar{p}'(\bar{r}',\omega)\frac{e^{ik|\bar{r}-\bar{r}'|}}{|\bar{r}-\bar{r}'|} = 4\pi k^2\frac{n\bar{p}'}{c}\sum j_\ell(kr')h_\ell^{(1)}(kr)\overline{X}_{\ell m}^*(\theta',\phi')\overline{X}_{\ell m}(\theta,\phi) \tag{30}$$

and the expansion coefficients for the dipole field

$$\overline{E}_{dip}\left(\bar{r},\omega_R\right) = \sum_{\ell,m}\frac{ic}{n_1^2\omega_R}a_E(\ell,m)\overline{\nabla}\times\left[h_\ell^{(1)}(k_1r)\overline{X}_{\ell m}(\theta,\phi)\right] + a_M(\ell,m)h_\ell^{(1)}(k_1r)\overline{X}_{\ell m}(\theta,\phi) \tag{31}$$

$$\overline{B}_{dip}\left(\bar{r},\omega_R\right) = \frac{1}{c}\sum_{\ell,m}a_E(\ell,m)h_\ell^{(1)}(k_1r)\overline{X}_{\ell m}(\theta,\phi) - \frac{ic}{\omega_R}a_M(\ell,m)\overline{\nabla}\times\left[h_\ell^{(1)}(k_1r)\overline{X}_{\ell m}(\theta,\phi)\right]$$

can be calculated from $\overline{B}_{dip} = \overline{\nabla}\times\overline{A}_{dip}$, which gives

$$a_E(\ell,m) = \bar{p}'\left(\bar{r}',\omega_0,\omega_R\right)\overline{V}_E(\ell,m), \quad a_M(\ell,m) = \bar{p}'\left(\bar{r},\omega_0,\omega_R\right)\overline{V}_M(\ell,m) \tag{32}$$

$$\overline{V}_E(\ell,m) = 4\pi k_1^2\nabla'\times\left[j_\ell(k_1'\bar{r}')\overline{X}_{\ell m}^*(\theta',\phi')\right] \tag{33}$$

$$\overline{V}_M(\ell,m) = 4\pi k_1^3\frac{1}{\mu_1^2}j_\ell(k_1'r')\overline{X}_{\ell m}^*(\theta',\phi')$$

Boundary fields:

$$\overline{E}_b(\bar{r}) = \sum_{\ell,m}\frac{ic}{n_1^2\omega_R}b_E(\ell,m)\overline{\nabla}\times\left[j_\ell(k_1r)\overline{X}_{\ell m}(\theta,\phi)\right] + b_M(\ell,m)j_\ell(k_1r)\overline{X}_{\ell m}(\theta,\phi) \tag{34}$$

$$\overline{B}_b\left(\bar{r},\omega_R\right) = \frac{1}{c}\sum_{\ell,m}b_E(\ell,m)j_\ell(k_1r)\overline{X}_{\ell m}(\theta,\phi) - \frac{ic}{\omega_R}b_M(\ell,m)\overline{\nabla}\times\left[j_\ell(k_1r)\overline{X}_{\ell m}(\theta,\phi)\right]$$

Scattered Raman fields:

$$\overline{E}_R\left(\bar{r},\omega_R\right) = \sum_{\ell,m}\frac{ic}{n_2\omega_R}c_E(\ell,m)\overline{\nabla}\times\left[h_\ell^{(1)}(k_2r)\overline{X}_{\ell m}(\theta,\phi)\right] + c_M(\ell,m)h_\ell^{(1)}(k_2r)\overline{X}_{\ell m}(\theta,\phi) \tag{35}$$

$$\overline{B}_R\left(\bar{r},\omega_R\right) = \frac{1}{c}\sum_{\ell,m}c_E(\ell,m)h_\ell^{(1)}(k_2r)\overline{X}_{\ell m}(\theta,\phi) - \frac{ic}{\omega_R}c_M(\ell,m)\overline{\nabla}\times\left[h_\ell^{(1)}(k_2r)\overline{X}_{\ell m}(\theta,\phi)\right]$$

The boundary conditions on the surface of the particle are:

$$\bar{n} \times \bar{E}_R = \bar{n} \times (\bar{E}_{dip} + \bar{E}_b), \quad \bar{n} \times \bar{H}_R = \bar{n} \times (\bar{H}_{dip} + \bar{H}_b). \tag{36}$$

This again is sufficient to calculate the expansion coefficients of the Raman scattered field, which yields (Druger, Arnold, and Folan, 1987; Lange and Schweiger, 1994):

$$c_E(\ell, m) = f_E(\ell) a_E(\ell, m) \quad c_M(\ell, m) = f_M(\ell) a_M(\ell, m) \tag{37}$$

$$f_E(\ell) = \frac{in_2^2/(\mu_1 k_1 a)}{(n_1^2/\mu_1) j_\ell(k_1 a)[k_2 a \cdot h_\ell^{(1)}(k_2 a)]' - (n_2^2/\mu_2) h_\ell^{(1)}(k_2 a)[k_1 a \cdot j_\ell(k_1 a)]'} \tag{38}$$

$$f_M(\ell) = \frac{i\mu_2/(k_1 a)}{\mu_1 j_\ell(k_1 a)[k_2 a \cdot h_\ell^{(1)}(k_2 a)]' - \mu_2 h_\ell^{(1)}(k_2 a)[k_1 a \cdot j_\ell(k_1 a)]'}$$

The quantities k_1 and k_2 have to be evaluated with ω_R for all frequency shifted fields.

The explicit calculation of the expansion coefficients of the boundary fields is not necessary but will be given here for completeness:

$$b_E = \frac{\dfrac{n_1^2}{\mu_1} h_\ell^{(1)}(k_1 a)[k_2 a \cdot h_\ell^{(1)}(k_2 a)]' - \dfrac{n_2^2}{\mu_2} h_\ell^{(1)}(k_2 a)[k_1 a \cdot h_\ell^{(1)}(k_1 a)]'}{\dfrac{n_2^2}{\mu_2} h_\ell^{(1)}(k_2 a)[k_1 a \cdot j_\ell(k_1 a)]' - \dfrac{n_1^2}{\mu_1} j_\ell(k_1 a)[k_2 a \cdot h_\ell^{(1)}(k_2 a)]' a_E} \cdot a_E(\ell, m) \tag{39}$$

$$b_M = \frac{\mu_2 h_\ell^{(1)}(k_2 a)[k_1 a \cdot h_\ell^{(1)}(k_1 a)]' - \mu_1 h_\ell^{(1)}(k_1 a)[k_2 a \cdot h_\ell^{(1)}(k_2 a)]'}{\mu_1 j_1(k_1 a)[k_2 a \cdot h_\ell^{(1)}(k_2 a)]' - \mu_2 h_\ell^{(1)}(k_2 a)[k_1 a \cdot j_\ell(k_1 a)]'} a_M(\ell, m)$$

14.3.4 TOTAL SCATTERED RAMAN FIELD

The total inelastically scattered light has to be calculated in principle by the incoherent integration over the contributions of all molecular dipoles.

$$\phi_{tot} \propto |\bar{E}_R(\bar{r}, \omega_R)|^2_{tot} = \int_{Particle} |E_R(\bar{r}, \bar{r}', \omega_R, \omega_0)|^2 dV' \tag{40}$$

This is obviously not practicable. To reduce the computing time to manageable quantities, the particle is divided into a sufficient number of subvolumes, and the radiation of each volume is represented by a single dipole with appropriate amplitude (Kerker et al., 1979).

In Figure 14.9(a), the $|E_t|^2$ distribution of the transmitted field in the equatorial plane of a micro particle is shown. The distribution was computed with the method of multipole expansion outlined above. The transmitted field depends strongly on position. Therefore, a large number of subvolumes have to be chosen to represent the inelastically scattered light with sufficient accuracy. This results

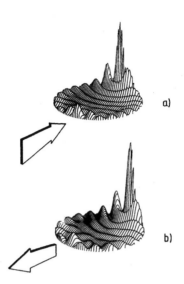

FIGURE 14.9 Square of the amplitude of the transmitted electrical field for an incident plane wave (upper diagram) and the distribution of dipoles uniformly distributed in the equatorial plane to the emission in the direction indicated by the arrow.

in still extreme computational times for larger particles, and no calculation of the intensity of Raman scattering for particles larger than $X = 20$ are known to the author (Kerker et al., 1979). Fortunately a number of applications of Raman scattering to particle analyses are possible without an explicit calculation of the scattered light intensity.

In Figure 14.9(b), the effect of the particle on the dipole radiation is shown. For this purpose a number of dipoles, all oscillating with the same amplitude, were uniformly distributed in the equatorial plane, then the contribution of each dipole to back scattering (emission in the direction shown by the arrow) was calculated. It can be seen that the contribution of different dipoles to the back direction is quite different depending on the position. There is a remarkably good agreement between results of geometrical optics shown in Figure 14.8 and the exact calculations.

The contribution of different regions of the equatorial plane to the Raman scattering into three different scattering directions (shown by arrows) is shown in Figure 14.10. It is a result of the superimposition of the focusing effect and the projection effect. With this kept in mind, the interpretation of Figure 14.10 is quite straightforward. In the case of forward scattering, the projection effect does not enhance the region of the main maximum of the transmitted field, but that of the smaller second maximum. This explains the appearance of two areas from the region where light is emitted, preferentially in the forward direction (scattering angle $\theta = 0°$). Using similar arguments, the relatively complicated situation reproduced in Figure 14.10(b) can be understood. Figure 14.10(c) shows that a relatively small region contributes predominantly to back scattering. This becomes clear because for back scattering the region where the transmitted field has its maximum (due to the focusing effect) coincides with the region from which the outgoing radiation is emitted predominantly in the backward direction (scattering angle $\theta = 180°$) caused by the projection effect.

14.3.5 SCATTERING CROSS SECTION

With the results of the foregoing section we can now calculate the scattering cross section of a particle containing an arbitrary number of molecules of a specific chemical component. For this purpose, we first calculate the differential scattering cross section $(\partial \sigma_A / \partial \Omega)_{par}$ for Raman scattering of a single molecule of species A at the location $\vec{r}'$ within the particle.

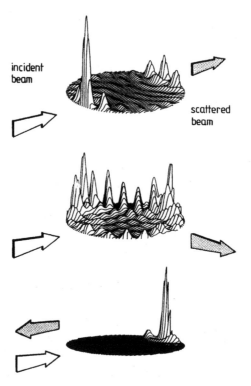

FIGURE 14.10 Contribution of different regions in the equatorial plane to Raman scattering in the direction indicated by the grey arrows. The white arrows indicate the direction of the incident plane wave.

$$\left(\frac{\partial\sigma_A}{\partial\Omega}\right)_{par}=\frac{1}{D_0}r^2\cdot\left(\frac{\partial\phi_A}{\partial\Omega}\right)_{par}=\frac{r^2}{E_0^2}\left|\overline{E}_R\right|^2 \tag{41}$$

where D_0 is the power density of the incident radiation. Representing $h_\ell^{(1)}(x)$ by its asymptotic form for $x \gg 1$ and expressing $\overline{E}_R$ by Equation (35), we get:

$$h_\ell^{(1)}(k_2r)=(-i)^{\ell+1}\frac{e^{ik_2r}}{k_2r},\quad \overline{\nabla}xh_\ell^{(1)}(k_2r)\cdot\overline{X}_{\ell m}(\theta,\phi)=(-i)^\ell\frac{e^{ik_2r}}{k_2r}k_2\left[\overline{n}x\overline{X}_{\ell m}(\theta,\phi)\right]$$

$$\left(\frac{\partial\sigma_A}{\partial\Omega}\right)_{par}=\frac{1}{k_2^2}\left|\sum_{\ell,m}(-i)^{\ell+1}c_E(\ell,m)\left[\overline{n}x\overline{X}_{\ell m}(\theta,\phi)\right]+c_M(\ell,m)\overline{X}_{\ell m}(\theta,\phi)\right|^2 \tag{42}$$

We now assume that the polarizability is a scalar and use the dimensionless quantities:

$$\overline{W}_E=\frac{\overline{V}_E}{4\pi k_1^3},\quad \overline{W}_M=\frac{\overline{V}_M}{4\pi k_1^3} \tag{43}$$

and get:

$$\left(\frac{\partial\sigma_A}{\partial\Omega}\right)_{par} = \frac{16\pi^2 k_1^6 \alpha^2}{k_2^2}\left|\sum_{\ell,m} f_E(\ell)\overline{W}_E(\ell,m)\left(\overline{E}_t(\overline{r}')/E_0\right)\left[\overline{n}x\overline{X}_{\ell m}(\theta,\phi)\right]\right.$$

$$\left. + f_M(\ell)\overline{W}_M\left(\ell_{\ell m}\right)\left(\overline{E}_t(\overline{r}')/E_0\right)X_{\ell m}(\theta,\phi)\right|^2 \tag{44}$$

The square of the polarizability α can by expressed by the total molecular scattering cross section σ_M using Equation (9) and we get:

$$\left(\frac{\partial\sigma_A}{\partial\Omega}\right)_{par} = 6\pi\left(n_1/n_2\right)^2\sigma_A \cdot p\left(k_0,k_R,n_1,n_2,\overline{r}'\right) \tag{45}$$

The molecule location factor p given by:

$$p\left(k_0,k_R,a,n_1,n_2,\overline{r}'\right) =$$

$$\left|\sum_{\ell,m} f_E(\ell)\overline{W}_E(\ell,m)\left(\overline{E}_t(\overline{r}')/E_0\right)\left[\overline{n}x\overline{X}_{\ell m}(\theta,\phi)\right] + f_M(\ell)\overline{W}_M(\ell,m)\left(\overline{E}_t(\overline{r}')/E_0\right)\overline{X}_{\ell m}(\theta,\phi)\right|^2$$

depends on the size parameters of the incident ak_0 and the scattered light ak_R, on the refractive indices n_1,n_2, and on the position $\overline{r}'$ of the specific molecule. A similar expression was given by Schweiger (1987) for the location factor averaged over the particle volume.

The total Raman scattering cross section for Raman scattering on a particle containing N_A molecules of chemical species A is then:

$$\left(\frac{\partial\sigma_A}{\partial\Omega}\right)_{tot} = \int_{particle}\left(\frac{\partial\sigma_A}{\partial\Omega}\right)_{par} n_A(r')dV' \tag{46}$$

The number density n_A of molecules of species A is given by:

$$n_A(\overline{r}') = \frac{N_0}{M_A}\rho_A(\overline{r}') = \frac{N_0}{M_A}\rho c_A(\overline{r}') \tag{47}$$

where N_0 is Avogadro's number, M_A is the molecular weight of species A, ρ is the mean particle density, and c_A the mass fraction. Introducing Equation (47) into Equation (46) yields with Equation (45):

$$\left(\frac{\partial\sigma_A}{\partial\Omega}\right)_{tot} = 6\pi\left(n_1/n_2\right)^2\sigma_A\frac{N_0}{M_A}\cdot\rho\int_{particle} p(\overline{r}')c_A(\overline{r}')dV'$$

$$= 6\pi\left(n_1/n_2\right)^2\cdot\sigma_A\left(\omega_0,\omega_R\right)\cdot N_A P\left(k_0\cdot k_R,n_1,n_2,c_A(\overline{r})\right)$$

$$P\left(k_0\cdot k_R,n_1,n_2,c(\overline{r}')\right) = \frac{1}{V\cdot\overline{c}_A}\int_{particle} p(\overline{r}')c_A(\overline{r}')dV' \tag{48}$$

where $N_A = N_0 \cdot \rho \cdot V/M_A$ is the total number of molecules of species A in the particle and $\bar{c}_A$ is the mean mass fraction of A. Schweiger (1991) measured the size dependence of Raman scattering on single homogeneous optically levitated particles and found a linear dependence on the particle volume. This means that P = const. If there is no concentration gradient, the Raman signal is proportional to the number of molecules in the particles.

From Equation (48), we see that for Raman scattering on spherical microparticles the ratio ϕ_A/ϕ_B of the intensities of two Raman lines due to component A and due to component B is given by:

$$\frac{\phi_A}{\phi_B} = \frac{\sigma_A(\omega_0, \omega_R)}{\sigma_B(\omega_0, \omega_R)} \frac{N_A}{N_B} \frac{P_A}{P_B} \tag{49}$$

If the Raman frequencies are not very different and the two components are equally distributed within the particle, the concentration profile factors P cancel, and the ratio is simply proportional to the mole fractions and the scattering cross sections.

14.3.6 MORPHOLOGY-DEPENDENT RESONANCE EFFECTS

One peculiarity of the scattering process on microparticles has attracted considerable interest, especially in recent years. This is the excitation of so-called structural or morphology dependent resonances (MDRs). At specific size parameters, the inelastically scattered light shows a sharp increase in magnitude. The physical reason for this resonance effect is that the microparticle can be considered as an optical cavity. If the incident light wave coincides with an eigenmode, the incident wave is coupled effectively to this eigenmode. The amplitude of this eigenmode is considerably enhanced compared to nonresonant excitation and the radiation of this eigenmode contributes preferentially to the scattering process. In the picture of geometrical optics surface waves are excited at resonance conditions. These waves are trapped within a particle by total reflection on the surface and arrive with the correct phase after each round trip. Figure 14.11 shows the distribution of the square of the amplitudes of the electric field in the equatorial plane off resonance (upper diagram) and on resonance (lower diagram) for particles with nearly the same size. The generation of regular field patterns near the surface in the resonance case is clearly visible. The grey scales are different in the two figures. The contribution of the transmitted field in the interior of the particle, which is practically identical in both cases, is, therefore, not visible in the lower diagram, because it is to weak compared to the resonant field near the rim of the plane. Thurn and Kiefer (1984) were the first to observe output MDRs on micro glass spheres.

Mathematically, resonances are caused because at specific (but complex) arguments the denominator of one of the expansion coefficients becomes zero. The expansion coefficient has a pole. As this is the case only for complex arguments, one speaks of *virtual eigenmodes*. For real frequencies the denominator cannot become exactly zero; however, it becomes small enough to cause in some cases an order of magnitude increase in the corresponding expansion coefficient for the transmitted or inelastically scattered fields (given by Equations (28) and (37)). Although the corresponding expansion coefficient for elastic scattering given by Equation (29) reaches its maximum at the same size parameter, the increase is by far not so dramatic, but resonance peaks are also clearly observable. The denominator has for complex arguments an infinite number of zeros. The resonance width increases with increasing order of zeros, and the amplitude decreases. Higher-order MDRs are often barely visible because they are very broad. Some low-order MDRs, on the other hand, are too narrow to be visible. The last holds especially for elastic scattering.

The present understanding of resonance phenomena in the interaction of light with microparticles was reviewed by Hill and Benner (1988) and by Hill and Chang (1992). Until recently most of the theoretical and experimental work on structural resonances was done on elastic scattering. In the last few years, the investigation of inelastic resonance effects, such as lasing or fluorescence,

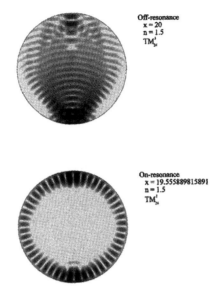

Off-resonance
x = 20
n = 1.5
TM_{24}^1

On-resonance
x = 19.555889815891
n = 1.5
TM_{24}^1

FIGURE 14.11 Distribution of the square of the electric field amplitude in the equatorial plane of a microparticle of resonance (upper plot) and at resonance of the TM_{24}-mode (lower plot). A plane wave hits the particle from the top; x is the size parameter; n is the relative index of refraction.

is in the center of ongoing research. If the resonance condition is fulfilled, the transmitted field can rise dramatically.

In inelastic scattering processes, the frequency of the scattered light is different from the incident light. MDRs can be excited in the incident fields as well as in the scattered fields. The corresponding size parameter has to be calculated at the wavelength of the incident wave; these MDRs are called *input resonances*. The resonances calculated at the wavelength of the inelastically scattered wave, are called *output resonances*. *Double resonances* are also possible. A double resonance is a resonance in the transmitted field and simultaneously in the Raman field (fluorescence field) (Schweiger, 1990a). If a resonance of the transmitted field is excited by an appropriate wavelength, a resonance in the Raman spectrum is very probably present simultaneously, because the Raman spectrum usually covers a certain wavelength range, so that one or several Raman frequencies fulfill the resonance condition. Owen et al. (1982a, b) discussed the excitation of MDRs for elastic, fluorescence, and Raman scattering on microparticles and glass fibers with diameters in the micrometer range.

The excitation of MDRs depends sensitively on the size parameter. The determination of the size parameter at which resonances are observed, is therefore, a very accurate method to determine the size of individual particles. Hill et al. (1985), Eversole et al. (1993), and Huckaby, Ray, and Das (1994) have shown this for elastically scattered light, and Tzeng et al. (1984) have used the morphology dependent peaks in fluorescence light to accurately determine the size of evaporating microparticles. Schweiger (1991) investigated the size dependence of Raman scattering by measuring the Raman scattered intensity from a single optically levitated microparticle. He used the resonances in the scattered field to accurately determine the particle size.

The excitation of input or output resonances causes peaks in the Raman spectrum of microparticles not present in bulk material. The interpretation of Raman spectrum of spherical or spheroidal particles is complicated by these MDR peaks. Careful provisions have to be made to avoid interpreting resonances as "true" Raman lines. Figure 14.12 shows an example of resonances in the Raman spectrum. The intensity of Raman scattered light was recorded at a fixed wavelength for a slowly evaporating DBP particle trapped by radiation forces in an Ar–ion laser beam. The continuous change of the particle size caused the consecutive excitation of input and output resonances. The width of the output resonances is determined by the wavelength dependence of

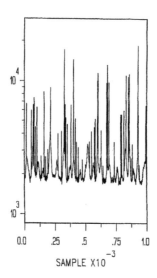

FIGURE 14.12 Amplitude of the Raman scattering at a fixed Raman shift on a microparticle with changing size. The size reduces with increasing sample number. The large narrow peaks are caused by input resonances, the smaller broader peaks by output resonances.

the resonance (natural or theoretical width), the slit function of the monochromator and the evaporation speed. The width of the input resonances is only determined by the natural width and the evaporation speed. Under the specific experimental conditions (Schweiger, 1991) used to record the Raman intensity reproduced in Figure 14.12, the narrow sharp peaks are input resonances.

14.4 APPLICATION OF RAMAN SCATTERING

14.4.1 TEMPERATURE MEASUREMENT

The shape of Raman bands can be affected by the temperature, as was discussed previously. This is especially the case for the rotational structure in gas phase Raman scattering. This is widely used for gas phase temperature determination in combustion processes, for example. Until recently, it was assumed that this technique cannot be applied to aerosols or sprays due to the strong elastic scattering on the particles, but with modern experimental equipment, the problem can be overcome in most cases (Grünfeld et al., 1994). We will discuss only the temperature measurement of the aerosol particles.

Aqueous particles are very suitable for temperature determination because the stretching mode of water shows an easily detectable change in shape with temperature, as can be seen from Figure 14.13. Vehring and Schweiger (1992) used this effect to measure the temperature of evaporating water droplets. The temperature of these particles as a function of the distance from the orifice of the generator, which was used to produce the particles, is shown in Figure 14.14. The experimental set-up for these measurements is given in Figure 14.15. This is a typical experimental set-up for Raman scattering on aerosols. The main components of the set-up are:

- High-power laser, usually an Ar–ion laser (typically 5W power at 514.5 nm) or a frequency-doubled Nd–YAG laser
- High quality, high aperture optics, to collect as much scattered light as possible
- Monochromator or spectrograph with high stray light rejection
- Highly sensitive light detector, such as a high-quality photo multiplier or a liquid nitrogen cooled CCD camera

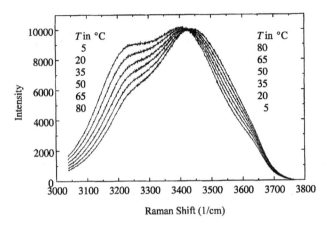

FIGURE 14.13 Effect of temperature change on the shape of the OH stretching band of water. The spectra are normalized at $\Delta v = 3425$ cm^{-1}.

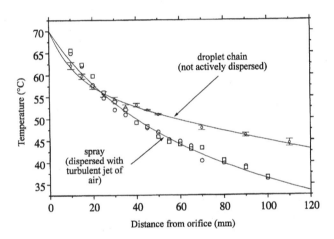

FIGURE 14.14 Temperature microparticle of water as a function of the distance from the orifice of the particle generator.

To improve the suppression of the elastically scattered light, a holographic Raman filter is often used in front of the entrance slit of the monochromator. If necessary, a polarization rotator is used to adjust the polarization of the incident beam perpendicular to the scattering direction because the scattering cross section is highest in this case. If an Ar–ion laser is used, a plasma filter is also recommended to remove the plasma light generated in the discharge tube from the laser beam.

Various techniques are used to generate the aerosol particles and investigate their behavior by Raman scattering, such as optical or electrodynamic trapping of single particles, or generation of a chain of highly monodispersed particles (see, for example, the review of Schweiger, 1990).

14.4.2 Concentration Measurements

14.4.2.1 Gas Phase Composition

The determination of the chemical composition of gaseous media by Raman scattering is a highly successful technique in combustion research. Measurements of the gas phase composition in two phase flows are complicated by the intense elastic scattering on the particle phase. Raman scattering

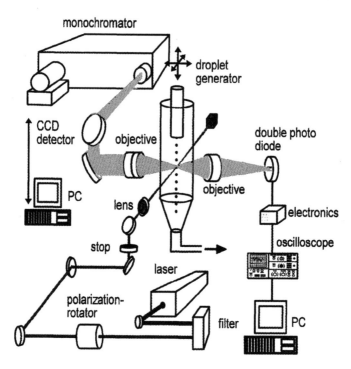

FIGURE 14.15 Typical set-up for Raman scattering on aerosols.

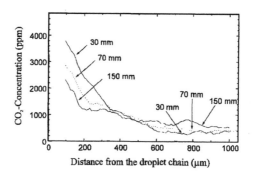

FIGURE 14.16 CO_2 concentration as a function of distance for evaporating water microdroplets saturated with CO_2. The concentration profiles are shown at three different distances from the droplet generator.

was applied in the past almost exclusively in single phase situations. However, for a complete analysis of heat and mass transport processes in aerosols, the measurement of the gas phase concentration is also necessary. This is possible with the experimental equipment available today. Figure 14.16 shows the gas phase concentration in the neighborhood of evaporating water microdroplets saturated with CO_2 as a function of distance from the droplets at three different distances from the orifice of the aerosol generator. The high spatial resolution was achieved with the 2D Raman technique described later.

14.4.2.2 Particle Composition

The most widely used application of Raman scattering in aerosol research is the determination of the chemical composition of aerosol particles. The first application of Raman scattering for this purpose was reported by Stafford, Chang, and Kindlmann (1977), but only in the last decade has this technique regained interest. Fung and Tang (1988a, 1988b, 1989, 1991, 1992) investigated the composition of single particles made from inorganic salt solutions. Buehler et al. (1991) analyzed the composition of various multicomponent aerosol particles trapped in an electrodynamic balance during evaporation. With the same technique, the chemical composition of individual toxic aerosol particles was investigated by Foss et al. (1993). Schweiger (1987) determined the composition of particles composed of DES (diethylsebacate) and DBP (dibutylphthalate).

Theoretical analysis has shown that the intensity of the Raman lines scattered by the particles is proportional to the number of molecules and, therefore, to the mole concentration in the particles. In addition, the intensity depends also on the shape of the particles and the concentration profile within the particle. Also, since no systematic investigation of the concentration profile factor is presently known, it can be assumed that the effect is small if the Raman frequencies of the various components are not very different and their concentration profiles are flat. Then the concentration profile factors in Equation (49) cancel each other, and the relative Raman intensities can be used to determine the relative concentrations within the particles without explicit knowledge of the concentration profile factors.

An important aspect is the detection limit. It depends on several circumstances, such as: Laser power, aperture angle of the scattering optics, scattering cross section of the chemical substances, sensitivity of the detector, detector noise, and background radiation. This background radiation can have several sources, such as insufficient suppression of light scattered elastically on the aerosol particles or on optical components, walls of the test chamber, etc. Another important source of background radiation is fluorescence or overlapping of the Raman bands of different substances. The last is the limiting factor in aqueous solutions, because the Raman spectrum of water extends from the laser line over more than 4000 wave numbers (cm^{-1}). The consequence is that there is always some residual Raman scattering of water in the wavelength ranch of the Raman spectrum of the substance of interest. The background in Figure 14.5 in the wave number range from 1200 to 1500 cm^{-1} is an example. The detection limit is, therefore, determined by the ratio of the Raman line of the specific component to the water background. After Fung, Imre, and Tang (1994), this detection limit is 2.5 millimolar for sulfate and nitrates.

Raman scattering was also applied to study the evaporation process of aerosol particles on droplet chains. Vehring et al. (1995) investigated not only the composition of evaporating particles but also the gas phase composition in the neighborhood of the particles. For this purpose, a 2D-Raman imaging technique was used. The region of interest is a part of a chain of evaporating droplets and the surrounding gas phase, which is imaged onto the entrance slit of the spectrograph. The laser beam passes through the region of interest perpendicular to the droplet chain. The imaging optics generates an image of the region of interest so that the image of the laser beam is oriented parallel to the entrance slit. In the exit plan of the instrument, a spectrally resolved image is generated. The spectrograph disperses the scattered light in the direction perpendicular to the laser beam, whereas the spatial resolution along the laser beam is preserved in the vertical direction. The evaluation of such a 2D-Raman image allows the determination of the gas phase concentration as a function of the distance from the droplet chain as well as the concentration in the liquid phase. Simultaneous to the local gas phase concentration, the concentration of CO_2 in the aerosol droplets was determined.

14.4.2.3   Chemical Reactions

Raman spectroscopy is a technique sensitive to particle composition. It can be used to monitor dynamic changes caused by chemical reaction on or in particles. Rassat and Davies (1992) observed the chemical reaction of a single electrodynamically levitated microparticle of calcium oxide and sulfur dioxide and water vapor. An example of observation of the polymerization of a single optically trapped particle of a photopolymer is shown in Figure 14.17. It shows part of the Raman spectrum of a photopolymer microdroplet, which was trapped optically before polymerization and kept in place until the polymerization process was completed (Esen, private communication).

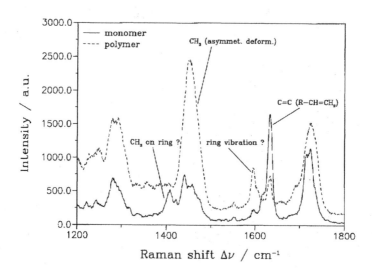

FIGURE 14.17 Part of the Raman spectrum recorded on a photopolymer, before polymerization (monomer) and after polymerization.

14.5 CONCLUSION

The theoretical analysis of Raman scattering on particles in the micrometer range is available in principle. The technique of multipole expansion, well known from elastic light scattering, is also suited to analyze Raman scattering. However, due to the incoherent nature of the linear Raman scattering process, the application of this technique causes extreme computation times. More effective methods would be highly desirable for practical applications.

The spectroscopic techniques have experienced a very rapid progress in the last few years. Detectors are now available which approach the theoretical detection limit, at least in the range of visible light. Although powerful lasers are available, very well suited for Raman spectroscopy, they are costly and voluminous. However, the rapid progress in the development of solid state lasers will change the situation favorably in the near future. After a period of very little change in dispersive instrumentation, much more compact, instruments are under development or already on the market. New components such as holographic or electrooptical filters are incorporated into these instruments.

As a summary, Raman scattering is a powerful technique for aerosol analyses, and its application in aerosol science will increase. This trend will be enhanced by the rapid progress in instrumentation.

ACKNOWLEDGMENT

The author would like to thank his collaborators, especially Thomas Kaiser, Stefan Lange, and Reinhard Vehring for their support in assembling this paper.

REFERENCES

Anderson, A., 1971, *The Raman Effect, Vol. 1: Principles*, Marcel Dekker, New York.
Anderson, A., 1973, *The Raman Effect, Vol. 2: Applications*, Marcel Dekker, New York.
Barber, P.W. and Chang, R.K., 1988, *Optical Effects Associated with Small Particles,* World Scientific, Singapore.
Barber, P.W. and Hill, S.C., 1990, Light scattering by particles: computational methods, *World Scientific.*

Bohren, G.F. and Huffman, D.R., 1983, *Absorption and Scattering of Light by Small Particles,* John Wiley & Sons.

Brame, G.E. and Graselli, J.G., 1976, 1977, *Infrared and Raman Spectroscopy,* Part A, B, C. Marcel Dekker, New York.

Brandmüller, J. and Moser H., 1962, Einführung in die Raman-Spektroskopie, Dr. Dietrich Steinkopf.

Buehler, M.F., Allen, T.M. and Davis, E.J., 1991, *J. Coll. Interf. Sci.,* 146: 79-89.

Chew, H., McNulty, P.J. and Kerker, M., 1976, *Phys. Rev. A.,* 13: 396-404.

Corney, A., 1977, *Atomic and Laser Spectroscopy,* Clarendon Press, Oxford.

Davis, E.J., Buehler, M.F. and Ward, T.L., 1990, *Rev. Sci. Instrum.,* 61: 1281-1288.

Dolish, F.R., Fately, W.G. and Bentley, F.F., 1974, *Characteristic Raman Frequencies of Organic Compounds,* John Wiley & Sons, New York.

Eversole, J.D., Lin, H.-B., Huston, A.L., Campillo, A., Leung, P.T., Liu, S.Y. and Young, K., 1993, *J. Opt. Soc. Am.,* B, 10: 1955-1968.

Fini, G., Mirone, P. and Patella, P., 1968, *J. Mol. Spectrosc.,* 28, 144-160.

Foss, W., Li, W., Allen, T.M., Blair, D.S. and Davis, E.J., 1993, *Aerosol Sci. Technol.,* 18: 187-201.

Freeman, S.K., 1974, *Applications of Laser Raman Spectroscopy,* John Wiley & Sons, New York.

Fung, K.H. and Tang, I.N., 1988a, *Chem. Phys. Lett.,* 147: 509-513.

Fung, K.H. and Tang, I.N., 1988b, *Appl. Optics.,* 27: 206-208.

Fung, K.H. and Tang, I.N., 1989, *J. Coll. Interf. Sci.,* 130: 219-224.

Fung, K.H. and Tang, I.N., 1991, *Appl. Spectrosc.,* 45: 734-737.

Fung, K.H. and Tang, I.N., 1992, *Appl. Spectrosc.,* 46: 1189-1193.

Fung, K.H., Imre, D.G. and Tang, I.N., 1994, *J. Aerosol Sci.,* 25: 479-485.

Graselli, J.G., Snavely, M.K. and Bulkin, B.J., 1981, *Chemical Applications of Raman Spectroscopy,* John Wiley & Sons, New York.

Grünefeld, G., Beushausen, V., Andersen, P., Hentschel, W., 1994, *Appl. Phys.,* 58: 333-342.

Herzberg, G., 1945, *Molecular Spectra and Molecular Structure Vol. II, Infrared and Raman Spectra of Polyatomic Molecules,* Van Nostrand Reinhold Corp., New York.

Herzberg, G., 1950, *Molecular Spectra and Molecular Structure Vol. I, Spectra of Diatomic Molecules,* Van Nostrand Reinhold Corp., New York.

Hill, S.C. and Benner, R.E., 1988, *Optical Effects Associated with Small Particles* (Edited by Barber, P.W. and Chang, R.K.) World Scientific, Singapore.

Hill, S.C. and Chang, R.K., 1992, Nova Science Publishers, New York.

Hill, S.C., Rushforth, C.K., Benner, R.E. and Conwell, P.R., 1985, *Applied Optics,* 24, 2380-2390.

Huckaby, J.L., Ray, A.K. and Das, B., 1994, *Applied Optics,* 33: 7112-7125.

Jackson, J.D., 1975, *Classical Electrodynamics,* John Wiley & Sons, New York.

Kaiser, Th., Voßmerbäumer, C., Schweiger, G., 1992, *Ber. Bunsenges. Phys. Chem.,* 96 Nr. 8: 976-980.

Kerker, M., 1969, The Scattering of Light, *Phys. Chem.,* Vol. 16, Academic Press.

Le Duff, Y. and Quillon R., 1973, *Advances in Raman Spectroscopy.,* 1: 428-435.

Lederman, S. and Sacks, S., 1984, *AIAA J.,* 22, 161-173.

Long, D.A., 1977, *Raman Spectroscopy,* McGraw-Hill, London.

Mie, G., 1908, *Ann. Phys.,* 25, 377-445.

Owen, J.F., Barber, P.W. and Chang, R.K., 1982a, *Microbeam Analyses* (Edited by Heinrich, K.F.J.), pp. 255-260, San Francisco Press, San Francisco.

Owen, J.F., Chang, R.K. and Barber, P.W., 1982b, *Aerosol Sci. Technol.,* 1, 293-302.

Rambau, R. and Schweiger, G., 1988, *Proc. European Aerosol Conf.,* Lund, Sweden, August 30–September 2.

Rassat, S.D. and Davies, E.J., 1992, *J. Aerosol Sci.,* 23: 165-180.

Ross, S.D., 1972, *Inorganic Infrared and Raman Spectra,* McGraw-Hill, New York.

Schrader, B. and Meier, W., 1974, 1975, 1976, *Raman/Infrarot-Atlas organischer Verbindungen,* Bd. 1,2,3, Verlag Chemie.

Schrader, B., 1986, *Physical and Chemical Characterization of Individual Airborne Particles* (Edited by Spurny, K.R.) Ellis Horwood Last, Chichester U.K.

Schrötter, H.W. and Klöckner, H.W., 1979, in *Raman Spectroscopy of Gases and Liquids* (Edited by Weber, A.) Springer-Verlag, New York.

Schweiger, G., 1990a, *Opt. Lett.,* 15: 156-158.

Schweiger, G., 1990b, *J. Aerosol Sci.,* 21: 483-509.

Schweiger, G., 1991, *J. Opt. Soc. Am. B.,* 8: 1770-1778.

Srivastrava, R.P. and Zaidi, H.R., 1979, in *Raman Spectroscopy of Gases and Liquids* (Edited by Weber, A.) Springer-Verlag, New York.

Stafford, R.G., Chang, R.K. and Kindlmann, P.J., 1977, in *NBS Special Publication 464*, Proc. 8th IMR Symp., 1976: 659-667.

Stratton, J.A., 1941, *Electromagnetic Theory,* McGraw-Hill, New York.

Tanabe, K., 1984, *J. Raman Spectrosc.,* 15, 248-251.

Thurn, R. and Kiefer, W., 1984, *J. Raman Spectrosc.,* 15: 411-413.

Tzeng, H.M., Wall, K.F., Loma, M.B., and Chang, R.K., 1984, *Opt. Lett.,* 9: 273-275.

van de Hulst, H.C., 1981, Dover Publications, New York.

Vehring, R. and Schweiger, G., 1992, *Soc. f. Appl. Spectrosc.,* 46: 25-27.

Vehring, R., Moritz, H., Niekamp, D. and Schweiger, G., 1995, wird veröffentlicht.

Zerda, T.W., Thomas, H.D., Bradley, M. and Jonas, J., 1987, *J. Chem. Phys.,* 86, 3219-3224.

15 Aerosol Time-of-Flight Mass Spectrometry

Christopher A. Noble and Kimberly A. Prather

CONTENTS

15.1 INTRODUCTION

Aerosols play a fundamental role in the physics and chemistry of our atmosphere. Due to their ubiquity throughout the environment, aerosols has been the focus of numerous studies, emphasizing either bulk or individual particle analysis. Recently, there has been an increase in the number of mass spectral methods aimed at a real-time characterization of aerosol samples on a single particle basis, with an emphasis on the analysis and modeling of atmospheric particles. As its title suggests, the focus of this chapter is on studies performed using aerosol time-of-flight mass spectrometry, detailing system development along with applications demonstrating the new types of information made available by a real-time aerosol analysis technique which measures both size and composition for single particles.

15.1.1 ATMOSPHERIC AEROSOLS

Atmospheric aerosols, produced by anthropogenic, biogenic, and geogenic sources are prevalent throughout our environment and, as a result, have an influence on the physical processes occurring in the atmosphere. With millions of tons of particulate matter emitted annually (Hinds, 1982), aerosols serve as the primary carrier of material in our atmosphere. Furthermore, submicrometer particles may act as cloud condensation nuclei (CCN) and ultimately affect atmospheric visibility and albedo (Quinn et al., 1995).

There are numerous examples of the role aerosols play in atmospheric chemical processes. Produced by heterogeneous chemical reactions in the atmosphere, gas phase acids (e.g., hydrochloric acid, nitric acid, and sulfuric acid) may be dissolved in water or ice and removed from the atmosphere through acid deposition. This acidifies surface water supplies, causing premature weathering of geological formations and buildings and affecting both plant and animal life (Seinfeld, 1986). In polluted urban environments, increased photochemical reactivity results in the production of organic particulate matter (Grosjean, 1992). In the upper atmosphere, polar stratospheric cloud (PSC) particles are central to the chemical reactions resulting in stratospheric ozone depletion (Molina, 1996).

Besides having both physical and chemical impacts on our atmosphere, particles also directly influence life and health. Since the 1970s, a number of epidemiological studies have shown correlations between high particulate mass concentrations and both mortality and morbidity (Dockery and Pope, 1994). In addition to directly affecting human health, some amount of atmospheric aerosols are viable biological particles. These range from small airborne viruses (~0.1 μm) to larger bacterial spores and pollen (~100 μm) (Duce et al., 1983).

15.1.2 SINGLE PARTICLE MASS SPECTROMETRY OF ATMOSPHERIC AEROSOLS

In the 1980s, several off-line mass spectrometry (MS) techniques were employed for the chemical analysis of single particles. One of the principle off-line MS methods is laser microprobe mass spectrometry (LMMS) (Kaufmann, 1986; Van Vaeck et al., 1994). A primary goal of a number of researchers using LMMS has involved the characterization of single atmospheric particles (Jambers et al., 1995; Jambers and Van Grieken, 1996). Using LMMS, Van Grieken and co-workers have performed numerous studies investigating the chemical characterization of individual particles from various environments, ranging from coastal (Wouters et al., 1990) and forest (Wouters et al., 1993) to urban areas (Wonders et al., 1994). LMMS has also been used to address the toxicity of coal mine dust (Tourmann and Kaufmann, 1993), soot from an oil shale retort (Mauney and Adams, 1984), and particles collected in the Antarctic (Hara et al., 1996). Other off-line mass spectral methods have been employed for single particle analysis, such as secondary ion mass spectrometry (SIMS) (Klaus, 1986; Bentz et al., 1995) and inductively coupled plasma mass spectrometry (ICPMS) (Kaneco et al., 1995).

Unlike off-line methods, real-time single particle mass spectrometry (RTSPMS) makes it possible to analyze particles on a real-time, continuous basis. For information on RTSPMS techniques, one should consult two excellent reviews by Stoffel(s) and Allen (1986) on early RTSPMS methods and Johnston and Wexler (1995) on current RTSPMS research. To date, several RTSPMS techniques have been used to sample ambient aerosols.

In early RTSPMS studies, Davis (1977a, 1977b, 1978) monitored various species, such as lithium, potassium, and lead, in ambient particles over time intervals up to one hour. Analysis of ambient marine aerosols was performed by Giggy, Friedlander, and Sinha (1989). Hinz, Kaufmann, and Spengler (1994, 1996) have reported compositional analysis of ambient air containing tobacco smoke particles, as well as ambient laboratory air particles using a size-selected analysis scheme. Using bipolar ion detection, they analyzed 0.8 μm particles found in ambient laboratory air and performed principal component analysis on a random sample of the entire data set. Sampling at Caribou, Colorado, Murphy and Thomson (1995) have reported the composition of single ambient particles along with estimated particle diameter. Most recently, Nobel and Prather (1996a, 1996b) made a detailed study of classifying individual particles in ambient samples at Riverside, California, based on both measured chemical composition and aerodynamic diameter. Using aerosol time-of-flight mass spectrometry (ATOFMS), these studies by Noble and Prather were the first demonstrations of real-time measurement of the size and composition of single aerosol particles of polydisperse samples.

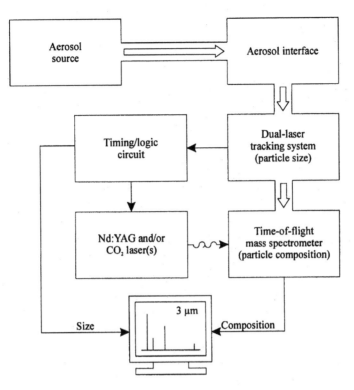

FIGURE 15.1 Schematic representation of time-of-flight aerosol beam spectrometer. Particles are accelerated across the pressure drop created by the nozzle. Scattered light is detected by PMTs as each particle passes through the lasers. The scattering pulses are used to start and stop an electronic timer.

15.2 METHODS

Aerosol time-of-flight mass spectrometry (ATOFMS) is a unique analytical method which incorporates two existing techniques for particle analysis: time-of-flight aerosol beam spectrometry (TOFABS) and time-of-flight mass spectrometry (TOFMS). The coupling of these two methods allows for real-time determination of both the size and chemical composition of individual particles in polydisperse aerosol samples (Figure 15.1). Both laboratory based and field portable ATOFMS systems have been developed.

15.2.1 TIME-OF-FLIGHT AEROSOL BEAM SPECTROMETRY

Originally patented in the 1970s, TOFABS was developed by Dahneke for determining the size distribution of an aerosol sample (Figure 15.2) (Dahneke, 1978; Schwartz and Andres, 1978; Israel and Friedlander, 1967). By expanding a carrier gas through a capillary or orifice into vacuum, a particle beam is created. Throughout the expansion, transfer of momentum occurs between the carrier gas and the suspended particles. This results in a separation of the suspended particles based on particle aerodynamic diameter (d_a). In the process of gaining momentum, the smaller particles achieve a higher final velocity than the larger, more massive, particles. Ultimately, the majority of the carrier gas is pumped away, while the particles continue as a focused beam. Because particles of different sizes travel at different velocities, the particle time-of-flight over a fixed distance will be a function of its aerodynamic diameter. Typically, particle time-of-flight is measured by detecting the scattered light as each particle passes through two continuous wave lasers oriented at 90° to the particle beam.

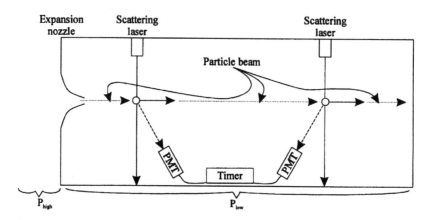

FIGURE 15.2 Schematic representation of a time-of-flight mass spectrometer, operating in the positive ion analysis mode. Ions are generated in the ion source and accelerated into the flight tube. Ions separate based on m/z, with smaller ions traveling more quickly.

By relying on various properties or characteristics of single particles, other methods are capable of providing near real-time size distributions of aerosol samples. Though TOFABS provides size analysis based on the aerodynamic properties of the particle, other instruments, such as a differential mobility analyzer (DMA), characterize the movement of charged particles through an electric field (Yeh, 1993). By relying on the optical properties, such as particle size, refractive index, laser wavelength, and angle of collection, an optical particle counter (OPC) both sizes and counts particles (Gebhart, 1993). Several techniques have been coupled with RTSPMS methods to provide complementary sizing information on single particles. Although ion signal integration (Allen and Gould, 1981; Giggy et al., 1989; Reents et al., 1995), light scattering intensity measurements (Kievit et al., 1992; Hinz et al., 1994; Carson et al., 1995; Murphy and Thomson, 1995; Salt et al., 1996), aerodynamic sizing (Sinha, 1984; Prather et al., 1994; Hinz et al., 1994; Yang et al., 1996), and observation through a telescope (Yang et al., 1995a) have all been attempted with RTSPMS systems, light scattering intensity measurements and aerodynamic sizing are the two methods which are being incorporated in most of these real-time aerosol analysis instruments.

15.2.2 TIME-OF-FLIGHT MASS SPECTROMETRY

Basic principles of TOFMS have remained unchanged since it was originally proposed by Stephens in 1942 (Figure 15.3) (Cotter, 1994; Price, 1994). Positioned in the ion source region of the mass spectrometer, a sample is volatilized and ionized. A potential energy is imparted to the resulting ions by an electric field in the source region. An electric field in the source accelerates the ions out of the ion source into a field free drift region or flight tube. Because the ions are roughly equivalent in kinetic energy, the velocity at which they travel is a function of the ion mass-to-charge ratio (m/z), commonly referred to as the ion mass for singly charged ions. Also, because the length of the flight tube is constant, and therefore, the distance the ions travel is constant, the ion time-of-flight is a function of the mass of the ion. Low mass ions have higher velocities than larger mass ions and, as a result, smaller ions have a shorter time-of-flight before striking an ion detector. The signals from all ion masses result in a mass spectrum which provides elemental and/or chemical information about the sample being analyzed.

In comparison to other mass spectrometric methods, TOFMS offers several capabilities which are desirable for real-time chemical analysis of aerosol particles (Skoog and Leary, 1992). Foremost, TOFMS measures a complete mass spectrum from a single desorption/ionization event. This is crucial to RTSPMS techniques because each particle is available only once for vaporization/ionization. Scanning instruments, such as quadrupole and magnetic sector mass spectrometers, are

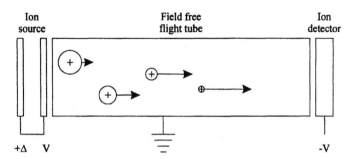

FIGURE 15.3 Block diagram of aerosol time-of-flight mass spectrometer. (Adapted from Nordmeyer, T. and Prather, K.A., 1994, *Anal. Chem.* 66: 3540-3542. With permission.)

able to measure only a limited number of ion masses per ionization event which, consequently, limits the amount of available chemical information. Second, mass analysis using TOFMS is performed in real-time, making *in situ*, continuous analysis possible. Furthermore, TOFMS has relatively high sensitivity, a theoretically unlimited mass range, and fast analysis time (on the order of 10^{-5}s). Combined with these qualities, further characteristics of TOFMS make it the ideal mass analyzer for a field-portable RTSPMS instrument. First, TOFMS is quite rugged and durable, making transportation of the instrument relatively easy. Second, TOFMS is simple. Mass separation is performed using only electric fields which do not need the delicate handling and maintenance required by MS systems utilizing magnetic fields.

15.2.3 AEROSOL TIME-OF-FLIGHT MASS SPECTROMETRY

A schematic diagram of the current aerosol time-of-flight mass spectrometer is shown in Figure 15.4 (Prather et al., 1994; Nordmeyer, 1995). The aerosol sample, initially at atmospheric pressure, is introduced into the system by free jet expansion through a nozzle. Passing through three skimmers, which separate differentially pumped regions, the sample is collimated into a narrow particle beam. This particle beam is directed through two low-power, continuous-wave argon laser beams which are separated by a known distance. Because these two laser beams are oriented at right angles to both the particle beam and each other, particles which are diverging from the center of the particle beam will not scatter light from both lasers. This effectively prevents the system from attempting to analyze particles which will not pass through the center of the ion source in the mass spectrometer. As a particle passes through the first argon laser beam (514 nm), scattered light is detected by a photomultiplier tube (PMT), which serves to start an electronic timing/logic circuit (Nordmeyer and Prather, 1994). On passing through the second argon laser beam (488 nm), scattered light is detected again, causing the timing/logic circuit to stop. This measured particle transit time serves two purposes. First, this transit is used, along with an appropriate calibration plot, to determine the particle aerodynamic diameter. Second, the transit time is used to determine the time at which the particle will arrive in the center of the ion source of the mass spectrometer, at which time the desorption/ionization laser(s) is fired. For conventional LDI, both CO_2 (10.6 µm) and Nd:YAG (266 nm for fourth harmonic) lasers may be used individually or together in two-step LDI. Under normal operating conditions, desorption/ionization efficiency is 20 to 50%. That is, for all particles for which a transit time is recorded, 20 to 50% of these particles also have mass spectra, providing the desired particle size and composition information. All data are stored on a personal computer for future analysis. Under optimal conditions, this can result in several hundred particles analyzed per minute.

Similar to other current RTSPMS projects, ATOFMS employs a time-of-flight mass analyzer for mass spectral analysis. The primary instrumental characteristic which distinguishes ATOFMS from other RTSPMS methods is the means by which individual particle size is determined. Current

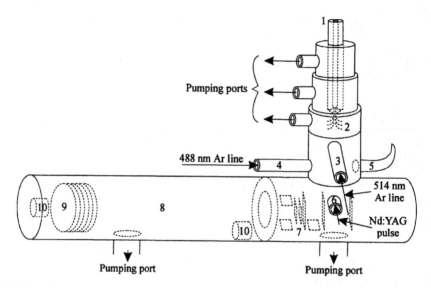

FIGURE 15.4 Schematic drawing of aerosol time-of-flight mass spectrometer. (1) Particle introduction. (2) Interface region. (3) Entry window for 514 nm argon laser. (4) Entry window for 488 nm argon laser. (5) Light horn to collect stray light. (6) Entry window for ND:YAG laser. (7) Ion optics of mass spectrometer. (8) Field free flight tube of mass spectrometer. (9) Reflectron. (10) Microchannel plate detector. (Adapted from Noble, C.A. and Prather, K.A., 1996a, *Environ. Sci. Technol.* 30:2667-2680. With permission.)

RTSPMS techniques incorporate ion signal integration, light scattering measurements, and both single-laser (Hinz et al., 1994) and dual-laser aerodynamic particle sizing (Sinha, 1991; Prather et al., 1994; Yang et al., 1996). Several researchers have reported that ion signal integration does not provide consistent results for either coarse particles or for chemically heterogeneous particles (Davis, 1978; Allen and Gould, 1981). As a result, this method of particle sizing is best applied to known fine particle samples. Because scattered light intensity is influenced by particle shape, size, refractive index, light wavelength, and angle of collection, this is a difficult method to couple with RTSPMS techniques for accurate particle size measurement (Salt et al., 1996). When this method is employed, it typically provides rough estimations of particle size rather than accurate measurements. While both single and dual-laser aerodynamic sizing methods are based on TOFABS, single-laser RTSPMS methods cannot analyze polydisperse samples in real-time. Because only one laser is used prior to LDI, a fixed time delay must be used prior to triggering the desorption/ionization laser. With a fixed time delay, only particles of one size may be analyzed at a time. By utilizing dual-laser particle sizing, ATOFMS may use a variable time delay before firing the desorption/ionization laser(s). Therefore, polydisperse aerosol samples can be analyzed in real-time.

For remote aerosol sampling, a new generation of the aerosol time-of-flight mass spectrometer has been developed which is based on the same basic principles as the original system. While still coupling dual-laser TOFABS and TOFMS, this new instrument is field transportable, with all equipment — ATOFMS system, lasers, vacuum pumps, power supplies, and computer — mounted on a single portable rack. Also, the TOFMS system has been redesigned with two mass analyzers which allow for simultaneous detection of both positive and negative ions from each particle.

15.3   SYSTEM CHARACTERIZATION

Because the primary goal of ATOFMS research involves measuring both the size and composition of single particles, some initial experiments have focused on comparing ATOFMS measurements with existing techniques. Several methods for particle size determination have been studied. Also, ATOFMS mass spectra of inorganic salts have been compared with published LMMS results.

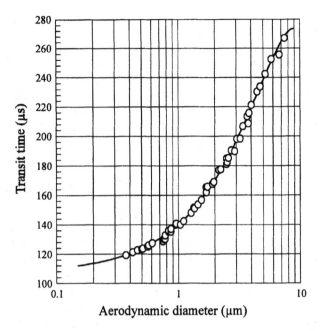

FIGURE 15.5   Empirical size calibration curve. (Adapted from Noble, C.A. and Prather, K.A., 1996a, *Environ. Sci. Technol.* 30:2667-2680. With permission.)

15.3.1   PARTICLE SIZING

Using monodisperse ammonium sulfate aerosol samples, a size calibration curve is empirically derived by measuring transit time for a known particle size (Figure 15.5) (Noble and Prather, 1996). Ammonium sulfate particles are used as the calibration aerosol because ammonium sulfate has a bulk density of 1.8 g/cm^3, which is the average density of ambient aerosol particles. A vibrating orifice aerosol generator (VOAG) is used to create particles with sizes greater than 1 μm, and a DMA is employed to select monodisperse particles with sizes less than 1 μm. Each data point in Figure 15.5 represents the average value of approximately 1000 measurements. Standard deviations for these values are typically less than 1.5% of the average value. The curve in this plot, a best fit third-order polynomial, is then applied to determine the particle size from measured particle transit time in a polydisperse sample.

As mentioned, several current RTSPMS methods employ light scattering measurements for complementary particle size determination. By utilizing one or both of the argon laser beams, ATOFMS is also capable of using light scattering intensity measurements to determine particle size (Salt et al., 1996). Light scattering intensity is determined as peak height, in volts, as detected by the PMT; particle diameter is measured as transit time, in microseconds, as recorded by the timing/logic circuit. By analyzing monodisperse samples, a comparison can be made between light scattering intensity measurements and aerodynamic sizing. For this comparison, three methods of measurement are employed. *Method 1 —* scattered light peak height intensity from the 514 nm argon beam is recorded for each particle passing through the 514 nm argon beam. *Method 2 —* scattered light intensity from the 514 nm argon line is measured only for those particles which scatter light from both argon beams. By requiring each particle to scatter light from both argon lines, Method 2 ensures that these particles are passing through the center of the 514 nm argon beam. As a result, variations in measured scattering intensity will not be the result of the particle passing through less intense incident radiation, as could be the case with Method 1. *Method 3 —* aerodynamic sizing is accomplished by measuring a transit time for each particle passing through both argon beams.

Monodisperse aerosols of sodium chloride (NaCl) and 2,4-dihydoxybenzoic acid (DHB) are created using a VOAG. These two compounds are chosen as representative inorganic and organic compounds. Frequency distributions of monodisperse DHB samples are shown in Figures 15.6a and 15.6b for Methods 1 and 2, respectively. It is apparent from these plots that light scattering intensity does not provide the reproducibility necessary to consistently measure particle size accurately. However, measured transit time is a unique function of aerodynamic diameter (Figure 15.6c). Similar results are seen when comparing these sizing methods for NaCl monodisperse aerosols. It is evident that aerodynamic sizing is more precise than light scattering intensity measurements for determining particle size in this size range.

For both monodisperse and polydisperse samples, a comparison is made between ATOFMS and a commercially available Aerosizer (Amherst Process Instruments) (Noble and Prather, 1996). Air flow from the aerosol source is split between both instruments during the sampling period. A graphical comparison of the size distributions measured for both monodisperse and polydisperse samples are shown in Figure 15.7a and 7b, respectively. Though the singlet peak in the monodisperse sample is the most prominent peak, the doublet and triplet peaks are also detected by both techniques. Not only does measured particle size correspond well, as evidenced by the similarity of size distributions for the monodisperse sample, but relative particle count at specific sizes also matches. This is demonstrated by similar peak heights in both size distributions. Though both instruments size particles using dual-laser TOFABS, the Aerosizer samples a higher number concentration than ATOFMS due to mechanical differences in these two instruments. This difference in sampling rate explains the smoothness of the Aerosizer curve compared to the ATOFMS curve because, over a limited sampling period, an Aerosizer samples many more particles than ATOFMS. Therefore, in comparing size distributions from both methods, the area under each curve is normalized to one.

Because ATOFMS utilizes two lasers for measuring aerodynamic diameter, it can be used to measure polydisperse samples in real-time (Nordmeyer and Prather, 1994). By using a dual-laser sizing configuration, variable time delays may be employed prior to triggering the Nd:YAG for LDI. Varying the delay time from the calculated optimal setting, it is demonstrated that the timing/logic circuit is triggering the Nd:YAG at the actual optimal delay time for each particle (Figure 15.8). By measuring that the highest desorption/ionization efficiency occurs at the calculated delay time regardless of particle size, this study shows that the timing/logic circuit is not imparting any size biasing effect. This is further demonstrated by comparing the size distribution of both sized and fully analyzed (those which are sized and produce mass spectra) particles (Figure 15.9).

15.3.2 Chemical Analysis

In an attempt to determine sulfur speciation based on ions present in a mass spectrum, a comparison between LMMS (Bruynseels and Van Grieken, 1984) — the off-line analog to RTSPMS methods — and ATOFMS has been performed for several sulfoxy salts (Salt, 1995). Both positive and negative mode spectra are compared for individual micrometer size particles composed of sodium sulfate (Na_2SO_4), sodium sulfite (Na_2SO_3), and sodium thiosulfate ($Na_2S_2O_3$). The oxidation states of sulfur are +6, +4, and +2, respectively, for each of these compounds.

Major mass/charge peaks which appear for both LMMS and ATOFMS in negative ion mode are 16 (O^-), 17 (OH^-), 48 (SO^-), 64 (SO_2^-), 80 (SO_3^-), 96 (SO_4^-), and 119 ($NaSO_4^-$). Figure 15.10 provides a graphical comparison of signal intensities for peaks at m/z 48, 64, 80, and 96. In order to make a relative comparison between peak intensities, each peak is normalized to the height of the SO_2^- peak (the dashed line in Figure 15.10). Though relative intensities for these peaks differ between both methods, a general trend may be observed for the intensities of the SO^- and SO_3^- ions. For both methods, the intensity of the SO^- and SO_3^- peaks increases based on the chemical composition of the particle ($Na_2SO_3 < NA_2SO_4 < NA_2S_2O_3$).

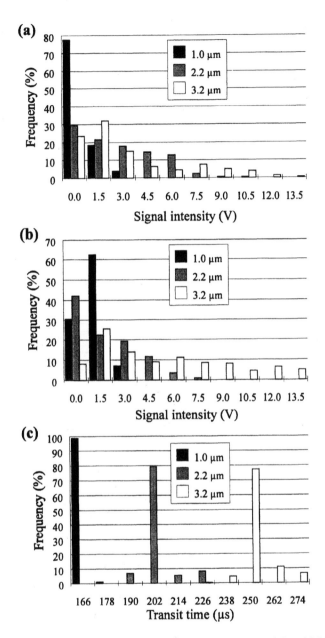

FIGURE 15.6 Frequency distributions for 1.0, 2.2, and 3.2 μm DHB particles. (a) Method 1 — scattered light intensity for all particles passing through 514 nm argon beam. (b) Method 2 — scattered light intensity for only those particles passing through the center of the 514 nm argon beam. (c) Method 3 — particle transit times. (Adapted from Salt, K. Noble, C.A., and Prather, K.A., 1996, *Anal. Chem.* 68:230-234. With permission.)

For positive ion detection, common peaks observed are 23 (Na^+), 46 (Na_2^+), 62 (Na_2O^+), 63 (Na_2OH^+), 101 (Na_3S^+), 110 ($Na_2SO_2^+$), 149 ($Na_3SO_3^+$), and 165 ($Na_3SO_4^+$). Results for positive ion detection are shown in Figure 15.11 for masses 101, 110, 149, and 165. The intensities of these peaks are normalized to $Na_3SO_4^+$ as indicated by the dashed line. While there is still not exact agreement between relative ion intensities, there does seem to be a greater similarity for the positive than for the negative ions. One observable trend for both techniques is that the relative intensity of $Na_3SO_3^+$ increases with decreasing sulfur oxidation state.

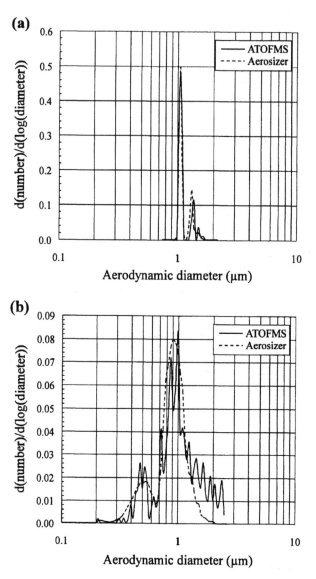

FIGURE 15.7 Size distributions comparing ATOFMS with an Aerosizer. (a) Monodisperse sample. (b) Polydisperse sample. (Adapted from Noble, C.A. and Prather, K.A., 1996a, *Environ. Sci. Technol.* 30:2667-2680. With permission.)

Although general trends were observed for both positive and negative ion modes, relative intensities between LMMS and ATOFMS do not correspond exactly. Several experimental differences between these methods may result in the observed differences in relative ion intensities. Most significantly, sample preparation was not identical for both techniques. For the LMMS study, sulfoxy salts were created by grinding salt grains down to the appropriate micrometer size. For the ATOFMS study, monodisperse aerosols are created from a solution of the dissolved salt and dried prior to analysis. Chemical changes may possibly occur in solution before each particle is dried and analyzed. Also, for LMMS, particles are placed on a substrate in the ion source for LDI; for ATOFMS, aerosols are introduced as a particle beam and desorbed/ionized in flight.

While these experimental differences may lead to differences in spectra, ATOFMS provides the reproducibility necessary for sulfur species determination when comparing Na_2SO_4, Na_2SO_3,

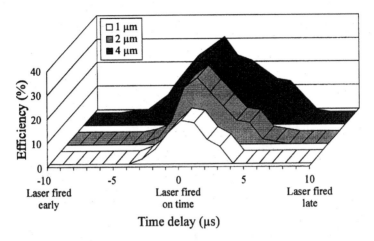

FIGURE 15.8 Desorption efficiency as a function of delay time for triggering the Nd:YAG laser. (Adapted from Nordmeyer, T. and Prather, K.A., 1994, *Anal. Chem.* 66: 3540-3542. With permission.)

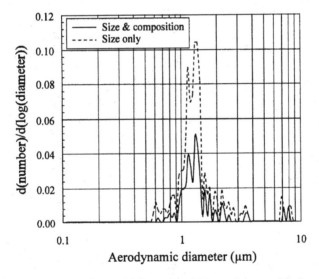

FIGURE 15.9 Size distributions for sized and fully analyzed (size and composition) particles. (Adapted from Noble, C.A. and Prather, K.A., 1996a, *Environ. Sci. Technol.* 30:2667-2680. With permission.)

and $Na_2S_2O_3$ in positive ion mode. For Na_2SO_4, the presence of a peak at m/z 165 with minor or no peaks at m/z 101, 110, and 149, are necessary. Peaks at m/z 149 and 165, with no peaks at m/z 101 and 110, indicate the presence of Na_2SO_3 in the original particle. Finally, a spectrum with peaks at m/z 101, 149, and 165 is indicative of a particle containing $Na_2S_2O_3$.

15.4 PARTICLE ANALYSIS

Following instrumental development and testing, ATOFMS has been used primarily to characterize ambient aerosols on a single particle basis in Riverside, California. To aid in the characterization of atmospheric aerosols, studies have been undertaken which focus on identifying and cataloging particles produced from potential aerosol sources.

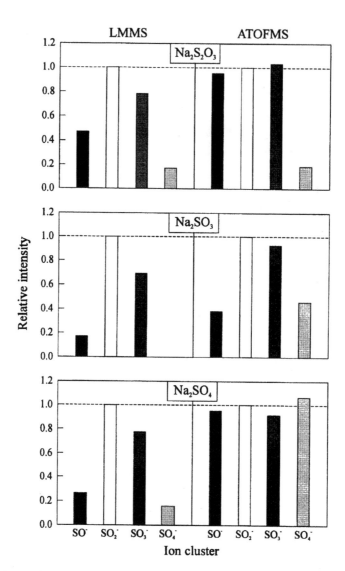

FIGURE 15.10 Comparison between signal intensities for LMMS and ATOFMS of major negative ion peaks from single sulfoxy particles. The y-axis has been normalized to the intensity of the SO_2^- ion cluster, which is indicated by the dashed line. (Adapted from Salt, K.L., 1995, *Characterization of Aerosol Particles Containing Sulfur and Nitrogen Species Using Aerosol Time-of-Flight Mass Spectrometry (ATOFMS)*, Doctoral Dissertation, University of California, Riverside. With permission.)

15.4.1 ATMOSPHERIC AEROSOLS

In initial atmospheric sampling experiments, based on the chemical species identified in a particle mass spectrum, chemically distinct classes are chosen such that the majority of all particles may be identified with one, and only one, category. This has led to the identification of organic, elemental carbon, marine (sea salt), inorganic oxide, and inorganic/organic particle categories. Several of these categories may be broken down further into more detailed classes. However, only five major particle categories were chosen for this experiment because the analysis and classification of all the individual mass spectra were performed manually. Work is currently under way for developing computer programs which will allow for automated mass spectral analysis and classification resulting in more refined chemical categories.

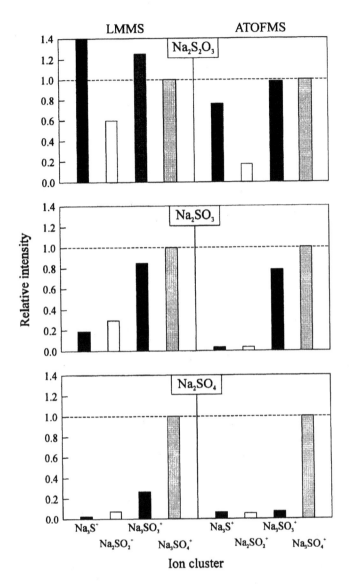

FIGURE 15.11 Comparison between signal intensities for LMMS and ATOFMS of major positive ion peaks from single sulfoxy particles. The y-axis has been normalized to the intensity of the $Na_3SO_4^+$ ion cluster, which is indicated by the dashed line. (Adapted from Salt, K.L., 1995, *Characterization of Aerosol Particles Containing Sulfur and Nitrogen Species Using Aerosol Time-of-Flight Mass Spectrometry (ATOFMS),* Doctoral Dissertation, University of California, Riverside. With permission.)

Several ambient sampling studies have been performed at Riverside, California (Noble and Prather, 1996a, 1996b). Sample spectra of the two most common types of carbon-containing ambient particles are shown in Figures 15.12a and 15.12b. Particles with similar compositions typically are found in the accumulation mode (0.1 μm < d_a < 1 μm). For particles smaller than 0.3 μm, an extrapolation of the size calibration curve (Figure 15.5) is employed to estimate the particle size. Spectra from this category are identified by the presence of hydrocarbon clusters extending out to C_5^+ or C_6^+. Related subcategories include mass spectral ion peaks due to minor inorganic components, such as metals and metal oxide clusters, ammonium, nitrogen oxide clusters, or sulfur oxide clusters. These organic particles are believed to be originally introduced into the atmosphere as products of combustion. Heterogeneous chemistry may explain the addition of ammonium, nitrogen oxides,

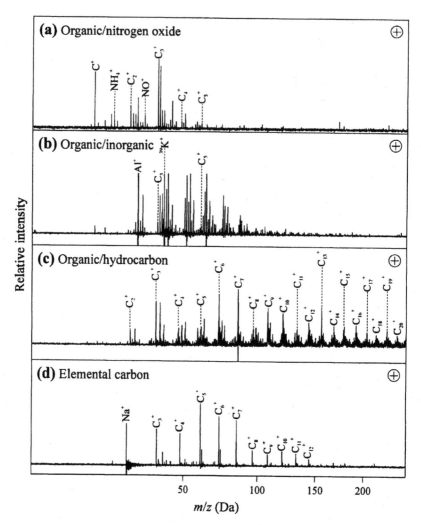

FIGURE 15.12 Sample positive ion LDI mass spectra of individual ambient carbon-containing particles. (a) Organic particle with peaks due to ammonium (NH_4^+ at m/z 18) and nitrate (NO^+ at m/z 30). (b) Organic particle with some inorganic components (Al^+ at m/z 27 and K^+ at m/z 39 and 41). (c) Organic particle with extended hydrocarbon envelopes. (d) Elemental carbon particle. (Adapted from Noble, C.A. and Prather, K.A., 1996a, *Environ. Sci. Technol.* 30:2667-2680. With permission.)

and sulfur oxides. Also shown in Figures 15.12c and 15.12d are examples of hydrocarbon and elemental carbon particle spectra. Hydrocarbon particles are identified by a hydrocarbon backbone extending out past C_{15}^+ in the mass spectrum. Mass spectra of elemental carbon particles contain mainly carbon peaks without the hydrogen-attached ion clusters or sodium. Both of these particle types are larger than the other organic particle types with sizes greater than ~0.8 μm and the mode centered at ~2 μm. Sources for these classes are also likely to be combustion processes.

One of the more common particle categories is the inorganic/organic category. Several example spectra of particles from this class are shown in Figure 15.13. This class qualitatively differs from the organic-inorganic type by the presence of a greater amount of inorganic components. The size range for this particle class is ~0.5 to 2.5 μm in diameter. Similar to the organic class, mass spectra for this particle type also have carbon peaks extending out to C_5^+ or C_6^+. However, there are also peaks at m/z 15 (CH_3^+), 18 (NH_4^+), 19 (H_3O^+), 30 (NO^+), and a variety of metal ions. Commonly observed metal ions in these mass spectra are Li^+, Na^+, Mg^+, Al^+, K^+, Ca^+, V^+, Cr^+, Mn^+, Fe^+, Co^+,

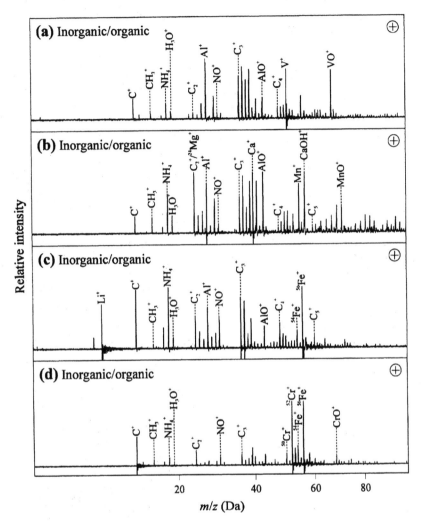

FIGURE 15.13 Sample positive ion LDI mass spectra of individual ambient inorganic/organic particles. All spectra contain peaks due to carbon clusters out to C_5^+, peaks at m/z 15, 17, 18, 19, and 30 and some metal cations. (a) Particle containing Al and V. (b) Particle containing Al, Ca, and Mn. (c) Particle containing Li, Al, and Fe. (c) Particle containing Cr and Fe. (Adapted from Noble, C.A. and Prather, K.A., 1996a, *Environ. Sci. Technol.* 30:2667-2680. With permission.)

Ni^+, Cu^+, Zn^+, Ge^+, Sr^+, Cd^+, Pt^+, and Pb^+. Aluminum, vanadium, and iron are the most common metals found in this particle type. A possible source for these particles is the combustion of a fuel that contains a significant amount of noncombustible metals.

Inorganic oxide particle spectra are typified by peaks from Na^+, Mg^+, Al^+, Si^+, K^+, Ca^+, Ti^+, Fe^+, and related metal/oxygen peaks (Figure 15.14). Because of particle composition coupled with the fact that these particles are commonly found in the coarse mode centered at 2 μm, these particles are thought to originate from dust or soil which has been suspended by wind.

Sample negative ion spectra of several different particle classes are shown in Figure 15.15. An example spectrum of a marine, or sea salt, particle is given in Figure 15.15c. Particles in this category range from ~1.5 to 4.0 μm in diameter and have chemical components similar to sea water. In negative ion mode, peaks due to Cl^- at m/z 35 and 37 are present. In positive ion mode, peaks from Na_2Cl^+ are evident at m/z 81 and 83. Because marine particles must pass through the polluted Los Angeles air basin to reach Riverside, marine particles sampled at Riverside often contain

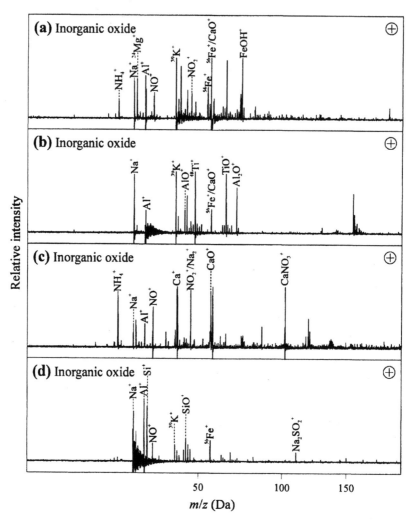

FIGURE 15.14 Sample positive ion LDI mass spectra of individual ambient inorganic oxide particles. (a) Particle containing Na, Mg, Al, K, and Fe. (b) Particle containing Na, Al, K, Ti, and Fe. (c) Particle containing Na, Al, and Ca. (d) Particle containing Na, Al, Si, K, and Fe. (Adapted from Noble, C.A. and Prather, K.A., 1996a, *Environ. Sci. Technol.* 30:2667-2680. With permission.)

nitrogen and sulfur oxides resulting from heterogeneous processes (Finlayson-Pitts and Pitts, 1989). Several ion peaks due to nitrogen and sulfur oxides are apparent in Figure 15.15c. Typically, negative ion spectra provide less information on the chemical composition of the particles than do positive ion spectra. Positive ion mode is useful for detecting metal cations and clusters, carbon and hydrocarbon clusters, ammonium ions, and some nitrate and sulfate clusters. Negative ion detection provides more detailed information on nitrogen and sulfur oxides, halide ions, and some carbon and hydrocarbon clusters.

Because size is measured on a single particle basis along with mass spectral information, size distributions of these chemically distinct categories can be presented. Figure 15.16 shows size distributions for both organic and inorganic particle classes. From these distributions, a division at ~1.4 μm is apparent, which represents the break between fine and coarse particle modes. The organic particles mainly fall below this size, while the inorganic particles are found above this size. One feature of interest is that the organic/inorganic class is larger in size than the pure organic particles

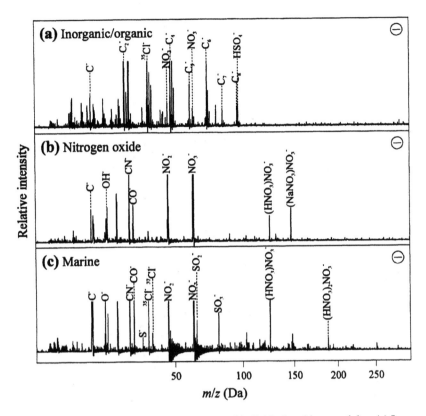

FIGURE 15.15 Sample negative ion LDI mass spectra of individual ambient particles. (a) Inorganic/organic particle. (b) Nitrogen oxide particle. (c) Marine particle. (Adapted from Noble, C.A. and Prather, K.A., 1996a, *Environ. Sci. Technol.* 30:2667-2680. With permission.)

and the inorganic/organic particles are slightly smaller than other inorganic types, possibly indicating a common source for the organic/inorganic and inorganic/organic categories.

In cases where specific chemical compounds are of interest, size distributions for all particles containing the compound(s) of interest may also be determined. Examples of this are shown in Figure 15.17 for four different compounds. During this sampling period, 70% and 25% of all mass spectra contained peaks at m/z 18 (NH_4^+) and 23 (Na^+), respectively. Sodium occurs mainly in particles larger than ~1.2 µm; ammonium occurs in particles from ~0.8 to 2.0 µm. In determining which particles contain either nitrogen oxides or sulfur oxides, several ion masses must be observed. Though the most common nitrate-indicating ion is m/z 30 (NO^+), other peaks also indicate the presence of nitrogen oxides in the particle: 46 (NO_2^+), 62 (NO_3^+), 92 ($Na_2NO_2^+$), and 108 ($Na_2NO_3^+$). Likewise, sulfur oxides are indicated by m/z 101 (Na_3S^+), 110 ($Na_2SO_2^+$), 149 ($Na_3SO_3^+$), and 165 ($Na_3SO_4^+$). It was found that 28% and 53% of particles sampled during this period contained sulfur oxides and nitrogen oxides, respectively. While nitrogen oxides are predictably found in both accumulation and coarse particles, sulfur oxides are observed only in the coarse mode. The lack of sulfur oxides in fine particles may be explained by the means in which sulfur oxide presence is determined. In positive ion mode, all ion masses for determining the presence of sulfur rely on the presence of sodium for forming cluster ions. As shown in Figure 15.17a, there is little sodium present in fine particles. Therefore, if the sulfur oxides are present in fine particles, there is not enough sodium to form ions clusters and be detected. For performing a more thorough analysis of aerosol systems, bipolar ion detection on a single particle basis is desired. As mentioned, this capability is available on the next generation of ATOFMS instrumentation.

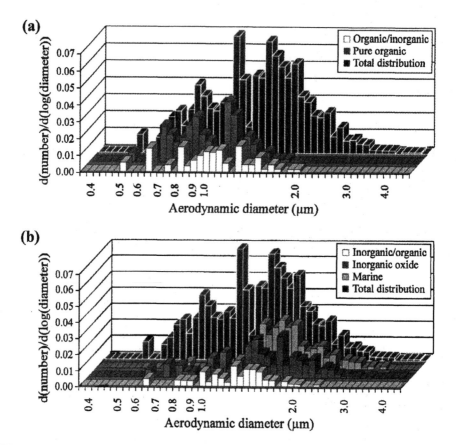

FIGURE 15.16 Particle size distributions sampled at Riverside, California, on 27 April 1995. (a) Size distribution for organic particle types. (b) Size distribution for inorganic particle types. (Adapted from Noble, C.A. and Prather, K.A., 1996a, *Environ. Sci. Technol.* 30:2667-2680. With permission.)

A unique sampling opportunity is presented by local fireworks displays on Fourth of July celebrations (Liu et al., 1996). Fireworks particles, from ignition and propulsion materials, oxidizers, and flame coloring compounds, are emitted into the atmosphere during each display. These particles have very distinct chemical signatures and are easily distinguished from typical ambient background particles. Commonly used materials include salts and pure metals containing magnesium, aluminum, potassium, copper, strontium, barium, and lead. During this sampling period (3 July 1995 to 7 July 1995), four chemically distinct classes were identified (Figure 15.18). Though other classes are detected, such as those containing Cu and Sr, the number concentration of these categories was significantly lower than the selected four classes. Of these four types, Class 1 is the most abundant, accounting for up to ~9% of all particles at times during the sampling. In Figure 15.19, the number concentration of fireworks particles is shown as a function of time. Though the highest number concentration of particles occurred at 0900 on 5 July, fireworks particles are detected throughout the entire sampling period. Fireworks particles detected prior to 4 July are probably due to small personal displays. Following the main episode, fireworks particles detected may be from personal displays or, more likely, from longer range transport from various displays throughout southern California.

15.4.2 Aerosol Sources

To assist in understanding atmospheric particles at the single particle level, the particles produced from several aerosol sources are currently being characterized using the ATOFMS technique. To

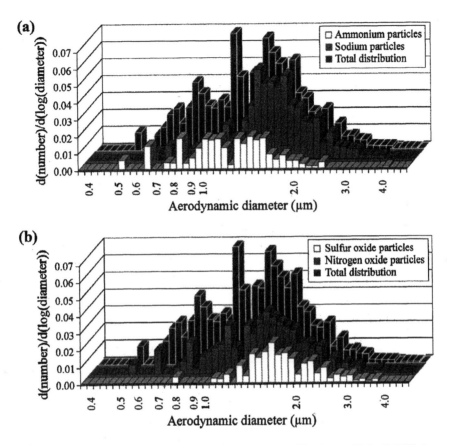

FIGURE 15.17 Particle size distributions sampled at Riverside, California, on 27 April 1995 demonstrating chemical speciation as a function of particle size. (a) Particles containing NH_4^+ and Na^+. (b) Particles containing sulfur oxide and nitrogen oxide ions. (Adapted from Noble, C.A. and Prather, K.A., 1996a, *Environ. Sci. Technol.* 30:2667-2680. With permission.)

date, the main aerosol sources studied are cigarette smoke, automobile exhaust, wood burning products, suspended soil, and household aerosols.

A significant source of indoor aerosols is cigarette smoke. Initial work to characterize cigarette smoke has focused on measuring changes in the size distribution and chemical composition over time (Morrical and Prather, 1996). Mass spectra of sidestream smoke — smoke which is not inhaled, but comes directly off the tip of a cigarette — are shown in Figures 15.20a and 15.20b. In Figure 15.20a, a peak from protonated nicotine $(C_{10}H_{14}N_2)H^+$ is evident at m/z 163. In both Figures 15.20a and 15.20b, the primary ion peak occurs at m/z 84, which corresponds to a loss of the pyridine ring from nicotine. The difference in the sidestream spectra shown in Figures 15.20a and 15.20b is primarily due to differences in LDI laser power. In Figure 15.20a, a relatively low laser power has been used for LDI, resulting in the parent ion of nicotine with little fragmentation. Higher laser power is used for the spectrum in Figure 15.20b, resulting in considerable fragmentation and no nicotine parent ion observed. Future incorporation of two-step LDI, using a CO_2 laser for desorption and a Nd:YAG for ionization, may allow the use of lower laser powers for better control of fragmentation processes (Kovalenko et al., 1992).

Sample spectra of other aerosol sources currently being studied using ATOFMS are shown in Figure 15.21. It is apparent that particulate emissions of cigarettes, diesel engines, wood burning, and soil vary significantly in chemical composition as indicated by the mass spectra. By characterizing various sources directly with the ATOFMS system, the chemical/size fingerprints of these

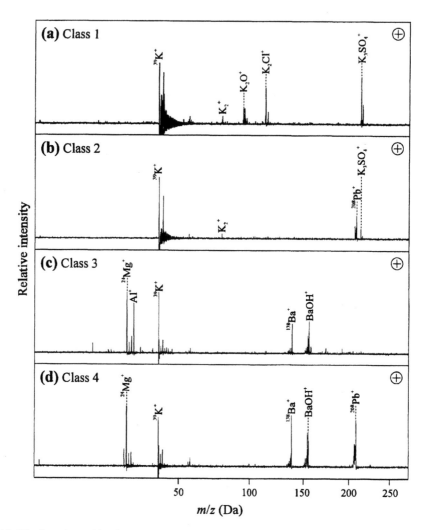

FIGURE 15.18 Sample positive ion LDI mass spectra of individual ambient fireworks particles. (a) Class 1 particle containing K. (b) Class 2 particle containing K, and Pb. (c) Class 3 particle containing Mg, Al, K, and Ba. (d) Class 4 particle containing Mg, K, Ba, and Pb. (Adapted from Liu, D.-Y., Rutherford, D., Kinsey, M., and Prather, K.A., 1996, *Anal. Chem.* 69:1808-1814. With permission.)

sources are determined (Silva and Prather, 1997). Ideally, this will allow for the determination of relative contributions of each source into the atmospheric aerosol budget. For example, the spectrum from diesel engine exhaust (Figure 15.21a) is very similar to a spectrum from an ambient particle sampled at Riverside, California (Figure 15.21c). Likewise, the spectrum from a suspended soil particle (Figure 15.21c) from a known location is similar to the inorganic oxide spectra shown in Figure 15.14.

15.5   CONCLUSION

Aerosol time-of-flight spectrometry is a unique analytical method which is capable of real-time aerosol characterization. By coupling dual-laser TOFABS with TOFMS, ATOFMS measures the size and composition of single particles in polydisperse samples. To date, ATOFMS is the only real-time instrumental method capable of obtaining both precise size and chemical composition information for individual particles in polydisperse aerosol samples.

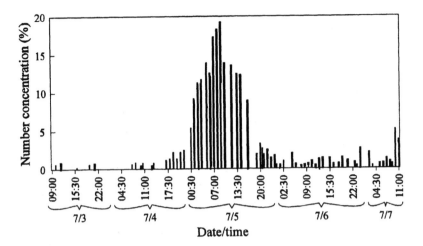

FIGURE 15.19 Number concentration of fireworks particles as a function of time. Number concentration is given as a percentage of the total number concentration detected during this sampling period. (Adapted from Liu, D.-Y., Rutherford, D., Kinsey, M., and Prather, K.A., 1996, *Anal. Chem.* 69:1808-1814. With permission.)

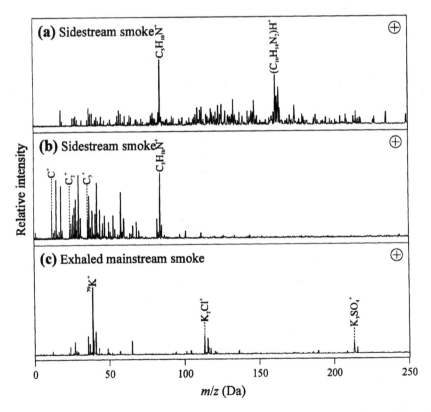

FIGURE 15.20 Sample positive ion LDI mass spectra of individual cigarette smoke particles. (a) Particle from sidestream smoke desorbed/ionized with relatively low laser power. (b) Particle from sidestream smoke desorbed/ionized with relatively high laser power. (c) Particle from exhaled mainstream smoke.

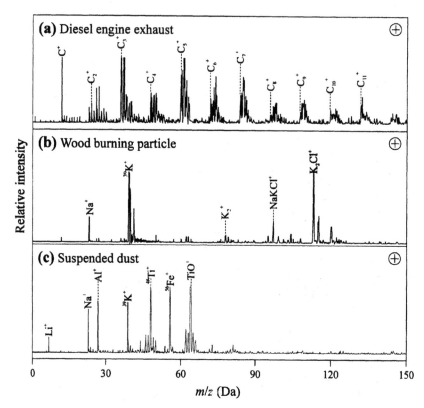

FIGURE 15.21 Sample positive ion LDI mass spectra of individual particles from various sources. (a) Particle from diesel engine exhaust. (b) Particle from burning wood. (c) Particle from suspended soil sample.

Initial research has focused on instrumental development, system characterization, and ambient aerosol analysis. Currently, work is concluding on the development, construction, and field testing of two portable bipolar ATOFMS systems which will be used for field sampling. Ongoing and future research planned for the ATOFMS systems include aerosol source characterization, continued ambient particle sampling, and modeling of heterogeneous chemical reactions and nucleation processes.

REFERENCES

Allen, J. and Gould, R.K., 1981, *Rev. Sci. Instrum.* 52:804-809.

Bentz, J.W.G., Goschnick, J., Schuricht, J., Ache, H.J., Zehnpfennig, J., and Benninghoven, A., 1995, *Fresenius J. Anal. Chem.* 353:603-608.

Bruynseels, F.J. and Van Grieken, R.E., 1984, *Anal. Chem.* 56:871-873.

Carson, P.G., Neubauer, K.R., Johnston, M.V., and Wexler, A.S., 1995, *J. Aerosol Sci.* 26:535-545.

Cotter, R.J., 1994, Time-of-flight mass spectrometry: basic principles and current state, in *Time-of-Flight Mass Spectrometry*, Cotter, R.J., Ed., American Chemical Society, Washington, Ch.2.

Dahneke, B, 1978, Aerosol beams, in *Recent Developments in Aerosol Science*, Shaw, D.T., Ed., John Wiley & Sons, New York, Ch. 9.

Davis, W.D., 1977a, *Environ. Sci. Technol.,* 11:587-592.

Davis, W.D., 1977b, *Environ. Sci. Technol.,* 11:593-596.

Davis, W.D., 1978, Continuous mass spectrometric analysis of environmental pollutants using surface ionization, in *Environmental Pollutants: Detection and Measurement*, Toribara, T.Y., Coleman, J.R., Dahneke, B.E., and Feldman, I., Eds., Plenum Press, New York, 395-411.

Dockery, D.W. and Pope, C.A., III, 1994, *Annu. Rev. Public Health* 15:107-132.

Duce, R.A., Mohnen, V.A., Zimmerman, P.R., Grosjean, D., Cautreels, W., Chatfield, R., Jaenicke, R., Ogren, J.A., Pellizzari, E.D., and Wallace, G.T., 1983, *Rev. Geophys. Space Sci.* 21:921-952.

Finlayson-Pitts, B.J. and Pitts, J.N., Jr., 1989, Particulate matter in the atmosphere: primary and secondary particles, in *Atmospheric Chemistry: Fundamentals and Experimental Techniques*, John Wiley & Sons, New York, Ch. 12.

Gebhart, J., 1993, Optical direct-reading techniques: light intensity systems, in *Aerosol Measurement: Principles, Techniques and Applications*, Willeke, K. and Baron, P.A., Eds., Van Nostrand Reinhold, New York, Ch. 15.

Giggy, C.L., Friedlander, S.K., and Sinha, M.P., 1989, *Atmos. Environ.* 23:2223-2229.

Grosjean, D., 1992, *Atmos. Environ.* 26A:953-963.

Hara, K., Kikuchi, T., Furuya, K., Hayashi, M., and Fujii, Y., 1996, *Environ. Sci. Technol.*, 30:385-391.

Hinds, W.C., 1982, Atmospheric aerosols, in *Aerosol Technology*, John Wiley & Sons, New York, Ch. 14.

Hinz, K.-P, Kaufmann, R., and Spengler, B., 1994, *Anal. Chem.* 66:2071-2076.

Hinz, K.-P., Kaufmann, R., and Spengler, B., 1996, *Aerosol Sci. Technol.* 24:233-242.

Israel, G.W. and Friedlander, S.K., 1967, *J. Colloid Interface Sci.* 24:330-337.

Jambers, W., De Bock, L., and Van Grieken, R., 1995, *Analyst* 120:681-692.

Jambers, W. and Van Grieken, R., 1996, *Trends Anal. Chem.* 15:114-122.

Johnston, M.V. and Wexler, A.S., 1995, *Anal. Chem.* 67:721A-726A.

Kaneco, S., Nomizu, T., Tanaka, T., Mizutani, N., and Kawaguchi, H., 1995, *Anal. Sci.* 11:835-840.

Kaufmann, R.L., 1986, Laser-microprobe mass spectroscopy (LAMMA) of particulate matter, in *Physical and Chemical Characterization of Individual Airborne Particles*, Spurny, K.R., Ed., Ellis Horwood Limited, Chichester, Ch. 13.

Kievit, O., Marijnissen, J.C.M., Verheijen, P.J.T., and Scarlett, B., 1992, *J. Aerosol Sci.* 23(Suppl. 1):S301-S304.

Klaus, N., 1986, Aerosol analysis by secondary-ion mass-spectrometry, in *Physical and Chemical Characterization of Individual Airborne Particles*, Spurny, K.R., Ed., Ellis Horwood Limited, Chichester, Ch. 17.

Kovalenko, L.J., Maechling, C.R., Clemett, S.J., Philippoz, J.-M., Zare, R.N., and Alexander, C.M. O'D., 1992, *Anal. Chem.* 64:682-690.

Liu, D.-Y., Rutherford, D., Kinsey, M., and Prather, K.A., 1996, *Anal. Chem.* 69:1808-1814.

Mauney, T. and Adams, F., 1984, *Sci. Total Environ.* 36:215-224.

Molina, M.J., 1996, *Angew. Chem. Int. Ed. Engl.* 35:1778-1785.

Morrical, B.D. and Prather, K.A., 1996, unpublished results.

Murphy, D.M. and Thomson, D.S., 1995, *Aerosol Sci. Technol.* 22:237-249.

Noble, C.A., Nordmeyer, T., Salt, K., Morrical, B., and Prather, K.A., 1994, *Trends Anal. Chem.* 13:218-222.

Noble, C.A. and Prather, K.A., 1996a, *Environ. Sci. Technol.* 30:2667-2680.

Noble, C.A. and Prather, K.A., 1996b, *Appl. Occup. and Environ. Hyg.*: in press.

Nordmeyer, T.E., 1995, *The Design and Construction of an Aerosol Time-of-Flight Mass Spectrometer for Ambient Aerosol Analysis*, Doctoral Dissertation, University of California, Riverside.

Nordmeyer, T.E. and Prather, K.A., 1994, *Anal. Chem.* 66:3540-3542.

Prather, K.A., Nordmeyer, T., and Salt, K., 1994, *Anal. Chem.* 66:1403-1407.

Price, D., 1994, Time-of-flight mass spectrometry: the early years as chronicled by the European time-of-flight symposia, in *Time-of-Flight Mass Spectrometry*, Cotter, R.J., Ed., American Chemical Society, Washington, Ch. 1.

Quinn, P.K., Marshall, S.F., Bates, T.S., Covert, D.S., and Kapustin, V.N., 1995, *J. Geophys. Res.* 100:8977-8991.

Reents, W.D., Jr., Downey, S.W., Emerson, A.B., Mujsce, A.M., Muller, A.J., Siconolfi, D.J., Sinclair, J.D., and Swanson, A.G., 1995, *Aerosol Sci. Technol.* 23:263-270.

Salt, K.L., 1995, *Characterization of Aerosol Particles Containing Sulfur and Nitrogen Species Using Aerosol Time-of-Flight Mass Spectrometry (ATOFMS)*, Doctoral Dissertation, University of California, Riverside.

Salt, K.L., Noble, C.A., and Prather, K.A., 1996, *Anal. Chem.* 68:230-234.

Schwartz, M.H. and Andres, R.P., 1978, On-line particle size determination in real time: an application of time-of-flight aerosol spectroscopy, in *Recent Developments in Aerosol Science*, Shaw, D.T., Ed., John Wiley & Sons, New York, Ch. 10.

Seinfeld, J.H., 1986, Acid rain, in *Atmospheric Chemistry and Physics of Air Pollution*, John Wiley & Sons, New York, Ch. 18.

Silva, P.J. and Prather, K.A., 1997, *Environ. Sci. Technol.* 31:3074-3080.

Sinha, M.P., 1984, *Rev. Sci. Instrum.* 55:886-891.

Sinha, M.P., 1991, *Proc. SPIE — Int. Soc. Opt. Eng. 1437 (Appl. Spectrosc. Mater. Sci.)*: 150-156.

Skoog, D.A. and Leary, J.J., 1992, Mass spectrometry, in *Principles of Instrumental Analysis (Fourth Edition)*, Saunders College Publishing, Fort Worth, Ch. 18.

Stoffel(s), J.J. and Allen, J., 1986, Mass spectrometry of single particles, *in situ*, in *Physical and Chemical Characterization of Individual Airborne Particles*, Spurny, K.R., Ed., Ellis Horwood Limited, Chichester, Ch. 20.

Tourmann, J.L. and Kaufmann, R., 1993, *Int. J. Environ. Anal. Chem.* 52:215-227.

Wonders, J.H.A.M., Houweling, S., De Bont, F.A.J., Van Leeuwen, H.P., Eeckhaoudt, S.M., and Van Grieken, R.E., 1994, *Int. J. Environ. Anal. Chem.* 56:193-205.

Wouters, L., Artaxo, P., and Van Grieken, R., 1990, *Int. J. Environ. Anal. Chem.* 38:427-438.

Wouters, L., Hagedoren, S., Dierck, I., Artaxo, P., and Van Grieken, R., 1993, *Atmos. Environ.* 27A:661-668.

Van Vaeck, L., Struyf, H., Van Roy, W., and Adams, F., 1994, *Mass Spectrom. Rev.* 13:209-232.

Yang, M., Dale, J.M., Whitten, W.B., and Ramsey, J.M., 1995, *Anal. Chem.* 67:1021-1025.

Yang, M., Reilly, P.T.A., Boraas, K.B., Whitten, W.B., and Ramsey, J.M., 1996, *Rapid Commun. Mass Spectrom.* 10:347-351.

Yeh, H.-C, 1993, Electrical technique, in *Aerosol Measurement: Principles, Techniques and Applications*, Willeke, K., and Baron, P.A., Eds., Van Nostrand Reinhold, New York, Ch. 18.

Section IV

Special Systems

16 Chemical Analysis and Identification of Fibrous Aerosols

Kvetoslav R. Spurny

CONTENTS

1-56670-040-X/99/$0.00+$.50
© 1999 by CRC Press LLC

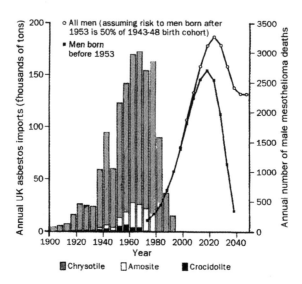

FIGURE 16.1 Import and industrial application of asbestos and the expected development of mesothelioma deaths in Great Britain. (From Peto, J., J. Hodgson, F. Matthews, et al. Continuing increase in mesothelioma mortality in Britain, *Lancet* 345:535-539 (1995). With permission.)

16.1 INTRODUCTION

Important information about an aerosol resides in the morphology (size, shape) and in the chemical composition of individual particles. To get this information, the existing, well-developed modern methods of single particle analysis are the most useful methodology.[1,2] In some cases (e.g., in the measurement and identification of fibrous aerosols), the methodology of single particle analysis is the methodology of choice. The anthropogenic fibrous aerosol existing in the working, indoor, and outdoor environments is always mixed with other types of aerosols and dusts. It is necessary to have methods that are able to differentiate between fibrous and nonfibrous (isometric, spherical) particles. The best way is to be in position to "see" directly (microscopical methods) the shape of individual particles or to detect the particle indirectly, e.g., by means of light scattering methods. Some other important aspects of such measurement are the biological effects of inhaled fibrous particles. These health effects are related to the particle shape and size and to several chemical markers of single particles (solubility, elemental composition, fiber surface activities, etc.). A combination of electron microscopical methods with single particle analysis is an optimal solution for the measurement and identification of fibrous aerosols. Fibrous dusts and aerosols consisting of asbestos and natural or man-made mineral vitreous fibers (MMVF), whether dispersed in workplace environments or the indoor and outdoor atmosphere, are toxic, fibrogenic, and carcinogenic.[3,4] Their measurement in air, personal dosimetry, and in some instances, real-time monitoring is necessary. Airborne asbestos is, therefore, a special case of fibrous aerosols. While asbestos fibers have many useful commercial and technological properties, there has been much concern regarding their ability to cause disease. There are three primary diseases that have been attributed to asbestos fiber exposure: asbestosis, mesothelioma, and lung cancer.

In the latter part of the 1990s, in several industrialized countries import and use of asbestos have been prohibited. Nevertheless, in most countries, mainly developing nations, asbestos is still used intensively, e.g., in the building industry as well as in several technological processes. However, in countries in which asbestos is banned, millions of tons of asbestos still remains in houses and buildings. Sanitation and removal activities will continue long after the year 2000. Also, the increase of asbestos related mortalities, mainly the mesothelioma mortalities, will continue. They were initiated by asbestos dust exposures 20 or more years ago (Figure 16.1).[5,6]

TABLE 16.1
Characteristics of Asbestos Minerals

Mineral type	Color	Asbestiform Variety	Nonasbestiform Variety	Chemical Formula
Serpentine	Light green to white	Chrysotile	Antigorite Lizardite	$Mg_3(Si_2O_5)(OH)_4$
Amphibole	Tan to brown	"Amosite"	Cummingtonite to grunerite	$Mg_7(Si_8O_{22})(OH)_2$ to $Fe_7(Si_8O_{22})(OH)_2$
	Blue	Crocidolite	Riebeckite	$Na_2Fe_5(Si_8O_{22})(OH)_2$
	White to tan	Anthophyllite	Anthophylite	$(Mg,Fe)_7(Si_8O_{22})(OH)_2$
	White to green	Tremolite	Tremolite	$Ca_2Mg_5(Si_8O_{22})(OH)_2$
	Green	Actinolite	Actinolite	$Ca_2(Mg,Fe)_5(Si_8O_{22})(OH)_2$

The replacement of asbestos by MMVF, starting intensively in the 1980s, did not solve the problem of health risk by fibrous aerosols.[7,8] The worldwide production of MMVFs lay over five megatons per year. The majority of these "new" fibrous materials have carcinogenic potencies less or much less than chrysotile and mainly crocidolite asbestos. Nevertheless, there do exist types of very persistent MMVFs, whose carcinogenicity is comparable with that of asbestos.

Asbestos is a term applied to several commercial minerals exploited for their tendency to produce long, thin fibers. The most important asbestiform minerals are the serpentine and the amphiboles (Table 16.1).[9]

Workplace exposure of airborne mineral fibrous aerosols is usually assessed by personal sampling in which a known volume of air is drawn through a membrane filter which is subsequently prepared for examination by phase-contrast optical microscopy (PCOM), and the fibers are counted following predetermined rules. Procedures that have been nationally and internationally standardized and tested by interlaboratory cooperation are available.[10-22] Furthermore, for workplace situations, a few real-time monitoring methods exist for the *in situ* detection of airborne mineral fibers. Nevertheless, they have not yet been standardized and are not generally fully accepted.[23,24]

When measurements of airborne mineral fibers are made outside, e.g., in the indoor or outdoor ambient air, it has been found that PCOM is inadequate. The concentrations of airborne mineral fibers to be measured are usually very much lower than those found in working places, and the fibers found are usually very much smaller, so that most cannot be seen by optical microscopy.

Quantification by PCOM can be considered as only an index of exposure. For this reason, better, and more sensitive and specific methods are in some instances necessary for measurements in workplaces. Electron microscopy is now the method of choice for the determination of mineral fiber concentrations, size measurements, and specific identification in almost all environmental situations. Scanning electron microscopy (SEM), transmission electron microscopy (TEM), identification techniques, such as energy-dispersive X-ray analysis (EDXA), and selective area diffraction (SAED) are currently well-proved, tested, and standardized procedures which can be fully recommended.[12-19] In spite of this, specific limitations, data conversion factors, data presentation, and interpretation methodologies are still under development and are not sufficiently conclusive and consistent. The same, unfortunately, is valid for a physically, chemically, mineralogically, and biologically clear and unambiguous definition of a "toxic fiber."

The former term, "respirable" fiber has now been replaced by "thoracic" fiber. Thoracic fibers are those capable of reaching any part of the respiratory system below the larynx, while respirable fibers include only those reaching the gas-exchange region of the lung. It has also been confirmed experimentally that fibers producing mesothelioma and lung cancer have different size characteristics.

In principle, the sampling method for fibrous aerosols should reflect potential health effects by collecting only fibers that reach the lung. The current technique for approximating the thoracic definition for fibers is to use an upper diameter limit of 3 or 3.5 μm. Recently realized theoretical

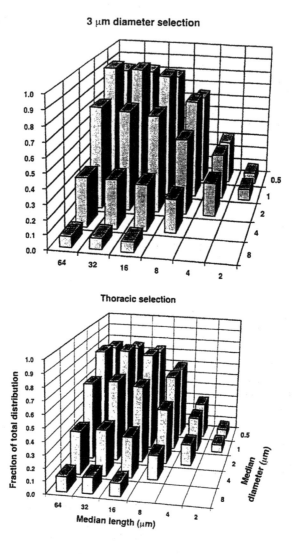

FIGURE 16.2 Calculated fractions of fibers counted by the PCOM method: for 3 μm diameter and for thoracic selection. (From Baron, A.P. Application of the thoracic sampling definition to fiber measurement, *Am. Ind. Hyg. Assoc. J.* 57:820-824 (1996). With permission.)

calculations could show that the 3 μm rule agrees with the thoracic definition within about ±25% for a wide range of possible size distributions. This agreement is well illustrated in the Figure 16.2.[25]

16.2   EARLY MEASUREMENT METHODS

A combination of inertial or thermophoretic fibrous particle separation with microscopical counting and examination of fibers had been used since about 1950, before filtration methods were established worldwide (Table 16.2).[26] The inertial separation (impaction) was realized in konimeters, impactors, and impingers. The konimeter, widely used in early studies of silica dust, was applied in the United States, South Africa, and Europe. Impingers, including the midget variety, were the standard collector for asbestos dusts in the United States until the mid-1960s. The great disadvantages of these methods are evident. Because of the uncertainty in the aerodynamic behavior of fibrous particles, the collection efficiencies could not be well predicted.[27-29] Mainly the collection of fine fibers was very

TABLE 16.2
Dust Measurements

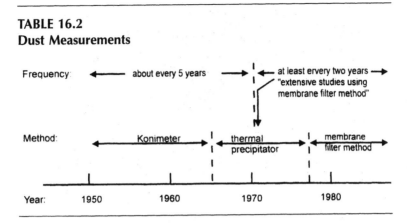

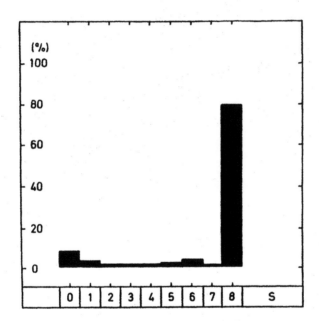

FIGURE 16.3 Separation efficiencies for fibrous amosite aerosol in an eight-stage cascade impactor and on a backing membrane filter. (From Farnham, J.E. and P. Roundtree, Health effects of synthetic vitreous fiber exposure, in G.A. Peters and B.J. Peters (Eds.) *Asbestos Pathogenesis and Litigation*, Butterworth Publ., Sevenoaks, UK, Vol. 13, pp. 111-139 (1996). With permission.)

poor. This is illustrated in Figure 16.3. Separation efficiencies for fine fibrous aerosols of a cascade impactor are plotted and compared with the amount of penetrated fibers (filter 8).[30]

During the 1980s, Zeis–Konimeter was used for the detection of asbestos fiber concentrations in some workplaces in the former East Germany.[31] In laboratory measurements, a sufficient correlation with the membrane filter method was reported. The detection limit was very poor. Only fiber concentrations over 1 fiber/ml could be detected.

A much better device for sampling fine fibrous aerosols was the thermal precipitator, also used for sampling fibrous dusts until the 1960s. The collected samples had shown an alignment of fibers in the laminar flow through the precipitator (Figure 16.4). All the "ancient" methods were also not very sensitive for fine particles and for low particle concentrations.

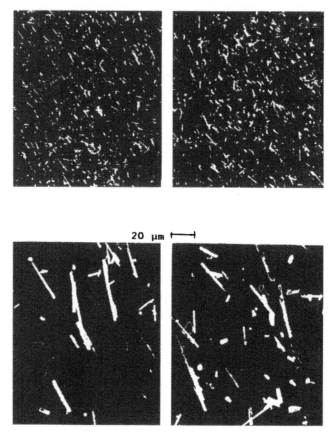

FIGURE 16.4   Scanning electron micrographs of amosite fibers sampled by a thermal precipitator.

16.3   SAMPLING BY FILTRATION

A convenient and widely used method for collecting fibers for counting, sizing, and identification is filtration. Filters can collect almost all sizes of fibrous particles. Different types of analytical aerosol filters are available. However, porous analytical aerosol filters, such as membrane filters (MFs), polycarbonate, or Nuclepore filters (NPFs), are the filters of choice for sampling fibers.

Air samples are collected using either an MF or an NPF.[32-34] Both are efficient, but they differ in the structure and separation properties. MF has a porous spongy structure and is the depth-type filter (Figure 16.5). Particles (fibers) are separated not only on the filter surface but also inside the approximately 100 μm thick filter. In comparison, the NPF is a thin (10 μm), smooth sheet with cylindrical pores (Figure 16.6). The particles are collected on the surface, mainly in the vicinity of pores. Filtration studies have shown that the collection and separation of fibrous particles in air filters differ from the collection and separation of spherical or isometric particles.[35]

The effects of fiber orientation in air and the alignment in laminar flows play important roles in the collection process, mainly for NPF with straight capillary pores (Figure 16.7). The theory of filtration of fibrous aerosols showed that the direct inception is the main separation mechanism.[36]

Fibrous particles in workplace air are mostly highly electrically charged. This electric charge can also influence positively the fiber collection efficiencies and the sampling process.[37-39] When sampling in workplaces with production of coarse mineral dust and fibrous aerosols, size-selective presamplers have to be used for collecting the thoracic fraction only.[40,41] All the techniques used for the evaluation of filter samples, by counting fibers are based on the assumption that the fiber distribution on the filter is random and homogeneous. Sampling airborne fibers on an MF or an

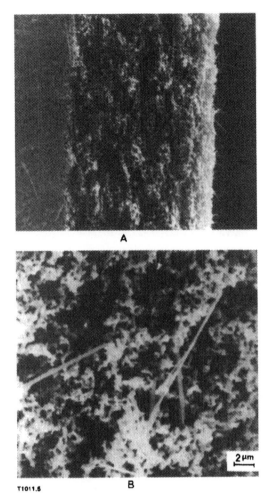

T1011.6

FIGURE 16.5 Scanning electron micrographs showing the cross section and surface (A) of a membrane filter as well as asbestos fibers "incorporated" in the inner filter structure (B).

NPF should follow a Poisson distribution. In order to realize such conditions, a double-backing filter (e.g., an MF with pore size 0.8 μm) must be used to minimize the flow variations across the filter.[42]

16.4 SAMPLING ON SURFACES

There are cases and situations in which fibers, mainly asbestos, exist in interfaces, e.g., air/solids. Such a practical case is the free or unattached deposition of asbestos fibers on the surfaces of corroded asbestos cement (AC) products (solid planar surfaces, such as roofs and facades). AC products are subject to weathering and corrosion just as other building materials, and the corroded AC products release asbestos fibers into the environment (including the general atmosphere). For this reason, the surface concentrations of such freely deposited fibers have to be measured. Special sampling equipment and procedures are necessary for such measurements. Such useful equipment has been developed and used in Germany.[43]

The fibrous emissions are collected on NPF or MF and then the number of fibers, their size distribution, and identities are evaluated by SEM and TEM. The principle and application of this instrument are shown in Figure 16.8. The sampling chamber (SC) is placed on the surface (e.g.,

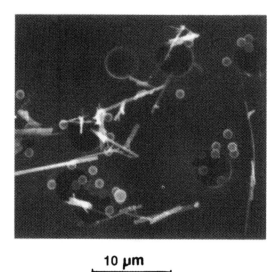

10 μm

FIGURE 16.6 Scanning electron micrograph showing the deposition of spherical and fibrous particles in the pores of an NPF.

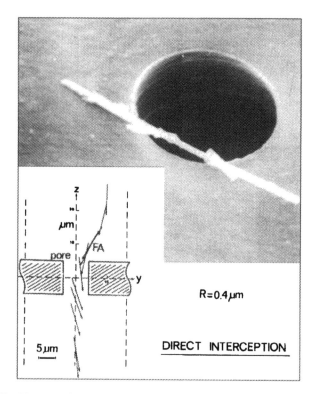

FIGURE 16.7 Electron micrograph (SEM) and schematic picture of fiber in one NPF pore.

asbestos roofing tiles). The contact between the instrument and the surface is sealed by means of polymer ribbons (D). Ambient air (L) flows over a corroded and weathered AC sheet (AZ) through the small slot (S). The sampling chamber can be maintained on the measured surface by means of the handle (H) as long as necessary. Asbestos fibers released from the corroded surface are then sampled on an analytical filter (MF). The fiber concentrations and fiber emissions are evaluated by

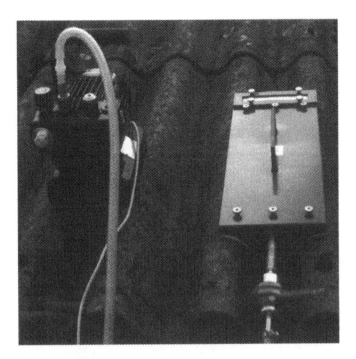

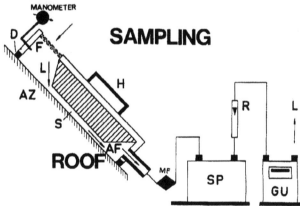

FIGURE 16.8 A photograph and schematic picture of the equipment for sampling emissions from surfaces.

an electron microscopical procedure (Figure 16.9). The other instruments are used to measure the flow rate (R) and the total air volume (GU). SP is a vacuum pump. The equipment and the procedure have been applied with very good success in practical environmental measurements.[43]

16.5 SPECIMEN PREPARATION

As already mentioned, all microscopic techniques are based on the collection of airborne particles by filtration of the air through an MF or an NPF filter. MF with a mean pore size of 0.8 μm and NFP with a mean pore size of 0.4 μm are used and have demonstrated satisfactory collection efficiency even for very fine fibers.[35]

Subsequently, a quantitative and representative specimen has to be prepared for PCOM, SEM, and TEM. Direct transfer techniques, i.e., techniques without filter destruction by ashing, dissolution, etc., are often preferred. In such cases, the individual particles appear on the final specimen in the same relative positions that they occupied on the original filter after air sampling. Furthermore,

FIGURE 16.9 Scanning electron micrograph of asbestos fibers sampled in the air over a corroded asbestos cement roof.

no physical, chemical, or morphological changes to the original airborne particles can occur. Some disadvantages and limitations of such direct transfer methods will be mentioned later.

PCOM is preferred for air sampling by an MF. A representative portion of the MF is then fixed and placed on a microscope slide. Standard techniques that clear the MF with a solvent to make it transparent and provide a refractive index suitably different from that of asbestos or other mineral fibers have been described and standardized.[10,20-22] Under such conditions, the fibers are clearly visible and can be sized and counted at magnifications of approximately ×450 (Figure 16.10).[44-48]

Air samples on MFs and NPFs can be used for analysis by both SEM and TEM. Air samples obtained on NPFs can be applied directly for SEM and EDXA evaluations without any further preparation. An example of such a technique is a procedure that was standardized in Germany.[49] The NPF (pore size 0.4 or 0.8 μm) must be coated with a very thin gold layer before sampling. The thickness of the gold coating on the exposed side of the NPF is approximately 40 nm. The coating is applied in a vaporizing unit or sputter coating unit. After sampling, the loaded filter is placed in the holder with a quartz glass mask and the sampled material is ashed in a cold ashing unit (oxygen plasma ashing). The cold ashing removes virtually all the organic material on the filter (e.g., soot particles). Randomly selected sections of the filter surface can then be taken directly, without further preparation, for SEM. The application of this procedure is shown schematically in Figure 16.11. The "pre-ashing" facilitates the counting by SEM considerably. In comparison with

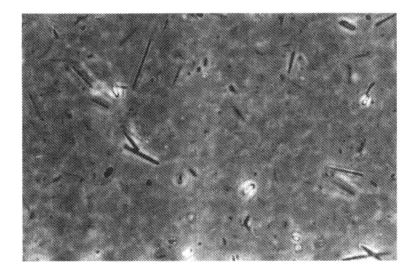

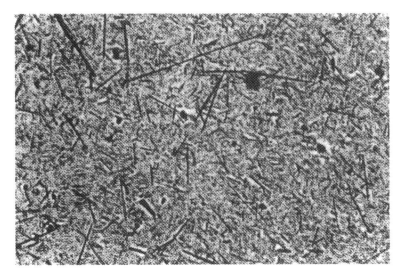

FIGURE 16.10 Optical microscope pictures (500×) of sampled asbestos fibers.

SEM, the specimen preparation for all modifications of TEM is more complicated and time consuming.[50-63]

Air samples are collected on either an MF or an NPF. The direct transfer specimen preparation techniques are different for these filter types. NPF-collected air samples are prepared according to standardized procedures.[52,53] The preparation steps are shown in the diagram in Figure 16.12a. A portion of the carbon-coated filter is placed on the TEM specimen grid and the NPF is dissolved away by treatment in a Jaffe washer,[50] shown schematically in Figure 16.12b. The polycarbonate matrix of the NPF is dissolved with chloroform.

The MF, made of the mixed esters of cellulose, cannot be prepared by means of the former procedure. An MF has a sponge-like texture (Figure 16.5), and carbon replication would yield a TEM specimen with an unacceptable amount of detail against which it would be difficult to detect small fibers. The MF sample must be converted into a thin film of plastic. The MF filter with the particles on its surface is subjected to a solvent treatment that causes the sponge texture to collapse into a thin, continuous film of plastic. This film can then be carbon coated and extracted. The series of steps are demonstrated in Figure 16.12c. A section of the filter can be placed on a small drop

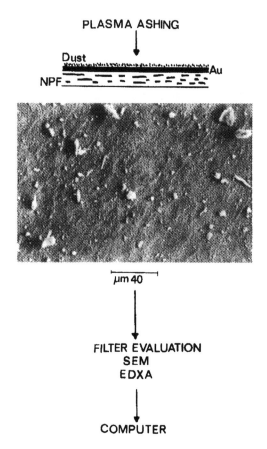

FIGURE 16.11   Flow chart showing the application of the German standardized VDI-procedure (SEM).

of a mixture of dimethylformamide, acetic acid, and water on a microscope slide, then heated on a slide warmer. In an alternative technique, a "hot block" device is used to expose the filter to acetone vapor.[55] Both methods recognize that some of the particles may become engulfed in the plastic during the filter-collapsing step, and that these particles will then not be transferred to the extraction replica. This problem is overcome by etching away a thin layer of the plastic in a plasma asher prior to the carbon coating step. In this way, any buried particles, close to the surface, are exposed again. At this point, the MF specimen is ready for examination by PCOM or SEM.[51] For SEM, the film surface must be coated with a gold film. The specimen preparation has to be continued for analysis by TEM. After coating with carbon, a portion of the filter area is treated in the Jaffe washer in a way similar to the NPF, except that the solvent used is either acetone or dimethylformamide. The sampled particles are then visible for the TEM observation and evaluation (Figure 16.13).

When the specimen preparations are completed, the microscopic counting, sizing, and fiber identification procedures can begin (Figure 16.14). Some investigations have mentioned that for SEM procedures, the MF techniques have several disadvantages and should not be recommended.[56] There also exist procedures using a combination of MF and NPF. The aerosol is sampled by MF. This sample is then ashed and the ash is resuspended in bidistilled water. For disturbing particle agglomerates, ultrasonic agitation is applied. After such a treatment, the low concentrated suspension is refiltered by NPF (pore size 0.2 μm) and specimens for SEM or TEM evaluation are prepared. Such procedures have the advantage in the ability to enhance the sensitivity; but also a great disadvantage — the ultrasonic agitation may significantly increase the fiber number concentrations and decrease the reproducibility of the whole procedure.[63-66] Ways to overcome these artifacts will be described later.

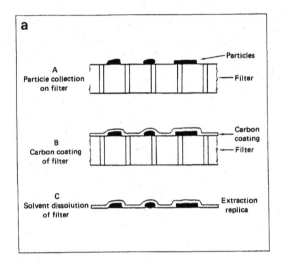

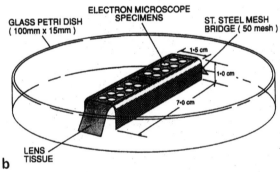

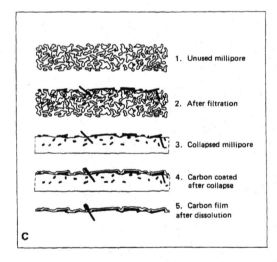

FIGURE 16.12 Schematic diagrams showing steps in the procedure for preparing a TEM specimen for an NPF (a), the design of the Jaffe Washer (b), and the procedure for preparing an MF sample (c). (From Chatfield, E.J. and M.J. Dillon, Some aspects of specimen preparation and limitations of precision in particulate analysis by SEM and TEM, *Scanning Electron Microscopy*, SEM-Inc., Chicago (1978) pp. 387-499. With permission.)

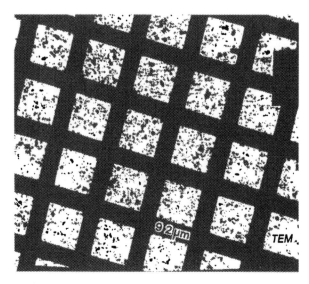

FIGURE 16.13 Transmission electron micrograph of a specimen grid and in air sampled particulates.

FIGURE 16.14 Very fine chrysotile fibers sampled on an NPF, observed by SEM (= REM) and TEM.

16.6 FIBER COUNTING, SIZING, AND IDENTIFICATION

16.6.1 PCOM Methods

A number of well-recognized standard methods for filter analysis by PCOM exist, including e.g., the 7400 National Institute for Occupational Safety and Health (NIOSH) method[19] and the European References Method.[20] These and other microscopic methods are all similar in that they count fibers on cleared MFs using PCOM at approximately ×450 magnification. This has provided a resolution of approximately 0.2 to 0.3 µm. Any piece of particulate matter meeting the prescribed definition of a fiber is counted, regardless of identity. Hence, one of the many limitations of PCOM for fiber analysis is its inability to ascertain the identity of the observed materials. PCOM provides an adequate index of exposure only if the fibers present in the air sample are in fact asbestos or other single types of fibers. In occupational environments, where, for example, asbestos is used, most of the fibers collected in an air sample are likely to be asbestos.

Conventional PCOM is based on counting the number of thoracic fibers (particles longer than 5 µm, with a length-to-diameter aspect ratio of 3:1 or more, but thinner than 3 µm) per unit graticule area. The counting is done either in 100 fields or until a total of 100 fibers are counted over at least 20 fields. The magnification used is ×450–500.

Another approach for estimating the fiber concentration on an MF is the "void-counting" procedure. This technique,[42,44] like the conventional method, is based on the assumption that the fiber distribution on the filter is random and homogeneous. A number of fields are scanned to determine whether fibers are or are not present in any given field. By taking the natural logarithm of the ratio of the number of fields scanned to the number of empty fields, the fiber concentration per field can be estimated. This procedure saves time during evaluation, and it is less strenuous to the operator. The void counting procedure requires that the deposit be uniform for accurate results.

Different counting automation methods were also described.[23,24] A modern technique for counting airborne fibers on MF uses an on-line evaluation by combining PCOM with a Macintosh computer.[67,68] The computer has special video display cards and software, which enable the microscopist to perform on-line determination of fibers on the display (Figure 16.15). Each fiber is marked by the microscopist with a mouse, and data are saved in the program. The void counting technique can also be used. It is also possible to measure the length and the diameter of fibers while counting. Data for several types of fibers have shown good agreement between counts by the conventional and on-line technique.

16.6.2 Electron Microscopy

Electron microscopy is a much more sophisticated methodology than the PCOM. Furthermore, excellent instruments are now available for such modern analytical procedures.

16.6.2.1 SEM Procedure

SEM is generally viewed as an intermediate analytical tool providing increased fiber visibility and moderate analytical capabilities compared with PCOM. It is a "screening" test procedure of high quality.[69-75] Nevertheless, the fiber visibility and analytical capabilities are poorer than with TEM. SEM is utilized to evaluate the physical parameters of mineral fibers, such as fiber size and shape, in samples of known identity. In addition, in combination with EDXA an approximate fiber identification can be achieved (Figure 16.16). It can be used to distinguish between organic and inorganic fibers and between silicate and nonsilicate fibers. Because of the lower costs for the instrumentation and for sample evaluation, SEM has also been standardized.

The German Verein Deutscher Ingenieure (VDI) Method[49] uses a direct sample evaluation at a magnification of approximately ×2000. Fibrous particles in air are sampled on a gold-coated NPF (pore size 0.4 to 0.8 µm). Before analysis, the organic portion of the sample is removed directly

FIGURE 16.15 A photograph of the microscope and the computer set-up for the evaluation of filter samples. (From Lundgren, L., S. Lundström, I. Laszlo, et al. Modern fibre counting — a technique with the phase-contrast microscope on-line to a Macintosh computer, *Ann. Occup. Hyg.* 39:455-467 (1995). With permission.)

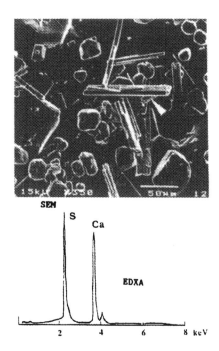

FIGURE 16.16 Electron micrograph (SEM) and element spectrum (EDXA) of airborne gypsum fibers.

from the gold-coated NPF surface using a special plasma ashing process (Figure 16.11). The individual fibers, in a randomly selected filter area, are then counted and classified according to fiber size and specific fiber types. EDXA is used to approximately identify the fiber types. The standardized method is used for detecting inorganic fibers with diameters (D_F) of less than 3 μm and lengths (L_F) of between 2.5 and 100 μm. The detectability and, in particular, the identifiability of fibers thinner than 0.2 μm is limited. Such limitations will be discussed later. Rules for fiber counting, sizing and identification are strictly standardized. Nevertheless, the problem of an exact determination of single fiber lengths, which will be mentioned later, is neglected.

FIGURE 16.17 A photograph of a modern transmission electron microscope (TEM).

16.6.2.2 TEM Procedure

Experience has shown that TEM, with the addition of SAED, is the system of choice for asbestos and other mineral fiber analysis. Modern TEM (Figure 16.17) has more than adequate resolution, and fibers can be identified using morphological signs, EDXA, and SAED. TEM has been developed for routine asbestos and mineral fiber air sample analysis. Widely accepted methods for analysis by TEM exist, which have been standardized and are applied in the workplace, indoors, and outdoors.[52-59] Most methods used the direct NPF procedure for air sampling and specimen preparation (Figure 16.12). In a routine asbestos or mineral fiber count, the fibers found on about 10 grid openings of a 200-mesh TEM grid (Figure 16.13) are counted, sized, and identified at magnification of about ×20,000. Mineral fibers are identified on the basis of shape (chrysotile is characterized by thin tubular fibrils with a hollow channel, Figures 16.18 and 16.19), "chemical" composition (EDXA), and the SAED pattern (Figure 16.19). Because of the costs, in some standardized procedures[52] different levels of analysis are proposed (Table 16.3). Each level of analysis provides increasing information and requires greater expertise, training, time, and costs. Nevertheless, TEM is a complex and sufficiently reliable method.[76-87]

16.7 TOTAL SAMPLE PREPARATION METHODS

Based on some known artifacts and imperfections of existing PCOM, SEM, and TEM procedures, which will be discussed in detail later, a sophisticated "Total Sample Preparation Methodology" for measuring fibrous aerosols was developed and described in Japan.[88] This methodology consists of two alternate procedures: parallel preparation and serial preparation (Figure 16.20). The parallel preparation method uses different portions of the same filter sample for measurements by PCOM (PCM), optical microscopy (OM), dispersion staining polarized microscopy (DS/PLM), SEM, and TEM. The serial preparation uses the same single filter portion for all measurements of OM, SEM, and TEM. After the first measurement of OM, the same sample is also observed by SEM, and successively retreated for TEM, respectively.

The MF sample (cellulose ester membrane, CE) is ashed by oxygen plasma on a glass slide. The evaluations by optical microscopical methods and by SEM are performed directly on the glass slide. Another SEM-version is illustrated in Figure 16.21. The glass slide with particles and fibers is carbon coated and is used for SEM+EDXA (EDX) analysis. This slide can be, after further sample development (Figure 16.22), used with an analytical transmission electron microscope (ATEM) as a direct transfer method.

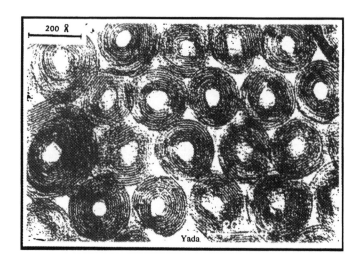

FIGURE 16.18 High-resolution electron micrograph of transverse section of chrysotile asbestos. (From Yada, K. Study of chrysotile asbestos by a high resolution electron microscope, *Am. Soc. Testing Materials,* ASTM 834:118-138 (1967). With permission.)

The last and most sensitive "indirect transfer method" (TEM 2) uses two steps in sample preparation (Figure 16.23). One half of the filter is attached to a clear glass slide using acetone vapor, with the dust side facing the glass slide, and then ashed in a low-temperature plasma asher. To suspend the residual mineral dust in isopropanol, the surface of the glass slide is wetted with a few droplets of isopropanol and shaved with a blade. The dust is immersed in a vessel with isopropanol by ultrasonification (30 sec). The suspension is filtered by NPF (pore size 0.2 μm). This procedure enables a preconcentration of the sampled fibers and consequently increases the detection limit appreciably. The effect of the 30-sec ultrasonification on the results — fiber sizes and concentration — was not mentioned.

16.8   LIMITATIONS OF PRECISION AND ACCURACY

All the methods described so far have advantages, disadvantages, and limitations. These are discussed below.

16.8.1   PCOM Methods

PCOM provides a theoretical resolution of approximately 0.2 μm. The detection limit is approximate because different filter types have different refractive indices and because of subjective counting problems. Therefore, the established limits in different publications lie between about 0.2 and 0.5 μm. In occupational environments, where asbestos or other natural or man-made mineral fibers are used, most of the fiber collected in an air sample is likely to be of the fiber type used as primary raw material in the production process or the technology involved. PCOM is most applicable in such situations.

In outdoor air and in nonoccupational atmospheric environments, a broad range of organic and inorganic materials meeting the dimensional fiber criteria will be collected, and few, if any, will be asbestos or other expected fiber types. A PCOM fiber count in these instances will provide little information about the asbestos or other fiber concentration in the sampled air. In addition, the limit of resolution, approximately 0.2 to 0.5 μm, is insufficient to observe all the fibers that may be present in the air sample collection filter. Chrysotile asbestos fibrils can be as small as 0.01 μm in diameter. These very thin fibers will not be included in a PCOM fiber count. In Figure 16.24, the

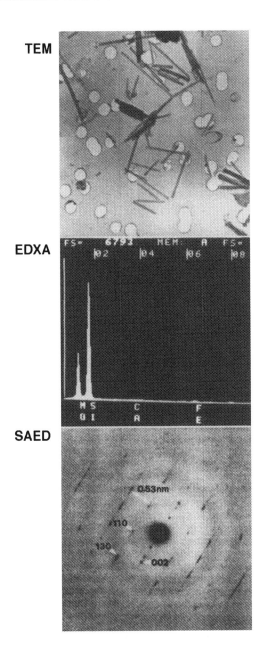

FIGURE 16.19 Examples of chrysotile fiber identification by shape (tubular morphology, TEM), crystallography (SAED), and characteristic element peaks (EDXA). (From Chatfield, E.J. ISO. International Standard Organization of Standardization, Analytical methods for determination of asbestos fibers in water and in air. Report ISO/TC 147/SC 2, W 618 and ISO TC 146/SC 3, W 61 (1991). With permission.)

"incapability" of PCOM in workplaces with asbestos is well demonstrated. The highest fiber concentrations for both fiber types — chrysotile and amosite — lie in the range of fiber diameters lower than 0.3 μm and are "invisible" by the PCOM evaluation procedure.

Hence, PCOM analysis only provides an index of exposure to those fibers optically visible. Fiber counts made by the PCOM method are subject to a great deal of variability. There is currently no absolute standard sample that can be used to certify the performance of an operator. The results

TABLE 16.3
Levels of Analysis for Airborne Asbestos Used in the EPA Methodology[52]

Level of Analysis	Fiber Identification Criterion	Applicability
Level I	Morphology; visual inspection of SAED pattern	Screening samples
Level II	Morphology; visual SAED; elemental analysis using EDXA	Regulatory action
Level III	Morphology; visual SAED; measurement of zone axis SAED from micrographs; elemental analysis using EDXA	Confirmatory analysis of controversial samples

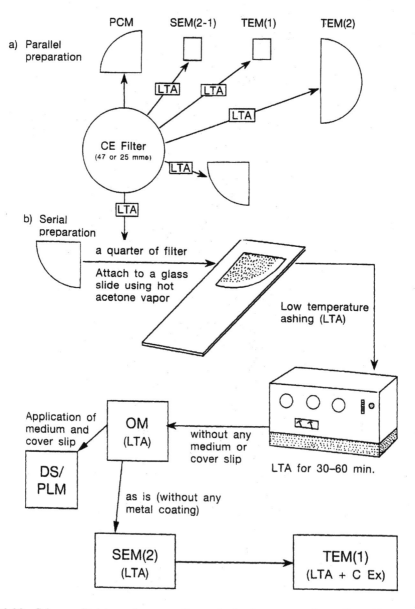

FIGURE 16.20   Schema of total sample preparation method using one membrane filter. (From Kohyama, N. and S. Kurimori, A total sample preparation method for the measurement of airborne asbestos and other fibers by optical and electron microscopy, *Industrial Health (Japan)* 34: 185–203 (1996). With permission.)

SEM(2-1)

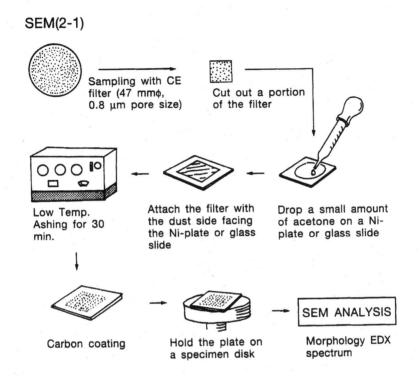

FIGURE 16.21 Flow chart of the sample preparation method for SEM (2-1). (From Kohyama, N. and S. Kurimori, A total sample preparation method for the measurement of airborne asbestos and other fibers by optical and electron microscopy, *Industrial Health (Japan)* 34:185-203 (1996). With permission.)

of different counters can differ within a relative standard deviation of 40% or greater. The American Industrial Hygiene Association started, with the sponsorship of the Asbestos Analyst Registry (AAR), a U.S. quality assurance program that evaluates individual analysis in the application of standard PCOM procedures. To qualify for listing in the AAR, an analyst must have received training, e.g., by the National Institute for Safety and Health (NIOSH). The theoretical detection limit for fiber concentrations lies in the range of 0.02 to 0.1 fibers per ml. This sensitivity is further limited by the background contamination of the filters and chemicals being used. For practical applications, the detection limit lies between 0.1 and 0.5 fibers per ml. For this reason, if the permissible exposure limit (PEL) in workplaces is 0.2 fibers per ml (or less in the future), the PCOM method will be insufficient.[53-58]

Nevertheless, in situations such as during the monitoring of the clearance of asbestos from buildings, the necessary detection limit could be lower than 0.03 fiber per ml, because the concentration of the total nonfibrous dust is low.

In many workplaces, other types of fibers are used in addition to asbestos. In such situations, and also in basic epidemiological studies in workplaces, the PCOM method is less suitable, and electron microscopy should be used. Many industrial hygiene analysts recommended that PCOM should be used in workplace applications as a preliminary check or screening test.[58]

16.8.2 SEM–EDXA PROCEDURE

SEM–EDXA is not intended to be an absolute method, but rather an intermediate option not encumbered with the time and expense requirements associated with TEM. The major limitations concern the limited resolution of SEM and associated fiber visibility and the inability of EDXA to adequately identify the fiber present. SEM is limited to fiber counting in the submicrometer size

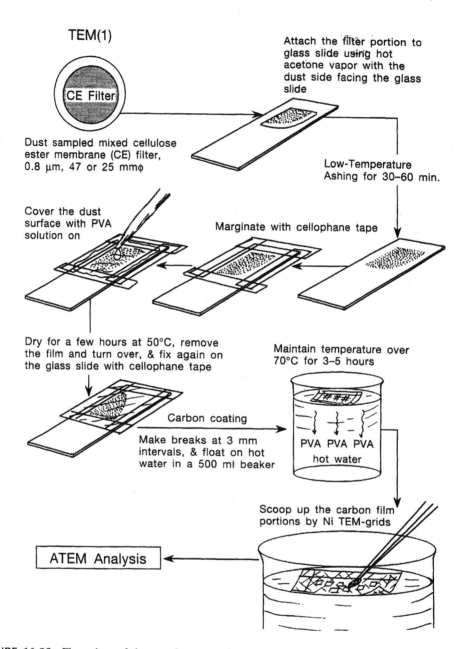

FIGURE 16.22 Flow chart of the sample preparation method for TEM (1). (From Kohyama, N. and S. Kurimori, A total sample preparation method for the measurement of airborne asbestos and other fibers by optical and electron microscopy, *Industrial Health (Japan)* 34:185-203 (1996). With permission.)

range. The diameter of the thinnest fibers of chrysotile (fibril) is approximately 0.03 μm or less. This is well below the visibility of optical microscopy (about 0.3 μm). Modern SEM, however, can achieve a visibility of about 5 nm. This capability can be realized only on photographs of idealized samples. In routine practice, it is much less. How the fiber visibility corresponds to the magnification used is demonstrated in Figure 16.25. SEM–EDXA provides rapid and positive identification of fibers with diameters over 0.1 μm. The quality of X-ray spectra produced is, to a large extent, a function of the diameter of the fiber being analyzed. Owing to high background excitation, the spectra produced from fibers with diameters less than 0.1 μm are difficult to evaluate. Gold-coated

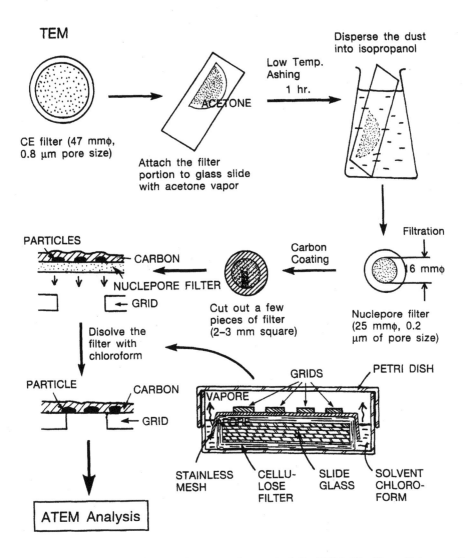

FIGURE 16.23 Flow chart of the sample preparation method for TEM (2). (From Kohyama, N. and S. Kurimori, A total sample preparation method for the measurement of airborne asbestos and other fibers by optical and electron microscopy, *Industrial Health (Japan)* 34:185-203 (1996). With permission.)

NPFs are also problematic for X-ray analysis. The gold peak overlaps with the silicon peak, which may cause additional uncertainty for small diameter fibers. As already mentioned, EDXA cannot completely distinguish between different types of asbestos or other mineral fibers. Information on the crystalline structure of the fibers, provided in TEM, is needed to identify the mineral species more completely. In addition, the tubular morphology of chrysotile (Figures 16.18 and 16.19), which is highly indicative, can be observed only using TEM. In a completely unknown situation, identification of fibers by EDXA alone is not normally possible. When, for example, an Mg–Si fiber also shows a Ca peak, it cannot be said whether it is an amphibole asbestos or chrysotile contaminated by calcium (as for fiber released from a corroded asbestos–cement).

Asbestos or other crystalline fiber identification made using SEM–EDXA is always open to question if it is not known whether other particle species of generally similar compositions are absent. For this reason, an optimistic pragmatism for using SEM as the "absolute" method should never override the necessary scientific objectivity.[54]

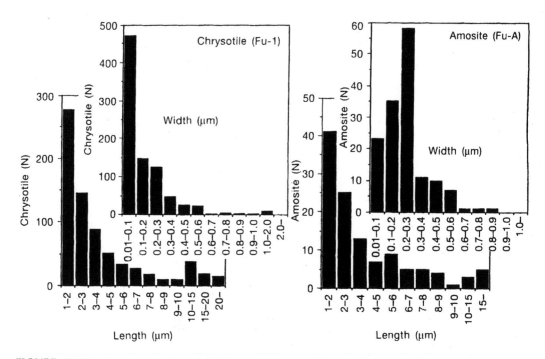

FIGURE 16.24 Examples for size distribution of airborne chrysotile and amosite fibers in workplace by TEM. (From Kohyama, N. and S. Kurimori, A total sample preparation method for the measurement of airborne asbestos and other fibers by optical and electron microscopy, *Industrial Health (Japan)* 34:185-203 (1996). With permission.)

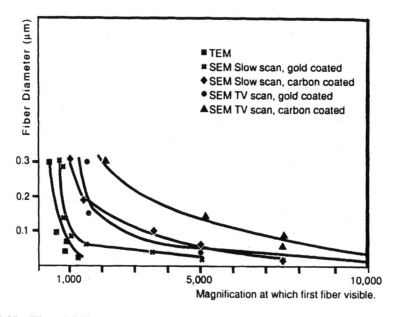

FIGURE 16.25 Fiber visibility on electron microscope screen for SEM and TEM. (From Breysse, P.N. Electron microscopic analysis of airborne asbestos fibers, *Crit. Rev. Anal. Chem.* 22:201-227 (1991). With permission.)

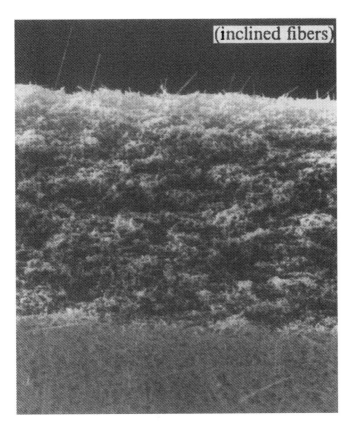

FIGURE 16.26 SEM micrograph showing asbestos fibers separated on the surface of a membrane filter with different orientation angles.

Another important artifact was observed and recently quantitatively described, when the SEM is used for fiber counting and sizing by a direct transfer method. In these procedures, fibrous aerosol is sampled on NPF. The NPF is coated by thin gold film before or, in most cases, after the sampling. The sample is then evaluated directly by SEM.

In several observations, it could be shown that the single fibers sampled on MF (Figure 16.26) and NPF (Figure 16.27) do not lie flat on filters.[89] In a new study,[90] variable inclinations of sampled fibers to the filter surfaces were measured quantitatively. Only about 60% of fibers lie flat on the filter surface, and many of the rest are balanced at high angles on the filter surface (Figure 16.28). These inclined fibers will appear foreshortened when the filter surface is viewed in conventional orientation during SEM evaluation. This could lead to an underestimation of fiber lengths. Some fibers may be so foreshortened that they do not appear to fulfill the criteria of a fiber. Moreover, it seems unlikely to derive widely applicable correction factors to allow for orientation effects.

In this regard, the direct SEM procedure has to be considered an "approximative" method. This method is, unfortunately, commonly used in animal inhalation studies to characterize the size distribution of the fibrous aerosols.[91] Because the carcinogenic effect of inhaled fibers is correlated to the fiber lengths, any underestimation of this parameter is be desirable.

16.8.3 TEM PROCEDURE

TEM analysis for asbestos, mainly for research purposes, started in the early 1970s. The PEL has decreased since 1970 from 12 to almost 0.1 fibers per ml, and a further decrease is under discussion.[58] For this reason, PCOM is not a satisfactory method for fiber detection in the workplace of the future. Furthermore, PCOM cannot detect fibers thinner than about 0.3 μm.

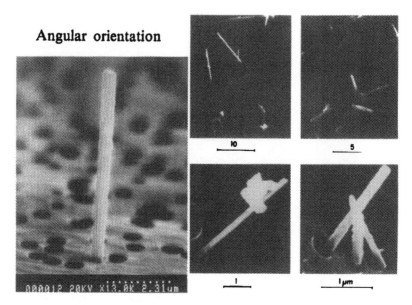

Angular orientation

FIGURE 16.27 SEM micrographs of fibers separated on the surface of two Nuclepore filters, showing different orientation angles.

SEM sizing of airborne fiber samples

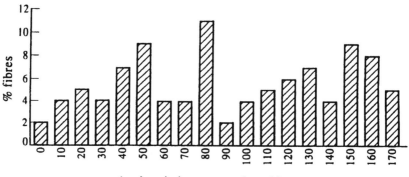

Angle relative to top edge of image

FIGURE 16.28 Angular orientation of 100 glass fibers within the plane filter. Angles measured relative to the top of the SEM screen. (From McIntosh, C. and A. Searl, Scanning electron microscope sizing of airborne fibre samples, *Ann. Occup. Hyg.* 40:45-55 (1996). With permission.)

Both SEM and TEM are capable of analyzing fibers (EDXA), to distinguish between silicates and nonsilicates, and to distinguish between asbestos and nonasbestos fibers. Nevertheless, fibers over 0.1 µm in diameter can be detected by SEM. The identification by TEM is approximately 0.0025 µm (fiber diameter). TEM also has a second technique available, SAED. Some recent publications, which will be discussed later, strongly suggest that both short and long mineral fibers are biologically active. For this reason, short fibers (shorter than 5 µm) are also a potential health risk and have to be counted in the analysis of air from all environments, including the workplace,[92,93] as well as in water, soil, and biological material. Another important reason is the basic scientific view, which should not consider any limitations. The most scientific and most sensitive method should be applied to measure the complete fiber array dispersed in the gas phase. Neither PCOM nor SEM can identify these short and thin fibers. The only logical option for analyzing air and other samples is, therefore, TEM.

The accuracy, sensitivity, detection limits, and exact specific identifications are also limited with TEM.[79-87] With optimized collection and fiber identification parameters, the accuracy and precision of the results are limited by specimen preparation and effects of statistical evaluation. The fiber sample must be uniform. The fibers are deposited randomly on the air sample filter, and the Poisson distribution defines the limits of precision. The fiber-counting rules and criteria have to be internationally standardized. It is also necessary to define all parameters of the analytical method. The filter collection efficiency for fibrous aerosols can also influence the results. NPFs with pore sizes 0.4 and 0.2 μm are, therefore, recommended as a standard. The specimen preparation technique must also be set up exactly and standardized.

Errors associated with TEM in chrysotile samples were investigated by interlaboratory tests.[81-86,94] The microscope can be a source of order-of-magnitude errors if it does not meet demanding requirements. By using a verified and standardized counting procedure, most operators work with a 50% error in counting fibers less than 1 μm in length. The accuracy is over 90% for fibers longer than 1 μm. The errors in counting fibers shorter than 1 μm are the reason such short fibers are no longer counted in many standardized procedures. For testing the microscope and the operator, analytical standards have to be used.[82] The detection limit of TEM can be in the region of 1,000 fibers/m^3 for all fibers and in the region of 100 fibers/m^3 for fibers longer than 5 μm.[53] The turnaround time for sample analysis by TEM is longer than in SEM and much longer than for PCOM. It depends on many factors and is in the range of 4 to 12 hours. The use of photographic negatives for counting and sizing fibers in combination with a microfilm reader facilitates the evaluation procedure and decreases the cost.[95]

16.9 COMPARISON

All the analytical methods available for examination of air sample filters with fibrous particles have limitations. As mentioned, PCOM can indicate a fiber dust index without any specific single particle identification. Furthermore, it detects only a small percentage of the fiber concentration that can exist in the measured air volume. Figure 16.29 illustrates the difference in the ability of PCOM and TEM. Often the part measured by PCOM is less than 5% of the total existing fibers. This situation is also well demonstrated in Figure 16.30. The fiber visibility in PCOM is very limited. Figure 16.30a shows a transmission electron micrograph of a single chrysotile fibril, in comparison with the diameter required for a fiber to be visible by PCOM (the circles show the visibility limits of 0.2 to 0.4 μm). Figure 16.30b shows a fiber length distribution of asbestos fibers dispersed in a workplace atmosphere. The shaded area represents the proportion of fibers that are detected when PCOM is used. Only those fibers longer than 5 μm are included in the fiber count, and the optical resolution limits the detectable diameter to about 0.2 to 0.3 μm. The relative standard deviation of PCOM can be as high as 60%. Figure 16.30c illustrates the distribution of fiber counts and how this is related to a fixed control limit.

Conversions between PCOM, SEM, and TEM have been performed on the basis of a contrast ratio. Figure 16.31 shows a comparison of fiber counts made by three methods for fibers longer than 5 μm and with a fiber diameter over 0.2 μm. It can be seen that there is a good correlation, but there is a trend for the PCOM fiber count to be considerably lower by about a factor of two.

The superiority of TEM over SEM has already been discussed. SEM size detection resolution is limited to fiber diameters over 0.1 μm because of contrast and identification problems. TEM produces total fiber counts that are greater than those from SEM, which in turn are greater than the PCOM counts. When the analysis is limited to fibers longer than 5 μm, however, SEM and TEM provide similar results. This similarity is to be expected.[97] There is a correlation between fiber length and diameter, with long fibers being thicker and hence more easily visible. The greater resolution of TEM may be necessary, however, when short fibers are counted. Such a case is shown in Figure 16.32. The TEM results are considerably greater than SEM counts.

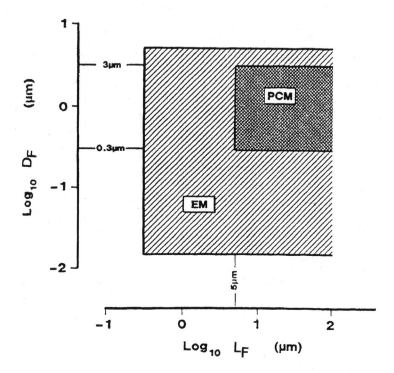

FIGURE 16.29   Relationship between the counting ability of the optical (PCM) and electron (EM) micro-scopical methods. (From Verma, D.K. and N.E. Clark, Relationship between phase contrast microscopy (PCM) and transmission electron microscopy (TEM) results of samples from occupational exposure to airborne chrysotile asbestos, *Am. Ind. Hyg. Assoc. J.* 56:866-873 (1995). With permission.)

Nevertheless, the agreement between TEM and SEM counts for fibers longer than 5 μm depends also on the type of fibers.[98] For example, TEM shows some discrimination in identifying chrysotile asbestos, whereas SEM favors amphiboles (amosite, crocidolite, and anthophyllite). Absolute and universal agreement, therefore, cannot be expected. For practical use, when indices of exposure to well known mineral fibers longer than 5 μm are evaluated, TEM and SEM may be comparable; by excluding SEM in a direct sample evaluation procedure.

16.10   DIRECT AND INDIRECT METHODS

Both SEM and TEM can evaluate specimens prepared by direct or indirect transfer methods.[79-87] Direct transfer (DT) was preferred in most published and standardized procedures. In the indirect transfer (IDT) procedures, cellulose ester MFs are used for fiber sampling. A potential advantage of the IDT methods is that during the air sampling period the filter can be overloaded and thus the detection limit can be significantly improved, mainly by an SEM evaluation procedure. Air samples collected on MFs are completely ashed, removing the filter material and any organic matter collected. The ashed material is then dispersed in very clean, fiber-free distilled water (pH ca. 4) and ultrasonically treated (for a short time with low ultrasonic energy). The suspension is imme-diately filtered by using an NPF (pore size 0.2 μm). The following specimen preparation steps for SEM and TEM are the same as in the DT procedures.

Many investigations and observations have shown differences in comparing DT and IDT methods. The differences were slight when only fibers longer than 5 μm were evaluated. However, the IDT methods overestimate the concentration of short fibers.[50-54] In measurements of workplaces, it could be shown that the effect of ultrasonic treatment by the IDT procedure is highly variable

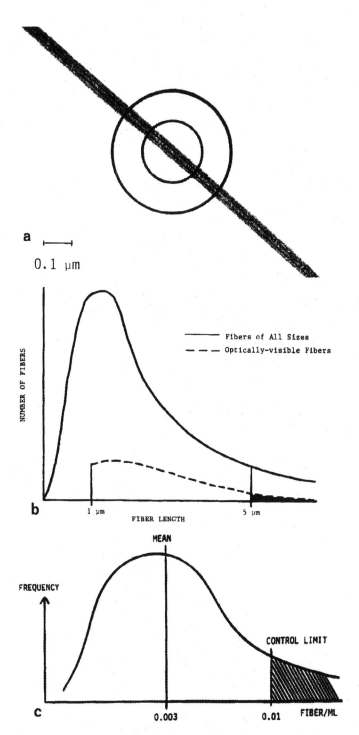

FIGURE 16.30 Limitations of the PCOM procedure: Transmission electron micrograph of a single fibril of chrysotile in comparison with the PCOM detection limits (circles) (a) From Chatfield, E.J. Fiber definition in occupational and environmental asbestos measurements, *Am. Soc. Testing Materials,* ASTM, 834:118-138(1984). With permission.); fiber length distribution of typical airborne asbestos dispersion (b) for fibers of all sizes and of optically visible fibers, and relationship between frequency distribution and control limit (c) (From Chatfield, E.J. Limits and detection and precision in monitoring asbestos fibers. Asbestos and Health Risk, *Air Poll. Control Assoc. J. APCA* (1987) pp. 79-90. With permission.)

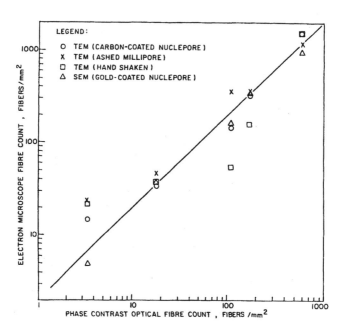

FIGURE 16.31 Comparison of PCOM fiber counts with fiber counts of optically visible fibers from TEM and SEM measurements. (From Chatfield, E.J. Limits and detection and precision in monitoring asbestos fibers. Asbestos and Health Risk, *Air Poll. Control Assoc. J. APCA* (1987) pp. 79-90. With permission.)

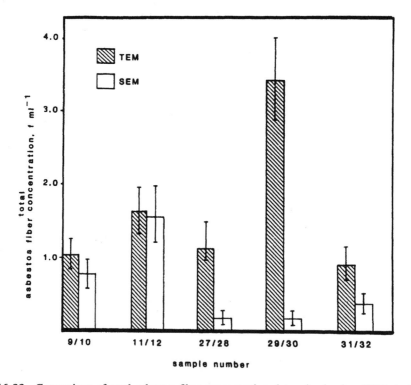

FIGURE 16.32 Comparison of total asbestos fiber concentration determined using SEM at ×10,000 and TEM at ×17,000 magnification. (From Breysse, P.N., J.W. Cherrie, J. Addison and J. Dodgson, Evolution of airborne asbestos concentration using TEM and SEM during residential water tank removal, *Ann. Occup. Hyg.* 33:243-256 (1989). With permission.)

and depends on the characteristics of the workplace where the samples were collected. When the ultrasonic treatment was excluded, no difference in fiber counts for fibers longer than 5 μm was observed between DT and IDT procedures.[65]

The ultrasonification effect was also observed when the IDT procedure was used in workplaces with man-made mineral fibers (e.g., rockwool and tungsten oxide).[99]

16.11   COUNT-TO-MASS CONVERSION

In an attempt to compare the mass concentrations that control the limits specified in terms of fiber number concentrations, some workers have assumed particular fiber sizes in order to convert these mass concentrations into numerical fiber counts and vice versa. Chrysotile fiber can be thought of as a cylinder, and the fiber mass (m_F) can then be calculated (the material density s = 2.55 g/cm^3):

$$m_F \text{ (ng)} = L_F \cdot D_F^2 \cdot s \cdot 10^{-3}$$

For the sample mass M_F we can obtain[57]:

$$M_F (\text{ng}) = 4 \cdot 10^{-3} \sum_{i=1}^{i=n} D_{Fi}^2 \cdot L_{Fi}$$

Nevertheless, experimental investigations have shown that the real sample mass can be two to three times higher than the "counted mass" M_F.[80] The conversion is of practical importance, for example, when stack emission concentrations have to be controlled. PCOM, SEM, and TEM can be applied for the measurements of fibrous emissions.[98,100] As the emission limits are prescribed as mass concentration (e.g., 0.1 mg/m^3), empirical conversion factors are recommended in different directives. For PCOM, the CEC (Council of the European Communities) recommends a conversion factor of 2 fibers/ml, corresponding to 0.1 mg/m^3 of asbestos powder (= 2×10^4 fibers/μg). Practical measurements in different factories have shown some variation between 0.4×10^3 and $4 \times 9 \cdot 10^3$ fibers/μg.[100]

16.12   MASS SPECTROMETRIC METHODS

The microscopical methods (PCOM, SEM, TEM, EDXA, SAED) used for routine identification of fibrous particles belong, in principal, among the group of modern physical methods for chemical analysis of individual particles.[1,2] Therefore, they do exist with other methods and procedures which can be applied for chemical identification of individual collected fibrous particles. These methods are usually not suitable for routine measurement of fibrous dusts and aerosols in environmental situations. Nevertheless, they can provide very important and useful complementary data and information about fibrous mineral microparticles for research purposes.

Such modern analytical methods, which have already been applied in the field of physical, chemical, and biological studies of fibrous mineral particles, include the laser microprobe mass spectrometry (LAMMS), secondary-ion mass spectrometry (SIMS), Raman microprobe spectrometry, infrared microscopy, and electron spectroscopy for chemical analysis (ESCA).[101-105]

LAMMS and SIMS have found application in single fiber analysis for scientific studies.[104,105] LAMMS is a combination of a laser beam with a time-of-flight (TOF) mass spectrometer. It utilizes a focused and pulsed laser beam to generate ions by laser ablation/ionization. Specimens for the LAMMS analysis are prepared in a way similar to that for SEM and TEM. In the LAMMS equipment, the specimen with fibrous particles is observed by means of an optical microscope or scanning electron microscope. The individual particles are chosen and then irradiated very briefly with the high-power pulsed laser (Nd:YAG). The absorption of the laser beam by a single fibrous particle produces heat for a short duration. The particle material is volatilized and expands into the vacuum. All ions generated are accelerated and mass separated by a TOF mass spectrometer.

The spectra obtained are characteristic for the chemical composition of such single microscopic fibers. Such spectra provide useful information about particle composition, contamination, corrosion, leaching, etc. Such mass spectra of a single fiber can also serve as useful fingerprints for different special fiber types (man-made mineral fibers, natural mineral fibers, such as zeolites, etc.).[104,105] Figure 16.33 shows examples of mass spectra for positive ions of a chrysotile (CH), wollastonite (W), and glass (G) fibers. The fingerprint differences are evident.

Secondary-ion mass spectrometry (SIMS) is based on ion beam sputtering. The specimen with fibrous particles is irradiated by the primary ion beam (O, Ar, Cs, Ga). The produced secondary ions are separated on the basis of their mass. Seyama et al.[105] have used SIMS for the analysis of asbestos fibers. It was possible to identify these fibers by the pattern of the positive secondary ion mass spectra. The image of asbestos fibers was measurable by using any major positive secondary ions. The smallest size of detectable asbestos fibers was about 5 μm and 1 μm in length and diameter, respectively. An example of such an analysis is shown in Figure 16.34.

16.13   IDENTIFICATION BY FLUORESCENT DYES

Asbestos and other mineral fibers can be stained with fluorescent dyes to enhance their visibility in a light microscope. Benarie tried to identify asbestos fibers by fluorochrome staining.[106] Awadalla et al. have successfully dyed chrysotile asbestos with triphenylmethane, which chelates with the $Mg(OH)_2$. The reaction was fast with basic fuchsin, malachite green, and methyl blue.[107]

The fluorescence technique may be used in some cases as a rapid screening, but it is not sufficiently specific for asbestos to provide unequivocal identification of fiber type.

16.14   ASBESTOS BULK ANALYSIS

There are cases in which the mass concentration of asbestos has to be determined directly. This can be the case in analysis of sampled or sedimented dusts in industry and mines, in materials, etc. Two analytical methods — the X-ray diffraction analysis and infrared spectroscopy — are preferentially used for the bulk analysis of asbestos-containing dusts and powders.

16.14.1   X-ray Diffraction Analysis

X-ray diffraction (XRD) techniques have been applied extensively to the qualitative and quantitative analysis of industrial dusts since about the 1960s.[108-117] The asbestos minerals each possess distinct crystalline structures, based upon the ordered arrangement of atoms, and are, therefore, amenable to analysis by XRD. X-ray powder diffraction data for the asbestos minerals show diffraction peaks, which make it possible to differentiate among almost all asbestos types (Figure 16.35).[115] The XRD is, therefore, capable of distinguishing among chrysotile, amosite, and crocidolite and is well suited to the quantitative determination of asbestos in mixtures.

Although the absolute limit of detection of asbestos is of the order of 10 μg using the direct on-filter (silver membrane filter) analysis of dust and an external standard, the limiting factor for the detection of asbestos in bulk samples is the lowest detectable proportion of asbestos in a mixture. For reliable quantitative determinations, the samples should contain more than 10% of asbestos.[112] Nevertheless, the matrix composition plays an important role. It could be shown, for example, that in a gypsum-based bulk material chrysotile concentrations in the range of 1% were also well measurable.[116]

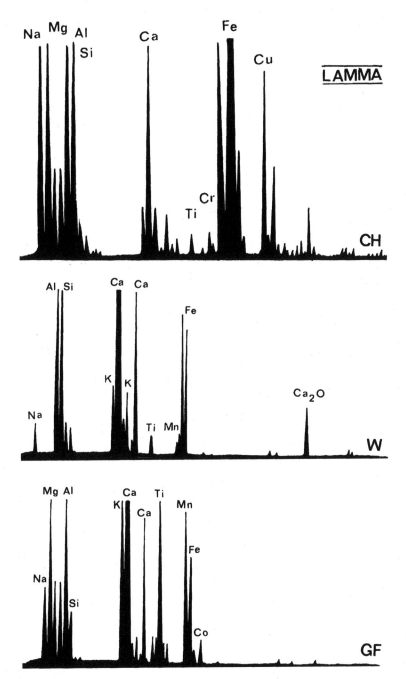

FIGURE 16.33 Example of mass spectrometric spectra (LAMMA) for positive ions: analysis of single fibers: chrysotile (CH), wollastonite (W), and glass fibers (GF).

16.14.1.1 Chrysotile Contamination

Chrysotile asbestos may be contaminated at the source by amphibole asbestos (e.g., tremolite, anthophyllite, vermiculite), which, in the event of long-exposure, might present a more serious health hazard to asbestos workers than would exposure to pure chrysotile. For this reason, the content of amphibole contaminants should be determined in chrysotile sampled in workplace air and in the raw chrysotile material. Routine analytical procedures, such as XRD, infrared spectrometry, and

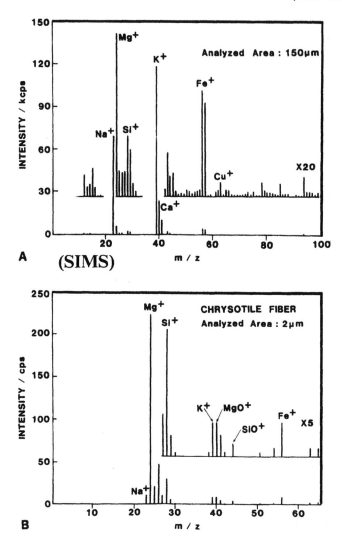

FIGURE 16.34 Positive secondary ion mass spectra of (A) chrysotile as bulk sample and (B) of single chrysotile fibers. (From Seyama, H., Y. Soma and M. Soma, et al. Application of secondary ion mass spectrometry for the analysis of asbestos fibres, *Fresenius Zschr. Anal. Chem.* 341:619-624 (1991). With permission.)

microscopic methods are insufficiently sensitive to determine amounts of amphibole much less than 1%. However, a method of chemical digestion of chrysotile samples has been tested and found successful for amphibole detection in the range of 0.01 to 0.5%. In the first step, the chrysotile sample is preheated to 600°C. The digestion, which affects the microstructural destruction of chrysotile, is carried out with dilute sulfuric acid. The amphiboles are resistant to this procedure and can then be determined by means of the routine XRD method.[118]

16.14.2 INFRARED SPECTROMETRY

The applicability of infrared (IR) spectrometry to the identification of minerals has been known since the 1950s. Parks[4,9] applied IR spectrometry to asbestos in 1971. He noted that the distinction between chrysotile and the amphibole group of asbestos minerals was much sharper than that between the individual amphiboles.

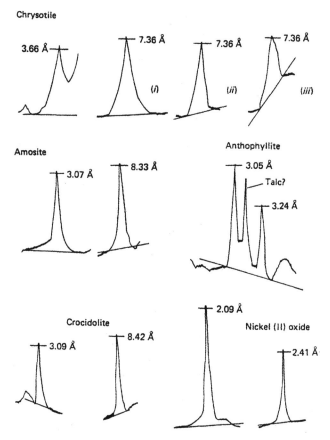

FIGURE 16.35   X-ray diffraction peaks for different types of asbestos and of nickel oxide. (From Puleda, S. and A. Marconi, Quantitative X-ray diffraction analysis of four types of amphibolic asbestos by the silver membrane filter method, *Int. J. Environ. Anal. Chem.* 36:209-220 (1989). With permission.)

A commonly used method of presenting the sample to the IR spectrometer is in the form of a pressed potassium bromide disc. The asbestos minerals are hydroxylated silicates and show two main absorption regions, one associated with stretching vibrations of Si–O groups in the structure at wave numbers of ca. 850 to 1150 cm^{-1}, and the other associated with stretching vibrations of OH groups — 3500 to 3700 cm^{-1}, In addition, the amphiboles show absorption bands at wave numbers less than 800 cm^{-1} (Figure 16.36). The method has the advantage of being a relatively rapid form of analysis and requires only a small amount of sample.[119-122]

16.15   EXPOSURE TESTS

Biological monitoring of mineral fibers is an alternative method for evaluating exposure or for detecting any early effects. Several biological monitoring tests of asbestos have been studied, such as for asbestos fiber concentrations in urine and asbestos bodies and fibers in sputum and in lung lavage fluids, etc.

16.15.1   URINE

TEM is the only method sufficiently sensitive and specific for the measurement and identification of fibers of all sizes in biological materials (e.g., in lung tissue) and fluids. Inhaled or ingested fibers appear to be able to pass through the broncheolar epithelium or gut membrane and then to migrate to various sites in the body.

Infrared spectroscopy

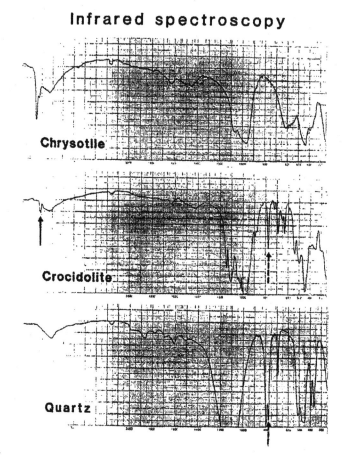

FIGURE 16.36 Infrared spectra for the differentiation among chrysotile, crocidolite, and quartz. (From Taylor, D.G., C.M. Nanadic and J.W. Crable, Infrared spectra for mineral identification, *Am. Ind. Hyg. Assoc. J.* 29:100-108 (1970). With permission.)

Urine samples from asbestos workers are filtered using an MF (pore size 0.45 μm). The ash of this sample is then dispersed in 1% acetic acid and the suspension is filtered again with an NPF (pore size 0.1 μm). The previously described TEM procedure is used for fiber counting, sizing, and identification (Figure 16.37). The detection limit is in the region of 100 fibers/ml. Concentrations found in urine samples of exposed asbestos workers are much higher than this level.[123]

16.15.2 Bronchoalveolar Lavage

Fibrous particles deposited in human lung tissue (Figure 16.38) as well as the secondarily produced asbestos bodies can be sampled in asbestos workers by means of a bronchoalveolar lavage. Asbestos bodies are inhaled asbestos fibers that have been coated with ferroprotein by macrophages. Bronchoalveolar lavage (BAL) is a simple and noninvasive technique that allows the recovery of large numbers of alveolar macrophages. BAL is primarily performed by bronchofibroscopy, washing a lung segment three times with 50 ml aliquots of normal saline solution. An average of 10 to 30 ml of the recovered fluid from the second and/or third sample are used for the measuring asbestos fibers by TEM and for counting and determining asbestos bodies by PCOM or SEM and TEM. Sodium hypochloride is often added to this liquid and the suspension is then filtered through an MF (pore size 0.45 μm). When the filter is dry, PCOM is used for counting and sizing fibers and asbestos bodies. For measurements by SEM and TEM, the usual specimens have to be prepared and/or NPF (pore size 0.2 μm) should be used (Figure 16.39). Concentrations of asbestos bodies

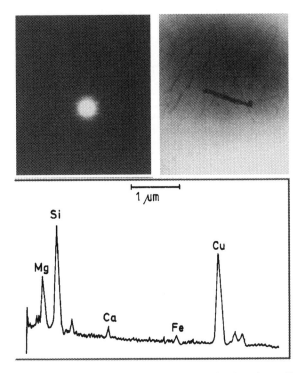

FIGURE 16.37 TEM-micrograph, EDXA, and SAED-analysis of a fine chrysotile fiber sampled in urine.

found in BAL fluid of asbestos exposed persons can be in a broad range between 1 to 1,000 asbestos bodies/ml. Good correlations can be found between exposure and the concentration of fibers and asbestos bodies.[124,125]

16.16 FIBER DEFINITIONS AND PHYSICOCHEMICAL CARCINOGENIC CHARACTERIZATION

16.16.1 CARCINOGENIC FIBER

For health protection, not only asbestos but all carcinogenic fibers should be measured and regulated. Therefore, there is a need to define a carcinogenic fiber and all the criteria of its carcinogenic potency.[126,127] Relationships exist between the physical (D_F, L_F, aspect ration L_F/D_F), chemical (durability, surface chemistry, surface electric charge), and biochemical (peroxidation) fiber properties and the expected tumor incidence.[128] There is a continuous transition from the noncarcinogenicity of fibers that are too short, too thick, or too soluble to have a possible carcinogenic potency, to fibers that have the optimum length and diameter and that are also very persistent.

The fiber size dependency is well defined, for example, in the *Pott hypothesis* shown in Figure 16.40).[127] The carcinogenicity factor C_f increases with increasing fiber length and decreases with the fiber diameter.

The minimum time that a fiber has to persist in the tissue to alter cells is still not clearly known. The time between the start of work in an asbestos or mineral fiber industry and diagnosis of asbestos — or mineral fiber-induced bronchial carcinoma is at least 5 to 10 years, and for mesothelioma, it is more than 10 years. A threshold of fiber persistence that has to be exceeded for induction of altered cells (Figure 16.41) is, therefore difficult to estimate. It can last months[129] or years.[130] At first approximation, a carcinogenic fiber should be longer than 1 µm and chemically persistent for more than 3 years. The traditional and conservative limit for the aspect ratio is >3:1. The latest publications and workshop discussions consider this no longer correct or acceptable.[92,131] For

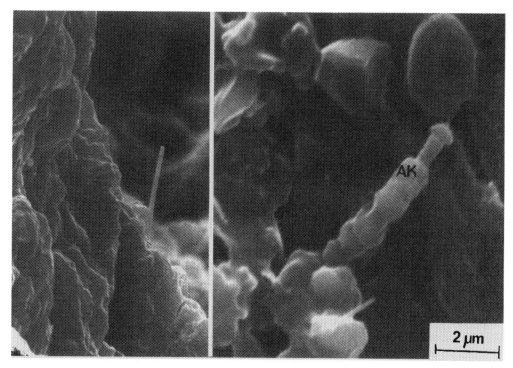

FIGURE 16.38 Electron micrographs (SEM) of the dried human lung tissue with incorporated asbestos fibers and asbestos body (AK).

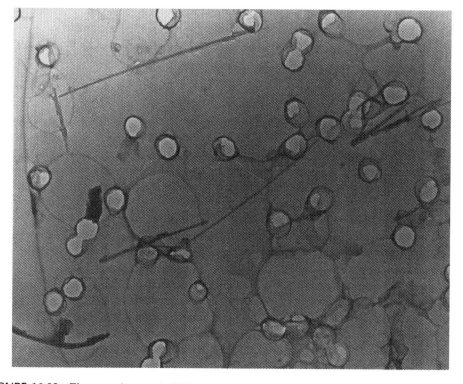

FIGURE 16.39 Electron micrograph (TEM) of very thin chrysotile fibers sampled by BAL procedure.

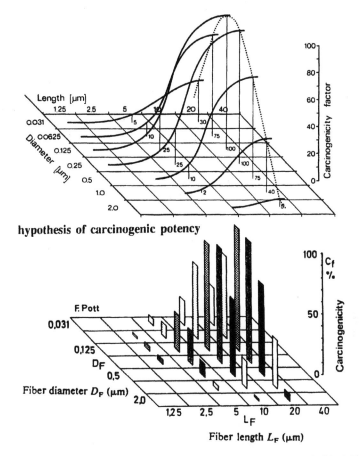

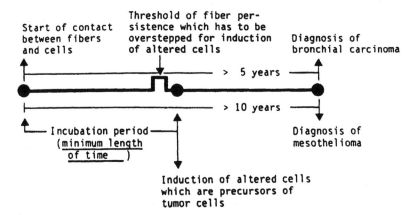

FIGURE 16.40   Chart depicting the hypothetical carcinogenic potency (C_f) of cylindrical fibers as a function of their lengths and diameters. (From Pott, F. Some aspects of the dosimetry of carcinogenic potency of asbestos fibers and other fibrous dusts, *Staub-Reinhalt Luft* 38:486-490 (1978). With permission.)

FIGURE 16.41   Schematic illustration of the exposure mechanism for respirable mineral fibers deposited in the lung tissue. (From Pott, F. The fiber as carcinogenic agent, *Zbl. Hyg.* B184:1-23 (1987). With permission.)

example, all asbestos structures have an aspect ratio above 5:1. Measurements have shown that asbestos fibers with aspect ratios of 3:1 are rarely found in the workplace, outdoors, or in indoor environment airborne samples.[131] In Figure 16.42, the results of such measurements are shown. The asbestos fibers have a median aspect ratio above 10:1 in workplace situations. In environmental

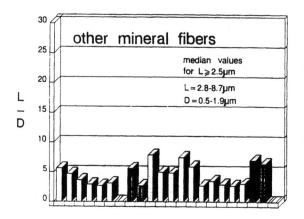

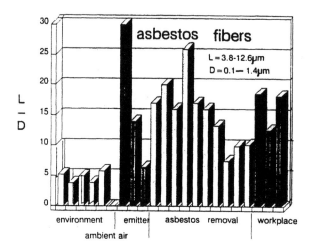

FIGURE 16.42   Results of 471 measurements: median length (L) compared with median diameters (D) of asbestos and other mineral fibers with L over 2.5 μm. (From Rödelsperger, K., U. Teichert, H. Marfels, et al. Measurement of inorganic fibrous particulates in ambient air and indoors with scanning electron microscope, in J. Bignon, J. Peto and R. Saracci (Eds.) *Non-Occupational Exposure to Mineral Fibres,* IARC Publ. Lyon, Vol. 90:361-366 (1989). With permission.)

situations, these ratios are above 6:1. On the other hand, non-asbestiform mineral fibers have aspect ratios in many cases of less than 5:1. Mineralogists are very dissatisfied with the existing asbestos fiber definitions used by hygienists and toxicologists.[132] They also agree that a shift to higher aspect ratios would radically decrease false-positive results without a decrease in sensitivity for true asbestos. The historical and conservative definition still used in occupational fiber counts clearly specifies limits on the fiber dimensions: mineral fibers longer than 5 μm with an aspect ratio above 3:1 and thinner than 3 μm. Such fibers are also designated as the "WHO (World Health Organization) Fibers."

Data published during the last few years suggest that both short (less than 5 μm) and long, thin, and chemically persistent mineral fibers are biologically active.[93] These studies have also borne out the observation that fibers less than 5 μm in length predominate in the lung tissue of asbestos-exposed workers. Short fibers may also constitute a majority of airborne dusts and aerosols. If it is prudent to assume that the short fibers are a potential health risk (also because their relative concentrations are high), then they must be counted during the analysis of air in all workplaces and environmental situations. For this reason, the only logical option for analyzing air samples properly is TEM.[47]

Fibers that have to be counted should be longer than 1 μm, thinner than 1 μm, and with an aspect ratio above 5:1. Of course, while these proposals are supported by many investigators, they have little support from the medical legislative community. From this point of view, other measuring procedures, PCOM, and SEM are still of practical importance. Nevertheless, they can be considered only as simple or good (SEM) screening tests. Legislation in the United States considers TEM as the only method of choice for indoor and outdoor measurements. Germany standardized and officially approved the SEM procedure.[49] In the majority of highly developed industrialized countries, TEM is the method of choice. In any case, the existing definition, which includes all fibers with aspect ratios at least 3:1, and longer than 5 μm, i.e., those seen by PCOM, is considerably behind the scientific knowledge and state-of-art in this field. It is very probable that the index of exposure obtained by PCOM may not represent the true exposure when the fiber size distributions change. PCOM is also much more affected by subjective effects than SEM and TEM. The results of different operators can vary considerably.

Considering the Pott fiber carcinogenicity model (Figure 16.40), it is clear that SEM and, mainly, TEM are able to determine almost completely the real fiber concentrations and their carcinogenic potency.

16.16.2 Physicochemical Characterization of Carcinogenic Fibers

The most important physical and chemical characterization parameters, which can be correlated to the carcinogenic potential of mineral fibrous aerosols are:

1. The measured or expected ambient air concentrations.
2. The fibrous particle size distributions. They are bivariate, because a fibrous particle is defined by its diameter (D_F) and length (L_F). Mathematically, such distributions can be described by a log-normal distribution function (Figure 16.43).[133,134]
3. The fiber durability, which can be defined and measured as the chemical solubility.
4. Fiber surface characterization: Zeta potential (surface electric charge) and the ability of catalytic production of reactive oxygen species (ROS) and metabolites (ROM).

16.16.2.1 Physicochemical Carcinogenic Factors

Using the Pott size-hypothesis-model, an integrated carcinogenic value can be obtained for a particular size distribution by weighing each fiber value C_f (Figure 16.44). The integral obtained approaches a fixed value, which represents the carcinogenic potency of the whole sample.[92]

Pott's hypothetical model of carcinogenic potency can be further combined with the Timbrell model, which accounts for the long-term retention of fibers in the lung.[135] An example of such a bivariate size distribution is shown in Figure 16.45. In our opinion, both of these graphical evaluation methods demonstrate an alternative and scientifically based index or definition of potential carcinogenicity for airborne fibrous dusts and aerosols.

16.16.2.1.1 Fiber Durability

For over 20 years, the biological activity of fibers has been linked to their dose, dimension, and durability. Fiber composition determines the chemical solubility, and consequently, the durability of fibers within the lung.[130]

Several methods have already been described and proved in the literature.[136] Simulated intra- and extracellular fluids are used as the liquid medium. The methods are dynamic ones. The 37°C fluid circulates with a flow rate of about 40 to 60 ml/day a flow cell (Figure 16.46). The tested fibers are sandwiched (B), mainly randomly oriented, between two perforated discs.[137] The duration of the entire test is about 60 to 120 days. At the end, the collected fluid is analyzed for silicon, and for boron and potassium. Si and B are determined by the IPC (inductively coupled plasma) method. The dissolution rates are measured in ng/cm².h, or as a half-life time, or often as time,

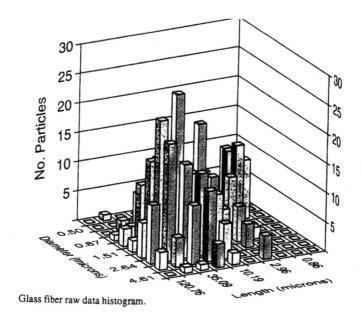

Glass fiber raw data histogram.

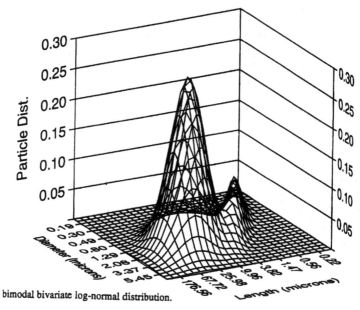

bimodal bivariate log-normal distribution.

FIGURE 16.43 Bimodal bivariate log-normal distribution of an aerosol containing fibrous and isometric particles. (From Moss, O.R., B.A. Wong, and B. Asgharian, Bimodal bivariate lognormal distributions in the application of inhalation toxicology, specific to the measurement of fiber and particle dosimetry, in U. Mohr (Ed.) *Toxic and Carcinogenic Effects of Solid Particles in the Respiratory Tract.* ILSI Press, Washington, D.C., pp. 623-628 (1994). With permission.)

which is necessary for a complete dissolution of a fiber with a diameter of 1 μm. For example, the dissolution rates obtained for different types of glass fibers varied from 1 to 50,000 ng/cm².h.

16.16.2.1.2   ROS and ROM

Interaction of fibers with lung macrophages leads to the release of many substances. Among them, reactive oxygen species and metabolites which include hydrogen peroxide, superoxide, and

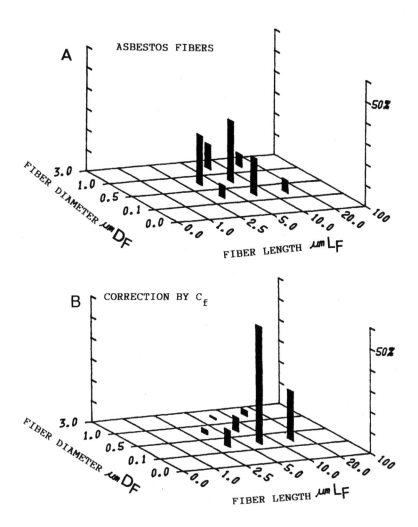

FIGURE 16.44 Diagrams illustrating the measured bivariate fiber size distributions obtained by SEM for all asbestos fibers (A) and for asbestos fibers with correlation to the fiber carcinogenic potency (B).

hydroxyl radicals are proposed to cause cellular and DNA damage.[138,139] There are well-developed methods for measuring the ROS and ROM release rate in cell suspensions *in vitro* as well as in individual cells.

The presence of transition metals, e.g., Fe, catalyze the production of highly reactive hydroxyl radicals (OH·) following the well-known Fenton reaction.[139-141] By using a simple leaching test the Fe-release, which correlates with the ability to form OH·, can be measured. A fiber probe is extracted by chelates and the leached amount of Fe is measured as a function of time (Figure 16.47).[142]

By means of ERS (electron spin resonance) spectrometry the production of OH· in suspensions of cells, liposomes, etc., can be determined.[143] As can be seen from Figure 16.48, the ability to produce OH· depends strongly on the fiber type.

The ability of different fiber types to produce ROMs, can be detected quantitatively also by means of reactions within individual cells.[144] Cell suspensions (e.g., lung macrophages) can be exposed to the tested fibers. Fibers incorporated in individual cells by phagocytosis initiate the H_2O_2 production. The released hydrogen peroxide is measured by means of a fluorescein dye (Figure 16.49) by using a laser cytometer (Figure 16.50).

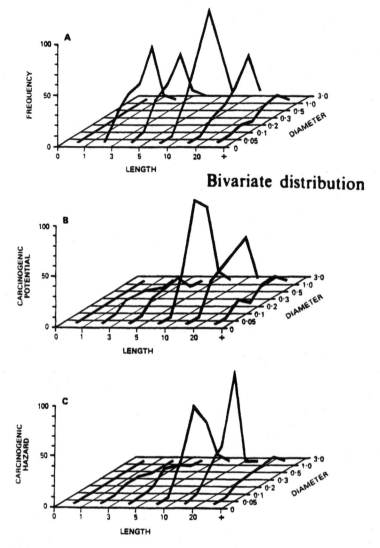

Bivariate distribution

FIGURE 16.45 Relative size distribution of (A) the air fiber sample (amosite), with (B) its carcinogenic potential (size $\times$ C$_f$), and (C) carcinogenic hazard (size $\times$ C$_f$ $\times$ "Timbrell"). (From Burdett, G.J., J.G. Firth, A.P. Rood, and P.R. Streeter, Application of fibre retention and carcinogenicity curves to fibre size distributions of asbestos, *Ann. Occup. Hyg.* 32:Suppl.1, 341-351 (1989). With permission.)

16.16.2.1.3   Results Interpretation

Having already characterized the fibrous powder or aerosol by fiber size distribution and solubility, the expected physicochemical carcinogenicity factor (designated as PhChC$_f$) can be obtained as a product of two partial parameters — PhC$_f$ (physical carcinogenicity factor) and ChC$_f$ (chemical carcinogenicity factor). It means that

$$\text{PhChC}_f = \text{PhC}_f \times \text{ChC}_f$$

The physical carcinogenic factor PhC$_f$ is a consequence of the fiber size distribution and is obtained by using the microscopically measured parameters (D$_F$ and L$_F$) and the fiber size hypothesis (the alternative form is presented in Figure 16.51a). It can be seen that TEM is superior in such evaluations to SEM, and SEM is superior to PCOM.

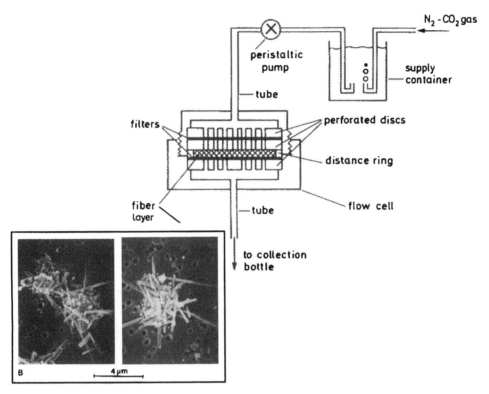

FIGURE 16.46 Schematic picture of the cell method for measuring chemical solubility of mineral fiber "sandwiches" (B). (From Scholze, H. and R. Conradt, An *in vitro* study of the chemical durability of siliceous fibres, *Ann. Occup. Hyg.* 31:683-692 (1987). With permission.)

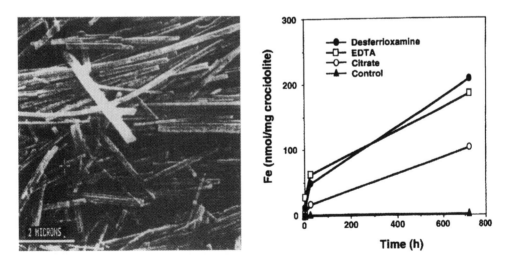

FIGURE 16.47 Leaching test for samples of mineral fibers: SEM-picture and leaching curves. (From Shen, Z.H., V.D. Parcler, and A.E. Aust, Mediated, thin layer cell; determination of redox-active iron the surface of asbestos fibers, *Anal. Chem.* 67:307-311 (1995). With permission.)

The chemical carcinogenic factor ChC_f is connected with the fiber solubility. Its dependency on the life time t is shown in Figure 16.51b. The parameter t represents the time which is necessary for the dissolution of a fiber with $D_F = 1$ μm.

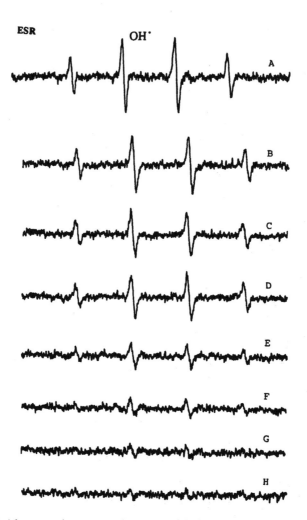

FIGURE 16.48 ESR (electron spin resonance) spectra of hydoxyl radicals in the presence of H_2O_2 and different mineral fibers: crocidolite (A,C), asmosite (B), tremolite (D), anthophyllite (E), chrysotile (F). Control for TiO_2 (G) and H_2O (H). (From Gulumian, M. and J.A. Van Wyk, Oxygen consumption, lipid peroxydation and mineral fibres, in R.C. Brown, J.A. Hoskins and N.F. Johnson (Eds.) *Mechanisms in Fibre Carcinogenesis,* Plenum Press, New York, NATO ASI Series A 223:439-446 (1991). With permission.)

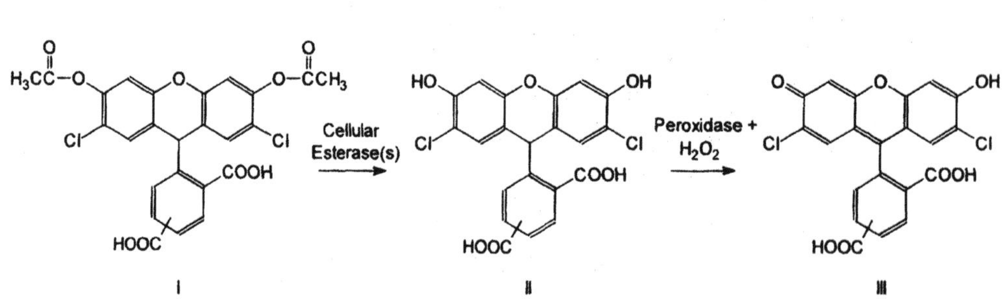

Forms of the fluorescin/fluorescein dye relevant to measuring the oxidative burst.

FIGURE 16.49 Reaction scheme for the fluorescence identification of the interaction of erionite fibers with individual cells. (From Hogg, B.D., P.K. Dutta, and J.F. Long, *In vitro* interaction of zeolite fibers with individual cells: Measurement of intracellular oxidative burst, *Anal. Chem.* 68:2309-2312 (1996).

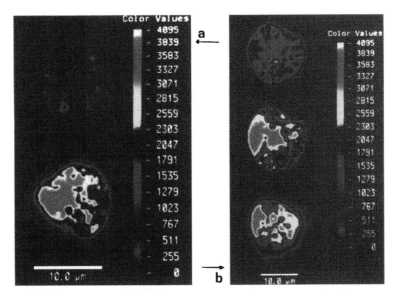

FIGURE 16.50 Fluorescence image of macrophage (a): top, control cell; bottom, cell exposed 30 min to erionite fibers. Fluorescence images of fiber-exposed cell taken along thin cross section along a vertical axis after 10 min of exposure (b). (From Hogg, B.D., P.K. Dutta, and J.F. Long, *In vitro* interaction of zeolite fibers with individual cells: Measurement of intracellular oxidative burst, *Anal. Chem.* 68:2309-2312 (1996). With permission.)

Mineral fibers with similar $PhChC_f$ and ROS and ROM production ability should be considered to have similar hazard potential. Such physicochemical characterization of fibrous mineral aerosols can substantially simplify the needed biological carcinogenicity testing methods.[136]

16.17   REAL-TIME MONITORING OF FIBROUS AEROSOLS

The basic philosophy, which enables the construction of such monitors, is based on the fact that fibrous particles may undergo alignment in laminar flows, as well as in magnetic and electric fields, and that the light scattering patterns for fibrous particles differ from those of spherical or isometric particles. This philosophy was explored at the beginning of the 1970s by Timbrell. He developed a technique for preparing permanently aligned fibers in magnetic field on glass slides. The samples were observed and evaluated by PCOM, and the light scattering pattern was measured. A relatively good differentiation among several fiber types was possible (Figure 16.52).[19,145-147]

One of the first "fiber monitors" was developed in Germany in the mid 1970s. Unfortunately, the equipment did not find commercial production. Fibrous particles were aligned by means of a laminar flow in the tubular nozzle. Intensities of the light scattered on single fibers were measured by two photomultipliers in a parallel (I_1) and in a perpendicular (I_2) direction to the air flow. By determining the ratio (I_2/I_1), it could be distinguished between fibers (ratio > 1) and isometric particles (ratio = 1).[148]

At the end of the '70s, Lilienfield et al.[149-151] described an instrument that monitored fiber concentrations in sampled air by laser light scattering from fibers oscillating in-phase with an electric field; the instrument has subsequently been employed for workplace measurements in the U.S., Japan, and elsewhere.[23] These fibrous monitors (FAM 1 and FAM 7400) are commercially available (MIE, Billerica, MA, USA). Figure 16.53 depicts the principle and the essential elements of the optical sensing configuration of both instruments.[151]

The operating principle of these instruments is based on a two-step concurrent sensing method. The first step consists of the alignment of fibers by passing them through a high-intensity electric

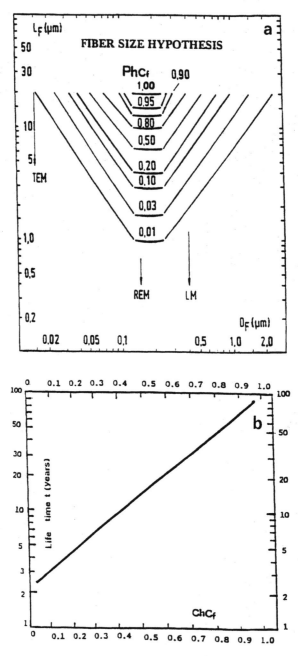

FIGURE 16.51 Diagrams for estimating the physical (PhC$_f$) and chemical (ChC$_f$) carcinogenic factors of tested mineral fibers.

field that also imparts an oscillatory motion to the fibers. The second step is the illumination of these fibers by a helium–neon laser and detection of the scattered light that is synchronous with the applied electric field. Each fiber is detected during its passage through the field oscillation/sensitivity region of the instrument by detecting the sequence of the light pulses produced by the electrically induced rocking motion of the fiber. The detection limit is approximately 0.2 fibers with diameter over 0.3 µm and with a length over 5 µm in 1 ml of air.

Another asbestos fiber analysis system, the M-88 Fiber Analyzer (Vickers Instruments, York, England) was developed in 1983 by Verril.[19] It works on the principle of magnetic field alignment/light

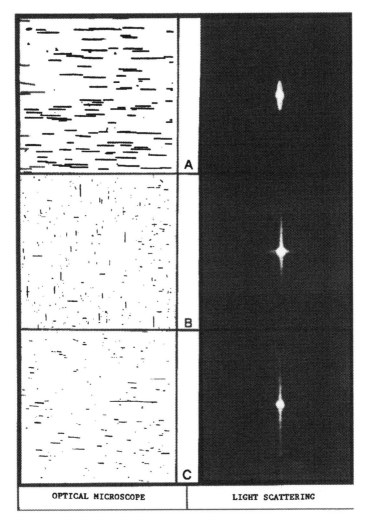

FIGURE 16.52 Phase contrast microscopy and light scattering of aligned fibrous samples: carbon fibers (A), amosite fibers (B), and crocidolite fibers (C). (From Timbrell, V. Desired characteristics of fibres for biological experiments, in P.V. Pelnar (Ed.) *Fibres for Biological Experiments,* Inst. of Occup. Hyg. and Environ. Health, Montreal (1973), pp. 89-122. With permission.)

scattering detection of asbestos fibers on filter samples. The next portable fiber monitor was described in 1992 by Rood et al.[152] Three separate physical techniques were combined. Air containing the fibers to be monitored is drawn into the instrument through a rectangular, slotted inlet, using a battery-operated pump. The flow acceleration at this inlet aligns the fibers. Corona charging then enables the fibers to be precipitated onto a standard microscope slide while retaining their alignment. The difference in light scattering along and at right angles to this deposit is used to determine the concentration of fibers. The components of this monitor are shown in Figure 16.54E, and its calibration curves are shown in Figure 16.54M. This monitor is also commercially available (Turnkey Instruments, London, UK).[153] The applicability and limitations of both monitors will be discussed later.

A new instrument, termed the aerosol shape analyzer (ASA), in which a rapid analysis of the transient spatial intensity distribution of laser light scattered by individual aerosol particles drawn from an ambient atmospheric environment, was developed and later commercially produced in the U.K. during the early 1990s.[154-159] The function of the ASA is illustrated in Figure 16.55. The instrument comprises a laser scattering chamber together with data capture, processing, and storage electronics, all within a single case, and a color monitor with integral touch screen control.

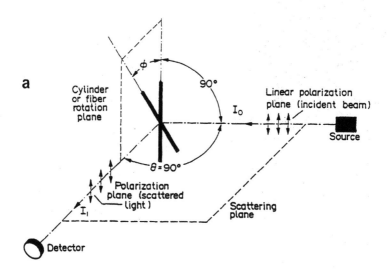

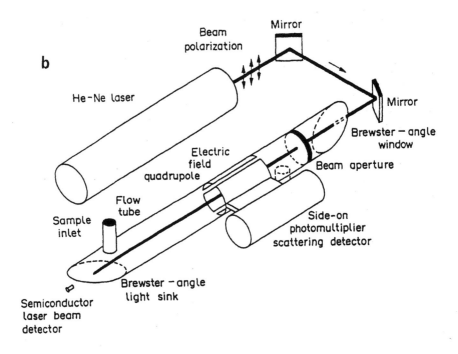

FIGURE 16.53 FAM-Monitor: Schematic picture of the measuring principle (a) and the optical configuration (b). (From FM-7400, Real time laser fiber monitor, Mie ATRC Comp. Billerica, MA, USA (1991). With permission.)

Light from a randomly polarized He–Ne laser enters the scattering chamber orthogonally to the plane of the figure and is directed via a 45° mirror onto an aerodynamically focused aerosol stream. Individual particles passing through the laser beam scatter light in all directions with an intensity distribution dependent on the particle shape and size. Forward scattered light at angles between 8° and 27° is focused via a lens onto a single photomultiplier tube. Light scattered at angles between 27° and 140° is incident on an ellipsoidal mirror which refocuses the light via a spatial filter to a collimating lens assembly. This light is incident upon three miniature photomultiplier tubes (E_1, E_2, E_3). Each

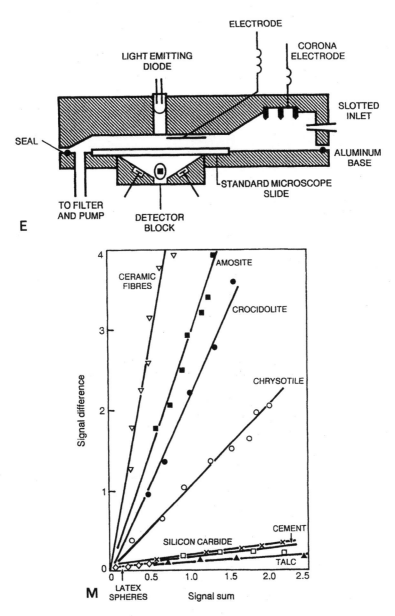

FIGURE 16.54 Portable Fiber Monitor: Section through top block (E) and calibration curves for different fibrous dusts (M). (From Rood, A.P., E.J. Walker and D. Moore, Construction of a portable fiber monitor measuring differential light scattering for aligned fibers, *Aerosol Sci. Technol.* 17:1-8 (1992). With permission.)

particle thus generates four simultaneous electrical pulses which are fed to the data processing electronics.

Particle laden air is drawn at a rate of approximately 1.5 L/min. The measured light scattering profile images indicate the particle shape. For spherical particles, the recorded profiles are symmetric (Figure 16.56). On the other hand, nonspherical particles show nonsymmetric scattering profiles. Examples of recorded laser scattering profiles on fibrous particles are shown in Figure 16.57. It can be seen that these profiles differ also in the correlation to the type of fibrous particles. Therefore, the application of the ASA for counting or for identification of fibrous particles on-line and in real time seems to be possible. Nevertheless, there are limitations concerning fiber size (≥ 1 µm) and fiber number concentration (≥ 10 fibers/L).[157]

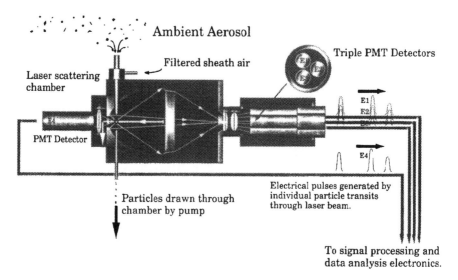

FIGURE 16.55 A functional schematic picture of the light scattering counter for fibrous aerosol. (From Kaye, P.H., K. Alexander-Buckley, E. Hirst, et al. A real-time monitoring system for airborne particle shape and size analysis, *J. Geophys. Res.* 101:19,215-221 (1996). With permission.)

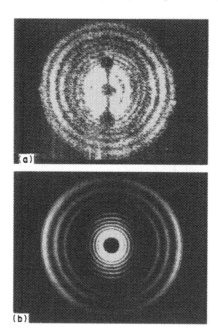

FIGURE 16.56 Experimental (a) and theoretical (b) scattering profile from a single droplet. (From Kaye, P.H., N.A. Eyles, I.K. Ludlow, and J.M. Clark, An instrument for the classification of airborne particles on the basis of size, shape and count frequency, *Atm. Environ.* 25A:645-654 (1991). With permission.)

A commercial version of the instrument is currently being marketed by a U.K. company, BIRAL (Bristol Industrial and Research Assoc. Ltd.).[156]

The ASA equipment was modified recently and a special instrument was developed and proved for the *in situ* and on-line airborne fiber counting and identification, designated SAFIRE.[159] In order to extract more subtle information about the morphology of a fibrous particle, the spatial intensity distribution of light scattered by the fiber was recorded in more detail. As illustrated in Figure 16.57, the chrysotile and crocidolite fiber profiles show some significant differences. The chrysotile fibers,

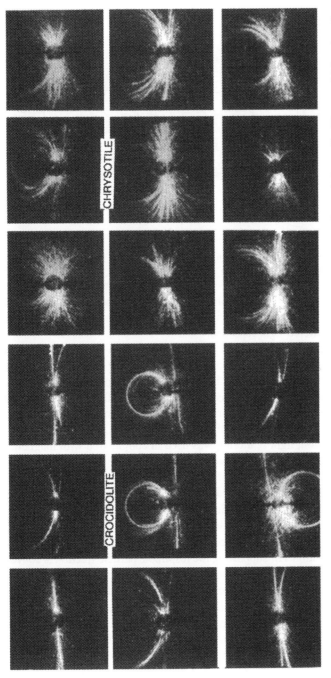

FIGURE 16.57 Laser scattering profiles recorded from individual airborne fibers: crocidolite and chrysotile. The acquisition time for each profile was 3 μs. (From Hirst, E., P.H. Kayer, and J.A. Hoskins, Potential for recognition of airborne asbestos fibres from spatial laser scattering profiles, *Ann. Occup. Hyg.* 39:623-632 (1995). With permission.)

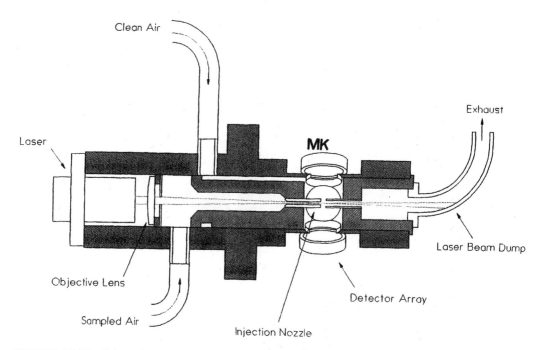

FIGURE 16.58 Schematic picture describing the function of the light scattering counting methods for individual fiber. (From Harley Sci. Ltd., Fibre Monitor, Fibrecheck FC 2, Newcastle upon Tyne, UK (1997).

being normally curved, cause scattering profiles other than those caused by the straight crocidolite fibers. In both chrysotile and crocidolite fibers, the fiber volume may be related to a first approximation of the total scattered light recorded.

Particle analysis rates in excess of 10^3 per second within a sample volume flow rate of 1 liter per minute are achievable, offering the possibility of detecting fiber concentrations at the recommended maximum exposure limit of 0.1 fibers/ml within a sampling period of a few seconds.

An alternative possibility for distinguishing asbestos fibers from other fiber types is to combine a light scattering procedure with the alignment of fibers in magnetic fields. The alignment of fibers in magnetic fields is a specific property of asbestos fibers.

In a recent publication, light scattering was used in conjunction with magnetic fiber alignment for a continuous on-line and real-time measurement of airborne fiber concentrations. Distinguishing asbestos from other airborne mineral fibers was well possible.*

Last but not least, another new laser-based "single fiber" detection instrument, FIBERCHECK, has also been commercially available in England since about 1995.[160] The diagram in Figure 16.58 shows how the equipment works. Air is sucked in through the inlet (at a rate of 2.0 L/min.) and passes through an air flow monitor device and a cyclone, which is designed to separate out the "respirable" particle fraction. Air supporting the respirable particle fraction is conveyed to the "optic module," where it is injected through a small nozzle and collected by a larger one. Coaxial with the nozzle is a focused beam of light from a semiconductor laser. Particles, supported by the air flow and illuminated by the laser, scatter light. As they pass the gap between the inlet and exhaust nozzle, this light can escape and is collected by an array of photo-detectors mounted radially around the axis of the system. The air is then removed from the system by a pump. Electronics are included to process the signals from the photodetectors.

---

* Z. Ulanowski, P.H. Kaye and E. Hirst. Respirable asbestos detection using light scattering and magnetic alignment. *J. Aerosol Sci.* Suppl 1. 29, S13-S14 (1998).

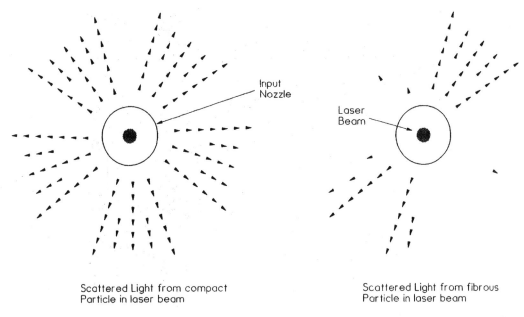

FIGURE 16.59 Schematic pictures illustrating the difference in the profiles of the scattered light from spherical and fibrous particles. (From Harley Sci. Ltd., Fibre Monitor, Fibrecheck FC 2, Newcastle upon Tyne, UK (1997). With permission.)

Fundamental to the operation of the device is the concept that spherical or compact particles will scatter light fairly symmetrically around the axis of illumination, while fibrous particles will show a marked preference, much more light being scattered "broadside" than end on, particularly when viewed at right angles to the axis (Figure 16.59). Also fundamental is that the nozzle design must ensure a turbulent air flow in the injection nozzle (Figure 16.58, MK) so that fibers are presented within the measurement region with a random orientation. The instrument is able to measure fiber concentrations ≥0.01 fibers/ml and provides statistical significance in its results after 15 mins. The British company Harley Scientific Ltd. in Newcastle-upon-Tyne produces this equipment (Figure 16.60).

Another very promising fiber aerosol analyzer (FAA) was developed recently in Germany. The measuring principle is based on the evaluation of the azimuthal light scattering pattern generated by single particles, which are illuminated by circularly polarized laser light. The instrument uses eight different optical detectors placed at a polar angle of 55° and in an azimuthal angular regime from 69° to 90°.

If a homogeneous, spherical scatterer (spherical aerosol particle) is uniformly illuminated by circularly polarized light, the azimuthal scattering pattern is uniform at all detector positions. For fibrous particles, however, the scattering intensity is nonuniform. Provided an alignment of the fiber axis perpendicular to the incident beam in an aerodynamic nozzle, the fiber length can be derived from the variability in the light scattering pattern at the azimuthal angles close to 90°. The diameter of the fiber follows from the mean value of all detectors.

The FAA can detect air concentrations of fibers with diameters >0.2 μm, which lay in the range of 0.1 fiber/ml and over.*

---

* H. Barthel, B. Sachweh, and F. Ebert. Measurement of airborne mineral fibers using a new differential light scattering device. *Meas. Sci. Technol.* 9, 210–220 (1998).

FIGURE 16.60   A photograph showing the commercial fibrous monitor FIBRECHECK FC-2. (From Harley Sci. Ltd., Fibre Monitor, Fibrecheck FC 2, Newcastle upon Tyne, UK (1997). With permission.)

16.17.1 CRITICAL EVALUATION OF FIBER MONITORS

Fiber monitors are used for the rapid on-line and real-time detection of airborne fibers in workplaces and in buildings during and after asbestos clearance. Nevertheless, the applicability of measured fiber concentrations for health risk evaluations seems still to be insufficiently conclusive. Some serious experimental comparisons of the FAM fibrous monitor with PCOM have been published.[161-163] Iles et al.[161] have compared the FAM with the membrane filter method (PCOM), mainly for chrysotile aerosols in the laboratory and in asbestos cement, friction material, and asbestos textile manufacturing factories. Correlation in laboratory trials was fairly good (Figure 16.61), but it was generally poor in workplaces, so FAM cannot be considered a substitute for the membrane filter method. FAM could be used for detecting changes in dust clouds of constant composition. However, by this criterion, it has little better than a nondiscriminating optical particle counter (e.g., the ROYCO instrument). These measurements were done in 1986. In 1990, Koyama et al.[163] made measurements of asbestos fiber concentrations in different indoor situations and under laboratory conditions. Parallel measurements with FAM and PCOM showed, in most instances, good agreement for counting asbestos fibers longer than 5 μm in concentrations between 0.1 and 2 fibers/ml. This agreement was very poor for fiber concentrations less than 0.1 fibers/ml. Studies by Phanprasit et al., done in 1988, were less successful. They made similar measurements and found a good correlation for chrysotile asbestos at concentrations between 0.1 and 7 fibers/ml. The reproducibility of these correlations was not satisfactory. For this reason, they concluded that the FAM does not provide air monitoring results that can be considered as an alternative to those of standard PCOM (membrane filter method).[162]

There is as yet very little information about the latest fiber monitors. It seems that these are more sensitive in counting lower fiber concentrations. The Fibrecheck monitor has already been compared with the standard membrane filter method (PCOM).[160] In the first approximation, the correlation was relatively good, mainly in the concentration range between 0.02 and 0.1 fibers/ml. In the higher fiber concentrations, the agreement was not so good. The Fibrecheck overestimated the measured concentrations in comparison with membrane filter method.

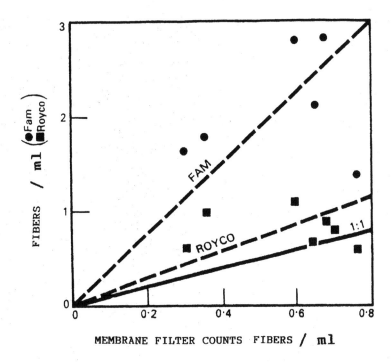

MEMBRANE FILTER COUNTS FIBERS / ml

FIGURE 16.61 Laboratory tests with chrysotile dusts. Comparison of PCOM (MF-method) with FAM and ROYCO. (From Iles, P.J., and T. Sheton-Taylor, Comparison of a fibrous aerosol monitor with membrane filter method for measuring airborne asbestos concentrations, *Ann. Occup. Hyg.* 30:77-87 (1986). With permission.)

16.18   REFERENCE FIBER STANDARDS AND CALIBRATION AEROSOLS

Reference standards of fibrous powders, dusts, and reference generators are very useful and a necessary aid in the measurement and analytical identification of fibrous aerosols in different atmospheric environments.

16.18.1   FIBER REFERENCE STANDARDS

There are different types of reference standards of fibrous materials (asbestos as well as natural and man-made fibers) available in research institutions. Such standards help microscope operators substantially in characterizing and identifying unknown environmental samples. For example, asbestos reference standards (chrysotile originated in Canada and Zimbabwe, amosite and crocidolite originated in South Africa, tremolite in California, and anthophyllite in Finland) are available from the IOM (Institute of Occupational Medicine) in Edinburgh, U.K.[164]

The reference fibrous materials, which had already been well characterized (PCOM, SEM, TEM, EDXA, SAED, etc.) by reputable laboratories, can then be used by operators for the preparation of calibration filter samples. Homogeneous, very diluted liquid suspensions of such fibrous powders, and/or diluted aerosols are filtered by MF or NPF, and those filters can be used as comparable standards. The filter standards representing different types of fibrous aerosols in workplaces, in indoor and outdoor atmospheres, can also be prepared and certified. For the purposes of electron microscopical measurement and analysis, such reference filter standards were prepared, for example, in the laboratory of the NIST (National Institute of Standards and Technology) Gaithersburg, MD, USA.[82]

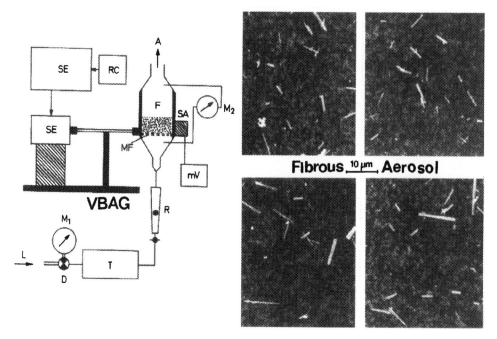

FIGURE 16.62 Schematic picture of the vibrating bed aerosol generator (VBAG) and scanning electron micrographs (right) of the dispersed amosite fibers.

16.18.2 THE GENERATION OF FIBROUS AEROSOLS

Among the several generation methods used for the production of reproducible fibrous aerosols,[165-167] two types of equipment seem to be of greatest interest: the vibrating bed generators and suspension generating equipment. The principle of a VBAG (vibrating bed aerosol generator) is shown in Figure 16.62.

After the dried fibrous material has been put through cylinder F, the vibration frequency and amplitude are set and the flow rate adjusted. The fiber concentration and size distribution of the generated aerosol are governed by the physical conditions in the fluidized bed, especially the frequency and amplitude of the mechanical vibrations. They appear to be functions of the vibration forces acting on the fiber agglomerates. The bed has a high porosity, and the fibrous powder is kept in a "fluidized and disintegrated" state. The bonding forces between individual fibers are thus very small, and fibers can be carried off by the flow of gas. The cohesion forces between fibers are to a first approximation, a function of fiber length, and hence greater vibration forces are needed for the dispersion of long fibers. It is, therefore, possible to control the aerosol concentration and fiber size quite accurately by adjusting the frequency and amplitude of the vibrations. The initial fiber size distribution of the powder in the bed is also, of course, of basic importance.

The VBAG is limited to the generation of fibers with a diameter of 0.5 μm and above.[166] For the generation of very finely dispersed fibrous aerosols, another generator type has to be used, e.g., a generator for dispersing liquid suspensions of very fine fibers.

The apparatus is shown schematically in Figure 16.63. The mixture of distilled water and methyl alcohol with very fine, e.g., chrysotile asbestos fibers, is dispersed in the generator DV. The solution in the container is treated ultrasonically (US). The dispersed microdroplets are dried in the columns SG (silica gel) and C (charcoal). The resulting fibrous aerosol is then discharged by a radioactive source (Kr –85). To get single fibers (AF) and almost no aggregates, the fibrous liquid mixture has to be sufficiently diluted.[167]

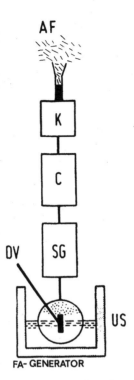

FIGURE 16.63 Schematic picture illustrating the generation of very fine dispersed fibrous aerosols. Description in the text.

16.19 CONCLUSION

The measurement, analysis, and identification of fibrous aerosols has made important progress in the last 10 years. The same can be said about developments in the field of toxicology of fibrous aerosols. Because of this, relatively exact determinations of fiber concentrations, size, identity, and carcinogenic potency are possible. Errors can be introduced during sampling and specimen preparation and by the microscopical and analytical equipment and operators. Therefore, before a laboratory is selected to perform the counting and identification of fibrous aerosol samples (by PCOM, SEM, and TEM), the capabilities of the personnel, equipment, and the standardized procedures should be checked and tested. Training, courses, workshops, and examinations for the operators are necessary. Use of PCOM and the light scattering fiber counters are suitable methods for screening detection of fibrous dusts in the workplace (used, for example, by factory industrial hygiene authorities). SEM is suitable for routine measurements in the workplace (used for example, by governmental industrial hygiene authorities). Nevertheless, there remains the caution relating to the direct filter evaluation procedure. SEM is a screening test of high quality for both indoor and outdoor measurements. TEM could be used for seasonal control (e.g., once a year) in the workplace and for indoor and outdoor measurements (done by public health authorities). This is also the only method for fiber analysis in liquids and in biological materials. Because the concentrations of fibrous aerosols are mostly low or very low and the single fibers can be very fine, with fiber diameters often less than 0.1 μm, automation procedures, both for filter sample evaluation and for satisfactory on-line counting and sizing of fibers in atmospheres are also not very promising in the future.

REFERENCES

1. Spurny, K.R. (Ed.) *Physical and Chemical Characterization of Individual Airborne Particles,* Ellis Horwood Ltd., Chichester, U.K. (1986).
2. Fletcher, R.A. and J.A. Small, Analysis of individual collected particles, in K. Willeke and P.A. Baron (Eds.) *Aerosol Measurement,* Van Nostrand Reinhold Publ., New York (1993), pp. 260-295.
3. Walton, W.H. The nature, hazard, and assessment of occupational exposure to airborne asbestos dust, *Ann. Occup. Hyg.* 25(2):117-247 (1982).
4. Selikoff, I.J. and D.H.K. Lee (Eds.) *Asbestos and Disease,* Academic Press, Inc., New York (1978).
5. Peto, J., J. Hodgson, F. Matthews, et al. Continuing increase in mesothelioma mortality in Britain, *Lancet,* 345:535-539 (1995).
6. Price, B. Analysis of current trends in US mesothelioma incidence, *Am. J. Epidemiology* 145:211-218 (1997).
7. Liddel, D. and K. Miller (Eds.) *Mineral Fibers and Health,* CRC Press, Boca Raton, FL, USA (1991).
8. Farnham, J.E. and P. Roundtree, Health effects of synthetic vitreous fiber exposure, in G.A. Peters and B.J. Peters (Eds.) *Asbestos Pathogenesis and Litigation,* Butterworth Publ., Sevenoaks, UK, Vol. 13, pp. 111-139 (1996).
9. Michaelis, L. and S.S. Chissick (Eds.) *Asbestos,* John Wiley & Sons, New York, USA (1979).
10. Heidermanns, G. Asbestgehaltbestimmung durch optische, chemische, röntgenographische und infra-rotspektroskopische Analysenverfahren, *Staub Reinhalt. Luft,* 33:66-70 (1973).
11. Harness, I. Airborne asbestos dust evaluation, *Ann. Occup. Hyg.,* 16:397-404 (1973).
12. Rickards, A.L. Estimation of submicron quantities of chrysotile asbestos by electron microscopy, *Anal. Chem.,* 45:809-811 (1973).
13. Chatfield, E.J. and H. Pullan, Measuring asbestos in the environment, *Canadian Res. Develop.,* December: pp. 23-27 (1974).
14. Spurny, K.R., J. W. Gentry and W. Stöber, Sampling and analysis of fibrous aerosol particles, in D.T. Shaw (Ed.) *Fundamentals in Aerosol Science,* John Wiley & Sons, New York, pp. 257-324 (1978).
15. Spurny, K. *Asbest-Analytik. Berichte der UBA-Umweltbundesamtes,* E. Schmidt Verlag, Berlin, 7/80:51-113 (1980).
16. Spurny, K.R., W. Stöber, H. Opiela and G. Weiss, On the evaluation of fibrous particles in remote ambient air, *Sci. Total Environ.,* 11:1-40 (1979).
17. Spurny, K.R. Sampling, analysis, identification and monitoring of fibrous dusts and aerosols, *Analyst* 119:560-590 (1994).
18. Spurny, K.R. Measurement of airborne asbestos fibers, in G.A. Peters and B.J. Peters (Eds.) *The Current Status of the Public Health Problem,* Butterworth Publ., Salem, USA, Vol. 9:75-130 (1994).
19. Baron, P.A. Measurement of asbestos and other fibers, in K. Willeke and P.A. Baron (Eds.) *Aerosol Measurement,* Van Nostrand Publ. Inc., New York, pp. 560-590 (1993).
20. HSE Methods for determination of hazardous substances: Asbestos fibers in air. Determination of personal exposure by European reference version of the membrane filter method. MDHS 39, HMSO, London (1984).
21. Kenny, L.C. Asbestos fiber counting by image analysis, *Ann. Occup. Hyg.,* 28:401-415 (1984).
22. Baron, A.P. *Asbestos analysis,* NIOSH Method 7400. Appl. Ind. Hyg. 2:R8-R10 (1987).
23. Kaga, A., Y. Inoue, N. Maehata and A. Yoshikawa, Automatic counting of asbestos fibers, in Masuda, S. and K. Takahashi (Eds.) *Aerosols,* Pergamon Press, Oxford, U.S. pp. 486-489 (1990).
24. Spurny, K. and C. Möhlmann, Direct-reading instruments for fibre measurement, *Gefarhrstoffe-Reinhalt. Luft.* 56:69-71 (1996).
25. Baron, A.P. Application of the thoracic sampling definition to fiber measurement, *Am. Ind. Hyg. Assoc. J.* 57:820-824 (1996).
26. Rödelsperger, K. and H.J. Woitowitz, Airborne fibre concentrations and lung burden compared to the tumor response in rats and humans exposed to asbestos, *Ann. Occup. Hyg.* 39:715-725 (1995).
27. Du Toit, R.S.J. and T.C. Gilfillan, Simultaneous airborne dust samples with konimeter, thermal precipitator and dosimeter in asbestos mines, *Ann. Occup. Hyg.* 20:333-340 (1977).
28. Du Toit, R.S.J., L.W. Isserow, T.C. Gilfillan, K. Robock and U. Teichert, Relationship between simultaneous airborne dust samplers taken with five types of instruments at South African asbestos mines and mills, *Ann. Occup. Hyg.* 27:373-387 (1983).

29. Hammad, Y.Y., J. Diem and H. Weils, Evaluation of dust exposure in asbestos cement manufacturing operations, *Am. Ind. Hyg. Assoc. J.* 40:390-395 (1979).

30. Asgharian, B., L. Zhang and C.P. Fang, Theoretical calculations of the collection efficiency of spherical particles and fibers in an impactor, *J. Aerosol Sci.* 28:277-287 (1997).

31. Kochan, F.K. Die Messung von Asbestfaserzahlkonzentrationen mit dem Zeiss-Konimeter und Membranfiltermethode, *Staub-Reinhalt. Luft* 47:146-150 (1987).

32. Spurny, K.R. Membranfilter in der Aerosologie, *ZBL für Biol. Aerosol-Forsch.* 12:1-39 (1965); 12/13:3-56 (1966); 13:3-56 (1967).

33. Spurny, K.R. Nuclepore-Siebfilter-Membranen, *Staub-Reinhalt. Luft* 37:328-334 (1977).

34. Spurny, K.R. and J.W. Gentry, Fractionation by graded Nuclepore filters, *Powder Technol.* 37:129-142 (1979).

35. Spurny, K.R. On the filtration of fibrous aerosols, *Sci. Total Environ.* 52:189-199 (1986).

36. Spurny, K.R. Zur Filtration faseriger Aerosole, *Filtrieren und Separieren (Germany)* 8:166-175 (1994).

37. Johnston, A.M., A.D. Jones and J.H. Vincent, The effect of electrostatic charge on the efficiencies of airborne dust samplers, *Am. Ind. Hyg. Assoc. J.* 48:613-621 (1987).

38. Baron, P.A. and G.J. Deye, Electrostatic effects in asbestos sampling I, II, *Am. Ind. Hyg. Assoc. J.* 51:51-69 (1990).

39. Vincent, J.H. *Aerosol Sampling,* John Wiley & Sons, Chichester, UK (1989).

40. Kenny, L.C., R. Aitken, C. Chalmers, et al. A collaborative European study of personal inhalable aerosol sampler performance, *Ann. Occup. Hyg.* 41:135-153 (1997).

41. Kaufer, E., J.C. Vigneron, J.F. Fabries, et al. The use of a new static device based on the collection of the thoracic fraction for the assessment of the airborne concentration of asbestos fibres by transmission electron microscopy, *Ann. Occup. Hyg.* 40:311-319 (1996).

42. Sahle, W. and G. Larsson, The usefulness of void-counting for fibre concentration estimation by optical phase contrast microscopy, *Ann. Occup. Hyg.* 33:97-111 (1989).

43. Spurny, K.R. On the release of asbestos fibers from weathered and corroded asbestos cement products, *Environ. Res.* 48:100-116 (1989).

44. Attfield, M.D. and S.T. Beckett, Void-counting in assessing membrane filter samples of asbestos fibres, *Ann. Occup. Hyg.* 27:273-282 (1983).

45. WHO, Reference methods for measuring airborne man-made fibres, *Environmental Health Report 4,* WHO, Copenhagen (1995).

46. Friedrichs, K.H. Particle analysis by light microscopy, in K.R. Spurny (Ed.) *Physical and Chemical Characterization of Individual Airborne Particles,* Ellis Horwood, Chichester, UK (1986), pp. 161-172.

47. ISO, Determination of the number concentration of airborne inorganic fibres by phase contrast optical microscopy (PCOM). Membrane filter method, *Int. Standard ISO/DIS 8672,* International Organization for Standardization, Geneva (1988).

48. HSE, Asbestos fibres in air. Light microscope method for use with the control of asbestos at work regulations, *Health and Safety Executive.* HES MDHS 39-3, London (1990).

49. VDI, Messen anorganischer faserförmiger Partikeln in der Aussenluft, REM Verfahren. VDI Richtlinie 3492 Blatt 1, 1-40 (1991).

50. Chatfield, E.J. and M.J. Dillon, Some aspects of specimen preparation and limitations of precision in particulate analysis by SEM and TEM, *Scanning Electron Microscopy,* SEM-Inc., Chicago pp. 387-499 (1978).

51. Burdett, G.J. and A.P. Rood, Membrane-filter. Direct-transfer technique for the analysis of asbestos fibers and other inorganic particles by transmission electron microscopy, *Environ. Sci. Technol.* 17:643-648 (1983).

52. Yamate, G., S.C. Agarwal and R.D. Gibson, Methodology for the measurement of airborne asbestos by electron microscopy, US-EPA, Environmental Protection Agency, Report 698-02-3266 (1984).

53. Chatfield, E.J. Asbestos measurement in workplace and ambient air atmospheres, in S. Basu and J.R. Milette (Eds.) *Electron Microscopy in Forensics.* Occup. and Environ. Health Service. Plenum Press pp. 149-186 (1986).

54. Chatfield, E.J. Limits and detection and precision in monitoring asbestos fibers. Asbestos and Health Risk, *Air Poll. Control Assoc. J. APCA* (1987) pp. 79-90.

55. NIOSH, Method 7402, TEM, National Institute for Occupational Safety and Health. Publ. 84, US Government Publ. Office, Washington, D.C. (1990).

56. Sahle, W. Lack of confidence in results obtained by SEM for airborne fiber concentration evaluation for dust sampled on porous membrane filters, *Ann. Occup. Hyg.* 34:101-105 (1990).

57. Chatfield, E.J. ISO. International Standard Organization of Standardization, Analytical methods for determination of asbestos fibers in water and in air. Report ISO/TC 147/SC 2, W 618 and ISO TC 146/SC 3, W 61 (1991).

58. Breysse, P.N. Electron microscopic analysis of airborne asbestos fibers, *Crit. Rev. Anal. Chem.* 22:201-227 (1991).

59. Johnson, L. Asbestos analysis registry, *The Synergist.* AIHA 2:5-6 (1991).

60. Ortiz, L.W. and B.L. Isom, Transfer technique for electron microscopy of membrane filter samples, *Am. Ind. Hyg. Assoc. J.* 35:423-431 (1974).

61. Höhr, D. Methodische Voraussetzungen zur transmissions elektronenmikroskopischen Bestimmung faserförmiger Partikeln, *Staub-Reinhalt. Luft* 44:337-341 (1984).

62. Boltin, W.R., B.H. Clark, L. Detter-Hoskin, and T. Kremer, Alternative instrumentation in the analysis for asbestos in various media, *American Laboratory,* 21:15-26 (1989).

63. Lee, R.J., T.V. Dagenhart, G.R. Dunmyre et al. Effect of indirect sample preparation procedures on the apparent concentration of asbestos in settled dusts, *Environ. Sci. Technol.* 29:1728-1736 (1995).

64. Spurny, K.R., W. Stöber, H. Opiela and G. Weiss, On the problem of milling and ultrasonic treatment of asbestos and glass fibers in biological and analytical applications, *Am. Ind. Hyg. Assoc. J.* 41:198-203 (1980).

65. Kaufer, E., M.A. Billon-Galland, J.C. Vigneron, et al. Effect of preparation methods on the assessment of airborne concentrations of asbestos fibres by transmission electron microscopy, *Ann. Occup. Hyg.* 40:321-330 (1996).

66. Lee. R.J., D.R. Van Orden and G.R. Dunmyre, Interlaboratory evaluation of the breakup of asbestos-containing dust particles by ultrasonic agitation, *Environ. Sci. Technol.* 30:3010-3015 (1996).

67. Lundgren, L., S. Lundström, I. Laszlo, et al. Modern fibre counting — a technique with the phase-contrast microscope on-line to a Macintosh computer, *Ann. Occup. Hyg.* 39:455-467 (1995).

68. Lundgren, L.S., Lundström and G. Sundström, et al. A quantitative method using a light microscope on-line to a Macintosh computer for the analysis of tremolite fibres in dolomite, *Ann. Occup. Hyg.* 40:197-209 (1996).

69. Goldstein, J.I. and H. Yankowitz, *Practical Scanning Electron Microscopy,* Plenum Press, New York, (1977).

70. Goldstein, J.I., D.E. Newbury, P. Echlin, et al. *Scanning Electron Microscopy and X-ray Microanalysis,* Plenum Press, New York, (1981).

71. Czarnecki, J., C. Loipführer and G. Renk, Mustererkennung zur Identifizierung faseriger Stäube in Rahmen des Immissionschutzes, *Staub-Reinhalt. Luft* 53:147-152 (1993).

72. Rödelsperger, K., R. Arhelger, B. Brückel and H.J. Woitowitz, Nachweis von Künstlichen Mineral-fasern aus Dämmstoffen in der Atemluft am Arbeitsplatz, *Staub-Reinhalt. Luft* 53:395-399 (1993).

73. Kaufer, E., T. Schneider and J.C. Vigneron, Assessment of man-made mineral fibre distribution by scanning electron microscopy, *Ann. Occup. Hyg.* 37:469-479 (1993).

74. Terry, K.W. Particle size distribution of airborne dust using a scanning electron microscopy, *Aerosol Sci. Technol.* 23:475-478 (1995).

75. Yada, K. Study of chrysotile asbestos by a high resolution electron microscope, *Am. Soc. Testing Materials,* ASTM 834:118-138 (1967).

76. Reimer, L. *Transmission Electron Microscopy,* Springer-Verlag, Heidelberg (1993).

77. Langer, A.M., A.D. Mackler and F.D. Pooly, Electron microscopical investigation of asbestos fibers, *Environmental Health Perspectives* 9:63-80 (1974).

78. Pooly, F.D. The identification of asbestos dust with an electron microscope microprobe analyzer, *Ann. Occup. Hyg.* 18:181-185 (1975).

79. Sebastien, P., A.M. Billon and X. Janson, et al. Utilisation du microscope electronique a trasmission pour la mesure des contaminations par l'amiante, *Arch. Mal. Prof. Med. Trav.* 39:85-96 (1978).

80. Pooly, F.D. and N.J. Clark, Quantitative assessment of inorganic fibrous particulate in dust samples with an analytical TEM, *Ann. Occup. Hyg.* 22:253-271 (1979).

81. Steen, D., M.P. Guillemin, P. Buffat and G. Litzistorf, Determination of asbestos fibers in air. TEM as a reference method, *Atm. Environ.* 17:2285-2297 (1985).

82. Small, J.A., E.B. Steel and P.J. Sheridan, Analytical standards for the analysis of chrysotile asbestos in ambient environment, *Anal. Chem.* 57:204-208 (1985).
83. Steel, E.B. and J.A. Small, Accuracy of TEM for the analysis of asbestos in ambient environment, *Anal. Chem.* 57:209-213 (1985).
84. Rood, A.P. and R. Streeter, Comparison of the size distribution of occupational asbestos and MMMF determined by TEM, *Ann. Occup. Hyg.* 32:361-367 (1988).
85. Takao, S., S. Tagami and T. Sakurai, Studies on the estimation of airborne asbestos in environments utilizing TEM, *J. Japan Soc. Air Pollut.* 24:214-226 (1989).
86. Turner, S. and E.B. Steel, Accuracy of TEM analysis of asbestos on filters. Interlaboratory study, *Anal. Chem.* 63:868-872 (1991).
87. Clark, N.E., D.K. Verma and J.A. Julian, Determination of the variability in elemental composition of asbestos fibres by analytical TEM, *Ann. Occup. Hyg.* 39:79-88 (1995).
88. Kohyama, N. and S. Kurimori, A total sample preparation method for the measurement of airborne asbestos and other fibers by optical and electron microscopy, *Industrial Health (Japan)* 34:185-203 (1996).
89. Spurny, K.R. Uncertainties and limitations of SEM method, in *Proceedings of the 4th Int. Colloquium on Dust Measurement Technique and Strategy, Asbestos Int. Assoc.,* AIA, London, pp. 356-361 (1982).
90. McIntosh, C. and A. Searl, Scanning electron microscope sizing of airborne fibre samples, *Ann. Occup. Hyg.* 40:45-55 (1996).
91. Muhle, H., F. Pott, B. Bellmann, et al. Inhalation and injection experiments in rats to test the carcinogenicity of MMMF, *Ann. Occup. Hyg.* 31:755-764 (1987).
92. Chatfield, E.J. Fiber definition in occupational and environmental asbestos measurements, *Am. Soc. Testing Materials,* ASTM, 834:118-138 (1984).
93. Lippy, B.C. and J.A. Boggs, Measuring airborne asbestos. The significance of small fibers detected under TEM, *J. Environ. Health,* 52:157-160 (1989).
94. Pang, T.S.W., F.A. Schonfeld and K. Patel, The precision of a method for the analysis of amosite asbestos, *Am. Ind. Hyg. Assoc. J.* 49:351-356 (1988).
95. Verma, D.K., N.E. Clark and J.A. Julian, Asbestos fiber characterization using analytical transmission microscope and a microfilm reader, *Am. Ind. Hyg. Assoc. J.* 52:113-119 (1991).
96. Verma, D.K. and N.E. Clark, Relationship between phase contrast microscopy (PCM) and transmission electron microscopy (TEM) results of samples from occupational exposure to airborne chrysotile asbestos, *Am. Ind. Hyg. Assoc. J.* 56:866-873 (1995).
97. Breysse, P.N., J.W. Cherrie, J. Addison and J. Dodgson, Evolution of airborne asbestos concentration using TEM and SEM during residential water tank removal, *Ann. Occup. Hyg.* 33:243-256 (1989).
98. Robertson, K.T., T.C. Thomas and L.R. Sherman, Comparison of asbestos air samples by SEM–EDXA and TEM-EDXA, *Ann. Occup. Hyg.* 36:265-270 (1992).
99. Sahle, W. and I. Laszlo, Airborne inorganic fibre level monitoring by transmission electron microscope (TEM): Comparison of direct and indirect sample transfer method, *Ann. Occup. Hyg.* 40:29-44 (1996).
100. Puleda, S. and A. Marconi, Study of the count-to-mass conversion factor for asbestos fibers in samples collected at the emission of three industrial plants, *Ann. Occup. Hyg.* 35:517-524 (1991).
101. Malisa, H. and J.W. Robinson, Analysis of Airborne Particulates by Physical Methods, CRC Press, FL, USA (1978).
102. Takeishi, K. Raman microprobe analysis of environmental asbestos particles, *Bunseki Kagaku (Japan)* 30:774-779 (1980).
103. Kaufmann, R. and P. Wieser, Laser microprobe mass analysis (LAMMA) in particle analysis, *NBS-Publications* 533:199-223 (1980).
104. Spurny, K.R., J. Schörmann and R. Kaufmann, Identification and microanalysis of mineral fibers by LAMMA, *Fresenius Zschr. Anal. Chem.* 308:274-279 (1981).
105. Seyama, H., Y. Soma and M. Soma, et al. Application of secondary ion mass spectrometry for the analysis of asbestos fibres, *Fresenius Zschr. Anal. Chem.* 341:619-624 (1991).
106. Benarie, M. Identification of asbestos fibers by fluorochrome staining, in *Aerosols in the Mining and Industrial Work Environment,* Ann Arbor, Ann Arbor Sci. Publ., USA (1983).
107. Awalda, F.T. and F. Habashi, Reaction of chrysotile asbestos with triphenylmethane dyes, *J. Material Sci.* 25:87-92 (1990).

108. Crable, J.W. and M.J. Knott, Application of x-ray diffraction to determination of chrysotile in bulk of settled dust samples, *Am. Ind. Hyg. Assoc. J.* 25:449-453 (1966).

109. Goodhead, K. and K.W. Mertindale, The determination of amosite and chrysotile in airborne dusts by an x-ray diffraction method, *Analyst* 94:985-989 (1969).

110. Keenan, R.G. and J.R. Lynch, Techniques for the detection, identification and analysis of fibers, *Am. Ind. Hyg. Assoc. J.* 29:587-597 (1970).

111. Nanadic, C.M. and J.W. Crable, Application of X-ray diffraction to analytical problems in occupational health, *Am. Ind. Hyg. Assoc. J.* 30:529-538 (1971).

112. Heidermanns, G., G. Riediger and A. Schütz Asbestbestimmung in industriellen Feinstäuben und Lungenstäuben, *Staub-Reinhalt. Luft* 36:107-111 (1976).

113. Taylor, M. Methods for the quantitative determination of asbestos and quartz in bulk samples using x-ray diffraction, *Analyst* 103:1009-1020 (1980).

114. Davis, B.L. Quantitative analysis of asbestos minerals by reference intensity x-ray diffraction procedure, *Am. Ind. Assoc. J.* 51:297-303 (1980).

115. Puleda, S. and A. Marconi, Quantitative x-ray diffraction analysis of four types of amphibolic asbestos by the silver membrane filter method, *Int. J. Environ. Anal. Chem.* 36:209-220 (1989).

116. Ruizhong, H., J. Block, J.A. Hriijac, et al. Use of x-ray powder diffraction for determining low levels of chrysotile asbestos in gypsum-based bulk materials, *Anal. Chem.* 68:3112-3120 (1996).

117. Caquineau, S., M.C. Magonthier, A. Gaudichet and L. Gomes, An improved procedure for the x-ray diffraction analysis of low mass atmospheric dust samples, *Eur. J. Mineral.* 9:157-166 (1997).

118. Addison, J. and L.S.T. Davies, Analysis of amphibole asbestos in chrysotile and other minerals, *Ann. Occup. Hyg.* 34:159-175 (1990).

119. Gadsen, J.A., J. Parker and W.L. Smith, Determination of chrysotile in airborne asbestos by infrared spectroscopic technique, *Atm. Environ.* 4:667-670 (1970).

120. Taylor, D.G., C.M. Nanadic and J.W. Crable, Infrared spectra for mineral identification, *Am. Ind. Hyg. Assoc. J.* 29:100-108 (1970).

121. Bagioni, R.P. Separation of chrysotile from minerals that interfere with its infrared analysis, *Environ. Sci. Technol.* 9:941-948 (1975).

122. VDI, Manuelle Asbest-Staubmessung in strömendem Reingas. IR-Methode, *VDI-Richtlinie* 3861, 1-14 (1990).

123. Guillemin, M.P., G. Litzidtorf and P.A. Buffat, Urinary fibers in occupational exposure to asbestos, *Ann. Occup. Hyg.* 33:219-233 (1989).

124. De Vuyst, P., P. Dumorker, E. Moulin, et al. Diagnostic value of asbestos bodies in bronchoalveolar lavage fluid, *Am. Rev. Respir. Dis.* 136:1219-1224 (1987).

125. Karjalainen, A., S. Anttila, T. Mäntylä, et al. Diagnostic value of asbestos bodies in bronchoalveolar lavage fluid in relation to occupational history, *Am. J. Ind. Med.* 26:645-654 (1994).

126. Pott, F. Problems in defining carcinogenic fibres, *Ann. Occup. Hyg.* 31:799-802 (1987).

127. Pott, F. Some aspects of the dosimetry of carcinogenic potency of asbestos fibers and other fibrous dusts, *Staub-Reinhalt Luft* 38:486-490 (1978).

128. Brown, R.C., J.A. Hoskins and N.F. Johnson, Mechanisms in fibre carcinogenesis, *NATO ASI Series A. Life Sci.* 223 (1991).

129. Pott, F. The fiber as carcinogenic agent, *Zbl. Hyg.* B184:1-23 (1987).

130. Eastes, W., J. Hadles and J. Bender, Assessing the biological activity of fibers: Insights into the role of fiber durability, *J. Occup. Health Safety* 12:381-385 (1996).

131. Rödelsperger, K., U. Teichert, H. Marfels, et al. Measurement of inorganic fibrous particulates in ambient air and indoors with scanning electron microscope, in J. Bignon, J. Peto and R. Saracci (Eds.) *Non-Occupational Exposure to Mineral Fibres,* IARC Publ. Lyon, Vol. 90:361-366 (1989).

132. Platek, S.F., R.P. Riley, and S.D. Simon, The classification of asbestos fibers by scanning electron microscopy and computer digitizing tablet, *Ann. Occup. Hyg.* 36:155-171 (1992).

133. Cheng, Y.S. Bivariate lognormal distribution for characterizing asbestos fiber aerosols, *Aerosol Sci. Technol.* 5:359-371 (1986).

134. Moss, O.R., B.A. Wong, and B. Asgharian, Bimodal bivariate lognormal distributions in the application of inhalation toxicology, specific to the measurement of fiber and particle dosimetry, in U. Mohr (Ed.) *Toxic and Carcinogenic Effects of Solid Particles in the Respiratory Tract.* ILSI Press, Washington, D.C., pp. 623-628 (1994).

135. Burdett, G.J., J.G. Firth, A.P. Rood, and P.R. Streeter, Application of fibre retention and carcinogenicity curves to fibre size distributions of asbestos, *Ann. Occup. Hyg.* 32:Suppl.1, 341-351 (1989).
136. Spurny, K.R. Testing the toxicity and carcinogenicity of mineral fibers, in G.A. Peters and B.J. Peters (Eds.) *Asbestos Health Effects, Treatment and Control.* The Michie Comp. Charlottesville, VA, USA (1995), pp. 169-215.
137. Scholze, H. and R. Conradt, An *in vitro* study of the chemical durability of siliceous fibres, *Ann. Occup. Hyg.* 31:683-692 (1987).
138. Gulumian, M. and J.A. Van Wyk, Hydroxyl radical production in the presence of fibers by a Fenton-type reaction, *Chemico-Biological Interaction* 62:89-97 (1987).
139. Kamp, D.W., P. Graceffa, and S.A. Weizman, The role of reactive oxygen species in asbestos-induced diseases, in G.A. Peters and B.J. Peters (Eds.) *Asbestos Risk and Medical Advances.* Butterworth Publ. Inc. (USA) Vol. 8:247-272 (1993).
140. Hardy, J.A. and A.E. Aust, The effect of iron binding the ability to catalyze DNA single-strand breaks, *Carcinogenesis* 16:319-325 (1995).
141. Hardy, J.A. and A.E. Aust, Iron in asbestos chemistry and carcinogenicity, *Chem. Rev.* 95:97-118 (1995).
142. Shen, Z.H., V.D. Parcler, and A.E. Aust, Mediated, thin layer cell; determination of redox-active iron the surface of asbestos fibers, *Anal. Chem.* 67:307-311 (1995).
143. Gulumian, M. and J.A. Van Wyk, Oxygen consumption, lipid peroxydation and mineral fibres, in R.C. Brown, J.A. Hoskins and N.F. Johnson (Eds.) *Mechanisms in Fibre Carcinogenesis,* Plenum Press, New York, NATO ASI Series A 223:439-446 (1991).
144. Hogg, B.D., P.K. Dutta, and J.F. Long, *In vitro* interaction of zeolite fibers with individual cells: Measurement of intracellular oxidative burst, *Anal. Chem.* 68:2309-2312 (1996).
145. Timbrell, V. Alignment of carbon and other man-made fibres by magnetic fields, *J. Appl. Phys.* 43:4839-4840 (1972).
146. Timbrell, V. Desired characteristics of fibres for biological experiments, in P.V. Pelnar (Ed.) *Fibres for Biological Experiments,* Inst. of Occup. Hyg. and Environ. Health, Montreal (1973), pp. 89-122.
147. Timbrell, V. Alignment of respirable asbestos fibres by magnetic fields, *Ann. Occup. Hyg.* 26:299-311 (1975).
148. König, R. and G.H. Seger, Spezielle Methoden zur Bestimmung der Faserkonzentration, *VDI-Bericht Nr.* 475 (1983).
149. Lilienfeld, P., P. Elterman, and P. Baron, Development of a prototype fibrous aerosol monitor, *Am. Ind. Hyg. Assoc. J.* 40:270-282 (1979).
150. Lilienfeld, P. Rotational electrodynamics of airborne fibers, *J. Aerosol Sci.* 16:315-322 (1987).
151. FM-7400, Real time laser fiber monitor, Mie ATRC Comp. Billerica, MA, USA (1991).
152. Rood, A.P., E.J. Walker and D. Moore, Construction of a portable fiber monitor measuring differential light scattering for aligned fibers, *Aerosol Sci. Technol.* 17:1-8 (1992).
153. Turnkey Instruments Ltd., London, FM 90 Respirable Fibre Monitor (1992).
154. Kaye, P.H., N.A. Eyles, I.K. Ludlow, and J.M. Clark, An instrument for the classification of airborne particles on the basis of size, shape and count frequency, *Atm. Environ.* 25A:645-654 (1991).
155. Kaye, P.H., E. Hirst, J.M. Clark and F. Michaeli, Airborne particle shape and size classification from spatial light scattering profiles, *J. Aerosol Sci.* 23:597-611 (1992).
156. Biral, Aerosol Shape Analyzer, Model B 1000. Bristol Industrial Res. Assoc. Ltd., Bristol, UK (1994).
157. Hirst, E., P.H. Kayer, and J.A. Hoskins, Potential for recognition of airborne asbestos fibres from spatial laser scattering profiles, *Ann. Occup. Hyg.* 39:623-632 (1995).
158. Kaye, P.H., K. Alexander-Buckley, E. Hirst, et al. A real-time monitoring system for airborne particle shape and size analysis, *J. Geophys. Res.* 101:19,215-221 (1996).
159. Hirst, E., P.H. Kaye, S. Saunders, A. Ferguson, and Z. Wang-Thomas, SAFIRE-A neutral network based instrument for *in-situ* airborne fibre recognition, 10th Aerosol Society Conference, UK, Svansea, *J. Aerosol Sci.* 28:331-342 (1997).
160. Harley Sci. Ltd., Fibre Monitor, Fibrecheck FC 2, Newcastle upon Tyne, UK (1997).
161. Iles, P.J., and T. Sheton-Taylor, Comparison of a fibrous aerosol monitor with membrane filter method for measuring airborne asbestos concentrations, *Ann. Occup. Hyg.* 30:77-87 (1986).
162. Phanprasit, W., V.E. Rose, and R.K. Oestenstad, Comparison of the fibrous aerosol monitor and the optical fiber count technique for asbestos measurement, *Appl. Ind. Hyg.* 3:28-33 (1988).

163. Koyama, H., Y. Konishi, and T. Takata, Measurement of asbestos concentrations in the atmosphere by a fibrous aerosol monitor, in S. Masuda and K. Takahashi (Eds.) *Aerosols,* Pergamon Press, Oxford, UK (1990), pp. 490-494.

164. Tyllee, B.E, L.S.T. Davies, and J. Addison, Asbestos reference standards-made available for analysis, *Ann. Occup. Hyg.* 40:711-714 (1996).

165. Spurny, K.R., C. Boose, D. Hochrainer, and F.J. Mönig, A note on the dispersing of fibrous powders, *Ann. Occup. Hyg.* 19:85-87 (1976).

166. Spurny, K.R. Fiber generation and length classification, in K. Willeke, *Generation of Aerosols and Facilities for Exposure Experiments,* Ann Arbor Sci. Publ., Ann Arbor, MI, USA (1980) pp. 257-298.

167. Spurny, K.R., J. Schörmann, and H. Opiela, Aerosols and protective clothing, *Am. Ind. Hyg. Assoc. J.* 51:36-43 (1990).

17 Detection and Analysis of Bacterial Aerosols

Kvetoslav R. Spurny

CONTENTS

17.1 INTRODUCTION

Bacterial aerosols belong to a great family of aerodisperse systems consisting of fine and coarse particles of variable biological origin, e.g., pollen, fungal spores, bacteria, viruses, protozoa, algae, etc., which are commonly designated as bioaerosols.[1,2] Bioaerosols are ubiquitous and occur in all sizes. They constitute an important part of atmospheric as well as of indoor and workplace aerosol.[3-5]

Bacterial aerosols are natural as well as man-made aerodisperse systems. Their particle sizes belong in the range of particles with transition physical behavioral properties. They also exist outdoors as well as indoors and can be natural or biotechnological in origin. From the point of view of measurement and identification, bacterial aerosols are considered to be the most complicated aerodisperse particulates. They are living objects and undergo physical, chemical, and biological changes which are time and space dependent, and are further connected with environmental properties. All these factors make measurement and identification difficult. The same is true for laboratory generation and investigation of bacterial aerosols.

The classical as well as the modern sampling and identification methods have been well summarized and evaluated in some new books and reviews.[1-6] Some progress was made by the improvement of the existing "classical" procedures,[4-6] and several new physical methods of modern analytical chemistry and biochemistry seem to be very promising.

Classical bacterial detection and identification are usually based on morphological evaluation of microorganisms as well as viability tests in or on various media, susceptibility to various phages and antibiotics, and the ability to metabolize various compounds, etc. Generally, no single test provides a definitive identification of an unknown bacterium. Consequently, classical methods are time consuming and cannot be developed as on-line, *in situ,* and real-time measurement and monitoring procedures. For this reason, one fruitful area of application is the development of rapid methods for detecting and identifying bacterial aerosols on the basis of fast physical and chemical characterization of single particles.[6-8] Physical methods for the detection of airborne microorganisms depends on measuring the number, size, and shape of the aerosol particles. These measurements, however, indicate only the presence of aerosol particles in a size and form range. For example, a sudden increase in the ratio of the number of particles in the size range of 0.5 to 5.0 μm could be indicative of artificially generated bacterial aerosols. Physical measurements, being nonspecific, makes no distinction between microbial particle and particles of other origins.

Chemical methods, being more specific than physical methods, are able to distinguish between bacterial and other aerosol particles, in some cases, even different bacterial species can be identified. Nevertheless, they still cannot fully replace classical culturing methods that enable concentrations of viable bacterial particles to be established and their species identified. Modern physical methods of chemical analysis are exciting and promising alternatives to classical identification, with the future surely confirming their promise. Chemical and physicochemical methods for detecting and identifying bacteria, and microorganisms in general, depends on chemical measurements of the biochemical components of unique chemical structures in microorganisms and, consequently, can distinguish biological from nonbiological materials. Commonly used for this purpose are tests for the presence of nucleic acids, adenosine triphosphate, and proteins. For example, biological compounds (e.g., proteins, nucleic acids, peptides, carbohydrates) react with certain dyes to produce shifts in the light adsorption characteristics of the latter. Measuring this shift by conventional spectroscopy can provide the basis of detecting biological compounds in an unknown sample. For example, acridine orange, which becomes bound to nucleic acids, produces fluorescence in the microbial particles under ultraviolet light.

Such methods are fairly rapid and provide information about the biological nature of the sample but have several limitations. Notably, none of the methods depends on the viability of the sample and, consequently, these methods provide no information on the living state of the microorganisms and provides little information for differentiating them.

Progress in analytical and colloid chemistry, based on theoretical and instrumental advances, has had important and positive influences on the development of modern microbiological analysis. In the last quarter century, there has been a revolution in the application of sensitive and increasingly informative methods of chemical analysis.[8] This has happened, to a large extent, because of advances in electronics, computers, and laser technology. These technologies have allowed practical application of methods in biology and microbiology. Detection sensitivity related to these methods is high, and therefore, the analysis of very small samples, and even of 1 μm single particle, is possible.[8,9] The list of these methods is a long one, ranging from atomic and molecular spectroscopy on the one hand, to separation and surface science on the other. In addition to infrared and nuclear spectroscopy, which began the revolution, we must add the various fluorescence spectroscopies, Raman and resonance Raman spectroscopy, and associated techniques such as coherent anti-Stokes Raman spectroscopy. Separation methods, e.g., gas chromatography and high pressure liquid chromatography, when coupled with mass spectrometry, have enhanced the revolution. These and many other methods have increased our capacity to do basic research.

Laser spectroscopical methods, chromatography, mass spectrometry, and various combinations of these methods have attracted more attention for rapid analysis of microorganisms. The composition of cell walls and DNA content differ markedly between bacterial species, and therefore, can be used for determination of microorganisms in different milieus, including air.[10]

17.2 SAMPLING AND SPECIMEN PREPARATION

Efficient separation of various sizes of microbes from the air and their collection into and onto a medium are based on the physical characteristics of the airborne particles and on the physical parameters of the sampler. They also depend upon the ambient conditions, e.g., wind speed, humidity, air temperature, etc. The applicability and performance of various bioaerosol and bacterial aerosol samplers has been investigated, tested and critically evaluated in several publications.[11-27] The majority of successful applications of chemical detection methods have been with liquids. The first task in sampling bacterial aerosols for chemical detection is to obtain a suitable quantity of preconcentrated microorganisms in a relatively small volume of liquid, i.e., in an aqueous suspension.

Microbes dispersed in air are subjected to different physical and toxic stresses.[28-33] Desiccation is experienced by all airborne microbes. Radiation, oxygen, ozone and its reaction products, and various air pollutants also decrease viability and infectivity through chemical, physical, and biological modifications predominantly to phospholipids, proteins, and nucleic acid moieties. Changes in water content, therefore, occur in all microorganisms while airborne and during sampling, and this represents the most fundamental potential stress. The act of sampling itself can induce significant viability losses, as can prehumidification before and during collection as lost water is slowly replaced. Viruses without structural lipids are also destabilized by desiccation, and viruses with structural lipids are least stable at high relative humidities. Different bacteria species have different optimum aerosol survival temperatures, which in most cases is about +10°C. Nonviable microbes may not be considered as dead because under certain conditions damaged moieties can be repaired.

Bacterial aerosols have a broad range of particle sizes, and selective sampling methods sometimes have advantages. Microbial particles often occur in the atmosphere as agglomerates, i.e., coagulates with other material such as dust particles or fog droplets, etc., and in a dusty atmospheric environment microorganisms may be carried on particles larger than 3 µm in diameter.[34]

Bulk samples of bacterial aerosol particles can be collected on a solid or into a liquid. Such samples can then be used directly or after further processing. Membrane and Nuclepore filters are the most useful tools for filtration sampling,[35] while cascade and virtual impactors, aerosol centrifuges, and thermal and electrostatic precipitators enable sample collection onto thin organic films.[8] In the past, gelatin porous filters were used successfully. These filters are soluble in water. Therefore, after air sampling, such a filter was dissolved, and the resulting water–gelatin suspension of sampled bacteria was directly used for evaluation by counting the bacterial colonies.[36]

Sampled particles can then be analyzed and evaluated either in bulk or in some cases as a collection of individual particles.[7,8] In the latter case, the analytical equipment (e.g., mass or micro-Raman spectrometer) has to be combined with a microscopical tool (LM, SEM, etc.). For several methods, e.g., fluorescence, luminescence, infrared and Raman spectroscopies, as well as for flow cytometry and electrochemical methods, samples suspended in liquids are desirable, and for these, impinger methods and mainly the "wet microcyclones" provide the most suitable samples.

Preparation and optimization of liquid suspension or solid samples for different analytical procedures follow collection of airborne material. For the application of gas chromatographic and mass spectrometric methods, the analyte must first be converted into the vapor phase.[10] In addition, vaporization of biological material must be preceded by fragmentation into smaller units by hydrolysis, and subsequent reaction with suitable compounds gives relatively volatile derivatives.[9,37,38]

The second method involves heating the sample in a nonoxidizing atmosphere (pyrolysis), whereby large molecules are fragmented to produce small units in the vapor phase. Fragmentation takes place at preferred junctions in the molecules, and the fragments contain information about their parent molecules. Several designs of pyrolyzers have been reported in the literature,[39-41] and are classified in terms of their mode of operation, either the continuous (e.g., furnace) or pulse. The pulse mode pyrolyzers are commonly used and include resistively heated pyrolyzers, Curie-point pyrolyzers, and laser pyrolyzers. Encouraging results have been obtained with Curie-point pyrolyzers.[42-44] In this method, a small sample (5 to 20 µg) is deposited directly onto a ferromagnetic wire, and high frequency inductive heating is employed to raise the temperature of the wire to its Curie temperature. It takes about 100 ms to reach equilibrium temperature, and typically, the sample is pyrolyzed in 1 to 10 s. Fully automated systems for Curie-point pyrolysis by gas chromatography and mass spectrometry are available.[42]

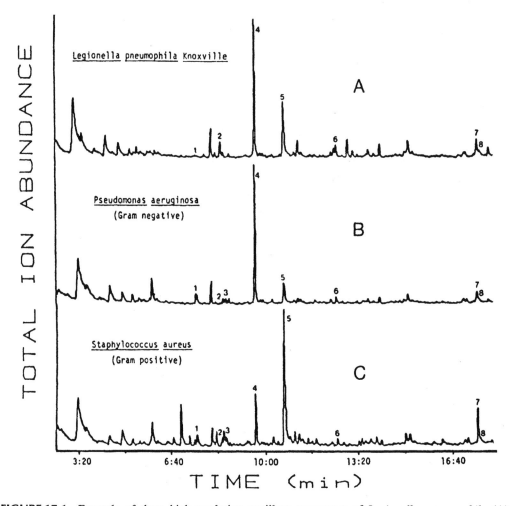

FIGURE 17.1 Example of three high-resolution capillary pyrograms of *Legionella pneumophila* (A), *Pseudomonas aeruginosa* (B) and *Staphylococcus aureus* (C). (From Fox, A. and Morgan, S.L. The chemotaxonomic characterization of microorganisms by capillary gas chromatography and gas chromatography–mass spectrometry, in W.H. Nelson (Ed.) *Instrumental Methods of Rapid Microbiological Analysis.* VCH Publishers, 1985, pp. 135-164. With permission.)

Laser pyrolysis has many desirable features,[8,42] including high temporal and spatial resolution, very fast heating rate, and ability to pyrolyze extremely small amounts of sample (e.g., 10^{-12} g or less), enabling single cell analysis.

17.3 DETECTION, ANALYSIS, AND IDENTIFICATION

17.3.1 BULK METHODS

17.3.1.1 Chromatography

Capillary gas chromatography (GC) and GC couplet with mass spectrometry (GCMS) are fast, specific, and sensitive techniques for characterizing microorganisms directly. Furthermore, analyzing structural components rather than metabolic products does not require that organisms be viable, and consequently, lengthy secondary (and possibly primary) culturing may be avoided.

Pyrolysis gas chromatography (PGC) can be used for detection and identification of microorganisms as a bulk method, and pyrograms of sampled microorganisms predominantly contain amino acids, sugars, carbohydrates, purine and pyrimidine bases, etc. PGC was first applied in the 1960s and demonstrated that pyrograms of microorganisms and cell fractions generally are very complex (Figure 17.1). Consequently, microorganisms have to be differentiated by two or three characteristic peaks in a phyrogram containing perhaps 200 or more components. In the past, this meant tedious visual comparisons of many complex pyrograms, but since the beginning of the 1970s, computerized pattern recognition techniques have been used to speed up the data analysis.[45-47]

Mass spectrometry (MS) is a critical component of this chemotaxonomic analytical approach because GC has nonselective detection (e.g., flame ionization) that does not permit unequivocal identification of chromatographic peaks. MS improves selectivity as well as allowing identification, or at least partial characterization, of the chemical nature of the constituents of interest. The combined technique of GCMS can separate and identify only volatile components, but unfortunately most macromolecular components of bacterial cells are not volatile. Therefore, bacterial samples have to be treated before GC, MS, and GCMS analysis. The most suitable treatment is thermal fragmentation, i.e., pyrolysis. Further important improvements in sensitivity and selectivity can be achieved by using the new technique, FTMS (Fourier transform mass spectrometry).[38] Further important improvements can be achieved by the application of the gas chromatography/ion-trap tandem mass spectrometry (tandem MS or GC/MS/MS).[48] This procedure provides information about the presence of different microorganisms. Several chemical markers, such as ergosterol, muramic acid, and 3-hydroxy fatty acids (Figure 17.2), are very useful for the determination of microorganism species. The low-picogram range of such markers can be identified.

17.3.1.2 Fluorescence and Luminescence Spectroscopy

Fluorescence- and luminescence-based identification techniques generate data only from some molecular components of the microorganisms. These are, therefore, more specific.[49] Primary and secondary fluorescence spectroscopical methods may be subdivided into direct and indirect methods. Primary fluorescence methods are those in which naturally fluorescent components of bacteria are examined. Consequently, only bacteria containing or producing some fluorescent pigment may be examined by primary fluorescence methods, but these become less specific when applied to a mixture of bacteria. Secondary fluorescence methods, in contrast, involve introducing a foreign fluorophore to the bacteria before identification. Such methods, therefore, do not depend on any natural fluorescence and can be applied more generally.

Immunofluorescence assays are of particular promise. Microbes that are inhaled by an animal host provoke an allergic response,[50] and the ensuing antibodies can be identified by specific fluorescence measurements, usually by addition of fluorescein.

Both fluorescence procedures may be used as direct and indirect methods. Direct analyses are those in which the bacterial cells are present during the analysis, unlike indirect methods in which the cells are absent.[51] Direct immunofluorescence assays are highly specific, and often the bacterium of interest can be detected in the presence of many similar bacteria. Some useful procedures were published recently.[52-55]

Chemiluminescence can be defined as light, usually visible, that is emitted as part of the energy released during certain exothermic reactions.[56-58] A variety of substances exhibit chemiluminescence, with the most suitable for bacteria being luminal. When bacteria are placed in a highly alkaline media (pH 13), they lyse and the hememoiety is released into the reaction mixture. The luminescence techniques can be used for the detection, but not for the identification, of microorganisms, but nevertheless, are very sensitive and permit the detection of about 10^3 microorganisms per milliliter.

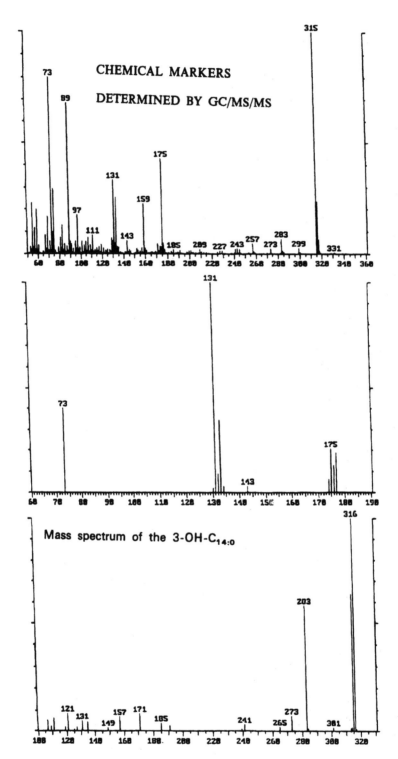

FIGURE 17.2   Mass spectrum of the 3–OH–$C_{14:0}$ fatty acid methyl ester/TMS (trimethylsilyl derivate) (top) fragment ion spectra (center and bottom). (From Saraf, A. and Larsson, L. Use of gas chromatography/ion-trap tandem mass spectroscopy for the determination of chemical markers of microorganisms in organic dust. *J. Mass Spectr.* 31, 389-396, 1996. With permission.)

Bioluminescence is caused by a sequence of reactions involving firefly luciferin, adenosine, triphosphate, luciferase, magnesium, and oxygen. Therefore, chemiluminescent techniques have been applied successfully in detecting bacterial aerosols.[59,60] The bioaerosol is sampled by some suitable collection method (e.g., wet cyclone air sampler), and the bacteria are collected into a solution. Iron porphyrins (e.g., heme) from bacteria catalyze the reaction between luminal and an oxidizing radical (e.g., perborate). Sodium hydroxide causes bacterial lysis, thereby releasing heme, and is also necessary for the luminal reaction:

$$\text{Luminol} + \text{ROOH} \xrightarrow[\text{hematin}]{\text{NaOH}} \text{aminophthalate} + \text{ROH} + N_2 + h\nu$$

The emitted light ($h\nu$) is proportional to the amount of heme released from the bacteria, provided the reaction takes place under fixed conditions. Light emission is measured with a photomultiplier tube, and the signal is registered by a recorder or a display. Standard calibration curves for different bacteria can be constructed, and a mean value curve from these curves can be used to determine amounts of bacteria in an unknown sample.

17.3.1.3   Infrared and Raman Spectroscopy

Infrared (IR) spectra applied to identifying bacteria was first reported in the 1950s.[49] Sampled bacteria are smeared onto an IR cell, and IR absorbance spectra are acquired using conventional instrumentation. The resultant spectra represent the chemical composition of the bacterium under investigation and are generally complex and broad band because of the large number of components in the spectra. These constraints limit the application of this technique, though Fourier transformation techniques (FTIR) are more promising.[61] Raman spectra are obtained by placing a sample in a beam of monochromatic radiation and measuring the intensity of the light scattered at longer wavelengths (or smaller values of frequency). Both Raman and IR can also be applied to single particle identification.

17.3.1.4   Biosensors

Electrochemical, immunochemical, and immunobiological methods are among very promising modern procedures which will probably influence the satisfactory solution of bacterial aerosol detection and analysis in the future.[62-66] Decreased electrical impedance of a culture occurs when actively metabolizing bacteria utilize large molecules in a culture medium and form smaller ion pairs. The sensitive instruments now available can usually detect actively metabolizing bacteria when 10^6 to 10^7 bacteria per milliliter are present in the culture liquid. In electrochemical methods, bacteria must be viable and their concentration in the liquid milieu relatively high. Application of these procedures to bacterial aerosols are, therefore, not very promising, unless concentrated samples are available.

The recent development of sensors using immunobiological techniques offers excellent selectivity through the process of antibody–antigen recognition and has revolutionized many aspects of chemical and biological sensor technologies.[63,64,67]

Immunosensors use different techniques, e.g., electrochemistry, radioactive labeling, piezoelectricity, spectroscopic methods, etc., while rapid and sensitive detection of toxins and pathogens has been accomplished using a light addressable potentiometric sensor (LAPS).[64] The detection step is preceded by the formation of a streptavidin-antibody/antigen/antibody-urease complex followed by its capture on a biotinylated nitrocellulose membrane. The LAPS is a semiconductor-based detector which monitors changes in pH over time due to conversion of urea to ammonia by urease in the nitrocellulose-bound toxin/pathogen antibody complex. Automated LAPSs have a continuous

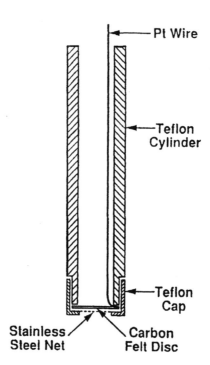

FIGURE 17.3 Schematic layout of the carbon felt electrode configuration. (From Rishpon, J., Gezundhajt, Y., Soussan, L. et al. Immunoelectrodes for the detection of bacteria, in *Biosensor Design and Application.* Mathewson, P.R. and Finley, J.W. (Eds.) ACS Symposium Series 511, 59-72, Am. Chem. Soc. Washington, D.C., 1992. With permission.)

tape of nitrocellulose on which analytes drawn in by the air sampler and reacted with the appropriate antibody mix are captured. The tape is automatically positioned under both the filtration unit and the LAPS unit. The detection limit is approximately 6000 organisms per sample.

It has been shown[68] that electrochemical immunosensors, which can detect the corresponding antigen, are extremely sensitive. In electrochemical enzyme immunoassays, an antigen or an antibody is ordinarily tagged with an enzyme and the enzymatic reaction is monitored by a potentiometric or amperometric electrode. The methods for the determination of bacteria are based on an enzyme-tagged immunoelectrochemical assay. Antibodies are immobilized on disposable carbon felt disc electrodes (Figure 17.3) and are used to capture antigens in test solutions. After a short incubation with the second antibody, which is labeled with the enzyme alkaline phosphatase, the activity of the enzyme electrode thus formed is measured (Figure 17.4). This enzyme reacts with the substrate, aminophenyl phosphate, and the product of this enzymatic reaction, p-aminophenol, is detected amperometrically. The use of rotating electrodes significantly shortens the incubation time and, together with a computerized electrochemical system, results in extremely high sensitivity. Concentrations as low as 10 cells ml^{-1} can be detected in less than 10 min (Figure 17.5).

Biosensors seem to be very promising tools for fast and sensitive detection of bacterial aerosols. They can be applied for the identification of very small samples. New techniques and instruments include fibre optic waveguides, surface plasmon resonance, vibrating mirrors, etc. Because of special bioreaction mechanisms, the bioassay methods deserve a separate review article.

A new sophisticated and very progressive method for remote sensing of bacterial aerosols using an immunoassay was published recently (see Lingler, F.S., Anderson, G.P., Davidson, P.T., et al., Remote sensing using an airborne biosensor. *Environ. Sci. Technol.* 32, 2461-2466, 1998). A fiber-optic biosensor capable of running four simultaneous immunoassays was integrated with an automated fluidics unit, a cyclone-type air sampler, a radio transceiver, and batteries on a small, remotely piloted airplane. The biosensor system is able to collect aerosolized bacteria in flight, identify them, and transmit the data to the operator on the ground. The method is, therefore, very useful for

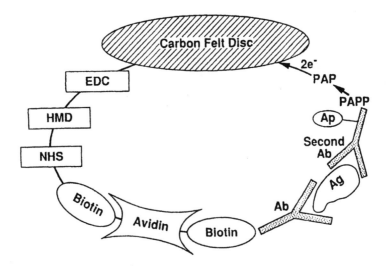

FIGURE 17.4   Scheme showing the immunoelectrode structure. (From Rishpon, J., Gezundhajt, Y., Soussan, L. et al. Immunoelectrodes for the detection of bacteria, in *Biosensor Design and Application*. Mathewson, P.R. and Finley, J.W. (Eds.) ACS Symposium Series 511, 59-72, Am. Chem. Soc. Washington, D.C., 1992. With permission.)

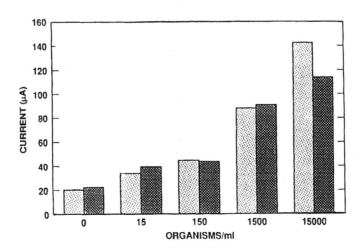

FIGURE 17.5   Detection of *Staphylococcus aureus*. Amperometric response of the immunoelectrode after incubation with different concentration of bacteria. The dark and light bars represent two sets of measurements. (From Rishpon, J., Gezundhajt, Y., Soussan, L. et al. Immunoelectrodes for the detection of bacteria, in *Biosensor Design and Application*. Mathewson, P.R. and Finley, J.W. (Eds.) ACS Symposium Series 511, 59-72, Am. Chem. Soc. Washington, D.C., 1992. With permission.)

warning of a biological warfare attack prior to human exposure. The detection limit lies in the range of 10^3 bacterial particles in 1 ml of the sampling fluid.

17.3.2   SINGLE PARTICLE DETECTION

Combining lasers with some classical spectrometric methods, such as atomic emissions and adsorption spectroscopy, molecular infrared and Raman spectroscopy, and fluorescence and mass spectroscopy, has greatly improved their performance such that the latter can now provide very fast and sensitive analyses and identification of very small amounts of elements or compounds, even in single airborne and waterborne particles.[8]

17.3.2.1 Microscopy

Light microscopy (LM), scanning electron microscopy (SEM), transmission electron microscopy (TEM), analytical electron microscopies (ASEM and ATEM), as well as very modern scanning near-field/atomic force microscopy (SNOAM), have been used for biological microanalysis,[69] as well as for counting and identifying single bacterial particles or colonies sampled from the atmospheric environment by impaction and filtration.[70] Nonspecific identifications or differentiations are then possible. Air particle sizes and forms, as well as their staining by different dyes, can be observed and measured. Specifically stained biological particles can be examined with automated LM. Fluorescence light microscopy (FLM) and fluorescence scanning microscopy (FSM) provide the most sophisticated methods for microscopical evaluation of airborne microorganisms. FLM and ethidium bromide as a selective fluorescence dye has provided a useful method for ambient measurement and for CFU (colony-forming units) estimation.[71,72] Applications for FSM are also promising because of high resolution and better automation facilities.[73] In FSM, the higher energy laser interacts with the specimen (e.g., Nuclepore filter with bacterial particles), which emits lower energy fluorescent light that passes back through the microscope optics, scanner, and dichroic to a sample achromatic lens that focuses it on an aperture. A detector (usually a photomultiplier tube) picks up this fluorescence and the output signal is amplified and digitized. An image is then formed by electronically storing and processing the single picture elements obtained from the scanned area.

SEM, ASEM, TEM, and ATEM have also been successfully applied in the identification of airborne microorganisms. Millipore and Nuclepore filters are suitable for sampling, and individual microbial particles may be counted on the filter specimen using magnifications of about 3000×.[74-79] ASEM and ATEM enable single microbial particles to be analyzed by the EDXA (energy dispersive X-ray analysis) procedures to give elemental spectra from each individual particle.[8,69] Newer analytical electron microscopes also include analysis of very light elements (e.g., carbon) and, by means of such fingerprinting methods, biological and mineral particles can be distinguished. In other investigations, combinations of LM and FM (fluorescence spectroscopy) with SEM have been used successfully.[74]

SNOAM is a relatively new procedure which has a spatial resolution of only a few tens of nanometers and sensitivity down to the single molecule level. It can have very useful applications in the bioanalytical field. This technique is analogous to STM (scanning tunnel microscopy) and to AFM (atomic force microscopy). SNOAM was successfully applied for simultaneous topographic and fluorescence imaging of biological samples in air and liquid. Optical resolution of this system is about 50 to 100 nm in the fluorescence mode. An example of the application of ANOAM in the investigation of *Escheria coli* bacteria is shown in Figure 17.6.[80] A distinction between mineral and biological particles is possible.

17.3.2.2 Flow Cytometry

In contrast to immunofluorescence assays, which use one highly specific reaction for identification of a bacterium, flow cytometry employs much less specific reactions to form a characteristic "response pattern."[81-83] In flow cytometry, a sample of bacteria is stained with one or more fluorescent dyes, after which the pattern of dye uptake for each bacterium is determined with a flow cytometer (Figure 17.7) In the case of the analysis of a bacterial aerosol, airborne microbial particles first have to be collected in liquids, e.g., by a wet microcyclone.

Most flow cytometric systems allow simultaneous measurement of cell volume and two or more fluorescent signals (Figure 17.8), and sophisticated flow cytometers employing mercury arc or laser light sources are commercially available.[81] Bacterial numbers, sizes, and identification patterns can be provided simultaneously. The derived histograms provide useful fingerprints for distinguishing species and this method, named mixed dye-fluorometry, is highly specific and may be used for multicomponent bacterial detection.

The ability of a microbial cell to reproduce itself on a nutrient agar plate constitutes the benchmark method for determining how many living cells may be contained in the sample of

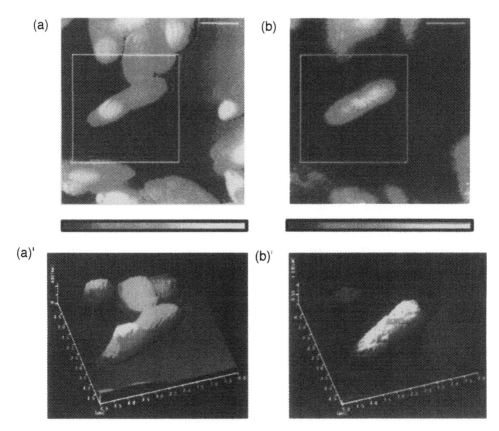

FIGURE 17.6   Topographic (a, a′) and near-field fluorescence (b, b′) images of *Escherichia coli* generated by green fluorescent protein. (From Tamiya, E., Shinichiro, I., Nagatani, N. et al. Simultaneous topographic and fluorescence imagings of recombinant bacterial cells containing a green fluorescent protein gene detected by a scanning near-filed optical/atomic force microscopy. *Anal. Chem.* 69, 3697-3701, 1997. With permission.)

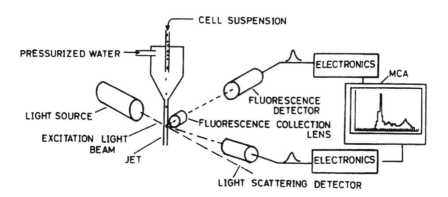

FIGURE 17.7   Schematic representation of a flow cytometer with detectors for fluorescence and light scattering measurement. (From Hadley, W.K., Waldman, F. and Fulwyter, M. Rapid microbial analysis by flow cytometry, in W.H. Nelson (Ed.) *Instrumental Methods of Rapid Microbiological Analysis,* VCH Publishers, 1985, pp. 67-89. With permission.)

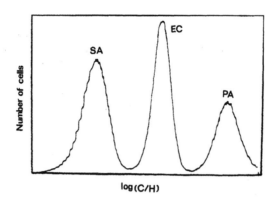

FIGURE 17.8   Distribution of the ratio of two fluorescence signals (C/H) over the cell population of *Staphylococcus aureus* (SA), *Escherichia coli* (EC) and *Pseudomonas aeruginosa* (PA) after double DNA staining. (From Hadley, W.K., Waldman, F. and Fulwyter, M. Rapid microbial analysis by flow cytometry, in W.H. Nelson (Ed.) *Instrumental Methods of Rapid Microbiological Analysis,* VCH Publishers, 1985, pp. 67-89. With permission.)

interest. However, there are also several "rapid" methods which also allow a speedier assessment of the "viable" microbial load in a sample. Flow cytometry is one of them. For example, by using the fluorescent dye rhodamine, "viable" and "nonviable" cells can easily be quantitatively distinguished in the flow cytometer by the extent to which they accumulate the dye.[84]

17.3.2.3   IR and Raman Spectroscopy

The Raman spectrum of a biological molecule is, in effect, a vibrational spectrum and modern Raman spectrometers (e.g., micro-Raman laser spectrometry) permit routine examination of solid or liquid microparticles.[85] High-quality Raman spectra in the visible region have been obtained from very small aggregates of chromobacteria, and useful spectra obtained from individual cells in bacterial mixtures. Argon laser excited resonance Raman spectrometry has been successfully applied to such measurements, and organisms present in the vicinity of the laser beam and in the beam itself (Figure 17.9) were observed and counted with ease by means of a television image via vidicon tube attached to the microscope. Figure 17.10 shows spectra of *R. palustris*. These results are most promising for the successful detection of bacterial aerosols, because the method permits analysis of airborne bulk samples as well as single microbial particles.[86-88] Also IR spectroscopy can be used in limited cases for bioparticle analysis.

17.3.2.4   Atomic Spectroscopy

One other modern analytical technique that can be exploited for nonspecific detection of single particles, including bio-particles, is atomic spectroscopy, particularly atomic emission spectroscopy (AES). Classical flame photometry,[89] as well as atomic absorption spectroscopy (AAS) and inductively coupled plasma spectroscopy (IPC), provide specific analysis of single aerosol particles.[90,91]

Applications of AES to atmospheric aerosol analysis started in the 1950s and was much improved in the 1980s, allowing the detection of elements such as K, Na, Li, Ca, and Ba in particles smaller than 1 μm. Organic particles (e.g., carbon and phosphorous, and some pyrolysis products) can be detected in the flame or in the gas plasma. Some new combinations (e.g., AES with GC and MS) further promote application of these techniques in microbiology as well as in the detection of bacterial aerosols.

Micro Raman spectroscopy of single particles

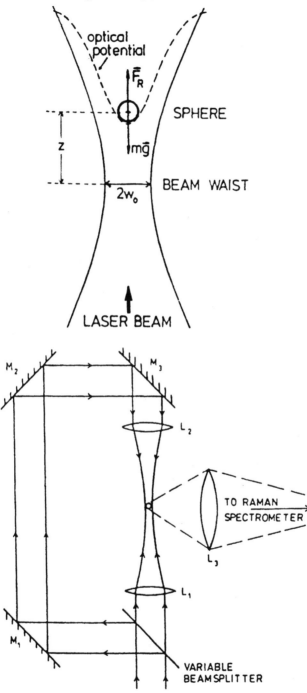

FIGURE 17.9 Schematic diagrams of spherical particles sitting in one or two (lower picture) laser beams. (From Kaprelyants, A.S. and Kell, D.B. Rapid assessment of bacterial viability and vitality by rhodamine 123 and flow cytometry. *J. Appl. Bacteriology* 72, 410-422, 1992. With permission.)

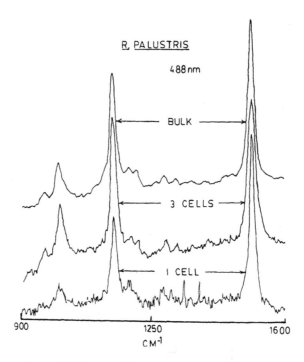

FIGURE 17.10 A comparison of bulk spectra from pure cultures and single cell detection spectra. (From Schweiger, G. Single microparticle analysis by Raman spectroscopy. *J. Aerosol Sci.* 20(8) 1621-1624, 1989. With permission.)

17.3.2.5 Fluorescence Bacterial Aerosol Particle Counters

Real-time discrimination between bacterial and other aerosols in the outdoor and indoor environments are possible by using fluorescence of several primary compounds of the microorganisms, namely amino acids tryptophan and tyrosine, reduced adenine dinucleotides, flavins, etc.[92] Combining the already well developed and commercially available optical particle counters and particle sizers with an aerosol-fluorescence spectrum analyzer, particle numbers and sizes, as well their fluorescence emissions produced by an irradiation with a UV laser, can be detected fast, on-line, *in situ,* and in real time. Several recently published studies[93-101] have made it possible to develop useful laboratory equipment including the first commercial equipment, such as, for example, fluorescence aerodynamic particle sizer (FLAPS), ultraviolet aerodynamic sizer spectrometer (UV-APS), etc.

Useful instruments for detection and discrimination between biological and nonbiological aerosols were developed and tested mainly during the 1990s. For fast fluorescence detection of airborne bacteria, flavins were first chosen as useful markers. All bacteria and bacterial spores contain flavins. Flavins have a broad excitation maximum near 450 nm and their fluorescence emissions peak between 515 and 565 nm. Therefore, argon lasers (488 nm), for example, could be used successfully for the excitation source. The working function of such an apparatus, the fluorescence particle counter (FPC), is illustrated in the block diagram in the Figure 17.11.[95]

The aerosol stream (particles) intersects the argon laser beam. The elastic scattering and fluorescence of each particle is detected, measured, and registered. The argon-ion laser excites the aerosols' fluorescence and elastic scattering. A spectrograph and a detector, two multiplier tubers, and processing electronics gather the necessary information about single airborne particles. Also other laser types, such as the KrF excimer laser (248 nm) and the Cr:LiSAF laser, tunable from 265 to 290 nm, have also been used in other equipment.[96,97]

Further progress in this field was achieved by developing a portable apparatus which combines the techniques of fluorescence measurement, commonly used in flow cytometry, and aerosol particle sizing to make possible two parameter measurements of individual particles in a sampled air flow.

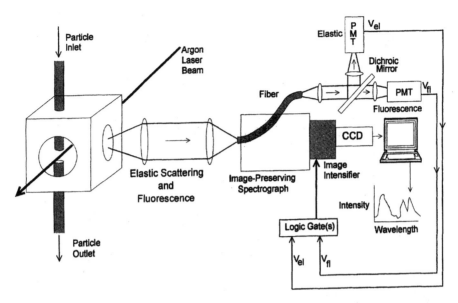

FIGURE 17.11 Block diagram of the conditional-sampling spectrograph detection system. (From Nachman, P., Chen, G., Pinnick, R.G. et al. Conditional-sampling spectrograph detection system for fluorescence measurements of individual airborne biological particles. *Appl. Optics.* 35(7)1069-1076, 1996. With permission.)

It can provide data to support three-dimensional histograms displaying accumulated particle counts vs. particle size and fluorescence. The output data are useful in revealing changes in concentration and specific airborne particle types in the sampled air, especially for rapidly ascertaining the likely presence and concentration of bacterial aerosols.

This apparatus, called the fluorescence aerodynamic particle sizer (FLAPS), can also distinguish between bacterial aerosol particles and inanimate material. Respirable particles (sizes between 0.5 and 15 μm) are sized by the time-of-flight method, initiated by an He–Ne laser, while biological-related fluorescence is elicited with excitation by He–Cd-UV laser at 325 nm.

The fluorescence characteristic is suggestive of biological properties inherent to the particle.[98] A commercial apparatus based on these investigations has been developed. Its name is the ultraviolet aerodynamic particle sizer (UV-PAS).[101] Its working scheme is illustrated in Figure 17.12, and the entire portable device is shown in Figure 17.13. Particle detection and characterization is realized by means of three measurements. In the first process, a time-of-flight signal is used to determine particle size and light-scattering intensity. The signal then triggers the second process, which targets the particle with a UV laser and measures any resulting fluorescence. The following section provides more detail on the unique method of operation. An example of resulting model measurement is illustrated in Figure 17.14. Time-dependent particle (P) counts and the percentage of fluorescence (F) are plotted.

17.3.2.6 Mass Spectrometry

MS has also proved to be a very promising modern method for chemical detection and identification of microorganisms, including bacteria.[102-117] For bacterial aerosols, the most suitable mass spectrometer is the pyrolysis mass spectrometer (PyMS). For mass spectrometric analysis of bacterial aerosols, laser pyrolysis has many advantages, including high temporal and spatial resolution, rapid heating rates, and the ability to pyrolyze extremely small amounts of sample (e.g., 10^{-12} g or less) so that single-cell analysis is possible. Laser pyrolysis mass spectrometers are available commercially and a quadruplet Neodymium–YAG laser is used for the laser microprobe mass analyzer (LAMMA).[107,108] The laser is focused to micron size and has a spatial resolution of this order. Special cascade impactors,[109] or filters (mainly Nuclepore filters[110]) are used for sampling airborne particles, including microorganisms.[111,112] The specimen in LAMMA is observed by light microscopy, and single particles are irradiated by the high-energy laser beam, pyrolyzed, and analyzed

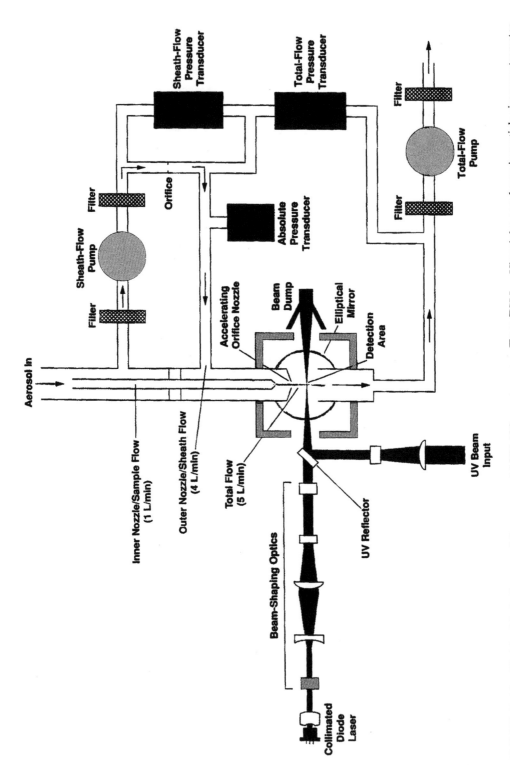

FIGURE 17.12 Working block diagram of the commercial UV-APS-spectrometer. (From TSI Inc. Ultraviolet aerodynamic particle sizer spectrometer. Information sheet, 1997. With permission.)

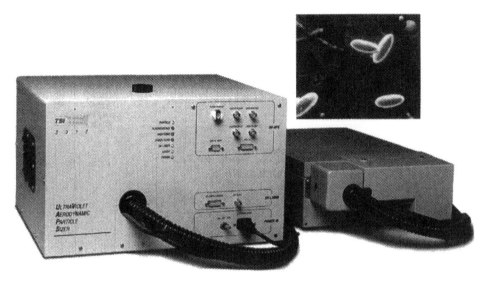

FIGURE 17.13 Photograph of the portable apparatus UV-APS spectrometer. (From TSI Inc. Ultraviolet aerodynamic particle sizer spectrometer. Information sheet, 1997. With permission.)

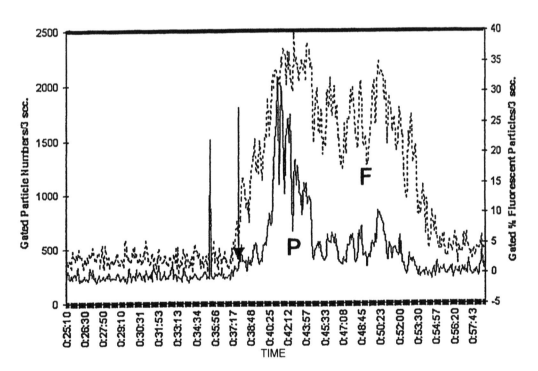

FIGURE 17.14 Example illustrating an on-line contemporal measurement of particle numbers/3 sec. and the correspondent percentage of fluorescence.

by the time-of-flight (TOF) mass spectrometer. Particle concentration, as well as mass spectra, of individual single bacterial particles are obtained. However, sample evaluation is time consuming. LAMMA-procedures are, therefore, more useful for calibrating the characterization of different types of bacteria.

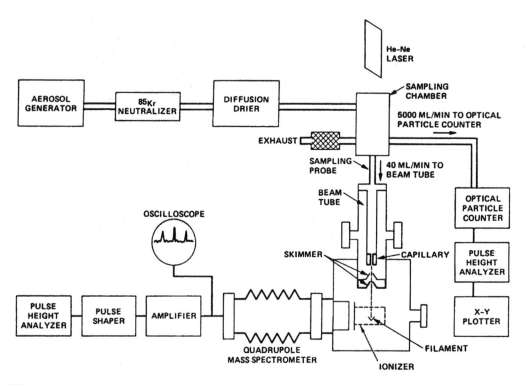

FIGURE 17.15 Schematic diagram of particle beam mass spectrometry apparatus. (From Sinha, M.P. Analysis of individual biological particles in air, in W.H. Nelson (Ed.) *Instrumental Methods for Rapid Microbiological Analysis.* VCH Publishers, 1985, pp. 165-192. With permission.)

Sinha[103] developed the first useful modification of PyMS. His particle analysis by mass spectrometry (PAMS) was an *in situ* and on-line procedure in which aerosol particles were introduced directly from the atmosphere in the form of a particle beam into the ion source of the mass spectrometer. This method enabled analysis of single particles on a continuous, real-time basis. The main components of the PAMS system were the particle beam generator and a quadruple mass spectrometer. Particle beams (analogous to molecular beams) are produced when the aerosol expands through a fine nozzle into a vacuum.[105,106,113] The particle beam enters the mass spectrometer chamber and individual particles are pyrolyzed. Two different pyrolysis systems have been examined and used. In the first, the accelerated aerosol particle impacts a V-shaped zone of refined rhenium filament (Figure 17.15). The filament is heated resistively and maintained at a constant temperature in the range of 200 to 1400°C. The particle striking the filament produces a plume of vapor molecules; these are ionized by electron impaction *in situ*. The second pyrolysis system uses multiple lasers, with the particles being pyrolyzed and ionized by a laser pulse while in flight in the aerosol beam.[103] The Nd–YAG laser used for this pyrolysis, volatilization, and ionization of the aerosol particles delivers 1 J of energy per pulse. In a further development, a "three-laser system" was used: one for measuring particle size, the second for volatization and pyrolysis, and the third for additional ionization.[113,114]

Bacterial particles in the aerosol beam are pyrolyzed into small mass fragments, and a burst of ions is produced from individual particles after volatilization and ionization by electron impaction in the ion source of the mass spectrometer (or in the additional laser). The mass spectrometer then measures the intensities of the different masses by scanning them in time. Intensities of different mass peaks are normalized to the most intense peaks in their respective spectrum (Figure 17.16). The similarity between spectra is due to the microorganisms having the same major chemical building blocks. However, some differences are apparent by visual inspection, while more objective

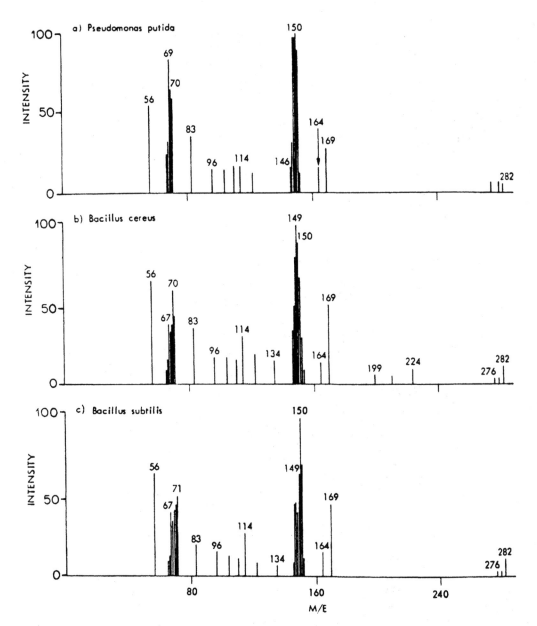

FIGURE 17.16   Mass spectra of three different bacterial particles. (From Sinha, M.P. Analysis of individual biological particles in air, in W.H. Nelson (Ed.) *Instrumental Methods for Rapid Microbiological Analysis.* VCH Publishers, 1985, pp. 165-192. With permission.)

comparisons of mass spectra and their reproducibility can be made by applying statistical procedures.[105,115] A library of mass spectral data of different microorganisms grown under different conditions should be compiled for identification by comparison. Further improvement in bacterial chemotaxonomic characterization and fingerprinting has been achieved by applying fast atom bombardment mass spectroscopy (FABMS). Positive and negative ion fast bombardment mass spectra may be obtained from lysed bacterial without extraction, because phospholipids and other polar lipids are selectively desorbed to provide molecular ions. FABMS may also serve as an aid in differentiating between microorganisms.[116,117]

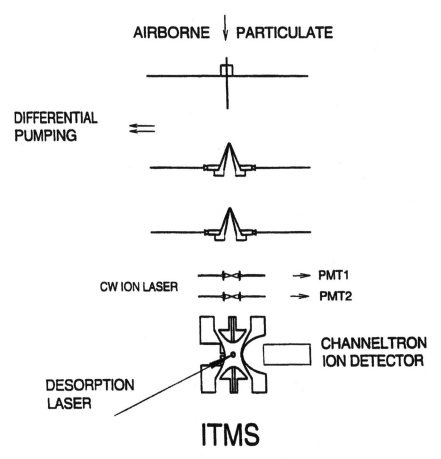

FIGURE 17.17 Schematic diagram of the experimental set up of the ITMS apparatus. (From Yang, MO, Reilly, P.T.A., Gieray, R.A., Whitten, W.B., and Ramsey, J.M. Complete chemical analysis of aerosol particles in real-time. *J. Korean Phys. Soc.* 30(2) 359-363, 1997. With permission.)

17.3.2.7 Recent Progress in Mass Spectrometric Bacterial Detectors

In several recently published papers, the usefulness of mass spectrometric methods for the analysis and identification of single aerosol particles, including bacteria, has been confirmed.[118-122] Similar to the fluorescence spectrometers, the new mass spectroscopic techniques are able to characterize an aerosol or bacterial particle in real time by size and composition.

An important improvement in sensitivity and identification has been achieved by using the tandem mass spectrometry of individual particles.[118,119] Figure 17.17 shows the principal diagram of such equipment. A laser desorption mass spectrometer (LDMS) combined with the MS/MS analyzer was used. Approximately 600 ml of air with particles enter the inlet system per minute. The particles are accelerated to nearly the speed of sound (150 to 400 m/s). Two parallel CW laser beams are focused on the particle beam. An ion trap mass spectrometer (ITMS) is used for the mass analysis.

In addition to the primary MS and MS/MS analysis, information about the particle size estimated by velocity measurement enhances the specificity of the microparticle characterization. Chemical detection limits appear to be well below femtograms per particle.

Individual bacterial aerosol particles are characterized by the mass spectrum (Figure 17.18). A further fragmentation by MS/MS enables a better fingerprinting for bacterial species. The bacterial aerosol usually consists of single cells as well as cell aggregates (Figure 17.19).

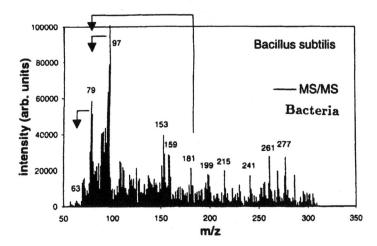

FIGURE 17.18 Average negative ion mass spectrum of *B. subtilis*. (From Yang, MO, Reilly, P.T.A., Gieray, R.A., Whitten, W.B., and Ramsey, J.M. Complete chemical analysis of aerosol particles in real-time. *J. Korean Phys. Soc.* 30(2) 359-363, 1997. With permission.)

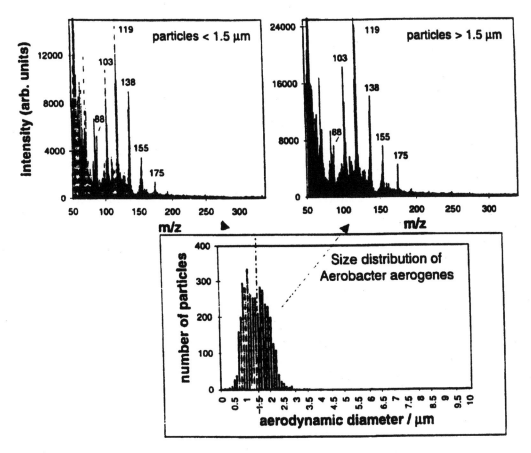

FIGURE 17.19 Size distribution and averaged positive ion mass spectra of *Aerobacter aerogens* for two particle size fractions. (From Yang, MO, Reilly, P.T.A., Gieray, R.A., Whitten, W.B., and Ramsey, J.M. Complete chemical analysis of aerosol particles in real-time. *J. Korean Phys. Soc.* 30(2) 359-363, 1997. With permission.)

A large number of ions are produced from a single airborne microorganism, and there are several prominent ions that appear in the mass spectrum. The presence of these ions may make possible a real-time single particle discriminator for microorganisms and particles of nonbiological origin. This means that it is possible to obtain a characteristic response similar to the response of a fluorescence particle counter.

Bipolar on-line laser mass spectrometry is a further important step on the road to the improvement of the mass spectrometric determination of single airborne particles.[120-122]

As has been shown, on-line mass spectrometry has been used successfully for the analysis of individual particles since the 1980s. The ability of a bipolar time-of-flight mass spectrometer results in an increase in accuracy in determining the chemical components of particles, when compared to classical off-line unipolar mass spectrometry.

The simultaneous detection of a positive ion spectrum and a negative ion spectrum for the same particle for each laser shot provides a fast and precise chemical characterization of single particles. A remarkable gain in analytical information can be obtained, and highly improved particle classification is possible.[120,121] Because of more complex information, differentiations among several bacterial species could be realized after further future investigations.

The bipolar on-line mass laser spectrometer was recently used in the development of a portable aerosol time-of-flight mass spectrometer (ATOFMS).[122] This instrument is a powerful new tool for providing temporal and spatial information on the origin, reactivity, and fate of aerosols, including airborne microorganisms. The portable ATOFMS is capable of analyzing the size and composition of individual particles from a polydisperse aerosol in real time. Particles are introduced into the instrument through a particle beam interface, sized by measuring the delay time between two scattering lasers, and compositionally analyzed using a dual polarity laser desorption/ionization time-of-flight mass spectrometer (Figure 17.20). The weight of this transportable apparatus is approximately 250 kg.

17.3.2.8 Secondary Ion Mass Spectroscopy (SIMS)

The well-known SIMS techniques are similar to MS and LAMMA. In principle, SIMS is a surface sensitive technique in which a solid surface under bombardment by so-called primary particles (i.e., atoms or molecules with energies in the KeV range) emits a number of secondary particles form the outermost molecular layers of the specimen. Emitted secondary particles comprise neutral atoms or molecules, positively or negatively charged atomic or molecular ions, excited particles, electrons, and protons. An analysis of the positive and negative secondary ions yields a mass spectrum. This method is very sensitive (detection limit of about 10^{-15} g or less). Bacterial aerosols collected by filters or cascade impactors and frozen may be directly analyzed as a bulk sample or as single particles to give mass spectra of the organic parts of microorganisms, thereby providing fingerprints for different species, or alternatively providing information on the presence of airborne microorganisms.[123,124] The method is sensitive enough to detect and analyze single micron-range bacteria.

Similarly, to the TOF–MS, a TOF–SIMS apparatus was developed and used for imaging and analyzing single cells or bacteria.[124] Furthermore, the new TOF–SIMS method can be used to obtain images of molecular species across a cell surface with the submicrometer ion probe beam. The method was successfully tested on a single-cell organism, *Paramecium*.

17.4   COMBINATIONS OF METHODS

In practical applications, it is necessary to distinguish between two important facets of the detection and identification of bioaerosols. First, the concentration and specification of harmful microorganisms have to be evaluated as indoor and outdoor pollutants; and the health risks should be estimated for workers or for the general population. In these cases, we usually do not need to know the results immediately. A combination of classical incubation methods combined, for example, with impaction

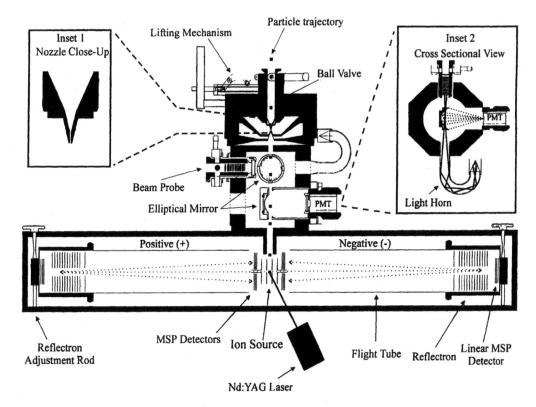

FIGURE 17.20 Schematic picture illustrating the particle beam interface and particle sizing region joined to the mass spectrometer region. (From Gard, E, Mayer, J.E., Morrical, B.D. et al. Real-time analysis of individual aerosol particles: Design and performance of a portable ATOFMS. *Anal. Chem.* 69, 4083-4091, 1997. With permission.)

sampling and some analytical tools are satisfactory and often the best solution. Furthermore, during the last decade the classical sampling methods for bacterial aerosols have been quantitatively well characterized and improved.

The second task deals with dangerous "applications" of bacterial aerosols in crime, war, and other activities. In such situations, for human protection purposes, we needed results in a very short time period (perhaps in less than 1 minute). We, therefore, need very fast, sensitive, and specific procedures. Such methods will also be necessary in the near future for the measurements of bioaerosols in workplaces and in the vicinity of biotechnological industries.

17.4.1 BIOLOGICAL WARFARE AGENTS

Biological warfare (WF) agents are derived from bacterial cultures either in the form of vegetative cells or spores. The major characteristic of these agents is that only about 10 viable microorganisms per person are required to cause an infection. The generation of bacterial aerosol clouds into the atmosphere is accomplished by several mechanical procedures, and the resulting products are bacterial particulate clouds with particle sizes of between 3 and 15 μm. The existing methodology for the detection, identification, and warning systems is based on the combination of methods characterized in earlier chapters.

In the mid-1970s, the already mentioned chemiluminescence wet chemistry technology, consisting of a two-step procedure (collection and detection modules[59]), was used in the U.S. This system measured the presence of biological iron in aerosol particles. Unfortunately, the method has been found to be nonspecific and unreliable.

The existing military protection strategies recommend the application of the fluorescence aerodynamic particle sizer as the first and fastest detection step. Particle size, concentration, and fluorescence are measured and detected at the same time, on-line and in real time (Figure 17.14). When the obtained parameters, characteristic size range (e.g., 3 to 15 μm), particle concentration, and fluorescence are increasing, there exists a potential danger of the presence of man-made harmful bacterial aerosols. In such a case, the use of a immunobiological method should follow. This method provides useful results in a few hours.

The third procedure belongs among the classical methods. The "unusual" aerosol is collected on filters which are stored dry for later microbiological evaluation. The results are available in days. All these sampling and measuring units can be contained in a mobile and easily transportable monitoring system. Based on the results of recent research, there are real options for further improvement of fast detection and identification technology. For example, fluorescence detectors could be extended for measuring the complete fluorescence spectra of single particles, thereby detecting the most dangerous bacterial aerosol groups (e.g., *Bacillus anthraxis,* etc.). By coupling biochemical and immunobiological reactions, the detection process might be marginally sped up.

Last, but not least, mass spectrometric detection and determination could become a very useful alternative or a complementary and rapid warning system for BW agents. A transportable environmental mass spectrometer (e.g., ATOFMS) is already available. The improved identification technologies, such as tandem and bipolar laser mass spectrometry, seem to be promising tools for a more precise identification of different airborne bacteria.

17.5 GENERATION OF BACTERIAL TEST AEROSOLS

The equipment as well as the detection and identification procedures have to be tested and calibrated. For this reason, useful and standardized methods for the generation of bacterial model aerosols are very necessary and desirable. There are a number of ways in which bacteria may be made airborne.[125-130]

Laboratory production of bacterial test aerosols is, however, aimed toward controllable reproduction of the required particle size, bacterial content, and level of concentration. Bacterial aerosols are produced from either liquid suspension or dried materials.[125,126]

Wet dispersion methods are very appropriate for bacteria.[126] Characteristics of the resulting bacterial aerosol depend on the dispersing shear forces and the sensitivity and agglomeration of the tested microorganisms. Usually, different generation methods have to be used for the production of model bacterial aerosols made of different microbial groups. The existing devices for the dispersion of liquid bacterial suspensions differ in their production abilities. For example, pneumatic nebulizers were found to produce considerably greater amounts of microbial fragments than the bubbling aerosol dispensers (Figure 17.21).[126] The concentrations and the amount of agglomerates can be controlled by additional arrangements, e.g., orifices. The aerosolization process produces stress on the suspended bacteria and influences the relative recovery of the aerosolized microorganisms. Therefore, the generation procedures have to be optimized, and the relative microbial recovery factor has to be measured and established for each generating system. Bacterial survival can be increased by several additional steps, e.g., by increasing and/or optimizing the air humidity,[126] by increasing the size of generating droplets (Figures 17.22),[127] etc.

17.6 CONCLUSION

When we studied membrane and gelatin filters in the 1960s, as well as several impaction sampling apparatuses for measuring bacterial aerosols, we found relatively poor reproducibility and broad variability in results.[131] Hygienists were, in these early days, the main users of the existing methods. There was little interest in the sampling and collection mechanisms, in the concentration variabilities, reproducibilities, etc. For more than 20 years, the field of basic investigation of bioaerosols,

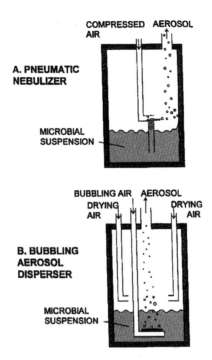

FIGURE 17.21   Schematics showing the two types of wet dispersion generators. (From Griffiths, W.D. and De Cosemo, G.A.L. The assessment of bioaerosols: A critical review. *J. Aerosol Sci.* 25(8) 1425-1458, 1994. With permission.)

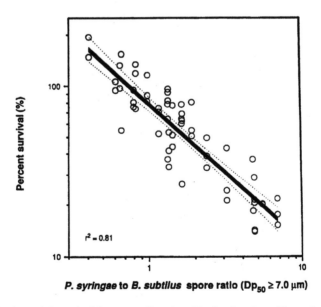

FIGURE 17.22   The bacterial survival increases directly with droplet size of bacteria. (From Reponen, T., Willeke, K., Ulevicius, V. et al. Techniques for dispersion of microorganisms into air. *Aerosol Sci. Technol.* 27, 405-421, 1997. With permission.)

including bacterial aerosols, escaped notice. A real renaissance in this research field dates from the 1980s, and relatively high interest began in the 1990s — mainly in the U.S., Canada, U.K., and the Scandinavian countries. The research results are of great theoretical and practical importance and are very promising.

REFERENCES

1. Fox, C.S. and Wathes, C.M. (Eds.) *Bioaerosols Handbook,* CRC/Lewis Publ. Boca Raton, U.S., 1995.
2. Ho. J.H. and Griffiths, W.D., (Eds.) Bioaerosols. *J. Aerosol Sci.* 25(8)1369-1613, 1994.
3. Nevalainen, A., Willeke, K, Liebhaber, F., et al., Bioaerosol sampling, in K. Willeke and P.A. Baron (Eds.) *Aerosol Measurements.* Van Nostrand Reinhold Publ., New York, 1993, pp. 471-492.
4. Matthias-Maser, S. and Jaenicke, R. Examination of atmospheric bioaerosol particles with radii >0.2 μm. *J. Aerosol Sci.* 25(8)1605-1613, 1994.
5. Navalainen, A., Pastuszka, J., Liebhaber, F., and Willeke, K. Performance of bioaerosol samplers. *Atm. Environ.* 26A, 531-540, 1992.
6. Terzieva, S., Donnelly, J., Ulevicius, V., et al. Comparison of methods for determination and enumeration of airborne microorganisms collected by liquid impingement. *Appl. Environ. Microbiology* 62(7)2264-2272, 1996.
7. Spurny, K.R. On the chemical detection of bioaerosols. *J. Aerosol Sci.* 25(8)1533-1547, 1994.
8. Spurny, K.R. (Ed.) *Physical and Chemical Characterization of Individual Airborne Particles.* John Wiley & Sons, Chichester, U.K., 1986.
9. Nelson, W.H. (Ed.) *Instrumental Methods for Rapid Microbiological Analysis.* VCH Publ. New York, 1985.
10. Nelson, W.H. (Ed.) *Moderate Techniques for Rapid Microbiological Analysis.* VCH Publ. New York, 1991.
11. Jacobson, A.L. Viable particles in the air, in A.C. Stern (Ed.) *Air Pollution.* Academic Press, New York, 1968, pp. 95-120.
12. Fincher, E.L. and Mallison, G.F. Intramural sampling of airborne microorganisms, in *Air Sampling Instruments.* Am. Conference of Govern. Industrial Hygienists, 1972, pp. E1-E6.
13. Macher, J.M. Positive-hole correction of multijet impactors for collecting viable microorganisms. *Am. Ind. Hyg. Assoc. J.* 50(11)561-568, 1989.
14. Mouilleseaux, A. Sampling methods for bioaerosols. *Aerobiologia.* 6, 32-35, 1990.
15. Fängmark, I., Wikström, L.E., and Henningson, E.W. Collection efficiency of a personal sampler for microbiological aerosols. *Am. Ind. Hyg. Assoc. J.* 52(12), 516-520, 1991.
16. Jensen, P.A., Todd, W., Davis, G.N., and Scarpino, P.V. Evaluation of eight bioaerosol samplers challenged with aerosols of free bacteria. *Am. Ind. Hyg. Assoc. J.* 53(10) 660-667, 1992.
17. Thorne, P.S., Kiekhaefer, Whitten, P., and Donham, K.J. Comparison of bioaerosol sampling methods in barns housing swine. *Appl. Environ. Microbiol.* 588(8) 2543-2551, 1992.
18. Kiefer, S. and Müller, W. Leistungsvergleich verschiedener Luftkeimsammler. *Tierärztl. Umschau* 47:495-500, 1992.
19. Juozaitis, A., Willeke, K., Grinshpun, S.A., and Donnelly, J. Impaction onto a glass slide or agar vs. impingement into a liquid for the collection and recovery of airborne microorganisms. *Appl. Environ, Microbiol.* 60(3), 861-870, 1994.
20. Henningson, E.W. and Ahlberg, M.S. Evaluation of microbiological aerosol samplers. A review. *J. Aerosol Sci.* 25(8)1459-1492, 1994.
21. Sommerville, M.C. and Rivers, J.C. An alternative approach for the correction of bioaerosol data collected with multiple jet impactors. Advances in Industrial and Environmental Hygiene. *Am. Ind. Hyg. Assoc. J.* 55(2), 127-131, 1994.
22. Terzieva, S., Donnelly, J., Ulevicius, V. et al. Comparison of methods for detection and enumeration of airborne microorganisms collected by liquid impingement. *Appl. Environ. Microbiology* 62(7), 2264-2272, 1996.
23. Nielsen, E.M., Breum, N.O., Nielsen, B.H., et al. Bioaerosol exposure in waste collection: A comparative study on the significance of collection equipment, type of waste and seasonal variation. *Ann. Occup. Hyg.* 41(3), 325-344, 1997.
24. Lacey, J. (Ed.) Sampling and rapid assay of bioaerosols *J. Aerosol Sci.* 28(3), 345-539, 1997.
25. Pahl, O., Phillips, V.R., Lacey, J., et al. Comparison of commonly used samplers with a novel bioaerosol sampler with automatic plate exchange. *J. Aerosol Sci.* 28(3), 427-436, 1997.
26. Griffiths, W.D., Stewart, I.W., Futter S.J., et al. The development of sampling methods for the assessment of indoor bioaerosol. *J. Aerosol Sci.* 58(3), 437-458, 1997.

27. Henningson, E.W., Lundquist, M., Larson, E., et al. A comparative study of different methods to determine the total number and the survival ratio of bacteria in aerobiological samples. *J. Aerosol Sci.* 28(3) 459-470, 1997.
28. Dennis, P.J. and Lee, V.J. Differences in aerosol survival between pathogenic and non-pathogenic strains of Legionella pneumophila serogroup 1. *J. Appl. Bacteriol.* 65, 135-141, 1988.
29. Cox. C.S. Airborne bacteria and viruses. *Sci. Prog. Oxf.* 73, 469-500, 1989.
30. Marthi, B., Fieland, V.P., Walter, M., and Seidler, R.J. Survival of bacteria during aerosolization. *Appl. Environ. Microbiol.* 56, 3463-3467, 1990.
31. Walter, M, Mathi, B., Fieland, V.P., and Ganio, L.M. Effect of aerosolization on subsequent bacterial survival. *Appl. Environ. Microbiol.* 56, 3468-3472, 1990.
32. Lindberg, C. and Horneck, G. Action spectra for survival and spore photoproduct formation of bacillus subtilis irradiated with short wavelength (200-300 nm) UV at atmospheric pressure and in vacuo. *J. Photochem. Photobiol. B. Biol.* 11, 69-80, 1991.
33. Thomson, C.M.A., Chanter, N., and Wathes, C.M. Survival of toxigenic pasteurella multocida in aerosols and aqueous liquids. *Appl. Environ. Microbiol.* 58(3), 932-936, 1992.
34. Simecek, J., Kneifova, J., and Stochl, V. Investigation of the microbial and dust air pollution. *Staub-Reinhalt. Luft.* 46(6), 285-289, 1986.
35. Hochrainer, D., Spurny, K.R., and Fabig, W. *Sampling of Microbial Aerosols,* Final Rep. (Fraunhofer-Society, Grafschaft, Germany), 1987, pp. 1-89.
36. Maier, K.H. and Vogel, K. Wasserlösliche Filter und ihre Anwendung in der Luftkeimuntersuchung. *Fachzschr. für das Laboratorium.* März, 119-127, 1965.
37. Irwin, W.J. (Ed.) *Analytical Pyrolysis.* Marcel Dekker, New York, 1982, p. 1-578.
38. Bucharan, M.V. and Hettich, R.L. Fourier transformation mass spectrometry of high-mass biomolecules. *Anal. Chem.* 65(5), 245A-258A, 1993.
39. Lattimer, R.P. Analytical pyrolysis. *J. Anal. Appl. Pyrolysis.* 17, 1-3, 1989.
40. Ericsson, I. and Lattimer, R.P. Pyrolysis nomenclature. *J. Anal. Pyrolysis.* 14, 219-221, 1989.
41. Kim, M.G., Inove, H., and Shirai, T. Development of a Curie-Point thermal desorption system and its application to the analysis of atmospheric dusts. *J. Anal. Pyrolysis.* 15, 217-226, 1989.
42. Sinha, M.P. Analysis of individual biological particles in air, in W.H. Nelson (Ed.) *Instrumental Methods for Rapid Microbiological Analysis.* VCH Publishers, Inc., 1985, pp. 165-192.
43. Snyder, A.P., Kremer, J.H., Meuzelaar, et al. Curie-Point pyrolysis atmospheric pressure chemical ionization. MS. *Anal. Chem.* 59, 1945-1951, 1987.
44. Chan, W.R., Kelbon, M., and Krieger-Brockett, B. Single particle biomass pyrolysis. *Ind. Engn. Chem. Res.* 27, 2261-2275, 1988.
45. Fox, A. and Morgan, S.L. The chemotaxonomic characterization of microorganisms by capillary gas chromatography and gas chromatography–mass spectrometry, in W.H. Nelson (Ed.) *Instrumental Methods of Rapid Microbiological Analysis.* VCH Publishers, 1985, pp. 135-164.
46. D'Agnostino, P.A., Provost, L.R., Anacleto, J.F., and Brooks, P.W. Capillary column gas chromatography-mass spectrometry and gas chromatography–tandem mass spectrometry detection of chemical warfare agents in a complex airborne matrix. *J. Chromatography.* 504, 259-268, 1990.
47. Matney, M.L. and Limero, T.F. Pyrolysis-gas chromatography/mass spectrometry analyses of biological particulates collected during recent space shuttle missions. *Anal. Chem.* 66, 2820-2828, 1994.
48. Saraf, A. and Larsson, L. Use of gas chromatography/ion-trap tandem mass spectroscopy for the determination of chemical markers of microorganisms in organic dust. *J. Mass Spectr.* 31, 389-396, 1996.
49. Rossi, T.M. and Warner, I.M. Bacterial identification using fluorescence spectroscopy, in W.H. Nelson *Instrumental Methods of Rapid Microbial Analysis.* VCH Publishers, 1985, pp. 1-50.
50. Alan, R. *Immunoassays: Chemical and Laboratory Techniques for the 1980s.* Liss, Inc., New York, 1980.
51. McElroy, W.D. and Seliger, H.H. *Bioluminescence in Progress.* Princeton University Press, Princeton, N.J., U.S.A., 1966, pp. 1-432.
52. Speight, S.E., Hallis, B.A., Bennett, A.M., and Benbough, J.E. Enzyme-linked immunosorbent assay for the detection of airborne microorganisms used in biotechnology. *J. Aerosol Sci.* 28(3), 483-492, 1997.

53. Rowell, F.J., Farrell, C., Nitescu, I., Cumming, R.H., and Steward, I.W. Rapid immunological assay methods for ceftazidime and alcalase in the workplace atmosphere. *J. Aerosol Sci.* 28(3), 493-500, 1997.

54. Nitescu, I., Cumming, R.H., Rowell, F.J., and Steward, I.A. Evaluation of a model for the detection of aerosol-containing protease in real-time using chromogenic substrates in the sampling of a cyclone and bubbler. *J. Aerosol Sci.* 28(3), 501-510, 1997.

55. Steward, I.W., Leaver, G., and Futter, S.J. The enumeration of aerosolized saccharomyces cerevisiae using bioluminescent assay of total adenylates. *J. Aerosol Sci.* 28(3), 511-524, 1997.

56. Cormier, M.J., Hercules, P.M., and Lee. J. *Chemiluminescence and Bioluminescence*. Plenum Press, New York, 1913.

57. Baeyen, S.W.R.G., Dekeukeleire, D., and Korkidis, K. *Luminescence Techniques in Chemical and Biochemical Analysis*. Marcel Dekker, New York. Practical Spectroscopy Series Vol. 12, 1991, pp. 1-654.

58. Neufeld, H.A., Pace, J.G., and Hutchison, R.W. Detection of microorganisms by bio- and chemiluminescence techniques, in W.H. Nelson (Ed.) *Instrumental Methods for Rapid Microbiological Analysis*. VCH Publishers, 1985, pp. 51-65.

59. Hallin, P., Linfors, G., and Sandström, G. A device for rapid detection of bacteria in air by chemiluminiscent technique. *5th Nord. Symp. in Aerobiology*, 1983, pp. 47-50.

60. Hallin, P., Linfords, G., and Sandström, G. Investigation of variation of the concentration of bacteria at outdoor testing with the use of detector for aerosol and bacteria. *FAO Report C 40201-22*. Natl. Defense Res. Inst. in Umea, Sweden, 1984, pp. 1-18.

61. Grader, G.S., Flagan, R.C., Seinfeld, H. and Arnold, S. Fourier transform infrared spectrometer for single aerosol particle. *Rev. Sci. Instrum.* 58(4), 584-588, 1987.

62. Hadley, W.K. and Yajko, D.M. Determination of microorganisms and their metabolisms by measurement of electrical impedance, in W.H. Nelson (Ed.) *Instrumental Methods of Rapid Microbiological Analysis*. VCH Publishers, 1985, pp. 193-209.

63. Smith, D.S., Harsan, M., and Nargessi, R.D. Principles of immunonoassay procedures, in E.L. Wehry (Ed.) *Modern Fluorescence Spectroscopy*. Plenum Press, New York, Vol. 3, Chap. 4, 1982.

64. Mackay, R.A., Godde, M.T., Stopa, P.J., and Zulich, A. Light addressable potentiometric sensor based detection of toxins and pathogens. *ACS 201st Natl. Meeting*, Atlanta, GA, Div. Environ. Chem. Preprint Papers 31, 351-354, 1991.

65. Van Emon, J.M. and Mumma, R.O. (Eds.) *Immunochemical Methods for Environmental Analysis*. (IMEA)ACS Symposium Series 442, Am. Chem. Society, Washington, D.C., 1990.

66. Mathewson, P.R. and Finley, J.W. (Eds.) *Biosensor Design and Applications*. ACS Symposium Series 550, Am. Chem. Society, Washington, D.C., 1992.

67. Vo-Dinh, T., Griffin, G.D., Sepaniak, M.J., and Alarie, J.P. Development of fibreoptic fluoroimmunosensor for chemical and biological analysis. *ACS 202nd. Natl. Meeting*, Atlanta, GA, Div. Environ. Chem. Preprint Papers, 31, 347-350, 1991.

68. Rishpon, J., Gezundhajt, Y., Soussan, L. et al. Immunoelectrodes for the detection of bacteria, in *Biosensor Design and Application*. Mathewson, P.R. and Finley, J.W. (Eds.) ACS Symposium Series 511, 59-72, Am. Chem. Soc. Washington, D.C., 1992.

69. Erasmus, D.A. *Electron Probe Microanalysis in Biology*. Chapman and Hall, London, 1978.

70. Leahy, T.J. and Sullivan, M.J., Validation of bacterial-retention capabilities of membrane filters. *Pharmaceutical Technol.*, 1978, pp. 7-12.

71. Fiser, Z., Hysek, J. and Binek, B. Quantification of airborne microorganisms and investigation of their interactions with non-living particles. *Int. J. Biometeorol.* 34, 189-193, 1990.

72. Palmgren, U., Ström, G., Malmberg, P. and Blomquist, G. The Nuclepore filter method: A technique for enumeration of viable and nonviable microorganisms. *Am. J. Ind. Med.* 10, 325-327, 1986.

73. Miller, W.I. and Forster, B. Fluorescence and confocal laser scanning microscopy: Applications in biotechnology. *Int. Laboratory*, March, 1991, pp. 18-26.

74. Eduard, W., Lacey, J. Karlson, K. et al. Evaluation of methods of enumerating microorganisms in filter samples for highly contaminated environment. *Am. Ind. Hyg. Assoc. J.* 51(8) 427-436, 1990.

75. Lacey, J., Karlson, K., Malmberg, P. et al. *Harmonization of Sampling and Analysis of Mold Spores*. Nordisk Ministerrad, Copenhagen, 1988, pp. 1-70.

76. Eduard, W., Sandren, P., Johansen, B.V., and Brunn, R. Identification and quantification of mold spores by SEM. *Ann. Occup. Hyg.* 32, 447-455, 1988.

77. Eduard, W.P. and Aalen, O. The effect of aggregation on the counting precision of mold spores on filter. *Ann. Occup. Hyg.* 32, 471-479, 1988.
78. Palmgren, U., Ström, G., Blomquist, G., and Malmberg, P. Collection of airborne microorganisms on Nuclepore filters: Estimation and analysis. *J. Appl. Bacteriology.* 61, 401-406, 1986.
79. Greene, V.W. Apartichrome analyzer. *Environ. Sci. Technol.,* 2, 104-109, 1968.
80. Tamiya, E., Shinichiro, I., Nagatani, N. et al. Simultaneous topographic and fluorescence imagings of recombinant bacterial cells containing a green fluorescent protein gene detected by a scanning near-field optical/atomic force microscopy. *Anal. Chem.* 69, 3697-3701, 1997.
81. Melamed, M.R., Mullaney, P.R. and Mendelson, M.L. (Eds.) *Flow Cytometry and Sorting.* John Wiley-Liss, New York, 1990.
82. Ruzicka, J. and Lindberg, W. Flow injection cytoanalysis. *Anal. Chem.* 64(9) 537A-544A, 1992.
83. Hadley, W.K., Waldman, F. and Fulwyter, M. Rapid microbial analysis by flow cytometry, in W.H. Nelson (Ed.) *Instrumental Methods of Rapid Microbiological Analysis,* VCH Publishers, 1985, pp. 67-89.
84. Kaprelyants, A.S. and Kell, D.B. Rapid assessment of bacterial viability and vitality by rhodamine 123 and flow cytometry. *J. Appl. Bacteriology* 72, 410-422, 1992.
85. Schrader, B. Micro-Raman fluorescence and scattering spectroscopy of single particles, in K.R. Spurny (Ed.) *Physical and Chemical Characterization of Individual Airborne Particles.* John Wiley & Sons, Chichester, U.K., 1986, pp. 358-379.
86. Hartman, K.A. and Thomas, G.J. The identification, interactions and structure of viruses by Raman spectroscopy, in W.H. Nelson (Ed.) *Instrumental Methods for Rapid Microbiological Analysis.* VCH Publishers, 1985, pp. 91-134.
87. Schweiger, G. Single microparticle analysis by Raman spectroscopy. *J. Aerosol Sci.* 20(8) 1621-1624, 1989.
88. Dalterio, R.A., Baek, M., Nelson, W.H. et al. The resonance Raman microprobe detection of single bacterial particles and cells for chromobacterial mixture. *Appl. Spectroscopy.* 41(2) 241-244, 1987.
89. Clark, N.J. A scintillation spectrometer with coincident counting for aerosol analysis and source attribution. *Atm. Environ.* 19(1), 331-339, 1985.
90. Kawaguchi, H., Fukasawa, N., and Mizuike, A. Investigation of airborne particles by inductively coupled plasma emission spectroscopy calibrated with monodisperse aerosols. *Spectrochemica Acta.* 41 B(12, 1277-1286), 1986.
91. Sneddon, J. Impaction-electrothermal AAS for direct and near real-time detection of metals. *Appl. Spectroscopy.* 43(6) 1100-1102, 1989.
92. Lakowicz, J.R. *Principles of Fluorescence Spectroscopy.* Plenum Press, New York, 1983.
93. Pinnick, R.G., Hill, S.C., Nachman, P. et al. Fluorescence particle counter for detecting airborne bacteria and other biological particles. *Aerosol Sci. Technol.* 23, 653-664, 1995.
94. Hill, S.C., Pinnick, R.G., Nachmann, P. et al. Aerosol-fluorescence spectrum analyzer: Real-time measurement of emission spectra of airborne biological particles. *Appl. Optics.* 34(30) 7149-7155, 1995.
95. Nachman, P., Chen, G., Pinnick, R.G. et al. Conditional-sampling spectrograph detection system for fluorescence measurements of individual airborne biological particles. *Appl. Optics.* 35(7) 1069-1076, 1996.
96. Pinnick, R.G., Hill, S.C., Chen, R.K. et al. Aerosol analyzer for rapid measurements of the fluorescence species of airborne bacteria excited with a conditionally fired pulsed 266 nm laser. *The Aerosol Association for Aerosol Research Conference,* Orlando, 1996, Abstracts, p. 271.
97. Seaver, M. and Eversole, J.D. Monitoring biological aerosol using UV fluorescence. *The Aerosol Association for Aerosol Research Conference,* Orlando, 1996, Abstracts, p. 270.
98. Hairston, P.L., Ho, J. and Quant, F.R. Design of an instrument for real-time detection of bioaerosols using simultaneous measurement of particle aerodynamic size and intrinsic fluorescence. *J. Aerosol Sci.* 28(3)471-482, 1997.
99. Chang, R. Biological aerosol detection by fluorescence spectra. *The Aerosol Association for Aerosol Research Conference,* Denver, 1997.
100. Hargis, P. Ultraviolet fluorescence spectroscopy of biological aerosols. *The American Association for Aerosol Research Conference,* Denver, 1997.
101. TSI Inc. Ultraviolet aerodynamic particle sizer spectrometer. Information sheet, 1997.

102. Sinha, M.P., Giffin, C.E., Norris, D.D. et al. Particle analysis by mass spectrometry. *J. Colloid Interface Sci.* 87(1) 140-153, 1982.

103. Sinha, M.P. Laser-induced volatilization and ionization of microparticles. *Rev. Sci. Instruments.* 55(6) 886-891, 1984.

104. Sinha, M.P. and Friedlander, S.K. Real time measurements of NaCl in individual aerosol particles by MS *Anal. Chem.* 57(9) 1880-1883, 1985.

105. Sinha, M.P., Platz, R.M., Friedlander, S.K., and Vilker, V.L. Characterization of bacteria by particle beam. MS. *Appl. Environ. Microbiol.* 49(6) 1366-1372, 1985.

106. Sinha, M.P. Analysis of individual biological particles in air, in W.H. Nelson (Ed.) *Instrumental Methods for Rapid Microbiological Analysis.* VCH Publishers, 1985, pp. 165-192.

107. Hillenkamp, F. and Kaufmann, R. Laser microprobe mass analysis (LAMMA), in M.L. Wolbarsh (Ed.) *Laser Applications in Medicine and Biology.* Plenum Press, New York, Vol. 4, 1982.

108. Kaufmann, R. Laser mass spectroscopy of particulate matter, in K.R. Spurny (Ed.) *Physical and Chemical Characterization of Individual Airborne Particles.* John Wiley & Sons, Chichester, 1986, pp. 226-250.

109. Wieser, P. and Wurster, R. Application of laser microprobe analysis to particle collectives, in K.R. Spurny (Ed.) *Physical and Chemical Characterization of Individual Airborne Particles.* John Wiley & Sons, Chichester, 1986, pp. 251-270.

110. Spurny, K.R. and Gentry, J.W. Aerosol fractionation by graded Nuclepore filters. *Powder Technology.* 24, 129-142, 1979.

111. Böhm, R. Sample preparation technique for the analysis of bacteria by LAMMA. *Fresenius J. Anal. Chem.* 308, 258-259, 1981.

112. Seydel, U. and Lindner, B. Qualitative and quantitative investigations of mycobacteria with LAMMA. *Fresenius J. Anal. Chem.* 308, 253-257, 1981.

113. Kievit, O., Marijnissen, J.C.M., Verheijen, P.T.J., and Scarlett, C.M.B. Some improvements on the particle beam generator. *J. Aerosol Sci.* 21(1) 685-688, 1990.

114. Marijnissen, J.C.M., Scarlett, C.M.B., and Verheijen, P.T.J. Proposed on-line aerosol analysis combining size determination and TOF mass spectrometry. *J. Aerosol Sci.* 19(7)1307-1310, 1988.

115. Kistemaker, P.G., Meuzelaar, H.L.C. and Posthumus, M.A. Microorganism analysis by MS, in C.G. Heden and T. Illeni (Eds.) *New Approaches of the Identification of Microorganisms.* John Wiley & Sons, London, 1975, pp. 179-191.

116. Heller, D.N., Cotter, R.J., Fenselau, C. and Oy, O.M. Profiling bacteria with fast atom bombardment spectroscopy. *Anal. Chem.* 59, 2806-2809, 1987.

117. Kissel, J. The Giotto particulate impact analyzer. *Phys. Blat.* 43(5), 131-133, 1987.

118. Yang, MO, Reilly, P.T.A., Gieray, R.A., Whitten, W.B., and Ramsey, J.M. Complete chemical analysis of aerosol particles in real-time. *J. Korean Phys. Soc.* 30(2) 359-363, 1997.

119. Gieray, R.A. and Whitten, W.B. Real-time detection of individual airborne bacteria. *J. Microbiol. Methods* 29, 191-199, 1997.

120. Hinz, K.P., Kaufmann, R., and Sprengler, B. Simultaneous detection of positive and negative ions from single airborne particles by real-time laser mass spectroscopy. *Aerosol Sci. Technol.* 24, 233-242, 1996.

121. Hinz, K.P., Kaufmann, R., Jung, R. et al. Characterization of atmospheric particles using bipolar on-line laser mass spectrometry and multivariate classification. *J. Aerosol Sci.* 28, Suppl. S305-S306, 1997.

122. Gard, E, Mayer, J.E., Morrical, B.D. et al. Real-time analysis of individual aerosol particles: Design and performance of a portable ATOFMS. *Anal. Chem.* 69, 4083-4091, 1997.

123. Klaus, N. Aerosol analysis by secondary mass spectroscopy, in K.R. Spurny (Ed.) *Physical and Chemical Characterization of Individual Airborne Particles.* John Wiley & Sons, Chichester, 1986, pp. 331-339.

124. Colliver, T.L., Brummel, C.L., Pacholski, M.L. et al. Atomic and molecular imaging at the single-cell level with TOF–SIMS. *Anal Chem.* 69-2225-2231, 1997.

125. Griffiths, W.D. and De Cosemo, G.A.L. The assessment of bioaerosols: A critical review. *J. Aerosol Sci.* 25(8) 1425-1458, 1994.

126. Reponen, T., Willeke, K., Ulevicius, V. et al. Techniques for dispersion of microorganisms into air. *Aerosol Sci. Technol.* 27, 405-421, 1997.

127. Lighthart, B. and Shaffer, B.T. Increased airborne bacterial survival as a function of particle content and size. *Aerosol Science and Technol.* 27, 439-446, 1997.

128. Rubel, G.O. Measurement of water vapour sorption by single biological aerosol. *Aerosol Sci. Technol.* 27, 481-490, 1997.

129. Madelin, T.M. and Johnson, H.E. Fungal and actinomycete spore aerosols measured at different humidities with an aerodynamic particle sizer. *J. Appl. Bacteriology.* 72, 400-409, 1992.

130. Cheng, Y.S. and Barr, E.B. Generation and characterization of biological aerosols for laser measurement, in *Inhalation Research Inst. Ann. Report,* Albuquerque, NM, 1995 and 1996.

131. Spurny, K. and Machala, O. Eine Bemerkung zur Wirksamkeit der Membranfilter für mikrobiale Aerosole. *Zschr. Biol. Aerosol-Forschunbg (Germany).* 13, 566-569, 1967.

Index

9 780367 399801